Advances in

VIRUS RESEARCH

VOLUME 38

ADVISORY BOARD

Advances in

VIRUS RESEARCH

Edited by

KARL MARAMOROSCH

Department of Entomology
Rutgers University
New Brunswick, New Jersey

FREDERICK A. MURPHY

Center for Infectious Diseases
Centers for Disease Control
Atlanta, Georgia

AARON J. SHATKIN

Center for Advanced Biotechnology and Medicine
Piscataway, New Jersey

VOLUME 38

ACADEMIC PRESS, INC.

Harcourt Brace Jovanovich, Publishers
San Diego   New York   Boston
London   Sydney   Tokyo   Toronto

This book is printed on acid-free paper. ∞

ACADEMIC PRESS, INC.
San Diego, California 92101

United Kingdom Edition published by
ACADEMIC PRESS LIMITED
24-28 Oval Road, London NW1 7DX

LIBRARY OF CONGRESS CATALOG CARD NUMBER: 53-11559

ISBN 0-12-039838-9 (alk. paper)

PRINTED IN THE UNITED STATES OF AMERICA
90 91 92 93 9 8 7 6 5 4 3 2 1

CONTENTS

Molecular Biology of Rotaviruses

A. R. BELLAMY AND G. W. BOTH

Molecular Biology of Varicella Zoster Virus

JEFFREY M. OSTROVE

Structure and Function of the RNA Polymerase of Vesicular Stomatitis Virus

AMIYA K. BANERJEE AND DHRUBAJYOTI CHATTOPADHYAY

Control of Expression and Cell Tropism of Human Immunodeficiency Virus Type 1

JEROME A. ZACK, SALVATORE J. ARRIGO, AND IRVIN S. Y. CHEN

Interferon-Induced Proteins and the Antiviral State

PETER STAEHELI

Expression of a Plant Virus-Coded Transport Function by Different Viral Genomes

JOSEPH G. ATABEKOV AND MIKHAIL E. TALIANSKY

Structural and Functional Properties of Plant Reovirus Genomes

DONALD L. NUSS AND DAVID J. DALL

Regulation of Tobamovirus Gene Expression

WILLIAM O. DAWSON AND KIRSI M. LEHTO

The "Merry-Go-Round": Alphaviruses between Vertebrate and Invertebrate Cells

HANS KOBLET

Emergence, Natural History, and Variation of Canine, Mink, and Feline Parvoviruses

COLIN R. PARRISH

ADVANCES IN VIRUS RESEARCH, VOL. 38

MOLECULAR BIOLOGY OF ROTAVIRUSES

A. R. Bellamy* and G. W. Both†

*Department of Cellular and Molecular Biology
University of Auckland
Auckland, New Zealand
and
†Laboratory for Molecular Biology
Division of Biotechnology
Commonwealth Scientific and Industrial Research Organization
North Ryde, Sydney, New South Wales 2113, Australia

I. Introduction

It is now over 10 years since the assignment of the rotaviruses to a separate genus of the family Reoviridae (Matthews, 1979). In the years that have followed the recognition of the separate status of this virus group, a considerable body of work has accumulated which clarifies the basic structure of the virus particle, the nature of the structural and nonstructural proteins encoded by the viral genome, and the

1

mechanisms involved in the intracellular replication of the virus. Previous reviews, particularly those by Holmes (1983), Estes *et al.* (1983), Cukor and Blacklow (1984), and Kapikian and Chanock (1985), have summarized most of this information. Much of the rapid expansion of knowledge about rotaviruses can be attributed to their clinical importance, which results from their identification as the single most important etiological agent causing acute gastroenteritis in the young (Cukor and Blacklow, 1984). Matsui *et al.* (1989) have reviewed clinical and immunological aspects of the virus and have detailed our understanding of the epitopes present on the important viral antigens.

The development of our knowledge of the rotavirus replication cycle has benefited from prior work on reovirus, the prototype strain of the Reoviridae (see, for example, reviews by Joklik, 1983; Zarbl and Millward, 1983; Ramig and Fields, 1983). Although precise information on some aspects of rotavirus replication is lacking, it is generally assumed that rotaviruses share with other members of Reoviridae a common mechanism for the replication of the genome. In this process the virion-associated RNA transcriptase, located in the central viral core, synthesizes single-stranded (positive sense) RNA transcripts that act both as mRNA and as the precursors for the synthesis of the double-stranded genomic RNAs. However, there are aspects of the replication cycle which set the rotaviruses apart from other members of the Reoviridae.

The most significant feature is the unique site of maturation for rotaviruses: the lumen of the rough endoplasmic reticulum (ER) (Estes *et al.*, 1983). Consequently, the assembly of rotaviruses involves the transport of viral components across the membrane of the ER, a process that involves a membrane budding event in which partially assembled virus particles are enveloped in a transient membranous vesicle. As detailed below, this unusual assembly pathway has provided the basis for a number of basic studies in cellular and molecular biology. Two viral glycoproteins are also localized to the rough ER prior to the budding event and these have provided unique opportunities for the study of signals involved in their localization to this intracellular compartment.

The relative speed with which information on the molecular biology of rotaviruses has accumulated in recent years is largely a result of the impact of recombinant DNA technology, which has enabled important genes from the virus to be cloned and sequenced. However, there are still gaps in our knowledge. It is our purpose in this chapter to highlight unique aspects of the rotavirus replication cycle, to review recent work on the molecular biology of group A rotaviruses, and to indicate those areas in which our knowledge is incomplete. We have excluded from this review information pertaining to the molecular biology of groups B, C, D, and E rotaviruses (Pedley *et al.*, 1983, 1986; Bremont *et al.*, 1988), in view of the relative paucity of information on these related, yet distinct, groups.

II. PARTICLE STRUCTURE AND COMPOSITION

A. General

The complete rotavirus virion is a double-shelled (ds) particle of 76.5 nm outer diameter with two layers of capsomeres enclosing the genome of 11 segments of double-stranded RNA (Rodger *et al.*, 1975; Newman *et al.*, 1975). The precise arrangement of both shells of capsomeres was recently determined by Prasad *et al.* (1988). Three-dimensional reconstructions were computed from electron-microscopic images of the virus embedded in a thin layer of amorphous ice. Although it had been shown previously that the inner capsomeres were arranged with T = 13 *l* skewed icosahedral symmetry (Roseto *et al.*, 1979; Metcalf, 1982), the organization of the outer layer had not been established. The model constructed by Prasad *et al.*, (1988) clarified much of what previously was in contention (see, for example, articles by Martin *et al.*, 1975; Esparza and Gil, 1978) and provided a wealth of information concerning the interrelationship of the inner and outer rotavirus shells.

In describing the structure and the morphogenesis of key elements of rotavirus particles, we refer here to virions as ds particles, to particles lacking the outer capsid layer as single-shelled (ss) particles, and to the particles which also lack VP6 (Bican *et al.*, 1982) as the central core. The ss particles are a normal intermediate of rotavirus infection and are produced in considerable quantity in the cell. The particle is exceedingly stable and is the morphological equivalent of the reovirus core produced by proteolytic digestion of the outer reovirus capsomeres (Shatkin and Sipe, 1968). Like the reovirus core, the rotavirus ss particles also contain the genome of segmented double-stranded RNA and the virion-associated transcriptase (Cohen, 1977; Mason *et al.*, 1980). Because the rotavirus genomic double-stranded RNAs are capped (Imai *et al.*, 1983a), the ss particles are also assumed to contain the other activities associated with modification of mRNA transcripts (i.e., 7-methyl G and 2'-*O*-methyl G methylases and guanylate transferase activities, as well as the nucleotide triphosphohydrolases), as is the case for the closely related reovirus (Furuichi *et al.*, 1975a,b).

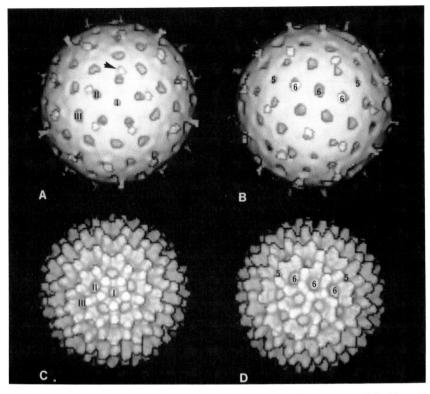

FIG. 1. Surface representations of the three-dimensional structures of double and single-shelled rotavirions along the icosahedral fivefold axis [(A) and (C)] and along the icosahedral threefold axis [(B) and (D)]. In (B) and (D) a pair of neighboring fivefold axes (designated as 5) and the 6-coordinated positions (designated as 6) relating them are shown to illustrate T = 13 *l* (laevo). Three types of channels [designated I, II, and III in (A) and (C), showing just one of each type] are found at all 5- and 6-coordinated positions spanning the outer and inner shell proteins. The protein spikes are situated at an edge of type II channels surrounding the fivefold positions in the double-shelled rotavirus. The spikes surrounding the central fivefold axis in (A) are pointing toward the viewer; one of these is arrowed. (Courtesy of B. V. V. Prasad.)

B. Single-Shelled Particle

Studies of images of freeze-etched and shadowed particles by Metcalf (1982) and by Roseto *et al.* (1979) showed that the capsomeres of the ss particle were arranged on a T = 13 *l* surface lattice. This was confirmed by the three-dimensional model of the ss particle of the virus (Prasad *et al.,* 1988), which showed, without a full icosahedral symmetry constraint to the computation, that 260 morphological units are located at all of the threefold axes around the 120 six-coordinated and 12 five-coordinated holes or channels in the shell. Reconstructions of both the ss and ds particles from this study are shown in Fig. 1. It is well established that the major surface protein of the particle is VP6 (Bican *et al.,* 1982) and that, when this protein is dissociated from the virus, it exhibits a tendency to form multimeric (probably trimeric) aggregates (Gorziglia *et al.,* 1985; Sabara *et al.,* 1987). Thus, the trimeric forms of capsomeres first observed under the electron microscope by Martin *et al.* (1975) and by Esparza and Gil (1978) almost certainly were aggregates of VP6 protein, and these would therefore correspond to the threefold clusters seen in the reconstructed images shown in Fig. 1A and B. Together, this information indicates that there are 780 molecules of VP6 per virion and that VP6 trimers are the basic morphological units of the surface lattice of the ss particle.

This conclusion is also supported by the arrangement of the VP6 in the aberrant tubular and sheet forms of rotavirus capsomeres found in fecal samples. Chasey and Labram (1983) demonstrated that these structures had a surface lattice of hexagonally arranged subunits. VP6 purified from ss particles by treatment with CaCl (Bican *et al.,* 1982) or LiCl (Ready and Sabara, 1987) also shows a pronounced tendency to form extended tubular forms, the walls of which are composed of a lattice of hexagonally arranged subunits. VP6 expressed in insect cells by recombinant baculoviruses exhibits a similar tendency to aggregate into tubular forms (Estes *et al.,* 1987). It is now clear that the basic hexagonal unit of all of these tubular and sheet forms is derived from clusters of VP6 trimers. Ready *et al.* (1988) processed images of the tubular form of VP6 and demonstrated that the trimers were arranged on a hexagonal lattice, but the resolution of the optical reconstructions was insufficient to reveal details of their molecular organization. However, they calculated the volume of the trimer and inferred that it was probably an elongated structure, a form consistent with the earlier description of the inner capsid protein as a trimeric truncated cone (Esparza and Gil, 1978).

The two-dimensional arrays of purified VP6 can also be induced to form sheets as well as tubes by varying the calcium ion concentration. Studies with these arrays have enabled further details to be deduced about the arrangement of the VP6 trimers (Berriman *et al.,* 1990) (Fig. 2a). The hexameric arrangement of the VP6 trimers can be seen on a 9.8-nm lattice, with some individual trimers and hexagonal clusters also present. Computer-processed reconstruction of the array resolves the packing of the asymmetric molecules making up the trimer (Fig. 2b). The central channel enclosed by the trimeric cluster is roughly 2 nm in diameter, while the larger channels enclosed by the hexameric arrangement of the trimers are approximately 6 nm in diameter. The

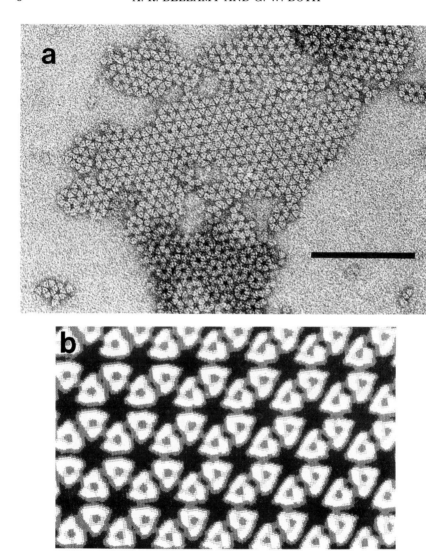

FIG. 2. (a) Electron micrograph of a single-layered VP6 paracrystal negatively stained with uranyl acetate. Regions of overlapping layers can be seen at the top and the bottom. Bar, 0.1 μm. (b) Fourier noise-filtered magnification of the array displayed in gray levels, with protein shown as white areas. The hexagonal lattice has a periodicity of 9.8 nm. (Courtesy of J. Berriman.)

model proposed by Prasad *et al.* (1988) for the virion shows that in the intact virion the larger channels (which are approximately 14 nm deep) extend through both the inner and outer layers of capsomeres. These authors speculate that the large channels might provide a route for both the entry of nucleotide triphosphates and for the egress of mRNA following transcription. Since VP6 has also been identified as the protein which acts as the ligand in interactions with the NS28 receptor (see Section V,B), it is tempting to speculate that the receptor–ligand interaction might involve the interaction of NS28 with the 2-nm central channel enclosed by the individual VP6 trimers.

Treatment of the intact virion with chelating agents removes the outer layer of capsomeres (probably by removing Ca^{2+} required for the stability of ds particles) and yields ss particles which are indistinguishable from those produced in infected cells (Cohen *et al.*, 1979). These have a buoyant density of 1.36 g/cm^3 on isopycnic CsCl gradients, in contrast to the intact virion, which bands at a density of 1.38 gm/cm^3. Particles prepared directly from infected cells or by treatment with chelating agents exhibit RNA transcriptase activity. In fact, the ability to synthesize mRNA *in vitro* over long periods is a remarkable feature of the ss particle, and this ability continues almost indefinitely, provided that the supply of ribonucleoside triphosphates is maintained. Thus, the rotavirus RNA transcriptase appears to exhibit activity similar to that shown by the equivalent reovirus enzymic system (Shatkin and Sipe, 1968; Skehel and Joklik, 1969), but there is little information concerning the identity of the rotavirus proteins involved in the transcription and capping of mRNA.

When ss particles are treated with chaotropic agents (e.g., high concentrations of $CaCl_2$), the outer layer of VP6 is removed to yield a central core that lacks RNA transcriptase activity. The transcriptase activity of the central core was reconstituted when VP6 reassociated with the particle after the concentration of the chaotropic agent was reduced (Bican *et al.*, 1982). Similarly, Sandino *et al.* (1986) reported that up to 70% of the transcriptase activity was restored when EGTA or EDTA was used to remove the high concentrations of Ca^{2+} used initially to dissociate VP6 from the particle. Restoration of transcriptase activity to the central core was also largely independent of the strain of virus used to prepare VP6, implying that the binding sites for the assembly of trimers on the surface of the inner nucleocapsid are conserved across rotavirus strains. From analysis of the rotavirus VP1 gene (Cohen *et al.*, 1989; see Section VII,A), it seems likely that VP6 itself is not the transcriptase activity and that it may form only part of a much more complex multienzyme system. The recent success in achieving expression of the proteins of the rotavirus central core using recombinant baculoviruses (Cohen *et al.*, 1989) could enable the individual enzymatic activities of the individual proteins in ss particles to be identified.

C. Double-Shelled Particle

The outer layer of capsomeres is also arranged in a $T = 13\ l$ surface lattice, but the surface of the ds particle is much smoother than the ss particle and consequently the symmetry is much less obvious. The lack of definition provided by images of the outer capsomeres frustrated early attempts to discern symmetry by conventional electron microscopy. However, the $T = 13\ l$ lattice can be inferred from the arrangement of the surface spikes first observed in the reconstructed images by Prasad *et al.* (1988) (Fig. 1A and B). The spikes are also quite easily observed when freeze-etched preparations of the virus are shadowed with platinum (Both *et al.*, 1989) or following negative staining by using low electron radiation doses (Berriman and Bellamy, 1990). In the reconstructed image by Prasad *et al.* (1988) (Fig. 1A and B), the spikes themselves are about 45 Å long and 35 Å wide and possess a knob at the distal end. The spike appears bifurcated in some of the reconstructed images (Fig. 1A and B), and such a structure is also evident in the negatively stained images presented in Fig. 3. Here, many of the spikes possess a clear "Y" or "lollipop" configuration.

Based on the particle symmetry and the calculation that the volume of the outer shell, including the spikes, is 4.5×10^7 Å^3, Prasad *et al.* (1988) estimated that there are 780 copies of VP7 and 60 copies of VP4 in the virion. Thus, it seems likely that the spikes and the smooth shell are composed of VP4 and the glycoprotein VP7, respectively. The assignment of VP4 to the spike must be regarded as provisional, pending further biochemical or immunological evidence, but it should be possible to identify the proteins involved unequivocally by decoration of the particle with Fab fragments of appropriate specific monoclonal antibodies. Perhaps the structure of the tips of the spikes will prove to be associated with some important aspect of their function, since other viral hemagglutinins also adopt an extended configuration (Wilson *et al.*, 1981; Furlong *et al.*, 1988).

III. Gene/Protein Coding Assignments

A summary of the nonstructural proteins and the structural polypeptides known to represent genuine components of the SA11 virus particle, together with their assignments to individual segments of double-stranded RNA, is presented in Table I. When comparing these

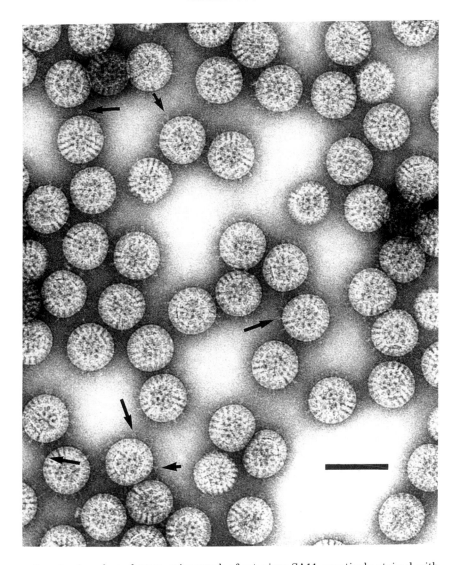

FIG. 3. Low-dose electron micrograph of rotavirus SA11 negatively stained with uranyl formate. The beam-sensitive protrusions, or spikes, can be seen in "Y" (small arrows) and "lollipop" (large arrows) conformations. Bar, 0.1 μm. (Courtesy of J. Berriman.)

data with those for other rotaviruses, it should be noted that the precise order of migration of the double-stranded RNA segments can vary for individual rotavirus strains.

There have been three major reasons for confusion in the literature concerning the identity of rotavirus proteins that represent genuine

TABLE I

ROTAVIRUS GENE CODING ASSIGNMENTS[a]

Genomic segment	Length (base pairs)	Protein	Open reading frame (amino acids)	Protein (molecular weight)[b]	Function	Virus strain	Comments/function
1	3302	VP1	1088	124,847	Central core	Bovine RF	Exhibits homology with RNA polymerases
2	2687	VP2	880	102,431	Central core	Bovine RF	Nucleocapsid protein?; contains "leucine zipper" sequence
3	2591	VP3	835	98,142	Central core	SA11	As for VP1
4	2362	VP4	776	86,775	Outer capsid	SA11	Hemagglutinin; cleaved to yield VP5* and VP8*
5	1611	NS53	495	58,484	Nonstructural	SA11	Possible metal binding domain
6	1356	VP6	397	44,816	Major protein in single-shelled particle	SA11	Subgroup-specific antigen; ligand for NS28
7	1104	NS34	312	36,072	Nonstructural	SA11	ORF starts at second AUG
8	1059	NS35	317	36,128	Nonstructural	SA11	—
9	1062	VP7 (glyco-protein)	326	37,198	Outer capsid or ER membrane	SA11	ORF starts at first AUG; type-specific antigen
10	751	NS28 (glyco-protein)	175	20,309	Nonstructural glycoprotein in ER membrane	SA11	Acts as a receptor for single-shelled particles
11	667	NS26	198	21,520	Nonstructural	SA11	Probably phosphorylated

[a]Based on data derived from cloned SA11 rotavirus genes in which sequences are available; otherwise, the data for the bovine RF strain are presented. Sequence references are gene 1, Cohen et al. (1989); gene 2, Kumar et al. (1989); gene 3, D. B. Mitchell and G. W. Both (unpublished observations); gene 4, Mitchell and Both (1988); gene 5, D. B. Mitchell and G. W. Both (unpublished observations); gene 6, Both et al. (1984b); gene 7, Both et al. (1984a); gene 8, Both et al. (1982); gene 9, Both et al. (1983a); gene 10, Both et al. (1983b); gene 11, Mitchell and Both (1988). For details of other sequences of cloned rotavirus genes, see text. ER, endoplasmic reticulum.

[b]All molecular weights are for the unmodified polypeptides.

structural components of the virus. First, during what might now be referred to as the "heroic age" of rotavirology, some early workers dealt with less than adequately purified virus, due to the early reliance on feces as a source material. This era predated the discovery that low concentrations of proteolytic enzymes enhanced the replication of rotaviruses in cell culture (Estes *et al.*, 1983). Second, the routine use of trypsin in virus propagation produced the complication that some cleavage products were incorrectly identified as primary gene products. Finally, problems were created by the conflicting terminology adopted for the various structural and nonstructural viral proteins and by the need to change terminology already in general use as conflicts in the numbering system became apparent (Liu *et al.*, 1988).

Despite these initial difficulties, several independent methods have been used to assign the individual gene products to particular RNA segments (Table I). The advent of cloned rotavirus genes and the use of appropriate *in vitro* transcription and translation systems (see, for example, Both *et al.*, 1983a; Mason *et al.*, 1983) have enabled the majority of the genes to be correctly assigned to their translation products. Information derived from genetic recombination among different rotavirus strains (Greenberg *et al.*, 1983a) was also particularly useful. The genetic approach was possible because rotaviruses share with reovirus the ability to undergo genetic recombination at a high rate when more than one strain is used to infect target cells (Ramig and Fields, 1983). When recombinants incorporating particular segments or groups of segments from one parental strain were analyzed, the immunological properties of the virus correlated with the presence of a particular coding segment of double-stranded RNA. This approach was first used successfully by Greenberg *et al.* (1983a) to assign VP6 to segment 6, the serotype protein VP7 to segment 9, and growth restriction to segment 4.

The key features that emerge from a consideration of these coding assignments are as follows. The rotavirus ss particle is composed of four structural proteins—VP1, VP2, VP3, and VP6—the latter being the major component (see Section II,B). VP6 is also the group-specific antigen which classifies most rotaviruses into one of two subgroups (Greenberg *et al.*, 1983b). The outer shell of the virion is composed of two proteins: VP4 (formerly termed VP3; Liu *et al.*, 1988) and VP7. VP4 is the hemagglutinin and VP7 is the type-specific antigen. Five of the rotavirus genomic segments code for nonstructural proteins, of which one, NS28, is a glycoprotein. There is now a significant body of evidence indicating that NS28 functions as a receptor during virion morphogenesis (see Section V,B), but the function of the other nonstructural proteins is unknown. However, cDNA clones are now available for all 11 of the segments, and all but VP3 have been sequenced. In fact, for segments which code for proteins of immunological significance, sequences are available for a large number of strains. The availability of these cloned gene copies has enhanced our understanding of the function of some of these gene products, but, as becomes apparent below, our knowledge is still incomplete.

IV. Viral Entry, Uncoating, and Morphogenesis

Despite the considerable progress that has been made toward elucidating the pathway of rotavirus morphogenesis, there are still many gaps in our knowledge. In this section we present an overview of the virus replication cycle; the molecular biology of specific proteins is discussed in subsequent sections. Briefly, a virus enters the cell and is uncoated to activate the virion transcriptase, which produces single-stranded mRNA transcripts. Following translation most proteins accumulate in viroplasmic inclusion bodies in the cytoplasm, but the two glycoproteins—VP7 and NS28—are directed to the ER membrane. RNA packaging and replication are proposed to occur in the inclusion bodies, which are also the sites of assembly of the ss particles. These particles bud through the membrane of the ER into the luminal space, becoming transiently enveloped in the process. Mature ds particles accumulate in cisternae of the ER, but the mechanism by which the envelope is lost and the outer capsid is acquired is unknown.

A. Entry and Uncoating

It is generally assumed that nonenveloped viruses (e.g., the members of Reoviridae) enter the cell by receptor-mediated endocytosis, involving incorporation of the incoming virion into an endocytic vesicle followed by localization of the particle to a low-pH compartment, where uncoating occurs (Dimmock, 1982). Petrie et al. (1982) and Quann and Doane (1983) observed rotavirus particles in vesicles and coated pits, suggesting that they entered by endocytosis. However, the interpretation of electron-microscopic evidence impinging on the possible mechanisms by which virus penetration is achieved is notoriously difficult, especially because it is not easy to demonstrate that the population of particles studied is actually the one that generates the productive infection.

More recent evidence suggests that the virus enters the cells by two routes: endocytosis and direct penetration of the cell membrane (Suzuki et al., 1985, 1986). It has been shown that inhibitors of endocytosis (e.g., dinitrophenol and sodium azide) and lysosomotropic agents (e.g., ammonium chloride and chloroquine) produce only a minimal effect on the infection of MA104 cells by rotavirus (Kaljot *et al.*, 1988), suggesting that conventional receptor-mediated endocytosis might not represent the major route of entry for productive infection. It has also been demonstrated that cleavage of the hemagglutinin VP4 plays a role in the virus entry, since trypsin-activated virus was internalized much more rapidly than was nonactivated virus. The release of ^{51}Cr from the cells, concomitant with virus entry, was interpreted as evidence that the particle directly penetrates the membrane (Kaljot *et al.*, 1988), a mechanism previously proposed for reovirus by Borsa *et al.* (1979). The direct transfer of rotavirus particles across the membrane would represent an entirely novel mechanism for entry, for which the molecular mechanism is presently obscure. It has been suggested that a region in VP4 which shows homology with the proposed fusion region of Sindbis virus could be involved in the entry of rotaviruses (Mackow *et al.*, 1988b). The sequence is located in VP5*, beginning 136 residues downstream from the amino terminus created by the cleavage of VP4. Several neutralizing monoclonal antibodies have been mapped to this region (Mackow *et al.*, 1988b).

B. *Location of Viral Proteins in the Cell*

Electron microscopy has identified two important features of rotavirus morphogenesis (Fig. 4). First, large viroplasmic inclusions are present in the cytoplasm of the cell, and, within these, assembled ss particles are clearly evident. Second, assembled virions are localized exclusively in the lumen of the rough ER, and the transfer of ss particles is achieved by a budding mechanism (Fig. 4) which results in the transient envelopment of these particles in a membranous vesicle. Such enveloped particles are known to accumulate in the presence of tunicamycin (Sabara *et al.*, 1982; Petrie *et al.*, 1983). Thus, viral proteins seem to be localized principally in the viroplasmic inclusions, the ER membrane, and the lumen of the ER.

Early work aimed at localizing the intracellular site of individual rotavirus proteins was initially carried out with monospecific polyclonal antisera coupled with electron microscopy and immunoperoxidase labeling techniques. VP2 and VP6 were localized to viroplasmic inclusion bodies (Chasey, 1980; Petrie *et al.*, 1982), but the resolution achieved in these early studies was not high. More recently, monoclonal antibodies have been used to localize proteins (Petrie *et al.*, 1984; Poruchynsky *et al.*, 1985; Richardson *et al.*, 1986; Kabcenell *et al.*, 1988). Using antibody-coated colloidal gold particles as specific

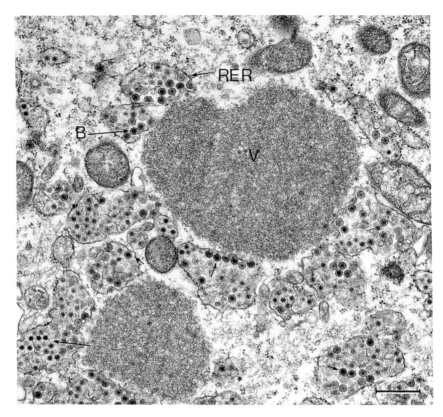

FIG. 4. Morphology of SA11 rotavirus-infected MA104 cells at 8 hours postinfection. Viroplasms (V) and the rough endoplasmic reticulum (RER) membrane-enveloped (short arrows) and mature particles (long arrows) are indicated. Budding particles (B) can also be seen. ×24,300. Bar, 500 nm. (Courtesy of M. S. Poruchynsky and P. A. Atkinson.)

postembedding immunocytochemical labels, the product of gene 11 and the nonstructural protein NS35 were shown to be located in viroplasmic inclusion bodies. VP6, the major inner capsid protein, was accessible to antibodies on some of the nonenveloped (i.e., ss) or deenveloped particles. However, the nonstructural glycoprotein, NS28, was localized to the cytoplasmic side of the ER membrane (Petrie et al., 1984). In fact, Kabcenell et al. (1988) found an annular localization for NS28 by immunofluorescent staining. This suggests it is located peripherally in the viroplasmic inclusions, consistent with its role as a receptor in budding (Au et al., 1988; Meyer et al., 1989), which is initiated where the ER membrane and the edge of the inclusion body are in contact.

VP7, the major outer capsid protein was localized among the membrane within the cisternae of the ER or associated with nonenveloped (i.e., mature) particles (Petrie *et al.*, 1984; Richardson *et al.*, 1986). The location of nonviral VP4 is less clear, although it can be detected in virus particles (Petrie *et al.*, 1984; Kabcenell *et al.*, 1988). By immunofluorescence it appeared to have a localization similar to those of VP7 and NS28, suggesting that it too could be associated in some way with the ER membrane, which is an active site of virus assembly (Kabcenell *et al.*, 1988). Proteins VP1, VP3, NS53, and NS34 have not been studied in this way, since appropriate immune sera or monoclonal antibodies are not yet available.

C. Transcription, Double-Stranded RNA Replication

Rotaviruses share with reovirus the ability to form reassortant viruses at high frequency when cells are simultaneously infected with two different strains of virus (Ramig and Fields, 1983; Kapikian and Chanock, 1985). This ability indicates that there must be a stage in viral infection when the plus-stranded segments are able to assort freely prior to the incorporation of an appropriate selection of 11 segments into the progeny particle. However, the mechanism by which this occurs is unknown. This problem applies also to reovirus and other viruses (e.g., Orthomyxoviridae) which also form reassortants at high frequencies. Developing an understanding of the mechanisms involved in reassortment remains one of the central areas of the molecular biology of the Reoviridae in which no significant progress has been made over the last 10 years.

However, some progress has been made in understanding the events involved in viral RNA synthesis. During rotavirus RNA replication, plus strands transcribed from the double-stranded segments serve as templates for minus strand synthesis (Patton, 1986). In one study of the kinetics of RNA replication (Stacy-Phipps and Patton, 1987), the levels of minus- and plus-stranded RNA were found to increase between 3 and 9–12 hours postinfection, at which time levels were maximal. However, the ration of plus- to minus-strand RNA changed during infection, and maximal levels of minus-stranded RNA synthesis occurred several hours prior to the peak for plus strand synthesis. Thus, the level of RNA replication did not correlate directly with the level of transcription.

Specific gene probes have also been used to monitor the levels of individual plus- and minus-stranded RNAs (Johnson and McCrae, 1989). Under conditions of high multiplicity of infection, it appeared that not all minus-strand species accumulated to the same level, the negative transcripts of genes *1, 4,* and *9* being present in greatest amounts. Incorporation of these species into infectious virus was inefficient, however. Plus-stranded species were produced in amounts that varied over a fivefold range, but the viral proteins they encoded were translated with vastly differing efficiencies. Genes *6* and *10* products were synthesized 36- and 54-fold more abundantly than the gene *11* protein, which was used as a reference standard, while the product of gene *1* was made at one-tenth of that level. Thus, some rotavirus mRNAs were translated much more efficiently than others, a feature previously noted for reovirus mRNAs (Shatkin and Kozak, 1983). This does not necessarily imply that some form of active translational control operates during infection. Rather, mRNAs inherently efficient in initiation could be preferentially translated as a result of the need for certain quantities of a particular viral protein during infection.

D. *Formation of Subviral Particles*

The assembly of rotavirus particles has been investigated by analysis of the complexes which can be isolated from lysates of infected cells. This work parallels earlier studies on the reovirus replicase particle by Zweerink (1974), Morgan and Zweerink (1975), and Zweerink *et al.* (1976). Lysates were fractionated on CsCl gradients or agarose gels to resolve subviral particles and these were then analyzed for protein and RNA content and for RNA replication and transcription activity (Helmberger-Jones and Patton, 1986; Patton and Gallegos, 1988; Gallegos and Patton, 1989). Particles with densities of 1.34 and 1.38 g/cm^3, which were presumed to be analogous to ss and ds particles, respectively, accounted for most of the transcriptase activity. The pellets from these gradients were enriched in complexes that had replicase activity, but this was partly due to the presence of single-stranded RNA in the complexes, since treatment with micrococcal nuclease reduced their density (Patton and Gallegos, 1988). Electrophoretic analysis showed that these particles were similar to ss particles in that they contained VP1, VP2, and VP6 (although this protein was underrepresented in comparison with ss particles), but in addition the nonstructural proteins NS34 and NS35 were also present. Once double-stranded RNA synthesis had occurred, however, these particles rapidly matured into ss particles (Patton and Gallegos, 1988).

More recently, subviral particles from rotavirus infected cells were resolved by electrophoresis under nondenaturing conditions in agarose gels (Gallegos and Patton, 1989). The complexes capable of replicating the 11 segments of double-stranded RNA were found to be heterogeneous in size and density. However, when their protein composition was analyzed, three types of replication intermediates were identified. These may correspond to particles described earlier in infected cells examined by electron microscopy (Esparza *et al.*, 1980). The smallest intermediate, referred to as the precore (45 nm, 160 S), was composed of structural proteins VP1 and VP3 and nonstructural polypeptides NS34, NS35, and NS53. The product of gene segment *11* (i.e., NS26) was also present. In comparison, the second intermediate (60 nm, 210 S) had also acquired VP2, the major structural protein of the central core. These particles appeared not to be the same as virion-derived central cores, since they did not comigrate during gel electrophoresis. The largest intermediate, a precursor to ss particles (75 nm, 320–390 S), was composed of VP1, VP2, VP3, and VP6 and nonstructural proteins NS34, NS35, and NS26. Analysis of these intermediates in rotavirus-infected cells pulse-labeled with ^{35}S-amino acids suggested that ss particles were assembled during infection by the sequential addition of VP2 and VP6 to a precore intermediate composed of VP1, VP3, NS34, NS35, NS26, and NS53 (Gallegos and Patton, 1989).

The use of nondenaturing methods, such as those used in the work detailed above, may eventually enable rotavirus precursor particles to be characterized and the steps in assembly to be clarified. However, some uncertainty remains as to the effect that the extraction procedures could have on particles that exist in the intact infected cell. Heterogeneity in particle composition may also be induced during extraction, and it is possible that proteins present in unrelated structures nevertheless comigrate on agarose gels and density gradients.

E. Budding of Single-Shelled Particles and Virus Maturation in the Endoplasmic Reticulum

The next step in the pathway of viral morphogenesis appears to be the entry of immature ss particles into the ER. During this process, or subsequent to it, the ss particles acquire an additional layer of outer capsomeres consisting of VP4 and VP7, thus yielding the progeny virions. The initial interaction between the ss particle and the membrane is mediated by the nonstructural glycoprotein NS28 and is dealt with in Section V,B. The means by which VP7 is directed to the ER and retained in that compartment for virus assembly is also the subject of separate discussion. Therefore in this section we point out subsequent parts of the proposed scheme for morphogenesis about which we know either nothing or remarkably little.

First, the location of VP4 in the cell is poorly understood. Immunofluorescence studies have suggested that this protein has a distribution in infected cells similar to those of NS28 and VP7, both of which are associated with the ER membrane. Thus, VP4 could also be membrane associated, but located on the cytoplasmic side, and could be acquired by ss particles during budding (Kabcenell *et al.*, 1988). Second, it is not understood how the membrane envelope acquired by ss particles during budding is subsequently removed. Rotavirus-infected cells treated with tunicamycin, which inhibits N-linked glycosylation, are known to accumulate membrane-enveloped subviral particles in the ER (Sabara *et al.*, 1982; Petrie *et al.*, 1983). The accumulation of membrane-enveloped particles cannot be due to any effect on the glycosylation of VP7 per se, since a mutant of SA11 which has lost the glycosylation site is nevertheless able to replicate normally (Estes *et al.*, 1982). Thus, it seems likely that NS28 is involved in the processes which attend removal of the envelope, given the known interaction of this protein with the ss particle (see Section V,B) and the fact that it is the only other vital glycoprotein which could be affected by tunicamycin treatment. Third, it is also not known how virus particles acquire VP7. Acquisition of soluble VP7 following removal of the membrane envelope was suggested by Petrie *et al.* (1984), but this mechanism seems incompatible with the integral membrane character of the protein (Kabcenell and Atkinson, 1985). VP7 could be acquired during the budding process if the protein excluded host membrane and protein components by forming patches in the membrane at the budding site. Fourth, mature rotavirus contains Ca^{2+} (Shahrabadi *et al.*, 1987), which is consistent with the instability of rotavirus particles in the presence of calcium-chelating agents (Cohen *et al.*, 1979; McCrae and Faulkner-Valle, 1981). When virus-infected cells were grown in calcium-deficient medium, only incomplete ss particles were produced and virus titers were dramatically reduced (Shahrabadi and Lee, 1986). Virus particles wrapped in fragments of ER membrane were also present, but budding of subviral particles into the ER was not observed (Shahrabadi *et al.*, 1987). Thus, the defect in virus production induced by low levels of calcium appears to occur during the assembly of the outer capsid layer, but it is not clear how calcium could be involved in the assembly process.

Finally, it has been suggested that rotavirus replication and assembly could occur in association with the cytoskeleton, since ss particles remained cell associated after extraction with Triton X-100 (Musalem and Espejo, 1985). A similar association of virus components with the cytoskeleton was observed during reovirus (Mora *et al.*, 1987) and bluetongue virus replication (Eaton *et al.*, 1987). An involvement with the cytoskeleton may therefore be a general feature of replication for the Reoviridae, but further investigation is required to establish whether this is the case.

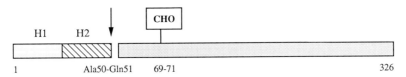

FIG. 5.   Features of the rotavirus SA11 VP7 protein. H1 and H2 are the hydrophobic signal peptide domains. Cleavage occurs between Ala 50 and Gln 51 (arrow) to generate mature VP7, which is glycosylated at Asn 69 [CHO (carbohydrate)].

V. ROLE OF SPECIFIC VIRAL PROTEINS IN MORPHOGENESIS

A. *The Major Glycoprotein VP7*

1. *Nature of the Carbohydrate*

The outer capsid glycoprotein VP7 was the first neutralizing antigen recognized for rotaviruses and is the principal antigen by which rotavirus serotypes are defined. The protein is the major component of the outer capsid and the only structural polypeptide which is glycosylated. Details of the domains of VP7 are shown in Fig. 5. Early investigations of the carbohydrate attached to VP7 were carried out by labeling the protein with [³H]mannose or [³H]glucosamine (Arias *et al.*, 1982; Ericson *et al.*, 1983). This showed that VP7 carbohydrate was sensitive to digestion with endoglycosidase H, indicating that it was of the high-mannose type, a finding confirmed by direct analysis of the oligosaccharides released from VP7 isolated from the bovine BDV 486 strain (Kouvelos *et al.*, 1984) and the simian SA11 virus (Both *et al.*, 1983c, Kabcenell and Atkinson, 1985; Kabcenell *et al.*, 1988). For the latter it was clear that the original Glc3Man9GlcNac2 moiety was processed predominantly to the Man8 and Man6 forms, but no complex carbohydrate was detected. Since complex carbohydrate is only added to proteins in transit through the Golgi apparatus (Hubbard and Ivatt, 1981), this is a clear indication that VP7 never reaches this organelle in the cell and that the molecule remains confined to the ER.

2. *Gene and Protein Structure*

The structure of the VP7 gene, first determined by cloning segment 9 from simian rotavirus SA11 (Both *et al.*, 1983a), has been confirmed by studies of many other strains (Elleman *et al.*, 1983; Arias *et al.*, 1984; Dyall-Smith and Holmes, 1984; Richardson *et al.*, 1984; Gunn *et al.*, 1985; Glass *et al.*, 1985; Mason *et al.*, 1985; Gorziglia *et al.*, 1986a, 1988c; Ruiz *et al.*, 1988; Green *et al.*, 1989; Reddy *et al.*, 1989; Hum *et al.*, 1989). At the 3' end there is complete conservation of eight bases between the various strains sequenced. At the 5' end there is even more extensive sequence conservation. Among eight isolates, representing six serotypes, there are no changes in the first 10 bases, and only nine changes from the basic *SA11* gene sequence are found in the different serotypes prior to position 72. However, the conserved RNA sequence overlaps the first of two potential inframe initiation codons in the gene which are also absolutely conserved. The 5' proximal codon has a weak consensus sequence for initiation; the second codon, a strong consensus (Kozak, 1984, 1986). Therefore, it is not clear whether this conserved sequence is maintained for reasons concerned with the initiation of protein synthesis or for some other reason (e.g., the conservation of signals involved in the packaging of the segment).

The longest open reading frame of the gene (bases 49–1026) predicts a protein of 326 amino acids. For the SA11 and OSU strains there is only a single glycosylation site, Asn–Ser–Thr at residue 69; this site is common to all strains except NCDV (Gunn *et al.*, 1985; Glass *et al.*, 1985). For other strains there are alternative Asn–X–Ser/Thr sites at residues 146–148 (unique to serotype 2), 238–240 (serotypes 1, 2, 6, and 8), and 318–320 (serotype 6); two of these are used in Wa (serotype 1) and bovine strain BDV 486 (Kouvelos *et al.*, 1984).

3. Location of the Signal Peptide

A plot of hydrophobicity across the protein shows that there are two prominent regions of hydrophobic amino acids near the amino terminus which have the characteristics of signal peptides (Gunn *et al.*, 1985). These have been referred to as the H1 and H2 domains (Fig. 5) (Whitfeld *et al.*, 1987). This correlates well with the observation that VP7 has a cleavable signal peptide (Ericson *et al.*, 1983), but raises questions as to which region actually functions as the signal sequence and, therefore, where cleavage occurs. These questions were addressed by introducing mutations into the *VP7* gene (Poruchynsky *et al.*, 1985; Whitfeld *et al.*, 1987). Mutations which deleted the regions coding for the first, the second, or both hydrophobic domains were constructed, and these genes were transiently expressed in COS cells using a simian virus 40 (SV40) expression vector. The presence of signal peptide function was monitored by whether the resulting protein became glycosylated. This work revealed that, in the absence of both hydrophobic regions, glycosylation did not occur. However, the presence of either of the hydrophobic domains was sufficient to permit glycosylation; that is, signal peptide function was present in both the H1 and H2 domains (Whitfeld *et al.*, 1987).

Other work demonstrated that the presence of the second hydrophobic domain alone was sufficient to permit the glycosylation and processing of VP7 (Stirzaker *et al.,* 1987; Poruchynsky and Atkinson, 1988). It is still not clear how frequently translation begins *in vivo* at the first weak initiation codon. Whether a minor amount of a protein with a subtly different function is made from the first initiation codon during infection is also not known. Nevertheless, authentic VP7 can be synthesized from the second initiation codon (see Section V,A) and presumably this one is used most of the time.

4. Processing of the Signal Peptide

It was reported that VP7 purified from virus was blocked at the amino terminus (Arias *et al.,* 1984). Thus, the site of signal peptide cleavage could not easily be determined by analysis of the viral protein. To circumvent this problem, Stirzaker *et al.* (1987) constructed modified forms of the *VP7* gene in which initiation occurred specifically at the first or second AUG codons. The mutated genes were then transcribed and translated *in vitro* in a reticulocyte lysate system in the presence of canine pancreatic microsomes to determine whether cleavage had occurred. The sizes of the processed products in each case were identical and therefore independent of whether translation began at the first or second ATG. This finding strongly suggested that cleavage occurred downstream from the second signal peptide, irrespective of whether one or both hydrophobic domains were present. A likely cleavage site between Ala50 and Gln51, identified from empirical rules (von Heijne, 1986), was mutated to determine whether this would prevent cleavage *in vitro.* Depending on whether both hydrophobic domains or just one (i.e., the H2) was present, changing Ala50 to valine prevented cleavage completely or resulted in another site being utilized, providing indirect evidence that Gln51 was the amino-terminal residue of mature VP7. Direct identification of the cleavage site was obtained by partial amino acid sequencing, which confirmed that both the native viral protein and the VP7 protein translated *in vitro* were processed to leave Gln51 at the amino terminus (Stirzaker *et al.,* 1987). In an earlier study (Chan *et al.,* 1986) it was suggested, on the basis of the presence of two VP7 species in purified virus, that the VP7 precursor was cleaved between the H1 and H2 domains. While the above evidence does not support that conclusion, the two studies are in agreement with respect to the existence of multiple (i.e., two or three) forms of VP7 in purified virus. However, the precise derivation of these species remains to be determined.

5. Retention of VP7 in the Endoplasmic Reticulum

The *in vitro* studies described above indicate that VP7 is directed to the ER by a cleavable signal sequence and then processed to remove the two most prominent hydrophobic regions in the protein. Furthermore, VP7 translated in the presence of microsomes becomes resistant to digestion with trypsin, indicating that it is translocated completely into the lumen of the ER (Kabcenell and Atkinson, 1985; Stirzaker *et al.*, 1987). The processed form of the protein is also resistant to extraction with high salt and sodium carbonate, and by this criterion it becomes an integral membrane protein after entering the ER (Kabcenell and Atkinson, 1985). This membrane-bound form of VP7 can be distinguished by immunological and biochemical techniques from the polypeptide already incorporated into virus particles, indicating that there are two pools of the protein present in the cell during infection (Kabcenell *et al.*, 1988). Based on the kinetics of the processing of oligosaccharides attached to viral versus membrane-bound VP7, it was suggested that the latter is the precursor of the former.

Considerable effort has been devoted to clarifying the mechanism by which VP7 is retained in the ER, since this protein is one of only a small number of membrane-associated proteins with this property. A deletion which removes amino acids 47–61 is sufficient to convert VP7 from a resident ER protein into a secreted polypeptide, which is then modified with complex carbohydrate as it passes through the Golgi apparatus (Poruchynsky *et al.*, 1985). Since VP7 never normally traverses the secretory pathway, it seems unlikely that the protein would carry a signal necessary for transport. Thus, the results obtained from deletion analysis argue strongly that a positive signal is required for retention in the ER and that transport is the default pathway. As noted above, the deletion of residues 47–61 caused secretion of VP7, while the deletion of amino acids 51–61 did not. The difference between these two mutations is that in the latter the H2 signal peptide is left intact, while in deletion 47–61 the last four residues of H2 have been removed. This suggests that the signal peptide itself is involved in both targeting VP7 to the ER and retaining it in that organelle. This has been tested by replacing the H2 region in the gene with sequences coding for the signal peptide from influenza hemagglutinin (HA), a membrane protein which is normally transported to the cell surface (Gething and Sambrook, 1981). VP7 derived from the HA precursor was secreted from the cells, indicating that the H2 signal peptide clearly plays a role in both targeting and retention (Stirzaker and Both, 1989).

The precise mechanism by which the signal region of VP7 acts as a retention sequence is still obscure, but two things are clear: First, the signal peptide exerts its effect despite being rapidly cleaved from the protein. Second, not all of the information necessary for retention is present in the H2 region. This was shown by splicing the H2 region onto another reporter molecule, the malaria S antigen; this modified molecule was still secreted (Stirzaker and Both, 1989). Another study which used α-amylase as a reporter molecule concluded that residues 62–111 of VP7 were necessary, but alone were not sufficient for retention (Poruchynsky and Atkinson, 1988). Collectively, the results indicate that the signal peptide and a region which lies within residues 62–111 of VP7 together are sufficient for retention of the protein in the ER, but the means by which the protein remains anchored in the membrane in the absence of obvious hydrophobic transmembrane domains remains a mystery.

6. VP7 as the Cell Attachment Protein?

When lysates prepared from rotavirus-infected cells were clarified by centrifugation and layered onto intact MA104 cells under conditions normally used for virus adsorption *in vitro,* the free form of VP7 bound to an unidentified component on the cell surface (Sabara *et al.,* 1985; Fukuhara *et al.,* 1988). Binding of the protein was competitively inhibited by the addition of increasing amounts of unlabeled intact homologous virus particles (Sabara *et al.,* 1985), whether or not they were treated with trypsin (Fukuhara *et al.,* 1988). Binding of VP7, or adsorption of virus particles to cells, was also inhibited by the addition of antiviral serum (Sabara *et al.,* 1985) or VP7-specific neutralizing monoclonal antibodies (Sabara *et al.,* 1985; Fukuhara *et al.,* 1988). A monoclonal antibody to VP4 also inhibited virus adsorption to a lesser extent, but a VP6-specific antibody and a nonneutralizing VP7 monoclonal had no effect (Sabara *et al.,* 1985). It was suggested that VP7 might be involved in attaching rotavirus particles to the cell surface. However, if this suggestion is confirmed, it would reflect a curious situation in view of the prominence of the viral hemagglutinin on the surface of the virion (Prasad *et al.,* 1988) and the fact that, for many other viruses, the hemagglutinating protein is the cell attachment protein (see Section VI).

B. The Nonstructural Glycoprotein NS28

1. General Biological Properties

NS28, the product of gene *10,* has also been called NCVP5 (Arias *et al.,* 1982). It was first shown by Petrie *et al.* (1982, 1984) that NS28 was localized to the ER membrane. More recently, Kabcenell *et al.* (1988) demonstrated, by immunofluorescence, using an NS28-specific monoclonal antibody, that the protein is localized to the periphery of the viroplasmic inclusion bodies found in the cytoplasm of the rotavirus-infected cells. These inclusion bodies have been identified as the sites of assembly of the ss particles (Poruchynsky *et al.*, 1985; Kabcenell *et al.*, 1988), which are then transferred into the lumen of the rough ER. NS28 is involved in this process, acting as a receptor for the budding particles (Au *et al.*, 1988; Meyer *et al.*, 1989). The interaction between NS28 and the ss particles is highly specific and involves the cytoplasmic domain of NS28 and the surface protein VP6 (Meyer *et al.*, 1989).

2. *Orientation of NS28 in the Endoplasmic Reticulum and Nature of the Carbohydrate*

Amino acid sequences for NS28 have been deduced from the nucleotide sequences of cDNA clones from various rotavirus isolates (Both *et al.*, 1983b; Okada *et al.*, 1984; Baybutt and McCrae, 1984; Ward *et al.*, 1985; Powell *et al.*, 1988). All segment *10* sequences possess three conserved inframe AUG codons, of which the first has a strong consensus sequence for initiation (Kozak, 1984), from which a protein of 175 amino acids can be translated. Two glycosylation sites are located in the first of three NH_2-terminal hydrophobic domains (Both *et al.*, 1983b): identified as H1, H2, and H3 (Fig. 6). One of these hydrophobic regions must provide signal peptide function, since the protein enters the ER membrane and is glycosylated, but the signal peptide must be uncleaved, since both these sites are present in the mature form of the protein. Furthermore, since the carbohydrate present on the molecule is exclusively of the high-mannose type (Both *et al.*, 1983b; Kabcenell and Atkinson, 1985), the protein clearly does not reach the Golgi apparatus, which is where the addition of more complex sugars occurs (Hubbard and Ivatt, 1981). Thus, NS28 shares with VP7 the distinction of being one of a small group of glycoproteins which are ER localized (Bergmann *et al.*, 1989).

The fact that asparagine residues at positions 8 and 18 are glycosylated also indicates that this region of the protein must be luminally oriented, in contrast to the COOH-terminal region of the protein, which is protease sensitive, indicating that it lies on the cytoplasmic side of the membrane (Ericson *et al.*, 1983; Kabcenell and Atkinson, 1985). Since there are three hydrophobic regions at the amino terminus of the protein and the first of these (containing the glycosylation sites) protrudes into the lumen, the possible options for the arrangement of NS28 in the membrane are somewhat limited. Bergmann *et al.* (1989) constructed a series of mutants of the gene and studied membrane insertion of the resulting translation products in a coupled transcription–translation system in the presence of canine pancreatic microsomes. These experiments confirmed that the H1 domain containing the glycosylation sites projects into the lumen and identified H2 as

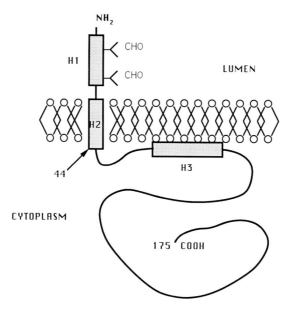

Fig. 6. A model for the topology of NS28 in the endoplasmic reticulum membrane. Hydrophobic domains H1, H2, and H3 are shown. Carbohydrate (CHO) is attached as asparagine residues 8 and 18. The polypeptide chain emerges from the membrane near amino acid 44.

a combined signal peptide/anchor sequence. The sensitivity of wild-type and mutant forms of NS28 to digestion with various proteases showed that the polypeptide emerged from the membrane near residue 44. Thus, H3 remains exposed on the cytoplasmic side of the membrane, leaving approximately 131 amino acids available for receptor–ligand interaction (Fig. 6).

Somewhat similar studies were undertaken by Chan *et al.* (1988). In this case H3 was identified as the membrane-spanning domain, due to its resistance to proteolytic attack. However, as pointed out by Bergmann *et al.* (1989), it seems possible that the refractory nature of the H3 region to proteolysis could be due to other factors; for example, if there were intermolecular disulfide bonds involving the cysteine residues present at positions 63 and 71, these might render this region of the protein resistant to proteolysis. No information is available concerning the possible multimeric state of NS28 in the membrane, but in view of its function as a receptor (see next section) it seems likely that in its native form NS28 could be a multimeric rather than a monomeric structure.

3. NS28 as a Receptor

A possible function for NS28 was first suggested by Petrie *et al.* (1983), who proposed that it might be a scaffolding protein that plays a role in virion morphogenesis. While such a role for NS28 cannot be excluded, definitive evidence is now available to show that NS28 acts as a receptor for ss particles prior to their transfer across the membrane into the lumen of the rough ER. Au *et al.* (1988) were the first to demonstrate that membranes prepared from SA11-infected cells were able to bind homologous ss particles in an *in vitro* binding assay. They also demonstrated that membranes prepared from the cells of the insect *Spodoptera frugiperda,* which had been infected with recombinant baculoviruses expressing the *NS28* gene, were able to bind ss particles, indicating that the presence of NS28 alone was sufficient for the receptor–ligand interaction. This work was confirmed and extended by Meyer *et al.* (1989), who used a series of recombinant vaccinia viruses to deliver NS28 and other rotavirus proteins to MA104 cells. Receptor binding was assessed using an *in vitro* assay, based on the binding of ^{125}I cores to the rough ER membranes extracted from the cells. The interaction of NS28 with ss particles was enhanced by the presence of Ca^{2+} and Mg^{2+}, and Scatchard analysis indicated a dissociation constant for the interaction on the order of $5 \times 10^{-11}\ M$. The affinity of a rotavirus ss particle for its receptor is therefore similar to that exhibited by other virus–receptor interactions (cf. rhinovirus at $5 \times 10^{-11}\ M$; Colonno *et al.,* 1988).

Meyer *et al.* (1989) identified VP6 as the ligand involved in binding, on the basis of the ability of this protein to block the interaction and by the use of a monoclonal antibody specific for this protein which also was able to block the interaction between NS28 and the ss particles. Single-shelled particles of rotavirus strain Wa, a member of subgroup II (Kapikian and Chanock, 1985), were able to block the interaction of SA11 ss particles with their homologous NS28 receptor (Meyer *et al.,* 1989). This finding indicates that the domain(s) of VP6 involved in binding must be conserved to a significant extent between subgroups I and II rotaviruses.

The interaction of NS28 with VP6 provides a model system that could be utilized further to investigate the events that occur when a virus particle interacts with a membrane-localized receptor during budding. Another attractive feature of the NS28–VP6 interaction is the relative ease with which the interaction can be monitored and the fact that the ss particle appears to be amenable to crystallographic analysis (B. Harris, I. Anthony, A. R. Bellamy, and S. C. Harrison, unpublished observations).

VI. VIRAL HEMAGGLUTININ VP4

A. Biological Properties

VP4 is the product of gene segment 4 and was formerly known as VP3 (Estes et al., 1983). It was renamed following identification of the product of gene segment 3 as a structural protein (Liu et al., 1988). VP4 is the hemagglutinin of the virus, since the ability of rotavirus strains to agglutinate red blood cells segregates with this gene segment (Estes et al., 1983; Kalica et al., 1983). Although VP4 is the minor component of the outer capsid of the virus, it is very important biologically. It is both a neutralizing antigen (Hoshino et al., 1985; reviewed by Matsui et al., 1989) and an important determinant of virulence and growth in cell culture (Greenberg et al., 1983a; Offit et al., 1986). These properties are similar to those described for the influenza hemagglutinin and the σ1 protein of reovirus. The parallels with the influenza virus hemagglutinin are close, given that both proteins hemagglutinate and must be cleaved for infectivity. There are also several binding sites for neutralizing monoclonal antibodies located on the HA1 subunit of hemagglutinin (Webster et al., 1982), and these are located around the proposed cell attachment pocket (Wiley et al., 1981). If the neutralizing monoclonal antibodies which recognize VP8* (Mackow et al., 1988a; Taniguchi et al., 1988) are analogous to those described for influenza HA1, they should also prevent virus adsorption; this could easily be tested.

In the presence of trypsin, the protein (88 kDa) is cleaved to produce polypeptides VP5* (60 kDa) and VP8* (28 kDa), which greatly enhances infectivity in vitro (Fig. 7) (Clark et al., 1981; Espejo et al., 1981; Estes et al., 1981). The precise mechanism by which this occurs is not known, but certain monoclonal antibodies against VP4 are capable of inhibiting the infection of preadsorbed virus, so long as internalization has not occurred (Fukuhara et al., 1988). Although the binding site of these two monoclonals was not specified, if they were directed to VP5*, they conceivably could function by blocking direct penetration of the virus (Kaljot et al., 1988) that might involve the putative fusion region (Mackow et al., 1988b); alternatively, they might interfere with the selective internalization of viral components (Fukuhara et al., 1988).

A change in the mobility of gene segment 4 was noted for the strain of SA11 known as 4F, and this virus has an altered phenotype (Pereira et al., 1984; Burns et al., 1989). Compared with the standard SA11 virus, 4F produces VP4, which is both more easily cleaved and apparently larger. The 4F strain also produces larger plaques, grows to

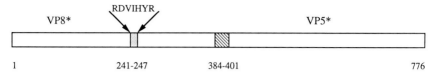

FIG. 7. Features of the rotavirus VP4 protein. VP4 is cleaved at residues 241 and 247 to yield VP8* and VP5*. The locations of amino acids 384–401, which show homology with the Sindbis virus fusion domain, are indicated.

higher titer, and is more stable (Burns *et al.*, 1989). These changes appear to be related, at least in part, to VP4 and emphasize the central importance of this protein to the biological properties of the virus.

B. Gene and Protein Structure

The structure of gene segment *4* has now been determined for nine human (Kantharidis *et al.*, 1987; Gorziglia *et al.*, 1988a), three bovine (Potter *et al.*, 1987; Kantharidis *et al.*, 1988; Nishikawa *et al.*, 1988), one porcine (Nishikawa and Gorziglia, 1988), and three monkey isolates (Lopez *et al.*, 1985; Lopez and Arias, 1987; Nishikawa *et al.*, 1988; Mackow *et al.*, 1988b; Mitchell and Both, 1988). An outline of the general structure of the protein is given in Fig. 7 and in Table I. The human genes code for proteins of 775 residues, while those of animal origin code for proteins of 776 amino acids. There is no signal peptide at the amino terminus of the protein so it is not glycosylated, despite the presence of numerous potential sites for the attachment of N-linked carbohydrate.

The amino terminus of viral VP4, like VP7, appears to be blocked, since the protein was not amenable to sequencing by Edman degradation. Similarly, the cleavage product VP8* was also blocked, indicating that it constituted the amino-terminal portion of the primary translation product. However, partial amino acid sequences were determined directly for the amino-terminal regions of two species of VP5*, and these were in almost complete agreement with the amino acid sequence deduced from the *VP4* gene of isolate SA11fm (Lopez *et al.*, 1985). This suggested that the trypsin cleavage sites in the protein are located carboxy-terminal to residues Arg241 and Arg247 and that six residues are removed during trypsin-activated cleavage (Lopez *et al.*, 1985). Arg241 is conserved in all VP4 sequences determined, and Arg247 varies in only three bovine isolates. In these cases an alternative arginine occurs at residue 246 (Lopez *et al.*, 1985; Kantharidis *et al.*, 1988). Considerable amino acid variation occurs between and adjacent to the cleavage sites (Lopez *et al.*, 1986; Gorziglia *et al.*, 1986b,

1988a; Nishikawa *et al.*, 1988; Kantharidis *et al.*, 1988), but whether this variation is a determinant of virulence, as it appears to be for influenza virus (Klenk and Rott, 1988), is not yet clear.

C. *Expression of VP4 to Study the Protein*

Several clones of VP4 genes are now available (Potter *et al.*, 1987; Mackow *et al.*, 1989; Mitchell and Both, 1989), and these will facilitate studies of the biology and antigenicity of VP4 similar to those already carried out for VP7. For example, if an appropriate assay could be developed, it might be possible to examine whether the putative fusion region identified in VP5* has any role in the penetration of the membrane by virus (Mackow *et al.*, 1988b; Kaljot *et al.*, 1988). The first studies with a partial clone of VP4 involved the expression in *Escherichia coli* of residues 42–387 (which includes the cleavage region) as a fusion protein with phage MS2 RNA polymerase (Arias *et al.*, 1987). The polypeptide produced induced antibodies in mice, which inhibited hemagglutination and boosted the neutralization response in seropositive mice. However, it was not possible to assess the biological activity of VP4 generated from this work. More recently, VP4 from the rhesus rotavirus has been expressed in insect cells using a recombinant baculovirus (Mackow *et al.*, 1989). About 5% of the total protein of the cells was VP4, and the expressed protein was antigenically identical to viral VP4, as judged by its reactivity with a panel of VP4-specific neutralizing monoclonal antibodies. The expressed protein was also able to bind erythrocytes, confirming its role as the hemagglutinin. Hemagglutinating activity of both the virus and the synthetic protein was inhibited by hyperimmune serum, VP4-specific monoclonal antibodies, and glycophorin, a finding which led to the inference that this protein might constitute the erythrocyte receptor for rotaviruses. Thus, VP4 expressed in the absence of the other rotavirus components with which it is normally associated in the virion appeared to be conserved structurally and functionally, indicating that expression of this gene in an appropriate eukaryotic system could yield large amounts of biologically active protein.

VII. OTHER STRUCTURAL PROTEINS

The role of the remaining rotavirus proteins in the events surrounding viral replication is poorly understood, despite the fact that only the VP3 gene now remains to be sequenced. In many cases our knowledge of the protein does not extend beyond the predicted sequence of the polypeptide. In other cases the sequence has provided a clue to the function of the polypeptide, but biochemical confirmation is needed. The details deduced from the gene structures are summarized in Table I. Here, we briefly summarize the biological features of the proteins.

A. Structural Proteins VP1, VP2, VP3, and VP6

By analogy with reovirus, and considering their location in the central core of the virus, VP1, VP2, and VP3 are likely candidates to provide the functions of mRNA synthesis and capping known to be present in the virus (Cohen, 1977; Imai et al., 1983a). VP1 is the most likely candidate to be the RNA polymerase of the virus, since it shows homology between residues 517 and 636, with consensus sequences that have been identified in putative RNA-dependent RNA polymerases from other viruses (Cohen et al., 1989). The protein was also expressed as a full-length nonfusion protein by using a recombinant baculovirus and was recognized by hyperimmune antiserum against purified rotavirus; it induced antibodies in guinea pigs, but these were nonneutralizing (Cohen et al., 1989).

Gene segment 2, encoding VP2, has been sequenced for the bovine RF (Kumar et al., 1989) and human Wa strains (Ernst and Duhl, 1989). Between amino acids 536 and 665, leucine occurs at every seventh residue over four hypothetical turns of an α helix. Such a motif is characteristic of the putative "leucine zipper" which is a feature of proteins which bind nucleic acids (Landschulz et al., 1988). Since VP2 has been shown to possess RNA binding activity (Boyle and Holmes, 1985) and is relatively abundant in the central viral core, it has been suggested that it might be a nucleocapsid protein (Kumar et al., 1989). VP2 is apparently myristylated, although the stoichiometry of labeling was not calculated (Clark and Desselberger, 1988). The protein also lacks the amino-terminal consensus sequence Gly–X–X–X–Ser/Thr present in many other myristylated proteins (Chow et al., 1987). Since other myristylated proteins attach to membranes (Henderson et al., 1983; Streuli and Griffin, 1987), it is possible that if myristylation proves to be a general feature of VP2, it could provide an anchor point for assembly of viral particles (Musalem and Espejo, 1985).

The remaining large protein of SA11, VP3 (molecular weight 88,000), is the product of gene segment 3. It was described only recently because it proved difficult to resolve from other viral proteins by gel electrophoresis (Liu et al., 1988). The protein was also translated poorly in vitro and is synthesized at low levels in the infected cell. VP3 can be detected in early replication complexes (Gallegos and Patton, 1989) and is a structural polypeptide found in the central core of the virus. Its function is unknown, but determination of the gene sequence is in progress.

VP6 is the major component of the ss particles and the subgroup-specific antigen of rotaviruses. Sequences have been determined for a variety of VP6 genes (Both et al., 1984b; Estes et al., 1984; Hofer et al., 1987; Gorziglia et al., 1988b), and many amino acid changes have been characterized, but in the absence of structural information it is not possible to identify amino acid changes that contribute to the antigenic differences between strains. However, the subgroup antigen specificity may be conferred by conformation epitopes which form via protein folding or subunit association, since subtypes I and II-specific monoclonal antibodies recognized only the trimeric form of VP6 (Gorziglia et al., 1988b). The properties of this protein with respect to multimer formation were discussed above. The presence of VP6 in ss particles is required for transcriptase activity, but, as noted above, it is not clear whether the protein participates directly in the transcription process or whether it is required indirectly to maintain the integrity of the central core. Clearly, there is also a role for VP6 in the NS28-mediated budding of subviral particles through the ER membrane (see Section V,B). VP6 also appears to myristylated at an unknown site (Clark and Desselberger, 1988), but how this relates to its function is unclear.

B. Nonstructural Proteins: NS53, NS34, NS35, and NS26

The role of the remaining nonstructural proteins in virus replication is poorly understood and there is still uncertainty as to whether NS26, the product of segment 11, forms part of the virion. Although it was previously classified as structural (Estes et al., 1983), it is now more generally thought to be nonstructural and is recorded as such in Table I. The situation can be clarified when a suitable antiserum becomes available via expression of the cloned gene. Gene sequences are now available for segments encoding NS53 (Bremont et al., 1987, D. B. Mitchell and G. W. Both, unpublished observations), NS34 and NS35 (Both et al., 1982, 1984a; Rushlow et al., 1988), and NS26 (Ward et al., 1985; Imai et al., 1983b; Mitchell and Both, 1988). Segments coding for NS53 and NS26 from different isolates vary in length, due to the presence of additional nucleotides in or near the 3' noncoding region (Nuttall et al., 1989; D. B. Mitchell and G. W. Both, unpublished observations). This accounts for the "short" and "supershort" electrophoretic profiles of some isolates. However, the origin of these extra sequences is unknown.

Equivalent genes encoding nonstructural proteins from different rotavirus strains are 74–92% and 76–97% conserved at the nucleotide and amino acid levels, respectively, suggesting that structural and functional constraints operating in the proteins could restrict the observed level of variation. However, a comparison of the sequences for NS53 from the bovine RF and SA11 viruses reveals much greater differences. The sequences are only 49% and 36% homologous at the RNA and protein levels, respectively. Nevertheless, a cysteine-rich region is conserved near the amino terminus of each protein. This region contains the general motif for a metal binding domain near the amino terminus of the protein (i.e., $Cys-X_{2-4}-Cys-X_{2-15}-a-X_{2-4}-$ a, where a is histidine or cysteine and X is any amino acid (Berg, 1986). This sequence

Bovine RF   42 C L D C C Q Y T N L T Y C R G C A L Y H V C Q W C S Q Y N R C 72
                 :   : :   :   :   : :   :   :
SA11   42 C . E C C . I A D . . H C Y . C S . P H . C K . C V . N R . C 72

could possibly form one or two metal binding domains, depending on the exact residues used. In addition the conserved sequence Cys–Lys–Trp–Cys at residues 325–328 could contribute to the metal binding domain if it were brought into juxtaposition by appropriate folding. It remains to be determined whether this putative metal binding domain actually functions as such and which metal ion, if any, is involved, although Zn^{2+} is an obvious candidate. An ability to bind Zn^{2+} has already been shown for the σ3 protein of reovirus (Schiff et al., 1988), and this protein and others containing zinc fingers are known to bind nucleic acids (Huismans and Joklik, 1976; Miller et al., 1985). Since NS53 is poorly expressed in the infected cell, these studies might be facilitated by expression of the cloned genes in an appropriate eukaryotic system.

The sequence of gene segment 11, which codes for the remaining nonstructural protein, NS26, has been determined from five rotavirus strains (Imai et al., 1983b; Ward et al., 1985; Mitchell and Both, 1988; Nuttall et al., 1989). The main features conserved in this protein are the unusually high serine content (e.g., for SA11 there are 38 serines in a total of 198 amino acids) and the presence of clusters of charged amino acids. The protein also appears to be modified during infection (Ericson et al., 1982) and, given its high serine content, the protein might also be phosphorylated. This gene also contains a potential overlapping reading frame which could code for a protein of 92–98 residues, depending on the strain (Mitchell and Both, 1988; Nuttall et al., 1989), but whether this reading frame is actually used in vivo remains to be determined.

VIII. Highlights and Future Directions

The past 10 years have seen remarkable progress in the development of our understanding of the molecular biology of rotaviruses. All of the gene segments have been cloned, and complete sequences for all but one of the genes are now available. However, our excitement with the rate at which information on the nature of rotavirus proteins has been obtained should not obscure the fact that we still understand little about the functions carried out by many of these proteins and the molecular mechanisms that underlie the intracellular replication of the virus.

Several areas can be highlighted as potential targets for further molecular analysis. First, by analogy with influenza virus and the picornaviruses, poliovirus, rhinovirus, and foot-and-mouth disease virus, for which crystallographic information is available (Wilson et al., 1981; Colman et al., 1983; Hogle et al., 1985; Rossman et al., 1985; Acharya et al., 1989), there is a clear need for information on the quaternary structure of rotavirus proteins and the three-dimensional structure of the virus itself. Structural information would enable the large body of sequence information on the VP4 and VP7 antigens to be correlated more rationally with the data on the antigenic epitopes of these proteins (reviewed by Matsui et al., 1989).

Second, the novel morphogenic pathway of rotaviruses provides a number of opportunities for further study of the processes that occur when an icosahedral ss particle is transferred across a membrane and, in the process, receives its outer layer of capsomeres. The central role of NS28 in the initiation of this budding event and the molecular nature of the interaction between this protein and its ligand, VP6, require further investigation. There are also intriguing questions relating to the mechanisms involved in the budding event itself and the problems of how the membrane envelope is removed and how the other outer capsid protein, VP7, is assembled onto the ss particles.

Third, rotaviruses, like reovirus, possess an integral virion-associated transcriptase which exhibits a remarkable stability in synthesizing the transcripts of all 11 genomic segments. Virtually nothing is known about which viral protein(s) constitutes this "molecular machine" or indeed how it functions.

Fourth, rotaviruses also share with reovirus the ability to form reassortants at high frequency when cells are infected simultaneously with more than one strain of virus. This feature of the viruses is of considerable importance for the generation of live candidate vaccine strains (Kapikian and Chanock, 1985) and is a process of considerable interest to molecular biologists. Although we can be reasonably certain that the reassortment events occur among precursor plus strands (Ramig and Fields, 1983), nothing is known of the molecular mechanism by which the correct set of 11 segments is selected and incorporated into the progeny particles. It is sobering to reflect that, despite being able to manipulate the entire rotavirus genome, at present the ability to incorporate these engineered genes into infectious virus to study the biological effect of mutations is lacking. This problem clearly needs further effort.

Finally, the rotaviruses also provide at least two proteins (i.e., VP7 and NS28) that are of more general interest to workers in the field of protein transport and targeting. Both proteins are localized exclusively in the membrane of the ER, and cloned genes are available for study. Few proteins characterized so far have these features. These two rotavirus proteins therefore provide useful model systems of general interest to cellular and molecular biologists, an advantage not anticipated when the study of rotaviruses was initiated, but an attractive feature nonetheless.

ACKNOWLEDGMENTS

We are grateful to the many workers in the field who provided us with copies of manuscripts prior to their appearance in the literature. We thank R. E. F. Matthews for helpful comments on the manuscript and A. McGill for assistance in preparing it. A. R. B. was supported by a grant from the Medical Research Council of New Zealand.

REFERENCES

Acharya, R., Fry, E., Stuart, D., Fox, G., Rowlands, D., and Brown, F. (1989). The three-dimensional structure of foot-and-mouth disease virus at 2.9 Å resolution. *Nature (London)* **337,** 709–716.

Arias, C. F., Lopez, S., Espejo, R. T. (1982). Gene protein products of SA11 simian rotavirus genome. *J. Virol.* **41,** 42–50.

Arias, C. F., Lopez, S., Bell, J. R., and Strauss, J. H. (1984). Primary structure of the neutralization gene of simian rotavirus SA11 as deduced from cDNA sequence. *J. Virol.* **50,** 657–661.

Arias, C. F., Lizano, M., and Lopez, S. (1987). Synthesis in *Escherichia coli* and immunological characterization of a polypeptide containing the cleavage sites associated with trypsin enhancement of rotavirus SA11 infectivity. *J. Gen. Virol.* **68,** 633–642.

Au, K.-S., Chan, W.-K., and Estes, M. K. (1988). Rotavirus morphogenesis involves an endoplasmic reticulum transmembrane glycoprotein. *In* "UCLA Symposium on Cell Biology of Viral Entry, Replication, and Pathogenicity" (R. Compans, A. Helenius, and M. Oldstone, eds.), pp. 257–267. Liss, New York.

Baybutt, H. N., and McCrae, M. A. (1984). The molecular biology of rotaviruses: VII. Detailed structural analysis of gene 10 of bovine rotaviruses. *Virus Res.* **1,** 533–542.

Berg, J. M. (1986). Potential metal-binding domains in nucleic acid binding proteins. *Science* **232,** 485–487.

Bergmann, C. C., Maass, D., Poruchynsky, M. S., Atkinson, P. H., and Bellamy, A. R. (1989). Topology of the non-structural rotavirus receptor glycoprotein NS28 in the rough endoplasmic reticulum. *EMBO J.* **8**, 1695–1703.

Berriman, J., and Bellamy, A. R. (1990). Manuscript in preparation.

Berriman, J., Meyer, J. C., Anthony, I., and Bellamy, A. R. (1990). Manuscript in preparation.

Bican, P., Cohen, J. Charpilienne, A., and Scherrer, R. (1982). Purification and characterization of bovine rotavirus cores. *J. Virol.* **43**, 1113–1117.

Borsa, J., Morash, B. D., Sargent, M. D., Copps, T. P., Lievaart, P. A., and Szekely, J. G. (1979). Two modes of entry of reovirus into L cells. *J. Gen. Virol.* **45**, 161–170.

Both, G. W., Bellamy, A. R., Street, J. E., and Siegman, L. J. (1982). A general strategy for cloning double-stranded RNA: Nucleotide sequence of the simian-11 rotavirus gene 8. *Nucleic Acids Res.* **10**, 7075–7088.

Both, G. W., Mattick, J. S., and Bellamy, A. R. (1983a). Serotype-specific glycoprotein of simian 11 rotavirus: Coding assignment and gene sequence. *Proc. Natl. Acad. Sci. U.S.A.* **80**, 3091–3095.

Both, G. W., Siegman, L. J., Bellamy, A. R., and Atkinson, P. H. (1983b). Coding assignment and nucleotide sequence of simian rotavirus SA11 gene segment 10: Location of glycosylation sites suggests that the signal peptide is not cleaved. *J. Virol.* **48**, 335–339.

Both, G. W., Mattick, J., Siegman, L., Atkinson, P. H., Weiss, S., Bellamy, A. R., Street, J. E., and Metcalf, P. (1983c). Cloning of SA11 rotavirus genes: Gene structure and polypeptide assignment for the type-specific glycoprotein. *In* "Double Stranded RNA Viruses" (R. W. Compans and D. H. L. Bishop, eds.), pp. 73–82. Elsevier, New York.

Both, G. W., Bellamy, A. R., and Siegman, L. J. (1984a). Nucleotide sequence of the dsRNA genomic segment 7 of simian 11 rotavirus. *Nucleic Acids Res.* **12**, 1621–1626.

Both, G. W., Siegman, L. J., Bellamy, A. R., Ikegami, N., Shatkin, A. J., and Furuichi, Y. (1984b). Comparative sequence analysis of rotavirus genomic segment 6—The gene specifying viral subgroups 1 and 2. *J. Virol.* **51**, 97–101.

Both, G. W., Stirzaker, S. C., Bergmann, C. C., Andrew, M. E., Boyle, D. B., and Bellamy, A. R. (1989). Analysis of rotavirus proteins by gene cloning, mutagenesis and expression. *Appl. Virol. Res.* **2**, in press.

Boyle, J. F., and Holmes, K. V. (1985). RNA-binding proteins of bovine rotavirus. *J. Virol.* **58**, 561–568.

Bremont, M., Charpilienne, D., Chabanne, D., and Cohen, J. (1987). Nucleotide sequence and expression in *Escherichia coli* of the gene encoding the nonstructural protein NCVP2 of bovine rotavirus. *Virology* **160**, 138–144.

Bremont, M., Cohen, J., and McCrae, M. A. (1988). Analysis of the structural polypeptides of a porcine group C rotavirus. *J. Virol.* **62**, 2183–2185.

Burns, J. W., Chen, D., Estes, M. K., and Ramig, R. F. (1989). Biological and immunological characterization of a simian rotavirus SA11 variant with an altered genome segment 4. *Virology* **169**, 427–435.

Chan, W.-K., Penaranda, M. E., Crawford, S. E., and Estes, M. K. (1986). Two glycoproteins are produced from the rotavirus neutralization gene. *Virology* **151**, 243–252.

Chan, W.-K., Au, K.-S., and Estes, M. K. (1988). Topography of the simian rotavirus nonstructural glycoprotein (NS28) in the endoplasmic reticulum membrane. *Virology* **164**, 435–442.

Chasey, D. (1980). Investigation of immunoperoxidase-labeled rotavirus in tissue culture by light and electron microscopy. *J. Gen. Virol.* **50**, 195–200.

Chasey, D., and Labram, J. (1983). Electron microscopy of tubular assemblies associated with naturally occurring bovine rotavirus. *J. Gen. Virol.* **64**, 863–872.

Chow, M., Newman, J. F. E., Filman, D., Hogle, J. M., Rowlands, D. J., and Brown, F. (1987). Myristylation of picornavirus capsid protein VP4 and its structural significance. *Nature (London)* **327**, 482–486.

Clark, B., and Desselberger, U. (1988). Myristylation of rotavirus proteins. *J. Gen. Virol.* **69**, 2681–2686.

Clark, S. M., Roth, J. R., Clark, M. L., Barnett, B. B., and Spendlove, R. S. (1981). Trypsin enhancement of rotavirus infectivity: Mechanisms of enhancement. *J. Virol.* **39**, 816–822.

Cohen, J. (1977). Ribonucleic acid polymerase associated with purified calf rotavirus. *J. Gen. Virol.* **36**, 395–402.

Cohen, J., Laporte, J., Charpilienne, A., and Scherrer, R. (1979). Activation of rotavirus RNA polymerase by calcium chelation. *Arch. Virol.* **60**, 177–186.

Cohen, J., Charpilienne, A., Chilmonczyk, S., and Estes, M. K. (1989). Nucleotide sequence of bovine rotavirus gene 1 and expression of the gene in baculovirus. *J. Virol.* **170**, 131–140.

Colman, P. M., Varghese, J. N., and Laver, W. G. (1983). Structure of the catalytic and antigenic sites in influenza virus neuraminidase. *Nature (London)* **303**, 41–44.

Colonno, R. J., Condra, J. H., Mitzutani, S., Callahan, P. L., Davies, M. E., and Murcko, M. A. (1988). Evidence of the direct involvement of the rhinovirus canyon in receptor binding. *Proc. Natl. Acad. Sci. U.S.A.* **85**, 5449–5453.

Cukor, G., and Blacklow, N. R. (1984). Human viral gastroenteritis. *Microbiol. Rev.* **48**, 157–179.

Dimmock, N. J. (1982). Initial stages of infection with animal viruses. *J. Gen. Virol.* **59**, 1–22.

Dyall-Smith, M. L., and Holmes, I. H. (1984). Sequence homology between human and animal rotavirus serotype-specific glycoproteins. *Nucleic Acids Res.* **12**, 3973–3982.

Eaton, B. T., Hyatt, A. D., and White, J. R. (1987). Association of bluetongue virus with the cytoskeleton. *Virology* **157**, 107–116.

Elleman, T. L., Hoyne, P. A., Dyall-Smith, M. L., Holmes, I. H., and Azad, A. A. (1983). Nucleotide sequence of the gene encoding the serotype-specific glycoprotein of UK bovine rotavirus. *Nucleic Acids Res.* **11**, 4689–4701.

Ericson, B. L., Graham, D. Y., Mason, B. B., and Estes, M. K. (1982). Identification, synthesis, and modifications of simian rotavirus SA11 polypeptides in infected cells. *J. Virol.* **42**, 825–839.

Ericson, B. L., Graham, D. Y., Mason, B. B., Hanssen, H., and Estes, M. K. (1983). Two types of glycoprotein precursors are produced by the simian rotavirus SA11. *Virology* **127**, 320–332.

Ernst, H., and Duhl, J. A. (1989). Nucleotide sequence of genomic segment 2 of human rotavirus Wa. *Nucleic Acids Res.* **17**, 4382.

Esparza, J., and Gil, F. (1978). A study on the ultrastructure of human rotavirus. *Virology* **91**, 141–150.

Esparza, J., Gorziglia, M., Gil, F., and Romer, H. (1980). Multiplication of human rotaviruses in cultured cells: An electron microscopic study. *J. Gen. Virol.* **47**, 461–472.

Espejo, R. T., Lopez, S., and Arias, C. (1981). Structural polypeptides of simian rotavirus SA11 and the effect of trypsin. *J. Virol.* **37**, 156–160.

Estes, M. K., Graham, D. Y., and Mason, B. B. (1981). Proteolytic enhancement of rotavirus infectivity: Molecular mechanisms. *J. Virol.* **39**, 879–888.

Estes, M. K., Graham, D. Y., Ramig, R. F., and Ericson, B. L. (1982). Heterogeneity in the structural glycoprotein (VP7) of simian rotavirus SA11. *Virology* **122**, 8–14.

Estes, M. K., Palmer, E. L., and Obijeski, J. F. (1983). Rotaviruses: A review. *Curr. Top. Microbiol. Immunol.* **105**, 123–184.

Estes, M. K., Mason, B. B., Crawford, S., and Cohen, J. (1984). Cloning and nucleotide sequence of the simian rotavirus gene 6 that codes for the major inner capsid protein. *Nucleic Acids Res.* **12**, 875–1887.

Estes, M. K., Crawford, S. E. Penaranda, M. E., Petrie, B. L., Burns, J. W., Chan, W.-K., Ericson, B., Smith, G. E., and Summers, M. D. (1987). Synthesis and immunogenicity of the rotavirus major capsid antigen using a baculovirus expression system. *J. Virol.* **61**, 1488–1494.

Fukuhara, N., Yoshie, O., Kitaoka, S., and Konno, T. (1988). Role of VP3 in human rotavirus internalization after target cell attachment via VP7. *J. Virol.* **62**, 2209–2218.

Furlong, D. B., Nibert, M. L., and Fields, B. N. (1988). Sigma 1 protein of mammalian reoviruses extends from the surfaces of viral particles. *J. Virol.* **62**, 246–256.

Furuichi, Y., Morgan, M., Muthukrishnan, S., and Shatkin, A. J. (1975a). Reovirus messenger RNA contains a methylated, blocked 5'-terminal structure: m7G(5')ppp(5')GmpCp–. *Proc. Natl. Acad. Sci. U.S.A.* **72**, 362–366.

Furuichi, Y., Muthukrishnan, S., and Shatkin, A. J. (1975b). 5'-Terminal m7G(5)ppp(5')Gmp *in vivo*: Identification in reovirus genome RNA. *Proc. Natl. Acad. Sci. U.S.A.* **72**, 742–745.

Gallegos, C. O., and Patton, J. T. (1989). Characterization of rotavirus replication intermediates: A model for the assembly of single-shelled particles. *Virology,* **172**, 616–627.

Gething, M. J., and Sambrook, J. (1981). Cell-surface expression of influenza haemagglutinin from a cloned DNA copy of the RNA gene. *Nature (London)* **293**, 620–625.

Glass, R. I., Keith, J., Nakagomi, O., Nakagomi, T., Askaa, J. J., Kapikian, A. Z., Chanock, R. M., and Flores, J. (1985). Nucleotide sequence of the structural glycoprotein VP7 gene of Nebraska calf diarrhea virus rotavirus: Comparison with homologous genes from four strains of human and animal rotaviruses. *Virology* **141**, 292–298.

Gorziglia, M., Larrea, C., Liprandi, F., and Esparza, J. (1985). Biochemical evidence for the oligomeric (possibly trimeric) structure of the major inner capsid polypeptide (45K) of rotaviruses. *J. Gen. Virol.* **66**, 1889–1900.

Gorziglia, M., Aguirre, Y., Hoshino, Y., Esparza, J., Blumentals, I., Askaa, J. J., Thompson, M., Glass, R., Kapikian, A. Z., and Chanock, R. M. (1986a). VP7 serotype-specific glycoprotein of OSU porcine rotavirus: Coding assignment and gene sequence. *J. Gen. Virol.* **67**, 2445–2454.

Gorziglia, M., Hoshino, Y., Buckler-White, A., Blumentals, I., Glass, R., Flores, J., Kapikian, A. Z., and Chanock, R. M. (1986b). Conservation of amino acid sequence of VP8 and cleavage region of 84-kD outer capsid protein among rotaviruses recovered from asymptomatic neonatal infection. *Proc. Natl. Acad. Sci. U.S.A.* **83**, 7039–7043.

Gorziglia, M., Green, K., Nishikawa, K., Taniguchi, K., Jones, R., Kapikian, A. Z., and Chanock, R. M. (1988a). Sequence of the fourth gene of human rotavirus recovered from asymptomatic or symptomatic infections. *J. Virol.* **62**, 2978–2984.

Gorziglia, M., Hoshino, Y., Nishikawa, K., Maloy, W. L., Jones, R. W., Kapikian, A. Z., and Chanock, R. M. (1988b). Comparative sequence analysis of the genomic segment 6 of four rotaviruses each with a different subgroup specificity. *J. Gen. Virol.* **69**, 1659–1669.

Gorziglia, M., Nishikawa, K., Green, K., and Taniguchi, K. (1988c). Gene sequence of the VP7 serotype specific glycoprotein of Gottfried porcine rotavirus. *Nucleic Acids Res.* **16**, 775.

Green, K. Y., Hoshino, Y., and Ikegami, N. (1989). Sequence analysis of the gene encoding the serotype-specific glycoprotein (VP7) of two new human rotavirus serotypes. *Virology* **168**, 429–433.

Greenberg, H. B., Flores, J., Kalica, A. R., Wyatt, R. G., and Jones, R. (1983a). Gene coding assignments for growth restriction, neutralization and subgroup specificities of the W and DS-1 strains of human rotavirus. *J. Gen. Virol.* **64**, 313–320.

Greenberg, H. B., McAuliffe, V., Valdesuso, J., Wyatt, R., Flores, J., Kalica, A., Hoshino, Y., and Singh, N. H. (1983b). Serological analysis of the subgroup protein of rotavirus, using monoclonal antibodies. *Infect. Immun.* **39**, 91–99.

Gunn, P. R., Sato, F., Powell, K. F. H., Bellamy, A. R., Napier, J. R., Harding, D. R. K., Hancock, W. S., Siegman, L. J., and Both, G. W. (1985). Rotavirus neutralizing protein VP7: Antigenic determinants investigated by sequence analysis and peptide synthesis. *J. Virol.* **54**, 791–797.

Helmberger-Jones, M., and Patton, J. T. (1986). Characterization of subviral particles in cells infected with simian rotavirus SA11. *Virology* **155**, 655–665.

Henderson, L. E., Krutzsch, H. C., and Oroszlan, S. (1983). Myristyl amino-terminal acylation of murine retrovirus proteins: An unusual post-translational protein modification. *Proc. Natl. Acad. Sci. U.S.A.* **80**, 339–343.

Hofer, J. M. I., Street, J. E., and Bellamy, A. R. (1987). Nucleotide sequence for gene 6 of rotavirus strain S2. *Nucleic Acids Res.* **15**, 7175.

Hogle, J. M., Chow, M., and Filman, D. J. (1985). Three-dimensional structure of poliovirus at 2.9 Å resolution. *Science* **229**, 1358–1365.

Holmes, I. H. (1983). Rotaviruses. *In* "The Reoviridae" (W. K. Joklik, ed.), pp. 359–423. Plenum, New York.

Hoshino, Y., Sereno, M. M., Midthun, K., Flores, J., Kapikian, A. Z., and Chanock, R. M. (1985). Independent segregation of two antigenic specificities (VP3 and VP7) involved in neutralization of rotavirus infectivity. *Proc. Natl. Acad. Sci. U.S.A.* **82**, 8701–8704.

Hubbard, S. C., and Ivatt, J. (1981). Synthesis and processing of asparagine-linked oligosaccharides. *Annu. Rev. Biochem.* **50**, 555–583.

Huismans, H., and Joklik, W. K. (1976). Reovirus-encoded polypeptides in infected cells: Isolation of two native monomeric polypeptides with affinity for single-stranded and double-stranded RNA, respectively. *Virology* **70**, 411–424.

Hum, C. P., Dyall-Smith, M. L., and Holmes, I. H. (1989). The VP7 gene of a new G serotype of human rotavirus (B37) is similar to G3 proteins in the antigenic C region. *Virology* **170**, 55–61.

Imai, M., Akatani, K., Ikegami, N., and Furuichi, Y. (1983a). Capped and conserved terminal structures in human rotavirus genome double-stranded RNA segments *J. Virol.* **47**, 125–136.

Imai, M., Richardson, M. A. Ikegami, N., Shatkin, A. J., and Furuichi, Y. (1983b). Molecular cloning of double-stranded RNA virus genomes. *Proc. Natl. Acad. Sci. U.S.A.* **80**, 373–377.

Johnson, M. A., and McCrae, M. A. (1989). Molecular biology of rotaviruses: VIII. Quantitative analysis of regulation of gene expression during virus replication. *J. Virol.* **63**, 2048–2055.

Joklik, W. K. (1983). The reovirus particle. *In* "The Reoviridae" (W. K. Joklik ed.), pp. 9–78. Plenum, New York.

Kabcenell, A. K., and Atkinson, P. A. (1985). Processing of the rough endoplasmic reticulum membrane glycoproteins of rotavirus SA11. *J. Cell Biol.* **101**, 1270–1280.

Kabcenell, A. K., Poruchynsky, M. S., Bellamy, A. R., Greenberg, H. B., and Atkinson, P. H. (1988). Two forms of VP7 are involved in the assembly of SA11 rotavirus in the endoplasmic reticulum. *J. Virol.* **62**, 2929–2941.

Kalica, A. R., Flores, J., and Greenberg, H. B. (1983). Identification of the rotaviral gene that codes for hemagglutination and protease-enhanced plaque formation. *Virology* **125**, 94–205.

Kaljot, K. T., Shaw, R. D., Rubin, D. H., and Greenberg, H. B. (1988). Infectious rotavirus enters cells by direct membrane penetration, not by endocytosis. *J. Virol.* **62,** 1136–1144.

Kantharidis, P., Dyall-Smith, M. L., and Holmes, I. H. (1987). Marked sequence variation between segment 4 genes of human RV-5 and simian SA11 rotaviruses. *Arch. Virol.* **93,** 111–121.

Kantharidis, P., Dyall-Smith, M., Tregear, G. W., and Holmes, I. H. (1988). Nucleotide sequence of UK bovine rotavirus segment 4: Possible host restriction of VP3 genes. *Virology* **166,** 308–315.

Kapikian, A. Z., and Chanock, R. M. (1985). Rotaviruses. *In* "Virology" (B. N. Fields, ed.), pp. 863–906. Raven, New York.

Klenk, H.-D., and Rott, R. (1988). The molecular biology of influenza virus pathogenicity. *Adv. Virus Res.* **34,** 247–275.

Kouvelos, K., Petric, M., and Middleton, P. J. (1984). Comparison of bovine, simian, and human rotavirus structural glycoproteins. *J. Gen. Virol.* **65,** 1211–1214.

Kozak, M. (1984). Selection of initiation sites by eukaryotic ribosomes: Effect of inserting AUG triplets upstream from the coding sequence for preproinsulin. *Nucleic Acids Res.* **12,** 3873–3893.

Kozak, M. (1986). Point mutations that define a sequence flanking the AUG initiator codon that modulates translation by eukaryotic ribosomes. *Cell (Cambridge, Mass.)* **44,** 283–292.

Kumar, A., Charpilienne, A., and Cohen, J. (1989). Nucleotide sequence of the gene for the RNA binding protein (VP2) of RF bovine rotavirus. *Nucleic Acids Res.* **17,** 2126.

Landschulz, W. H., Johnson, P. F., and McKnight, S. L. (1988). The leucine zipper: A hypothetical structure common to a new class of DNA binding proteins. *Science* **240,** 1759–1764.

Liu, M., Offit, P. A., and Estes, M. K. (1988). Identification of the simian rotavirus SA11 genome segment 3 product. *Virology* **163,** 26–32.

Lopez, S., and Arias, C. F. (1987). The nucleotide sequence of the 5′ and 3′ ends of rotavirus SA11 gene 4. *Nucleic Acids Res.* **15,** 691.

Lopez, S., Arias, C. F., Bell, J. R., Strauss, J. H., and Espejo, R. T. (1985). Primary structure of the cleavage site associated with trypsin enhancement of rotavirus SA11 infectivity. *Virology* **144,** 11–19.

Lopez, S., Arias, C. F., Mendez, E., and Espejo, R. (1986). Conservation in rotaviruses of the protein region containing the two sites associated with trypsin enhancement of infectivity. *Virology* **154,** 224–227.

Mackow, E. R., Shaw, R. D., Matsui, S. M., Vo, P. T., Benfield, D. A., and Greenberg, H. B. (1988a). Characterization of homotypic and heterotypic VP7 neutralization sites of rhesus SA11 gene 4. *Nucleic Acids Res.* **15,** 691.

Mackow, E. R., Shaw, R. D., Matsui, S. M., Vo, P. T., Dang, M.-N., and Greenberg, H. B. (1988b). Characterization of the rhesus rotavirus VP3 gene: Localization of amino acids involved in homologous and heterologous rotavirus neutralization and identification of a putative fusion region. *Proc. Natl. Acad. Sci. U.S.A.* **85,** 645–649.

Mackow, E. R., Barnett, J. W., Chan, H., and Greenberg, H. B. (1989). The rhesus rotavirus outer capsid protein VP4 functions as a hemagglutinin and is antigenically conserved when expressed by a baculovirus recombinant. *J. Virol.* **63,** 1661–1668.

Martin, M. L., Palmer, E. L., and Middleton, P. J. (1975). Ultrastructure of infantile gastroenteritis virus. *Virology* **68,** 146–153.

Mason, B. B., Graham, D. Y., and Estes, M. K. (1980). In vitro transcription and translation of simian rotavirus SA-11 gene products. *J. Virol.* **33,** 1111–1121.

Mason, B. B., Graham, D. Y., and Estes, M. K. (1983). Biochemical mapping of the simian rotavirus SA11 genome. *J. Virol.* **46,** 413–423.

Mason, B. B., Dheer, S. K., Hsaio, C.-L., Zandle, G., Kostek, B., Rosanoff, E. I., Hung, P. P., and Davis, A. R. (1985). Sequence of the serotype-specific glycoprotein of the human rotavirus Wa strain and comparison with other human rotavirus serotypes. *Virus Res.* **2,** 328–336.

Matsui, S. M., Mackow, E., and Greenberg, H. B. (1989). The molecular determinant of rotavirus neutralization and protection. *Adv. Virus Res.* **36,** 181–214.

Matthews, R. E. F. (1979). The classification and nomenclature of viruses: Summary of results of meetings of the International Committee on Taxonomy of Viruses in The Hague, September 1978. *Intervirology* **11,** 133–135.

McCrae, M. A., and Faulkner-Valle, G. P. (1981). Molecular biology of rotaviruses: I. Characterization of basic growth parameters and patterns of macromolecular synthesis. *J. Gen. Virol.* **39,** 490–496.

Metcalf, P. (1982). The symmetry of reovirus. *J. Ultrastruct. Res.* **78,** 292–301.

Meyer, J. C., Bergmann, C. C., and Bellamy, A. R. (1989). Interaction of rotavirus cores with the nonstructural glycoprotein NS28. *Virology* **171,** 98–107.

Miller, J., McLachlan, A. D., and Klug, A. (1985). Repetitive zinc-binding domains in the protein transcription factor IIIA from *Xenopus* oocytes. *EMBO J.* **4,** 1609–1614.

Mitchell, D. B., and Both, G. W. (1988). Simian rotavirus SA11 segment 11 contains overlapping reading frames. *Nucleic Acids Res.* **16,** 6244.

Mitchell, D. B., and Both, G. W. (1989). Complete nucleotide sequence of the simian rotavirus SA11 VP4 gene. *Nucleic Acids Res.* **17,** 2122.

Mora, M., Partin, K., Bhatia, M., Partin, J., and Carter, C. (1987). Association of reovirus proteins with the structural matrix of infected cells. *Virology* **159,** 265–277.

Morgan, E. M., and Zweerink, H. J. (1975). Characterization of transcriptase and replicase particles isolated from reovirus infected cells. *Virology* **68,** 455–466.

Musalem, C., and Espejo, R. T. (1985). Release of progeny virus from cells infected with simian rotavirus SA11. *J. Gen. Virol.* **66,** 2715–2724.

Newman, J. F. E., Brown, F., Bridger, J. C., and Woode, G. N. (1975). Characterization of a rotavirus. *Nature (London)* **258,** 631–633.

Nishikawa, K., and Gorziglia, M. (1988). The nucleotide sequence of the VP3 gene of porcine rotavirus OSU. *Nucleic Acids Res.* **24,** 11847.

Nishikawa, K., Taniguchi, K., Torres, A., Hoshino, Y., Green, K., Kapikian, A. Z., Chanock, R. M., and Gorziglia, M. (1988). Comparative analysis of the VP3 gene of divergent strains of the rotaviruses simian SA11 and bovine Nebraska calf diarrhea virus. *J. Virol.* **62,** 4022–4026.

Nuttall, S. D., Hum, C. P., Holmes, I. H., and Dyall-Smith, M. L. (1989). Sequences of VP9 genes from short and supershort rotavirus strains. *Virology* **171,** 453–457.

Offit, P. A., Blavat, G., Greenberg, H. B., and Clark, C. F. (1986). Molecular basis of rotavirus virulence: Role of gene segment 4. *J. Virol.* **57,** 46–49.

Okada, Y., Richardson, M. A., Ikegami, N., Nomoto, A., and Furuichi, Y. (1984). Nucleotide sequence of human rotavirus genome segment 10, an RNA encoding a glycosylated virus protein. *J. Virol.* **51,** 856–859.

Patton, J. T. (1986). Synthesis of simian rotavirus SA11 double-stranded RNA in a cellfree system. *Virus Res.* **6,** 217–233.

Patton, J. T., and Gallegos, C. O. (1988). Structure and protein composition of the rotavirus replicase particle. *Virology* **166,** 358–365.

Pedley, S., Bridger, J. C., Brown, J. F., and McCrae, M. A. (1983). Molecular characterization of rotaviruses with distinct group antigens. *J. Gen. Virol.* **64,** 2093–2101.

Pedley, S., Bridger, J. C., Chasey, D., and McCrae, M. A. (1986). Definition of two new groups of atypical rotaviruses. *J. Gen. Virol.* **67,** 131–137.

Pereira, H. G., Azeredo, R. S., Fiahlo, A. M., and Vidal, M. N. P. (1984). Genomic heterogeneity of simian rotavirus SA11. *J. Gen. Virol.* **65**, 815–818.

Petrie, B. L., Graham, D. Y., Hanssen, H., and Estes, M. K. (1982). Localization of rotavirus antigens in infected cells by ultrastructural immunocytochemistry. *J. Gen. Virol.* **63**, 457–467.

Petrie, B. L., Estes, M. K., and Graham, D. Y. (1983). Effects of tunicamycin on rotavirus morphogenesis and infectivity. *J. Virol.* **46**, 270–274.

Petrie, B. L., Greenberg, H. B., Graham, D. Y., and Estes, M. K. (1984). Ultrastructural localization of rotavirus antigens using colloidal gold. *Virus Res.* **1**, 133–152.

Poruchynsky, M. S., and Atkinson, P. (1988). Primary sequence domains required for the retention of rotavirus VP7 in the endoplasmic reticulum. *J. Cell Biol.* **107**, 1697–1706.

Poruchynsky, M. S., Tyndall, C., Both, G. W., Sato, F., Bellamy, A. R., and Atkinson, P. A. (1985). Deletions into an NH_2-terminal hydrophobic domain result in secretion of rotavirus VP7, a resident endoplasmic reticulum membrane glycoprotein. *J. Cell Biol.* **101**, 2199–2209.

Potter, A. A., Cox, G., Parker, M., and Babiuk, L. A. (1987). The complete sequence of bovine rotavirus C486 gene 4 cDNA. *Nucleic Acids Res.* **15**, 4361.

Powell, K. F. H., Gunn, P. R., and Bellamy, A. R. (1988). Nucleotide sequence of bovine rotavirus genomic segment 10: An RNA encoding the viral non-structural glycoprotein. *Nucleic Acids Res.* **16**, 763.

Prasad, B. V. V., Wang, G. J., Clerx, J. P. M., and Chiu, W. (1988). Three-dimensional structure of rotavirus. *J. Mol. Biol.* **199**, 269–275.

Quann, C. M., and Doane, F. W. (1983). Ultrastructural evidence for the cellular uptake of rotavirus by endocytosis. *Intervirology* **20**, 223–231.

Ramig, R. F., and Fields, B. N. (1983). Genetics of reoviruses. In "The Reoviridae" (W. K. Joklik, ed.), pp. 197–228. Plenum, New York.

Ready, K. F. M., and Sabara, M. (1987). In vitro assembly of bovine rotavirus nucleocapsid protein. *Virology* **157**, 189–198.

Ready, K. F. M., Buko, K. M. A., Whippey, P. W., Alford, W. P., and Bancroft, J. B. (1988). The structure of bovine rotavirus nucleocapsid protein (VP6) assembled in vitro. *Virology* **167**, 50–55.

Reddy, D. A., Greenberg, H. B., and Bellamy, A. R. (1989). Rotavirus serotype IV: Nucleotide sequence of genomic segment nine of St. Thomas 3 strain. *Nucleic Acids Res.* **17**, 449.

Richardson, M. A., Iwamoto, A., Ikegami, N., Nomoto, A., and Furuichi, Y. (1984). Nucleotide sequence of the gene encoding the serotype-specific antigen of human (Wa) rotavirus: Comparison with the homologous genes from simian SA11 and UK bovine rotaviruses. *J. Virol.* **51**, 860–862.

Richardson, S. C., Mercer, L. E., Sonza, S., and Holmes, I. H. (1986). Intracellular location of rotaviral proteins. *Arch. Virol.* **88**, 251–264.

Rodger, S. M., Schnagl, R. D., and Holmes, I. H. (1975). Biochemical and biophysical characterization of diarrhea viruses of human and calf origin. *J. Virol.* **16**, 1229–1235.

Roseto, A., Esciag, J., Delain, E., Cohen, J., and Scherrer, R. (1979). Structure of rotaviruses as studied by the freeze-drying technique. *Virology* **98**, 471–475.

Rossman, M. G., Arnold E., Erikson, J. W., Frankenberger, E. A., Griffith, J. P., Hecht, H.-J., Johnson, J. E., Kamer, G., Luo, M., Mosser, A. G., Reuckert, R. R., Sherry, B., and Vriend, G. (1985). Structure of a human common cold virus and functional relationship to other picornaviruses. *Nature (London)* **317**, 145–153.

Ruiz, A. M., Lopez, I. V., Lopez, S., Espejo, R. T., and Arias, C. F. (1988). Molecular and antigenic characterization of porcine rotavirus YM, a possible new rotavirus serotype. *J. Virol.* **62,** 4331–4336.

Rushlow, K., McNab, A., Olson, K., Maxwell, F., Maxwell, I., and Stiegler, G. (1988). Nucleotide sequence of porcine rotavirus (OSU strain) gene segments 7, 8, and 9. *Nucleic Acids Res.* **16,** 367–368.

Sabara, M., Babiuk, L. A., Gilchrist, J., and Misra, V. (1982). Effect of tunicamycin on rotavirus assembly and infectivity. *J. Virol.* **43,** 1082–1090.

Sabara, M., Gilchrist, J. E., Hudson, G. R., and Babiuk, L. A. (1985). Preliminary characterization of an epitope involved in neutralization and cell attachment that is located on the major bovine rotavirus glycoprotein. *J. Virol.* **53,** 58–66.

Sabara, M., Ready, K. F. M., Frenchick, P. J., and Babiuk, L. A. (1987). Biochemical evidence for the oligomeric arrangement of bovine rotavirus nucleocapsid protein and its possible significance in the immunogenicity of this protein. *J. Gen. Virol.* **68,** 123–133.

Sandino, A. M., Jashes, M., Faundez, G., and Spencer, E. (1986). Role of the inner protein capsid on in vitro human rotavirus transcription. *J. Virol.* **60,** 797–802.

Schiff, L. A., Nibert, M. L., Co, M. S., Brown, E. G., and Fields, B. N. (1988). Distinct binding sites for zinc and double-stranded RNA in the reovirus outer capsid protein sigma 3. *Mol. Cell. Biol.* **8,** 273–283.

Shahrabadi, M. S., and Lee, P. W. K. (1986). Bovine rotavirus maturation is a calcium-dependent process. *Virology* **152,** 298–307.

Shahrabadi, M. S., Babiuk, L. A., and Lee, P. W. K. (1987). Further analysis of the role of calcium in rotavirus morphogenesis. *Virology* **158,** 103–111.

Shatkin, A. J., and Kozak, M. (1983). Biochemical aspects of reovirus transcription and translation. *In* "The Reoviridae" (W. K. Joklik, ed.), pp. 79–106. Plenum, New York.

Shatkin, A. J., and Sipe, J. D. (1968). RNA polymerase activity in purified reovirus. *Proc. Natl. Acad. Sci. U.S.A.* **61,** 1462–1469.

Skehel, J. J., and Joklik, W. K. (1969). Studies on the *in vitro* transcription of reovirus RNA catalysed by reovirus cores. *Virology* **39,** 822–831.

Stacy-Phipps, S., and Patton, J. T. (1987). Synthesis of plus- and minus-strand RNA in rotavirus-infected cells. *J. Virol.* **61,** 3479–3484.

Stirzaker, S. C., and Both, G. W. (1989). The signal peptide of rotavirus glycoprotein VP7 is essential for its retention in the ER as an integral membrane protein. *Cell (Cambridge, Mass.)* **56,** 741–747.

Stirzaker, S. C., Whitfeld, P. L., Christie, D. L., Bellamy, A. R., and Both, G. W. (1987). Processing of rotavirus glycoprotein VP7: Implications for the retention of the protein in the endoplasmic reticulum. *J. Cell Biol.* **105,** 2897–2903.

Streuli, C. H., and Griffin, B. E. (1987). Myristic acid is coupled to a structural protein of polyoma virus and SV40. *Nature (London)* **326,** 619–622.

Suzuki, H., Kitaoka, S., Konno, T., Sato, T., and Ishida, N. (1985). Two modes of human rotavirus entry into MA104 cells. *Arch. Virol.* **82,** 25–43.

Suzuki, H., Kitaoka, S., Sato, T., Konno, Y., Iwasaki, Y., Numazaki, Y., and Ishida, N. (1986). Further investigation on the mode of entry of human rotavirus into cells. *Arch. Virol.* **91,** 135–144.

Taniguchi, K., Maloy, W. L., Nishikawa, K., Green, K. Y., Hoshino, Y., Urasawa, S., Kapikian, A. Z., and Chanock, R. M. (1988). Identification of cross-reactive and serotype 2-specific neutralization epitopes on VP3 of human rotavirus. *J. Virol.* **62,** 2421–2426.

von Heijne, G. (1986). A new method for predicting signal sequence cleavage sites. *Nucleic Acids Res.* **14,** 4683–4690.

Ward, C. W., Azad, A. A., and Dyall-Smith, M. L. (1985). Structural homologies between RNA gene segments 10 and 11 from UK bovine, simian SA11 and human Wa rotaviruses. *Virology* **144,** 328–336.

Webster, R. G., Laver, W. G., Air, G. B., and Schild, G. C. (1982). Molecular mechanisms of variation in influenza viruses. *Nature (London)* **296,** 115–221.

Whitfeld, P. L., Tyndall, C., Stirzaker, S. C., Bellamy, A. R., and Both, G. W. (1987). Location of signal sequences within the rotavirus SA11 glycoprotein VP7 which direct it to the endoplasmic reticulum. *Mol. Cell. Biol.* **7,** 2491–2497.

Wiley, D. C., Wilson, I. A., and Skehel, J. J. (1981). Structural identification of the antibody-binding sites of Hong Kong influenza haemagglutinin and their involvement in antigenic variation. *Nature (London)* **289,** 373–378.

Wilson, I. A., Skehel, J. J., and Wiley, D. C. (1981). Structure of the haemagglutinin membrane glycoprotein of the influenza virus at 3 Å resolution. *Nature (London)* **289,** 366–373.

Zarbl, H., and Millward, S. (1983). The reovirus multiplication cycle. *In* "The Reoviridae" (W. K. Joklik, ed.), pp. 107–196. Plenum, New York.

Zweerink, H. J. (1974). Multiple forms of ss–dsRNA polymerase activity in reovirus-infected cells. *Nature (London)* **247,** 313–315.

Zweerink, H. J., Morgan, E. M., and Skyler, J. S. (1976). Reovirus morphogenesis: Characterization of subviral particles in infected cells. *Virology* **73,** 442–453.

ADVANCES IN VIRUS RESEARCH, VOL. 38

MOLECULAR BIOLOGY OF VARICELLA ZOSTER VIRUS

Jeffrey M. Ostrove

Medical Virology Section
Laboratory of Clinical Investigation
National Institute of Allergy and Infectious Diseases
National Institutes of Health
Bethesda, Maryland 20892

I. Introduction

Herpesvirus infections occur in a diverse group of animals, from amphibians and teleosts to primates, including humans. Six distinct herpesviruses that infect humans have been isolated, and some of the diseases they cause are well characterized. These human viruses are distributed in three subfamilies of the family Herpesviridae: Alpha-, Beta-, and Gammaherpesvirinae. Herpes simplex viruses types 1 and 2 (HSV-1 and -2) are the best-studied members of the subfamily Alphaherpesvirinae, extensive biochemical and genetic analyses having been performed both *in vitro* and *in vivo*. The complete DNA sequence of HSV-1 was recently published (McGeoch *et al.*, 1988a). HSV-1 and -2 are renowned for causing recurrent mucocutaneous lesions in humans (for reviews, see Corey and Spear, 1986a,b; Straus *et al.*, 1984). Human cytomegalovirus, a member of the subfamily Betaherpesvirinae,

45

is the largest and genetically most complex member of the herpes-virus family, causing a range of illnesses. Severe infections occur in the immunocompromised host and *in utero,* resulting in diseases of newborns. Epstein–Barr virus, a member of the subfamily Gam-maherpesvirinae, is predominately a B-lymphotrophic virus associated with infectious mononucleosis, or glandular fever, and two common cancers: African Burkett's lymphoma and nasopharyngeal carcinoma (de-Thé, 1982). The recently described human herpesvirus type 6 causes exanthem subitum, or roseola, a common childhood disease (Salahuddin *et al.,* 1986; Takahashi *et al.,* 1988; Yamanishi *et al.,* 1988). Whether this virus is associated with other disease states has yet to be verified.

The focus of this review is varicella zoster virus (VZV), a member of the subfamily Alphaherpesvirinae. The virus causes varicella, or chickenpox, and, on reactivation from a latent state, zoster, or shin-gles.

The origin of the name "chickenpox" is uncertain. It might have its roots in the French word *"chiche,"* or "chickpea," relating to the size and central dimpling of the pock seen on the skin; another possibility is that it is a derivative of "gican," which sounds like "chicken" and is the Old English term for "itch" (Scott-Wilson, 1978). "Varicella" is a variation on the word "variola," meaning smallpox, which is a severe disease that was often confused with chickenpox (i.e., varicella) until Heberden distinguished them in the late 18th century.

The name used to describe the clinically distinct VZV reactivation infection, "zoster," is derived from the Greek word meaning "girdle," while the term "shingles" is from the Latin *"cingre"* (Christie, 1974). This describes the grids or griddle appearance of the vesicular lesions of the skin characterizing zoster, or shingles (Taylor-Robinson and Caunt, 1972).

The first reports of chickenpox are attributed to Persian physician Rhazes, who, in the ninth century, described a pustular skin disease that conferred no protection against smallpox (Bett, 1934). Over 1000 years later, von Bokay (1909) proposed the idea that the chickenpox infection in children could be related to herpes zoster exposure. Lipschütz in 1921 first described histologically that the vesicular le-sions present in chickenpox and in zoster appeared similar.

Varicella is acquired primarily by inhalation of an airborne virus. Replication of virus within the mucosa of the upper respiratory tract leads to initial waves of viremia, which result in the seeding of virus throughout the body. Viral replication in the basal layers of the epi-dermis causes the rash.

The clinical disease of chickenpox, or varicella, presents acutely

with fever and malaise. Following this prodrome the rash, often starting on the face and the scalp, rapidly progresses from macular lesions
to vesicles. The entire progression can sometimes take as little as 8–12
hours. Multiple rounds of vesicle formation occur over a 3- to 4-day
period. Gradually, the lesions dry, and resulting crusts fall off in 1–3
weeks. Varicella usually resolves uneventfully, but in newborns, immunocompromised children, and adults complications are common
(Arvin, 1987; Straus *et al.*, 1988). Treatment of varicella with anti-
virals (e.g., acyclovir) is usually not indicated for the normal host
(Straus *et al.*, 1988).

II. Biology of Varicella Zoster Virus

Weller (1953) first isolated VZV from the vesicle fluid of a patient
with chickenpox. In 1958 Weller and Witton isolated virus from varicella and zoster patients and demonstrated identity between these viruses which cause distinct infections, leading them to suggest the
name "varicella zoster virus." Infectious virus particles recovered
from varicella and zoster vesicles can replicate in selected primary or
continuous human cells, such as lung fibroblasts (Weller, 1953; Rapp
and Benyesh-Melnick, 1963), thyroid cells (Caunt, 1963), and brain
and ganglion cells (Gilden *et al.*, 1978; Wigdahl *et al.*, 1986). VZV can
also be grown on a number of continuous cell lines, such as HeLa
(Weller *et al.*, 1958); MeWo, a human melanoma cell line (Grose and
Brunell, 1978); Vero, an African green monkey kidney cell line (Caunt
and Shaw, 1969); and guinea pig embryo cells and chemically transformed guinea pig cells (Edmond *et al.*, 1981). The virus grows slowly,
but cytopathic changes occur 4–14 days postinoculation, depending on
the virus titer and cell culture. The cytopathic effect of replicating
VZV is characterized by rounded swollen cells that become refractile
to light. The cytopathic effect spreads in a spindle-shaped pattern
along the monolayer. The cells within the central areas of spreading
infection often detach from the monolayer, resulting in a plaque. At
all times, however, the virus remains highly cell-associated, with few,
if any, free infectious particles being released into the culture medium. Infected cells exhibit eosinophilic intranuclear inclusions characteristic of the herpesvirus family (Andrewes, 1964).

VZV spreads by direct cell–cell contact. This is in contrast to virus
growth *in vivo*, in which cell-free VZV can be readily documented in
aspirates of the vesicular lesions. In cell culture it is extremely difficult to obtain high titers of cell-free virus. Titers greater than 10^6
plaque-forming units (pfu) per milliliter are rare and, hence, this has

slowed our understanding of the growth and molecular biology of VZV. Only a few cell lines producing any cell-free virus have been described. The MeWo cell line (Grose *et al.*, 1979), produces approximately 10^2 pfu/ml culture supernatant. After sonic disruption 10^5 pfu/ml could be achieved. Various methods of concentrating the virus, including ultra-centrifugation or dialysis against hydrophilic compounds, usually result in a loss of titer. Treatment with 8% polyethylene glycol (PEG 6000) was used to precipitate virus and to increase the titer by 50-fold (Grose *et al.*, 1979).

It is not clear why high titers of infectious VZV are not released into the media of tissue culture cells infected with the virus. This is in stark contrast to HSV, in which 10^8–10^9 pfu/ml of culture media can be achieved readily. The first electron-microscopic studies of replicating virus demonstrated viral assembly of nucleocapsids in the nucleus of the infected cell, the virus obtaining its envelope by budding through the nuclear envelope (Cook and Stevens, 1968, 1970). Further electron-microscopic studies have noted VZV in cytoplasmic vacuoles, but it is not known whether these are lysosomal vesicles (Cook and Stevens, 1968, 1970; Gershon *et al.*, 1973). Recently, Jones and Grose (1988) have proposed that VZV obtains its membrane by budding though the Golgi apparatus, so the true process of VZV maturation within the cell is still not clear.

Quantitation of the particle–infectivity ratio from VZV-infected human thyroid cells (Shaw, 1968) disrupted by sonication revealed 8×10^{11} virions, 18% of them complete with core and envelopes, but only 10^5 infectious units per milliliter detected. This particle–infectivity ratio of approximately 10^7 is quite different from that reported for HSV-infected cells (Watson *et al.*, 1963), in which HSV-1 grown in HeLa cells had a particle–infectivity ratio of approximately 10. Of additional interest are the reports by Shiraki and Takahashi (1982), which visualized 10^9 virion particles per milliliter of VZV-infected human embryonic lung cell culture supernatant and noted zero infectivity. Only on sonication of the cells was infectivity obtained. The particle–infectivity ratio of their preparation was approximately 10^6, consistent with previous reports. The reason for this high ratio needs further study. Whether there are important structural differences in the envelopes of VZV that cause it to be so labile or whether virion assembly is an inefficient process is not known. Studies have not been performed to determine whether a final processing event is required for infectivity, as is the case for influenza virus, which requires a proteolytic cleavage event.

It is of interest to note that vesicle fluid from cases of varicella or zoster do have reasonable titers of cell-free virus, with 10^2 pfu/10–30 μl of aspirate. Seidlin *et al.* (1984) detected between 10 pg and 10 ng of VZV DNA in vesicle aspirates. This represents greater than 10^7 VZV DNA molecules with only 100 pfu. One possible explanation of the low infectivity was originally proposed by Gershon *et al.* (1973), who found, on subcellular fractionation, that VZV virions copurified with vacuoles that contained lysosomal enzymes. These acid hydrolases and other enzymes could be responsible for inactivation of the virus as it buds into the cytoplasm from the nucleus. The discovery of mannose 6-phosphate on VZV (Edson *et al.*, 1987) might be significant, in that the virus binds to the mannose 6-phosphate receptor and traffics the virion into lysosomes, where it is destroyed.

In contrast to the lytic nature of VZV growth in numerous cell types, VZV infection of the nonpermissive primary hamster embryo cells has been reported to result in morphological and oncogenic transformation (Gelb *et al.*, 1980). These transformed cells appeared to grow to higher densities in cell culture and exhibited virus-specific antigens. Fc receptors were expressed on their cell surface. Subcutaneous injection of these cells into inbred hamsters induced aggressive fibrosarcomas. The tumor-bearing animals developed antibodies against VZV antigens. After passage of these transformed cells in culture, VZV DNA could not be detected either by dot blot or Southern hybridization. Attempts to repeat these experiments (Gelb and Dohner, 1984) have been unsuccessful, so the ability of VZV to morphologically transform cells is still in doubt.

Animal Models

While VZV grows in tissue culture cells of human, monkey, and guinea pig origin, the human is still the best system in which to study the pathogenesis of VZV disease. The virus does not replicate efficiently in animals. Early attempts to experimentally infect animals were made by Rivers in 1926, who detected nuclear inclusions in the cells of monkeys infected intratesticularly with the tissues of human varicella lesions. Rivers concluded that there was limited viral replication, but no disease. Reagan *et al.* (1953) reported a generalized VZV infection of *Cerci saeus* monkeys following multiple inoculations with blood, vesicle fluid, and throat washings of human patients, but this was not considered an adequate animal model. More recently, replication of VZV in monkeys was examined by Asano *et al.* (1984), who reported a detailed study of the immunogenicity of wild-type and attenuated (Oka vaccine strain) strains of VZV in rhesus monkeys. Animals were subcutaneously inoculated with 10^4 pfu of virus. None of the animals displayed signs or symptoms of disease. Blood samples, nasal washings, and cerebrospinal fluid samples were negative for VZV growth. Most animals mounted a serum antibody response, although the titers varied markedly.

Infection of anthropoid apes by the human VZV has been reported. A female chimpanzee, an orangutan, and a male gorilla who were in close contact with the general public developed a varicellalike illness that was described by Heuschele (1960). In a more detailed report White *et al.* (1972) suggested the acquisition of varicella infection by a gorilla following exposure to the infected 24-year-old wife of a zookeeper, who had uncomplicated varicella. Fifteen days later typical signs of chickenpox occurred in her 9-year-old child, and 1 day later an 8-month-old 20-lb female gorilla residing in their home developed a varicellalike eruption and a slight fever. Over the next few days the rash progressed, with approximately 100 vesicles on the chest and the abdomen and with additional vesicle formation on the arms, back, face, and head. By 2 weeks all major signs of the disease were gone. A 14-month-old female orangutan living in the same household and who had daily contact with the others did not display any signs or symptoms of the disease. Virus was recovered from the gorilla's lesions, and both the gorilla and the orangutan developed complement-fixing antibodies against VZV by 5 weeks postinfection. This episode demonstrates direct human-to-ape transmission and the apparent differential species response in the formation of disease. Most recently, Provost *et al.* (1987) described the experimental infection of the common marmoset by VZV, but the animal developed no overt clinical signs of disease, although an immunological response was detected.

Many early claims of a small animal model for VZV infections could not be substantiated (Cole and Kuttner, 1925). However, studies in the guinea pig have shown the most promise. Myers *et al.* (1980) inoculated weanling guinea pigs intranasally, intracranially, or subcutaneously with VZV grown for 12 passages in fetal guinea pig tissue culture cells. Some animals infected by the intranasal route had a documented viremia, with nasal shedding of virus. This virus was transmitted to cagemates, and a high percentage of infected animals exhibited a serological response to VZV. Virus also could be recovered acutely from the blood, nasal washings, liver, spleen, dorsal route ganglia, spinal cord, and cerebral cortex of the brain. Using ^{32}P-labeled cloned *Hin*dIII-A, B-, or C- DNA restriction endonuclease fragments, VZV nucleic acid could be detected in the brain, liver, blood, and nasal washings (Myers *et al.*, 1985). Infected guinea pigs manifest minimal, if any, cutaneous lesions suggestive of varicella. This could be due to the inability of VZV to replicate efficiently at the elevated temperature of 39.3°C (Gold, 1965; Vaczi *et al.*, 1963), the basal temperature of a guinea pig.

Dunkel *et al.* (1989) reported that Hartley guinea pigs inoculated by intrastromal injection of VZV developed a diffuse punctate keratitis of the cornea. Ocular surface involvement progressed through day 6, with subsequent development of systemic and neurological VZV signs of infection (e.g., pneumonitis, fifth cranial nerve palsy, head and body tremors, and variable hindlimb paralysis). Virus could be recovered from the trigeminal ganglia, brain, and lungs.

Despite the many studies of VZV infections of the guinea pig (Matsunaga *et al.*, 1982; Walz-Cicconi *et al.*, 1986), a complete model for the disease has yet to be developed. In such a model one should be able to demonstrate a viremia and disseminated skin lesions which resolve spontaneously. This should be followed by a period of latency with no virus detection, and eventually one would hope to reactivate latent VZV with a resulting dermatomal zosterlike infection.

III. Virus Structure

Historically, the preliminary classification of viruses has often depended on knowledge of the virion structure. Viruses have been grouped together not necessarily by similar diseases they cause, but by the structure of the virus particle. In this regard the earliest visualization of the virus associated with chickenpox was carried out in 1911 by Aragao. Vesicle fluid from patients was studied using light microscopy, and "elementary bodies" 0.125–0.175 μm in diameter were seen. With the development and application of electron microscopy in the 1940s came Ruska's (1943a,b) description of the morphology of the virion as being round or polygonal with a central core. The relationship between VZV and herpes zoster virus was investigated by Rake *et al.* (1948), who, by electron microscopy, demonstrated the similarity in virion structure from varicella and zoster vesicle fluid.

Almeida *et al.* (1962), using phosphotungstic acid for negative staining, revealed the VZV to be enveloped, with an overall diameter of approximately 200 nm. The envelope showed 8-nm projections from its surface and surrounded a central body (i.e., a capsid) which had a subunit configuration. Disrupted virions that were penetrated by phosphotungstic acid revealed a 100-nm nucleocapsid with capsomers on axes of fivefold symmetry. Each capsid is now known to be made up of 162 hexagonal capsomers, which are assembled into an icosahedron structure with 5 : 3 : 2 axial symmetry (Fig. 1). The average diameter of the VZV capsid was close to the 95-nm diameter determined by Wildy *et al.* (1960) for HSV, the overall structure of the virions of varicella looking nearly identical to that of HSV. In most electron-microscopic studies of VZV, many particles are seen, a high percentage of which

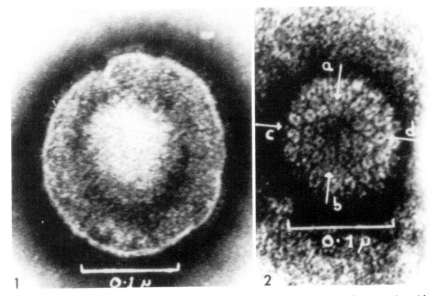

FIG. 1. Electron micrograph of VZV negativity stained with phosphotungstic acid. Panel 1 shows the complete enveloped virion while panel 2 shows nucleocapsids with resolution of the capsomers. Reprinted with permission from Almeida *et al.* (1962).

lack a central core and appear to be incomplete. This observation presomably relates to the high particle–infectivity ratio.

IV. VZV DNA

Early studies demonstrated that 5-halogenated pyrimidines (e.g., 5-iododeoxyuridine), the first clinically useful antiherpetic drugs used for the treatment of herpes simplex ocular disease (Kaufman, 1962; Kaufman *et al.*, 1962), inhibit the synthesis of DNA- but not RNA-containing viruses (Salzman, 1960; Herrman, 1961; Hanparian *et al.*, 1963). Using a plaque assay for VZV developed by Rapp and Benyesh-Melnick (1963), 5-iododeoxyuridine was shown to inhibit both HSV and VZV plaque formations to a similar extent. These early studies also described the kinetics of spread from one infected cell to another, with VZV requiring 8–16 hours postinfection to detect a focus of infected cells by immunofluorescence (Rapp and Vanderslice, 1964). These studies indicated that VZV replication required DNA synthesis and that DNA is the most likely genetic material for VZV.

Over ten years after Weller first isolated VZV, Melnick *et al.* (1964) classified it as a cell-associated herpesvirus of humans. Ludwig *et al.* (1972) described the first characterization of VZV DNA. DNA was isolated from purified virions grown in human embryonic lung fibroblasts. The buoyant density of the DNA, as determined by analytical ultracentrifugation, was calculated to be 1.705 g/cm^2. This corresponds to a guanine plus cytosine (G + C) content of 46%. Heat denaturation of the DNA showed an increase in density consistent with the DNA's being double stranded in nature. Identical results were obtained using DNA isolated from virus purified from both varicella and zoster cases, providing further evidence that these immunologically related viruses were the same (Weller *et al.*, 1958; Ludwig, 1972; Richards *et al.*, 1979).

The density and the percentage of G + C content of VZV were significantly lower than those reported for HSV-1 and -2, with densities of 1.726 g/cm^3 (67% G + C) and 1.728 (69% G + C), respectively (Goodheart *et al.*, 1968; Kieff *et al.*, 1971). HSV-1 and -2 DNA could be separated readily from chromosomal DNA based on their increased buoyant density, but the buoyant density of VZV DNA was sufficiently close to that of cellular DNA to require other techniques to separate and purify it for study. Dumas *et al.* reported in 1980 that trypsin treatment of VZV-infected cells resulted in the release of cell-free viral particles from which DNA could be readily isolated. Sucrose gradient centrifugation of these VZV-infected cell lysates yielded a lower-density enveloped virus and a higher-density nucleocapsid. Virions were lysed with sodium dodecyl sulfate (SDS) and pronase, and the DNA was purified on a linear sucrose gradient. Yields were approximately 2 μg/5 × 10^7 infected cells. Purified DNA was infectious when transfected into human embryonic lung cells by the calcium phosphate precipitation method (Graham and Van der Eb, 1973). Approximately 80–140 pfu/μg of VZV DNA were obtained.

Similar results were obtained with DNA isolated from virus cultured from varicella or zoster patients. Treatment at 56°C for 30 minutes, which inactivates the virus, had no effect on the transfection efficiency, but DNase I treatment destroyed infectivity. The purified DNA was visualized by electron microscopy, and length measurements were used to calculate a molecular weight of 80 million (Dumas *et al.*, 1980). It is interesting to note that when Davison and Scott (1986a) published the complete DNA sequence of VZV, it contained approximately 125,000 base pairs (bp) and a molecular weight of 83 million. It also has a G + C content of 46.02%, extremely close to that predicted by earlier studies.

V. Genomic Organization

The structural organization of the VZV genome was first proposed by Dumas *et al.* (1981) when reporting restriction endonuclease maps of VZV. Analyses of the relative molarity of all of the DNA restriction fragments separated by agarose gel electrophoresis demonstrated the presence of 0.5- and 2-M fragments. The arrangement of the fragments along the genome was determined, and terminal fragments were identified by their disappearance after λ exonuclease digestion.

Further analysis of the VZV genome by restriction enzymes, electron-microscopic studies of denatured genomic DNA, and DNA sequence analysis resulted in the genomic map as we know it (Straus *et al.*, 1981, 1982; Ecker and Hyman, 1982; Davison and Scott, 1986a). Figure 2 illustrates that the genome is considered operationally to be composed of two covalently joined segments: a 105-kilobase (kb) unique long (U_L) region and a 5.4-kb unique short (U_S) region. Both of these unique segments are flanked by inverted repeats. The terminal and internal repeats of the U_L region (TR_L and IR_L, respectively) are 88.5 bp in length, while TR_S and IR_S are about 6.8 kb in length. IR_S and TR_S can be readily visualized by electron microscopy, as they form a terminal lariat structure following denaturation of the DNA (Fig. 3) (Ecker and Hyman, 1982; Straus *et al.*, 1982; Gilden *et al.*, 1982). The 0.5 M and 2 M bands described above, which map in the short repeat region (Dumas *et al.*, 1981; Straus *et al.*, 1982; Ecker and Hyman, 1982; Gilden *et al.*, 1982), are due to isomerization of the entire short segment (Fig. 4). This presumably occurs during replication of the viral DNA.

Hence, the genome exists largely as two isomers with U_S in one of two orientations (Fig. 4). Unique submolar bands, less than 0.5 M, arise from the isomerization of the long segment and can be detected at a low frequency, leading to the conclusion that U_L and its repeats can invert, or the genome can circularize approximately 5% of the time. These unique fragments, which map to the joint between the long and short segments, have been detected by Southern hybridization (Kinchington *et al.*, 1985). A novel joint fragment has been cloned which contains sequences from the short repeat contiguous with those from the left end of the long repeat (J. M. Ostrove and S. E. Straus, unpublished observations). Such fragments could only be derived from circularization of the genome or isomerization of the long segment (Davison, 1984; Kinchington *et al.*, 1985; Ruyechan *et al.*, 1985; Hayakawa and Hyman, 1987). Straus *et al.* (1981) reported the visualization by electron microscopy of low-abundance circular forms of VZV, and this was later confirmed by Kinchington *et al.* (1985). Although the biological significance of this observation is unclear, a replication

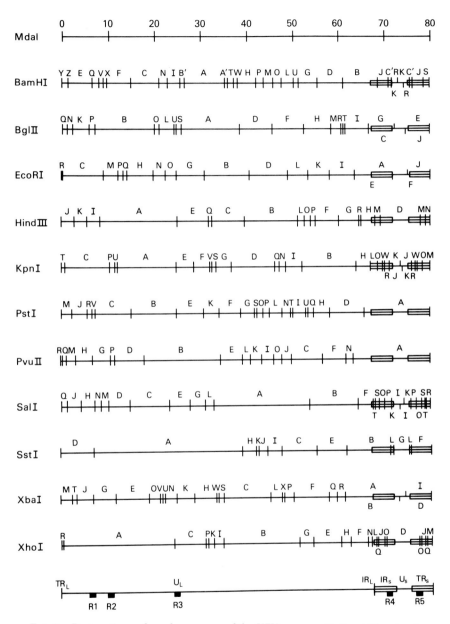

FIG. 2. Restruction endonuclease maps of the VZV genome. (Bottom) The structure of the genome, with TR_L and IR_L being the terminal and internal repeats of the unique long (U_L) segment and IR_S and TR_S the internal and terminal repeats of the unique short (U_S) segment. R1–R5 are the locations of the repeat elements, R4 and R5 being identical (R4/5).

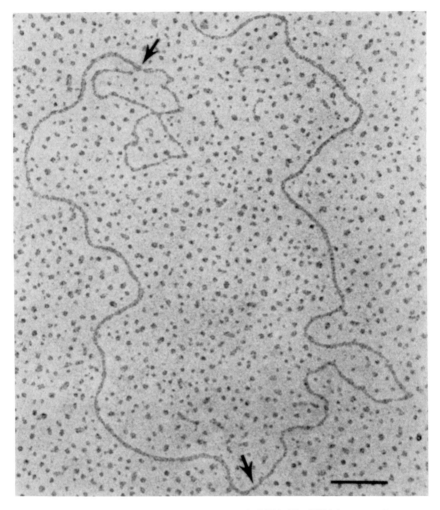

Fig. 3. Electron micrograph of denatured VZV DNA. The DNA between the arrows is double-stranded due to the hybridization of IR_S and TR_S. The single-stranded loop represents the U_S segment of the genome. Bar, 0.2 μm. Reprinted from Ecker and Hyman (1982).

model proposed by Davison (1984) involves circularization as an initial step in the process.

VZV, therefore, differs from HSV in that the long and short segments of the HSV genome isomerize at equal frequencies, resulting in four equimolar isomers of the genome. While VZV potentially has four isomers, approximately 95% of the genome exists in only two isomeric forms.

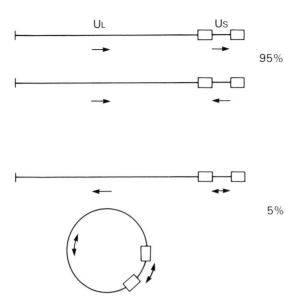

Fig. 4. VZV isomers. The two predominant forms of VZV contain the unique long (U_L) segment and its repeats in a single orientation, while the unique short (U_S) segment and its repeats are detected in an equimolar ratio in either of the two possible orientations, as shown by the arrows. A small percentage of the molecules exist either in a circular form or with the long segment inverted.

While the overall G + C content of the VZV genome is 46%, Ruyechan *et al.* (1985) reported an unequal distribution of G + C-rich regions within the genome. In the short repeat segments of VZV, they determined this distribution to be approximately 64%. The DNA sequence analysis of this region by Davison and Scott (1985) demonstrated it to have an overall 59% G + C content, in close agreement to Ruyechan *et al.* (1985). The repeat regions of HSV-1 have also been shown to contain a significantly higher G + C content (Delius and Clements, 1976). The asymmetric distribution of guanine and cytosine residues is of unclear significance.

VI. Restriction Endonuclease Maps

The construction of restriction endonuclease maps for VZV DNA has been useful for the elucidation of the genome structure, for molecular epidemiological studies, and for providing the foundation for the understanding of the molecular biology of VZV through the cloning and sequencing of the entire VZV genome (Dumas *et al.*, 1981; Ecker and Hyman, 1982; Straus *et al.*, 1982, 1983; Davison and Scott, 1983; Gilden *et al.*, 1983; Mishra *et al.*, 1984). The first restriction endonuclease digests of VZV were reported by Oakes *et al.* (1977), but restriction maps were not reported until 1981, when Dumas *et al.* described maps for the enzymes *Xba*I, *Pst*I, and *Bgl*II. Maps for *Bam*HI, *Bgl*I, *Eco*RI, *Hin*dIII, *Hpa*I, *Kpn*I, *Pvu*II, *Sal*I, *Sma*I, *Sst*I, and *Xho*I soon followed (Ecker and Hyman, 1982; Straus *et al.*, 1982, 1983; Davison and Scott, 1983; Gilden *et al.*, 1983; Mishra *et al.*, 1984). It was not until the publishing of the complete DNA sequence of VZV by Davison and Scott (1986a) that fine detailed restriction maps could be derived. Evidently, the earlier published maps often neglected to account for smaller internal fragments. Figure 2 shows detailed restriction maps of VZV DNA for many common enzymes.

Restriction enzyme digestion of large numbers of clinical isolates has led to interesting observations. Comparison of VZV strains isolated from varicella or zoster patients, from different vesicles on the same individual, or from a varicella isolate and subsequent zoster isolate from the same individual (Iltis *et al.*, 1977; Pichini *et al.*, 1983; Straus *et al.*, 1984) show no major differences in their restriction enzyme patterns. These studies of the viruses isolated from these two clinical syndromes help provide the molecular evidence for the Hope-Simpson (1965) hypothesis, which states that following varicella the virus remains latent in an individual, but reactivates later in life, resulting in zoster. Zweerink *et al.* (1981) purified VZV DNA from two different cases of varicella and subjected it to digestion with six different enzymes. Small but reproducible alterations in the migration of a few of the restriction fragments were noted. These differences represent stable changes between these strains, and no changes were seen in a single strain grown in WI 38 diploid fibroblast tissue culture cells, even after 46 passages *in vitro*. This study suggested that restriction endonuclease analysis of VZV strains might be a valuable epidemiological tool.

Martin *et al.* (1982) isolated VZV DNA from several clinical isolates and compared them with VZV strain Oka, the current live VZV vaccine strain (Gershon, 1987; Gershon *et al.*, 1988). Using *Bam*HI, *Bgl*I, and *Hpa*I, they were able to distinguish the Oka vaccine strain from wild-type strains. Further analysis by that group demonstrated an additional *Bgl*I site within the *Bam*HI-D fragment of Oka, as well as its parent, but not in any of the other strains isolated (Adams *et al.*, 1990). This restriction site polymorphism has been effectively exploited in studies of vaccine recipients. It permitted a clear determination of whether clinical disease in a vaccine recipient is due to the live

Oka vaccine strain or imperfect immunity followed by exogenous reinfection (Gelb *et al.*, 1987).

Mapping of regions of the genome in which restriction fragment-
length polymorphism occurs was reported by Straus *et al.* (1983).
Using 17 isolates and multiple restriction enzymes, four regions of the
genome were found by agarose gel electrophoresis to display altered
fragment mobility. Two of these regions mapped in the inverted repeats (since one change is repeated twice: R4/R5) and the two others
mapped at approximately 0.16 (R2) and 0.35 (R3) map units (Fig. 2).

Since then the mapping of the variable-sized fragments from many
different strains has been reported (Straus *et al.*, 1983; Hayakawa *et
al.*, 1984, 1986), and the molecular basis for the variation in fragment
sizes has been determined. Hondo *et al.* (1987) described both loss and
gain of restriction endonuclease sites within the VZV genome. While
this type of heterogeneity is detected, the most common cause of size
variation within the VZV genome is the reduplication or deletion of
small repeat elements that are G + C rich. Five repeat elements (R1–
R5) have been mapped, and their DNA sequences have been determined (Fig. 2) (Casey *et al.*, 1985; Davison and Scott, 1986a; Kinchington *et al.*, 1986).

Of these five repeating elements, only R2, R3, and R4/R5 were
mapped to variable-sized restriction fragments and are due to different numbers of repeats within the fragments (Table I). R1 is a repeat-

TABLE I

REPEAT ELEMENTS IN VZV

		Base pairs
Repeat 1 within ORF 11		
A	GGACGCGATCGACGACGA	18
B	GGGAGAGGCGGAGGA	15
C	GGACGCGATTGACGACGA	18
D	GGACGCGGCGGAGGA	15
X	GGA	3
Repeat 2 within ORF 14		
A	GCGGGATCGGGCTTTCGGGAAGCGGCCGAGGTGGGCGCGACG	42
B	GCGGGATCGGGCTTTCGGGTAGCGGCCGAGGTGGGCGCGACG	42
X	GCGGGATCGGGCTTTCGGGTAGCGGCCGAGGT	32
Repeat 3 within ORF 22		
A	GCCCGCGCA	9
X	GCCC	4
Repeat 4/5 between ORFs 62 and 63		
A	CCCCGCCGATGGGGAGGGGGCGCGGTA	27
X	CCCCGCCGATG	11

ing element mapping in open reading frame (ORF) 11 and varies among three different VZV strains sequenced (Kinoshita *et al.*, 1988). R1–R3 are found within coding regions, and R4/R5 maps in an intragenic region (Casey *et al.*, 1985; Davison and Scott, 1986a). R1 is composed of the repeat elements ABBABBCBBCBBBBABDABX, where A and C are 18-bp elements differing at nucleotide 10 by a C-to-T transition in the Dumas strain. Element D has the first 7 bp of A and C, followed by eight different base pairs. Element B has 3 bp changes in the first six bases and is then identical to element D for the remaining 9 bp. X is a partial copy of A, B, and D (Table I). In strains H-S1, H-N3, and YS structures of ABBABBABBABABAB, ABBABBCBBDAB, and ABBABBA*BBDABABAB (A* has a G-to-A alteration at base 17) were reported by Kinoshita *et al.* (1988). With all of these strain differences, the nucleotide sequence is still a multiple of 3 bp, causing no frameshifts in the ORF.

R2 consists of nearly perfect 42-bp tandem repeats (which can vary by 1 bp at position 21). This maps in the coding region of ORF 14, which codes for glycoprotein V (gpV) (Kinchington *et al.*, 1986). The Scott strain of VZV contains seven 32/42 copies while the Webster strain has three 32/42 copies of R2. Variations in the sizes of gpV transcripts and protein have been detected (P. R. Kinchington, P. Ling, W. T. Ruyechan, and J. Hay, personal communication), indicating that these variations become incorporated into gpV with proteins of corresponding greater or lesser size, depending on the strain of virus. The biological effects caused by variations in the size of this glycoprotein are not clear.

The VZV fragment *Bam*HI-A contains a 9-bp repeated element known as R3. It is the most difficult to clone and unstable region of the VZV genome. Greater than 100 copies of R3 may be found in the VZV sequence, but since it is so unstable on cloning into *Escherichia coli* vectors, the published sequence of VZV (Davison and Scott, 1986a) could lack up to 500 bp of DNA sequence in this region. How these putatively missing sequences relate to ORF 22 protein or any additional ORFs that may map in this area is not known. It is unlikely, though, that additional ORFs exist, due to sequence homology and the spatial arrangement of VZV genes compared to HSV-1 (McGeoch *et al.*, 1988a).

The final repeat element recognized R4/R5 maps in IR_S and TR_S sequences. Different strains contain a variable number of this 27-bp repeat (Casey *et al.*, 1985; Davison and Scott, 1986a; Hondo and Yogo, 1988). R4/R5 shares DNA sequence homology with a repeat element identified in a similar genome location in HSV-1 DNA (Casey *et al.*, 1985). Both map adjacent to the viral origin of DNA replication, termed "Ori_S" (see Section VIII,C).

VII. VIRALLY ENCODED PROTEINS

Studies aimed at characterizing VZV-encoded proteins began shortly after VZV was determined to be morphologically similar to HSV (Almeida *et al.*, 1962). Kapsenberg (1964) suggested the possibility that VZV and HSV share common antigens. These early studies demonstrated that increases in the titer of complement-fixing antibodies directed against VZV occurred following an HSV infection (Kapsenberg, 1964; Svedmyr, 1965). Similar studies were carried out by Ross *et al.* (1965), who demonstrated an approximately fourfold increase in complement-fixing antibody to HSV in 48% of varicella and 28% of zoster patients. By the late 1960s it was demonstrated clearly that HSV and VZV share immunological cross-reactivity, as determined by complement fixation and fluorescent antibody tests (Schmidt *et al.*, 1969; Ross *et al.*, 1965; Schaap and Huisman, 1968). Over the ensuing 10 years, numerous papers were published in which the immunological relationships of VZV to other primate and nonprimate herpesviruses were discussed (Harbour and Count, 1979; Felsenfeld and Schmidt, 1975, 1977).

In 1978 Wolff published the first attempt to identify and catalog VZV virion proteins. To do so, $[^{35}S]$methionine was added to VZV-infected human embryonic fibroblasts. Twelve hours later the cells were scraped from the monolayer and virions were purified by sucrose (10–60%) density gradient centrifugation. The integrity of enveloped virions was verified by electron microscopy. Radiolabeled virion proteins were analyzed by SDS–polyacrylamide gel electrophoresis (SDS–PAGE), and 31 bands ranging in size from 18 to 240 kDa were resolved. Fourteen of the 31 bands could be immunoprecipitated with human sera reactive with VZV (Wolff, 1978).

Over the next few years further SDS–PAGE analyses of the VZV polypeptides led to the identification of approximately 33 proteins, ranging in size from 16 to 280 kDa. As determined by the incorporation of $[^{14}C]$glucosamine, 10–13 of these were glycosylated (Wolff, 1978; Asano and Takahashi, 1979, 1980; Shemer *et al.*, 1980; Shiraki *et al.*, 1982; Zweerink and Neff, 1981; Grose and Friedrichs, 1982; Grose *et al.*, 1981).

Categorizing VZV proteins according to their time of synthesis in the infectious cycle has been difficult. A characterization scheme similar to that for HSV, identifying immediate early or α genes which do not require viral protein synthesis for expression, early, or β, genes which are expressed prior to DNA replication, and late, or γ, genes which follow DNA replication, was impossible to carry out with any great precision. Such kinetic analysis had been performed for HSV-1 (Honess and Roizman, 1974, 1975), since studies on temporal gene expression were feasible due to the availability of high-titered stocks of virus. Cells easily could be infected with 10–50 pfu per cell of HSV-1, while studies with VZV still used cell-associated virus or titers of cell-free virus infecting only a small percentage of cells (0.005–0.3%).

Lopetequi *et al.* (1985) infected human embryonic fibroblasts at a multiplicity of infection of 0.3 pfu per cell, using a titered stock of VZV at 1.6×10^6 pfu/ml. Immunoprecipitations were performed, using a hyperimmune monkey serum against VZV. Cells were treated with cycloheximide (50 µg/ml) for 12 hours and labeled with [^{35}S]methionine in the presence of actinomycin D to prevent further transcription. Under these conditions one should be able to detect immediate early proteins. Seven putative immediate early polypeptides with molecular weights of 180K, 145K, 135K, 118K, 44K, 42K, and 33K were resolved by SDS–PAGE.

Early proteins were analyzed using a similar strategy for infection, but in the presence of 20 µg/ml phosphonoacetic acid as an inhibitor of VZV DNA synthesis. After 12 hours postinfection cells were washed and incubated with cycloheximide and phosphonoacetic acid for an additional 6 hours to accumulate early protein mRNA. Cells were then washed and labeled with [^{35}S]methionine for 2 hours in the presence of phosphonoacetic acid. Under these conditions six early proteins were detected at molecular weights of 135K, 118K, 68K, 44K, 42K, and 35K. To group VZV proteins into the late class of gene products, cells were infected as above, but incubated for 36 hours until 80% of the cells demonstrated a cytopathic effect. Immunoprecipitated proteins with molecular weights of 180K, 123K, 84K, 81K, 68K, 48K, 44K, 37K, and 30K were detected. Unfortunately, these experiments, with such low multiplicity of infection, are not able to clearly resolve specific VZV polypeptides and identify the kinetic class of the protein.

Other attempts to map the immediate early genes of VZV were made by Shiraki and Hyman (1987), who used multiple (i.e., four) rounds of infection of the same human embryonic lung cell monolayer. This technique still only infected 13.6% of their cells. Cycloheximide treatment of the cells to block protein synthesis was performed to accumulate immediate early transcripts. After 24 hours cells were washed and placed in media containing actinomycin D (10 µg/ml) to block   transcription,   and   were   then   incubated   with [^{32}P]orthophosphate or [^{35}S]methionine for 3 hours. Immunoprecipitation with anti-VZV sera revealed four putative immediate early genes at 185, 69, 43, and 34 kDa. Although three of these protein sizes (185, 43, and 34 kDa) were in close agreement with some of the immediate early proteins suggested by Lopetequi *et al.* (1985), the authors state that "not all four proteins were seen with both radiolabels in all five gels" (Shiraki and Hyman, 1987). With such low titers of VZV, these classic drug-block experiments, used successfully to map temporal classes of genes for other members of the herpesvirus family, might not be possible.

A. VZV Glycoproteins

Various approaches have been used to identify and characterize VZV glycoproteins. The first studies to clearly define the sizes and the time of expression of VZV glycoproteins were conducted by Grose (1980). Human melanoma cells were infected with VZV and radiolabeled with the sugar precursors [^{3}H]glucosamine and [^{3}H]fucose. Cell extracts were immunoprecipitated with rabbit anti-VZV hyperimmune sera absorbed with uninfected cells and serum from a patient with zoster. Five glycoproteins were immunoprecipitated by both sera and were called gp118, gp98, gp88, gp62, and gp45 (Grose, 1980, 1981; Grose *et al.*, 1981). Asano and Takahashi (1980), using [^{14}C]glucosamine and [^{35}S]methionine, identified seven glycoproteins (with molecular weights of 105K, 90K, 84K, 72K, 70K, 60K, and 56K). Numerous groups catalogued VZV glycoproteins, but it was not until the use of specific antisera (e.g., monoclonal antibodies or antibodies directed against purified VZV glycoprotein) that the number of glycoproteins and the relationship between the different molecular weight forms of each began to be sorted out. Murine monoclonal antibodies were generated following the inoculation of mice with either purified VZV virions (Forghani *et al.*, 1982, 1984; Keller *et al.*, 1984) or VZV-infected cell extracts (Grose *et al.*, 1983; Okuno *et al.*, 1983). The majority of the murine monoclonal antibodies fell into three groups, the members of each recognizing one set of glycoproteins. These monoclonal antibodies were able to immunoprecipitate and/or react in Western blot analysis. The sizes and the names assigned to each of the glycoproteins varied from laboratory to laboratory, until a common nomenclature was agree on at the 10th International Herpesvirus Workshop in Ann Arbor, Michigan (Table II) (Davison *et al.*, 1986).

At first, it was not completely clear why one monoclonal antibody could immunoprecipitate numerous proteins. It was presumed that different degrees of glycosylation, as well as a proteolytic processing and posttranslation modification, resulted in families of related polypeptides. The exact number of glycoprotein genes coded for by the virus was not known until Davison and Scott (1986a) sequenced the entire VZV genome. This demonstrated that there were five ORFs— ORF 68, 31, 37, 67, and 14—which correspond to gpI–V (Table II). All

TABLE II

VZV-Encoded Glycoproteins

Glycoprotein	Gene	Molecular weight[a]	HSV-1 homolog
I	68	90,000–98,000, 80,000–83,000, 55,000–60,000, 45,000	gE
II	31	120,000–140,000, 57,000–65,000	gB
III	37	105,000–118,000	gH
IV	67	45,000–55,000	gI
V	14	95,000–105,000	gC

[a]Molecular weight ranges of mature and processed forms of the glycoproteins are as determined by SDS–PAGE and reported by Davison et al. (1986).

[b]From P. R. Kinchington, P. Ling, W. T. Ruyechan, and J. Hay, (personal communication).

of the previously characterized monoclonal antibodies mapped to gpI, gpII, or gpIII. The general characteristics of the amino acids that make up the ORFs include an amino-terminal hydrophobic domain for the translation of mRNA in membrane-bound ribosomes. An extensive hydrophobic region followed by basic residues mapping proximal to the carboxyl termini would make up the membrane anchor sequence. Antibodies against each of the ORFs gpI–V have demonstrated that they code for a viral glycoprotein.

gpI, a major viral glycoprotein, maps to ORF 68 in the U_S region of the genome. Its primary amino acid sequence codes for a protein of 70 kDa. This protein was first mapped by Ellis et al. (1985), using an E. coli expression library. Monoclonal antibodies directed to gpI detected expression of a polypeptide that mapped to the U_S segment of the genome. This fragment was used to hybrid select and in vitro translate a protein that reacted with antibodies to gpI.

Ostrove et al. (1985) mapped a 3.7-kb transcript to this region that might code for gpI. gpI is the predominant VZV glycoprotein found in isolated membrane fractions from infected cells (Grose, 1987). gpI contains both N- and O-linked sugars (Montalvo et al., 1985; Namazue et al., 1985; Edson et al., 1985), as determined by treatment with various glycosylases and inhibitors of glycosylation. In addition, evidence for the addition of fatty acids using [^{3}H]palmitic acid has been obtained by Namazue et al. (1989). Multiple species of gpI, ranging in size from 45 to 98 kDa, are detected and presumably represent different processed forms of gpI.

Analysis of the primary amino acid sequence of VZV gpI and comparison of it with that of HSV-1 demonstrate amino acid homology with glycoprotein E (gE) (Davison and McGeoch, 1986). Both sequences contain conserved amino- and carboxy-terminal hydrophobic domains, but the positions of potential glycosylation sites, Asp–X–Ser/Thr, are not conserved. These proteins share between 25% and 49% amino acid homology in various stretches in the carboxy-terminal half of the protein.

gpII has been mapped to ORF 31, which can code for a polypeptide of 98 kDa. ORF 31 is located at map position 0.47. gpII is a disulfide-linked glycoprotein which migrates in a size range between 57 and 140 kDa, depending on the experimental conditions (Okuno et al., 1983; Forghani et al., 1984; Grose et al., 1984; Keller et al., 1984; Vafai et al., 1984; Namazue et al., 1985; Montalvo and Grose, 1987). Under nonreducing conditions Grose et al. (1984) demonstrated [^{3}H]fucose-labeled gpII immunoprecipitates as peptides of 140, 118, 98, 62, and 45 kDa. When the samples are analyzed on reducing SDS–PAGE, the 140-kDa gpII species is reduced, and a 66-kDa species is found (Fig. 5). This 140-kDa glycoprotein is thought to be a heterodimer derived from two polypeptides of approximately 66 kDa (Grose et al., 1984).

Detailed studies performed by Keller et al. (1986) on gpII, utilizing hybrid selection followed by in vitro translation, demonstrated that a 100-kDa polypeptide could be precipitated with monoclonal antibodies as well as convalescent sera against gpII. This protein is encoded by 2.6-kb ORF 31. Confirmation that gp140 is proteolytically cleaved and is present as a disulfide-linked heterodimer came from amino-terminal sequence analyses of reduced gp66 that was derived from a highly purified nonreduced preparation. Amino acid sequence analysis revealed sequences consistent with two termini: one the native amino terminus from positions 9–20, the other mapping to amino acid positions 432–443. These results were compared to the amino acid sequence derived from the DNA sequence. An endoproteolytic cleavage event between amino acids 431 and 432 was proposed. This reaction is thought to be mediated by a cathepsin-related acid-thiol protease, due to the presence of two arginine residues at positions 430 and 431 (Keller et al., 1986).

The primary amino acid sequence of VZV gpII shares a high degree of homology to the HSV-1 gB gene. Direct alignment of the amino acids shows approximately 45% homology, and hydropathicity profiles are similar (Keller et al., 1986; Bzik et al., 1984; Pellett et al., 1985). This high degree of homology could imply that these glycoproteins share similar functions. An important difference between VZV gpII and its HSV-1 homolog, gB, is that there is no evidence that gB is proteolytically cleaved and, hence, some structural differences are clearly present.

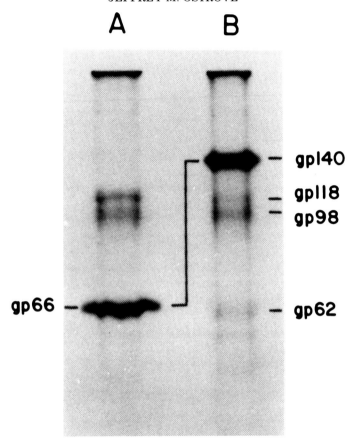

Fɪɢ. 5. SDS–PAGE analysis of gpII. [³H]Fucose-labeled VZV-infected cell polypeptides enriched for gpII were fractionated following treatment with 5% 2-mercaptoethanol (lane A) or with no reducing agent added (lane B). Reprinted from Grose *et al.* (1984).

Edson *et al.* (1985) clearly demonstrated that gpII and gB cross-react using both polyclonal and monoclonal sera directed against HSV. Two monoclonal antibodies that recognized HSV-1 gB were able to neutralize VZV in a complement-dependent fashion.

gpIII, the third and least abundant of the three major VZV glycoproteins, has been mapped to ORF 37. Keller *et al.* (1987) purified gpIII to over 95% homogeneity by affinity chromatography, using monoclonal antibodies that reacted with gpIII. Amino-terminal protein sequence analysis of gpIII led to an unambiguous amino acid sequence. Using this amino acid sequence, a pool of degenerate oligonucleotides capable of expressing codons for the amino acid sequence Lys–Ser–Tyr–

Val–Thr–Pro–Thr was synthesized. This degenerate oligonucleotide pool was used to hybridize to Southern blots of HindIII fragments of VZV DNA. Analysis of the DNA sequence of this HindIII fragment was able to predict an ORF that had amino acids that matched the amino termini of purified gpIII. This mapped to position 0.54 on the genome. The ORF encoding gpIII has a primary polypeptide sequence capable of producing a protein of 94 kDa. There are 11 potential sites for N-linked glycosylation.

Numerous monoclonal and polyclonal antisera against gpIII have been developed, the majority of which immunoprecipitate a single glycoprotein of approximately 118 kDa. Many of these antibodies were shown to contain neutralizing activity.

gpIV was first predicted not by the reaction with monoclonal antibodies or the identification of a specific radiolabeled glycoprotein band on a gel, but by analyses of the genomic sequence. This area of the genome was shown to code for an ORF with amino acids that included a hydrophobic amino-terminal and hydrophobic carboxyl-terminal region. ORF 67 coding for a 39-kDa protein contained such properties (Davison, 1983). Davison et al. (1985) synthesized a seven-amino acid peptide corresponding to the carboxy terminus of ORF 67. Rabbits were immunized and antisera were raised that reacted weakly to the [^{35}S]methionine or [2-^{3}H]mannose-labeled VZV-infected cells. Two labeled VZV glycoproteins of molecular weights 55,000 and 45,000 (i.e., gp55 and gp45) were immunoprecipitated.

Comparison of amino acid sequence homology with HSV-1 has identified that VZV ORF 67 shares homology with HSV-1 gI. Okuno et al. (1983) developed a monoclonal antibody that reacts with an intracellular 55-kDa glycoprotein and a secreted form at 45 kDa. The relationships of these glycoproteins with that identified by Davison et al. (1985) are unknown. P. R. Kinchington, W. T. Ruyechan, B. Moss, and J. Hay (unpublished observations) have expressed gpIV in a vaccinia virus expression system and raised antibodies against the native protein. This serum neutralizes VZV.

The existence of gpV, like gpIV, was also first considered on the basis of DNA sequence analysis. Kinchington et al. (1986), while characterizing and sequencing the R2 repeat, realized that the 42-bp repeats lie within a sequence capable of encoding a glycoprotein. The ORF 14 hydropathy plot was consistent with this. Five potential N-linked glycosylation sites were also identified. The potential variability in the number of the 42-bp repeats within VZV strains predicted gpV polypeptides ranging in size from 62 to 66 kDa. This has been verified by P. R. Kinchington, W. T. Ruyechan, and J. Hay (personal communication).

Hybridization studies of this area of the genome have demonstrated that the repeats are transcribed and that RNA isolated from different strains vary in size according to the numbers of repeat elements (P. Ling, W. T. Ruyechan, and J. Hay, personal communication). In addition, antisera raised against a synthetic peptide immunoprecipitated a glycoprotein of predicted size. ORF 14 has been cloned into a vaccinia expression system, and large amounts of gpV, as well as polyclonal sera directed against it, have been derived (P. R. Kinchington, J. Hay, and W. T. Ruyechan, unpublished observations).

Based on the genomic position of ORF 14 relative to HSV-1 and homology studies with gpV, amino acids between positions 108 and 546 share approximately 21% homology with HSV-1 gC.

In summary, VZV appears to code for five glycoproteins which are products of ORF 14 (gpV), ORF 31 (gpII), ORF 37 (gpIII), ORF 67 (gpIV), and ORF 68 (gpV). All have homologs in HSV-1, and they are gE, gB, gH, gI, and gC, respectively. In addition, ORFs 31 and 37 share homology with putative Epstein–Barr virus glycoproteins coded for by BALF4 and BXLF2, respectively (McGeoch et al., 1986) and with cytomegalovirus ORF 37 (Cranage et al., 1988). Antisera directed against any of these glycoproteins have the ability to neutralize VZV infectivity.

B. Virally Encoded Enzymes

1. Deoxypyrimidine Kinase

Studies of the biochemical consequences of VZV infection in tissue culture cells and the search for antiviral compounds led to the identification, characterization, and purification of a number of virus-specific enzyme activities. Dobersen et al. (1976) reported that the enzymatic basis for selective inhibition of VZV by 5-halogenated analogs of deoxycytidine (e.g., 5-bromodeoxycytidine) are due to the induction of 5-bromodeoxycytidine kinase activity. This deoxypyrimidine kinase (dPK) activity seemed to be associated with a protein that had a molecular weight of 70,000 and was more thermostable than the similar enzymatic activity induced by HSV-1 or -2, but less so than the human fibroblast (i.e., human embryonic lung cell) enzyme. Similar observations were also reported by Ogino et al. (1977), Hackstadt and Mallavia (1978), and Cheng et al. (1979). These studies verified the molecular weight of the enzyme activity to be approximately 70,000 under nondenaturing conditions. VZV dPK can phosphorylate bromodeoxyuridine, iododeoxycytidine and uridine, deoxycytidine and deoxythymidine. All of the ribonucleotide and deoxyribonucleoside triphosphates, except for deoxythymidine triphosphate, could serve as phosphate donors in the kinase reaction, ATP being the best, with a K_m of 16 μM (Cheng et al., 1979).

Stable cell lines expressing the VZV dPK were first described by Yamanishi et al. (1981). They infected Ltk⁻ cells with VZV. Under selective pressure of medium containing aminopterin and thymidine, cell clones were isolated. These cells expressed high levels of dPK activity, which could be inhibited by antisera directed against VZV-infected African green monkey kidney cell extracts (Yamanishi et al., 1980; Ogino et al., 1982).

Although a number of cell lines expressing VZV dPK were prepared, it was not until the work by Sawyer et al. (1986) that the gene and the transcript for VZV dPK were mapped. In their studies Ltk⁻ cells were transfected with recombinant DNA clones spanning 75% of the VZV genome. Two restriction fragments, BamHI-H and EcoRI-D, proved capable of converting Ltk⁻ to Ltk⁺ in selective hypoxanthine–aminopterin–thymidine medium. These restriction fragments overlapped by 2.2 kb and mapped between 0.50 and 0.52 map units on the VZV genome. Probes homologous to the region of overlap hybridized to a 1.8-kb VZV-specific mRNA transcript that was expressed in these biochemically transformed cells.

By DNA sequence analysis and S1 nuclease mapping, Davison and Scott (1986a) confirmed this and localized the 5′ and 3′ ends of the dPK transcript. The VZV dPK gene which maps to ORF 36 has a number of interesting features at both its 5′ and 3′ ends. This ORF can encode a protein of 37.8 kDa, one which shares amino acid homology with the thymidine kinase gene of HSV-1. The enzymatic activity of dPK is found as a homodimer with a native molecular weight of approximately 70,000 (Lopetequi et al., 1983; Shiraki et al., 1985). The 5′-untranslated region of the mRNA is 410 bp in length, with a consensus TATTAAA mapping approximately 33 bp upstream from the start of transcription. Within this untranslated region three additional TATA elements were found, yet no smaller transcripts have been detected by Northern blot hybridization (unpublished observations). Two ATG methionine codons are present in the 5′-untranslated leader sequence, but each is followed by a stop codon. This type of organization could play a role in the efficiency of translation (Kozak, 1984).

The 5′ promoter elements of the VZV dPK differ from that noted in HSV-1 by McKnight et al. (1981). VZV lacks the Sp1 binding sequences, but contains a number of G-rich boxes which could play a role in a transcription (Gelman and Silverstein, 1987; Inchauspe and Ostrove, 1989a).

The 3′ end of the ORF 36 sequence contains two putative poly(A) addition signals, an ATTAAA at 65,978 bp, and an AATAAA at 66,125 bp, but 3' end mapping by Davison and Scott (1986a) places the poly(A) addition site at 65,859 bp at the sequence AGTAAA. A similar AG-TAAA poly(A) addition sequence was also proposed for ORF 28, the VZV-encoded DNA polymerase.

The mode of action for many of the antiviral compounds used clinically or experimentally derive their specificity from their selective phosphorylation by the viral-encoded dPK. Acyclovir (ACV), for example, is phosphorylated to ACV monophosphate by the viral enzyme and then to the triphosphate by cellular enzymes. ACV triphosphate either inhibits the DNA polymerase directly or is incorporated into viral DNA, resulting in chain termination. ACV is an antiherpetic drug that is used extensively for the treatment of recurrent HSV and VZV infections (Elias *et al.*, 1977; Straus *et al.*, 1985). It inhibits VZV DNA replication at 3 μM, but inhibits cellular DNA synthesis at 300 μM.

ACV-resistant mutants of VZV were first described by Biron *et al.* (1982), who passaged VZV in human fibroblasts in the presence of increasing drug concentrations or continuous exposure to 100 μM ACV, a dose approximately 30 times the ED_{50}. Of five ACV-resistant mutants isolated, three were negative and two positive for the phosphorylation of ACV. The latter two might represent mutations in the VZV-encoded DNA polymerase. Similarly, Shiraki *et al.* (1983) isolated eight ACV-resistant mutants, six of which exhibited reduced thymidine kinase (i.e., dPK) activity.

Through the work by Kit (1985), amino acid homology among the thymidine kinase genes of six members of the herpesvirus family could be aligned and a putative ATP and thymidine (nucleoside)-binding region could be determined (Sawyer *et al.*, 1988) (Fig. 6).

The molecular basis of ACV resistance in one series of mutants was determined by Sawyer *et al.* (1988). Northern blot hybridization, using a probe internal to the *dPK* gene demonstrated that each of the mutants and their wild-type parents synthesized the 1.8-kb dPK mRNA. The *Pst*I-P restriction fragment which contains the *dPK* gene was cloned from each strain, and the DNA sequence was determined. Single-base pair changes between each mutant and its wild-type parent were noted. Two of the mutants (KB3 and 101) had identical G-to-A transitions, resulting in a premature stop codon at amino acid 225. Mutant *40a2* showed a T-to-C transition at amino acid 154, resulting in a L-to-P substitution. This dramatic change maps nine amino acids downstream from the putative nucleoside binding region (Fig. 6) of the enzyme. VZV strain 7-1-3 was the first ACV-resistant isolate recovered in a clinical setting from a patient with acquired immunodeficiency syndrome (AIDS) (Pahwa *et al.*, 1988). It was found to contain

A

```
VZV DUMAS (12-29).....V L R I Y L D G A Y G I G K T T A A..
VZV ELLEN (12-29).....V L R I Y L D G A Y G I G K T T A A..
PSEUDORABIES (3-20)...I L R I Y L D G A Y G T G K S T T A..
HSV-1 (49-66)........L L R V Y I D G P H G M G K T T T T..
HSV-2 (49-66)........L L R V Y I D G P H G V G K T T T S..
MARMOSET HERPESVIRUS..I L R V Y L D G P H G V G K S T T A..
  (10-27)
```

B

```
VZV DUMAS (129-146).....D R H P I A S T I C F P L S R Y L..
VZV ELLEN (129-146).....D R H P I A S T I C F P L S R Y L..
PSEUDORABIES (108-124)..D R H P V A A T V C F P L A R F I..
HSV-1 (162-178)........D R H P I A A L L C Y P A A R Y L..
HSV-2 (162-178)........D R H P I A S L L C Y P A A R Y L..
MARMOSET HERPESVIRUS....D R H A V A S M V C Y P L A R F M..
  (130-146)
```

FIG. 6. Identification of amino acid sequence homology in the (A) ATP binding site and (B) nucleoside (thymidine) binding site of five different herpesviruses. Reprinted with permission from Sawyer *et al.* (1988).

an R-to-Q substitution at residue 130, which is within the nucleoside binding site. This change, while destroying ACV-phosphorylating activity, still appears to allow the virus to retain some thymidine kinase activity (K. K. Biron and J. A. Fyfe, personal communication).

Two 5-bromodeoxyuridine-resistant (BUDR-R) thymidine kinase-deficient mutants of VZV were sequenced by Mori *et al.* (1988). These mutants contained changes that caused either the direct introduction of a stop codon at amino acid residue 69 (strain Oka-BUDR-R-oz) or a 2-bp deletion at position 376, causing a frameshift and premature termination at amino acid 162. These changes are all consistent with a nonfunctional dPK.

2. *VZV DNA Polymerase*

The earliest indication that VZV encoded its own DNA polymerase was from the work of May *et al.* (1977), who demonstrated that phosphonoacetic acid was able to inhibit the replication of VZV *in vitro*. By that time, it had already been well documented that phosphonoacetic acid inhibited the replication of herpesviruses (Shipkowitz *et al.*, 1973) and that its target of inhibition is the viral DNA polymerase (Mao and Robishaw, 1975).

Miller and Rapp (1977) and Mar *et al.* (1978) partially purified the VZV DNA polymerase and demonstrated differences from cellular polymerases α and β in the optimal pH, synthetic template specificity, and the effect of $(NH_4)_2SO_4$ on the reaction. Comparative studies of the DNA polymerases purified from HSV-1 and -2, cytomegalovirus, and VZV were reviewed by Mar and Huang (1979).

The gene for the VZV DNA polymerase was predicted from analysis of the DNA sequence of the genome (Davison and Scott, 1986a) and was shown to map to ORF 28. ORF 28 is predicted to code for a protein of 134 kDa, one that shares extensive amino acid homology with HSV-1 DNA polymerase (McGeoch et al., 1988a). Becker (1988) revealed that there was a 57% similarity among first 1000 amino acids of the DNA polymerase, and 49% homology among the remaining 194 amino acids. Conserved domains among different viral DNA polymerases, such as vaccinia virus, adenovirus, bacteriophage 29, and the herpesviruses (e.g., VZV, HSV-1 and -2, Epstein–Barr virus, and cytomegalovirus) have been described (Larder et al., 1987; Earl et al., 1986; Gentry et al., 1988).

3. Ribonucleotide Reductase

Another enzyme that is conserved among HSV-1 and -2, Epstein–Barr virus, and VZV is ribonucleotide reductase (RR), which catalyzes the conversion of ribonucleotides to their corresponding deoxyribonucleotides. This enzyme is present in prokaryotic and eukaryotic cells and could determine the rate-limiting step in DNA replication (Thelander and Reichard, 1979). Herpesviruses induce a ribonucleotide reductase that differs from the cellular enzymes in that they are not sensitive to allosteric regulators such as thymidine (Cohen, 1972; Spector et al., 1987). The VZV-encoded enzyme, as is the enzyme from almost all sources (Thelander and Reichard, 1979), is composed of two nonidentical subunits, the gene products of VZV ORFs 18 and 19. ORF 18 encodes the small subunit of RR, a protein of 306 amino acids (35 kDa), and is approximately 25 amino acids smaller than the corresponding subunit of HSV-1 RR (Davison and Scott, 1986a). ORF 19 codes for the large subunit of RR and contains 775 amino acids that constitute a polypeptide of 87 kDa, significantly smaller than the 136-kDa polypeptide of HSV-1. In general, a high degree of conservation exists in the subunits of RR (Nikas et al., 1986) between HSV and VZV, especially in the small subunit (Davison and Scott, 1986a). Such homology is further revealed by the 1986 report that antibodies synthesized against a synthetic peptide mapping to the small subunit of HSV-1 RR could immunoprecipitate the corresponding protein from VZV-infected cells (Dutia et al., 1986).

Little is known about the gene expression of RR, but two transcripts that are 3′ coterminal and share a polyadenylation site seem most likely (Davison and Scott, 1986a). Transcripts of 5.0 and 1.2 kb have been mapped for HSV-1 RR subunits (McLauchlan and Clements, 1982). Reinhold *et al.* (1988) mapped two transcripts of 3.55 and 1.45 kb to this region of the VZV genome, which might code for these enzymes. A mutant virus containing deletions in HSV-1 RR has been constructed (Goldstein and Weller, 1988), but still maintains the ability to replicate *in vitro*. It is possible that this enzyme plays a role in the pathogenesis and latency of herpesviruses (Jacobson *et al.*, 1989).

4. Thymidylate Synthetase

Following the completion of the DNA sequence of VZV, computer searches for similarities of VZV with sequences of many proteins and enzymes within the data base were performed. One such analysis revealed that the product of VZV ORF 13 contains a high degree of homology to both prokaryotic and eukaryotic thymidylate synthetases (TSs). These enzymes catalyze the reductive methylation of deoxyuridylate to thymidylate as the *de novo* means for synthesizing thymidine nucleotides. ORF 13 is 301 amino acids in length and is predicted to code for a protein of 35 kDa. Thompson *et al.* (1987) were able to express VZV TS activity in a strain of *E. coli* devoid of TS activity. Due to a deletion in the gene *thyA,* the bacteria could not grow in minimal agar. Addition of VZV sequences encoding TS to *E. coli* has since allowed this strain to grow. TS activity in VZV-infected fetal lung cells was then shown to be distinct and novel from the eukaryotic fetal lung TS enzyme, migrating to a different position on SDS–PAGE.

5. Kinases

By computer analysis alone, but still lacking biochemical verification, two VZV genes with counterparts in HSV-1 were shown to share homology with members of the protein kinase family of proteins encoded by eukaryotic cells and by retroviruses. VZV ORF 66, which is capable of coding for a protein of 44 kDa, has homology to HSV-1 gene *US3* (53 kDa). Homologies to the tyrosine kinase activity of the yeast CDC 28 PK (Loring and Reed, 1984), cAMP- and cGMP-dependent protein kinases, and the retroviral v-*mos*, v-*src*, v-*erbB*, and v-*abl* were detected (McGeoch and Davison, 1986).

More recent studies by Smith and Smith (1989) have shown that the VZV ORF 47, HSV-1 UL13, and Epstein–Barr virus BGLF4 resemble serine/threonine kinases. Homology is detected within six conserved regions of these enzymes (Leader, 1988). The presence of these specific activities and their role in VZV replication will undoubtedly be determined. A unique protein kinase activity has been found by Montalvo and Grose (1986) in VZV-infected cells. They showed that it was not related to the casein kinase family, but was able to phosphorylate VZV gpI *in vivo* and *in vitro*. Edson *et al.* (1987) also reported that VZV gpI was phosphorylated, and both groups demonstrated phosphothreonine and phosphoserine present on gpI. Montalvo and Grose (1986) indicated that the protein kinase activity was induced in VZV-infected cells, but whether it is the product of gene *47* or *66* or another viral or cellular gene is unknown.

6. VZV-Encoded DNase

Cheng *et al.* (1979) noted that DNase activity was sevenfold greater in VZV-infected cells than in mock-infected cells. The electrophoretic mobility of this DNase activity was similar in both infected and noninfected cells, so it was not clear that this was a novel enzyme activity. Nuclease activity had been reported in HSV-1-infected cells (Keis and Gold, 1963). An enzyme with 5' and 3' exonuclease activity (Strobel and Francke, 1980) was mapped to 0.17 map units on the HSV-1 genome by hybrid arrest translation of active enzyme (Preston and Cordingley, 1982). McGeoch *et al.* (1986) sequenced the region of the HSV-1 genome that contained the exonuclease gene. Direct comparison of the amino acids predicted from the VZV sequence with the HSV-1 sequence indicated that VZV gene *48* codes for the DNase activity. It shares 29% identity with HSV-1 UL12 (McGeoch *et al.*, 1988a). No further biochemical studies have been performed on the VZV enzyme.

C. Structural Proteins

Besides the five glycoproteins encoded by the VZV genome, little is known about the structural proteins of the virus. As discussed previously, 16–32 nonglycosylated proteins could be resolved by SDS–PAGE analyses of [^{35}S]methionine-labeled purified virions (Grose, 1981; Shemer *et al.*, 1980; Zweerink and Neff, 1981; Asano and Takahashi, 1979, 1980; Shiraki *et al.*, 1982). A prominent nonglycosylated protein of 155 kDa was detected by SDS–PAGE in virion preparations of VZV; this was called the major capsid protein. A protein of approximately this size has been a common feature among many of the herpesviruses studied (Killington *et al.*, 1977). Costa *et al.* (1984) demonstrated that a 6.0-kb HSV-1 mRNA localized between 0.23 and 0.27 map units on the HSV-1 genome could be translated *in vitro* and immunoprecipitated with antibodies to the major capsid protein of HSV-1. The DNA sequences of this region of HSV-1 were determined by Davison and Scott (1986b), and a protein with a predicted molecular weight of 149,000 was identified. Studies of sequence similarities between HSV-1 and VZV revealed that VZV ORF 40 encodes the major capsid protein. A 6.5-kb VZV RNA transcript mapped to the region might correspond to this protein (Ostrove *et al.*, 1985).

VIII. Gene Expression

Understanding of the temporal expression of VZV genes is still in its infancy. Only in the past few years have we developed a better understanding of viral and cellular proteins that could play a role in determining the sequential expression of viral genes. With the direct mapping of a number of viral proteins and the publishing of the complete sequence of the VZV genome, we have been able to ask questions about the regulation of gene expression. The low titers of cell-free virus have hampered the classic drug-block experiments, so many laboratories are studying isolated cloned genes and their expression (Felser *et al.*, 1988; Inchauspe and Ostrove, 1989b; Inchauspe *et al.*, 1989).

Analyses of the complete DNA sequence of VZV (Davison and Scott, 1986a) determined that the genome could code for 70 proteins distributed along both strands of the DNA (Fig. 7). Since three genes are found in the repeat elements IR_S and TR_S, two copies of each exist; hence, 67 unique VZV genes are predicted. This determination was performed by a computer program designed to determine the potential ORFs (Blumenthal *et al.*, 1982) and their translation into amino acid sequence (Taylor, 1986). It must be stressed that this type of analysis is a reasonable indication of, but in no way necessarily predicts, all VZV ORFs and their precise structures. Detailed transcription mapping and biochemical characterization of the protein product of each gene are necessary to determine how closely the predicted ORFs reflect what is expressed in VZV-infected cells.

Ostrove *et al.* (1985) published the first transcription maps of the VZV genomes. Using overlapping cloned restriction fragments as probes in Northern blot hybridizations, 58 distinct transcripts ranging in size from 0.8 to 6.5 kb were detected in the total RNA extracted from VZV-infected cells. Similar studies were carried out by Maguire and Hyman (1986), who separated poly(A)$^+$ and poly(A)$^-$ RNA. They detected 41–67 poly(A)$^+$ transcripts, many of whose size and location agreed with those reported by Ostrove *et al.* (1985). Maguire and Hyman described VZV-specific nonpolyadenylated cytoplasmic transcripts of 3.7 and 1.9 kb in size mapping to map positions 0.62–0.67. The significance of these poly(A)$^-$ transcripts is not known, but this is the only region of the genome where Davison and Scott (1986a)

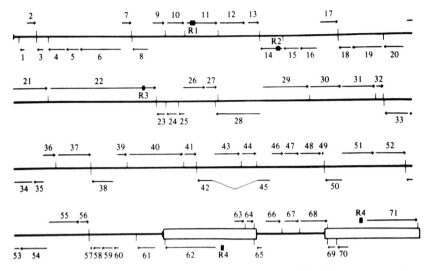

FIG. 7. Open reading frame analysis of the complete VZV DNA sequence. Sixty-seven unique ORFs have been predicted, where 69, 70, and 71 are identical to 62, 63, and 64. ORFs 42 and 45 may represent two exons of a gene predicted by its homology to HSV-1. Repeat elements (R) are also indicated. Reprinted with permission from Davison and Scott (1986a).

predicted the presence of a spliced RNA transcript mapping to ORFs 42 and 45; prediction was based on homology to HSV and primary DNA sequence.

A more detailed transcript map was reported by Reinhold et al. (1988), who utilized single-stranded RNA probes to determine the direction of transcription and to better define the limits of the transcript. In total 77 RNAs were detected, most of which agree fairly well with the ORFs proposed by the DNA sequence (Fig. 8).

The mapping of individual ORFs to specific genes is also being refined as the function of individual genes is identified. Examples include the dPK enzyme encoded by gene 36, which is expressed as a 1.8-kb RNA transcript (Sawyer et al., 1986; Davison and Scott, 1986a) and VZV "IE175," which is equivalent to HSV-1 ICP4 homology encoded by a 4.3-kb transcript mapping to gene 62 (Felser et al., 1988).

A. trans-Activation

Reports on the ability of HSV-1 infection to activate the expression of the enzyme thymidine kinase in cell lines biochemically transformed with the viral dPK gene have indicated that HSV-1 must code for trans-activating proteins (Leiden et al., 1976; Kit and Dubbs, 1977;

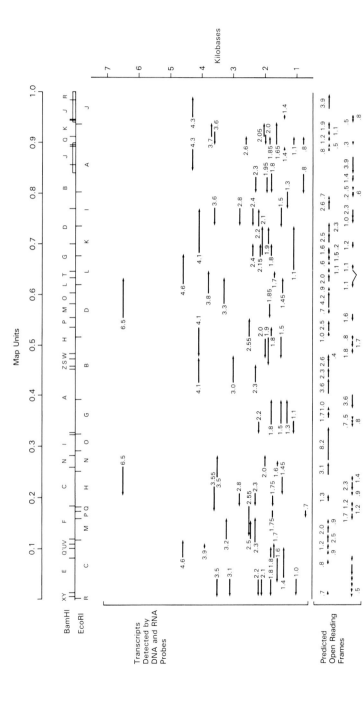

Fig. 8. Map location of mRNA species detected in VZV infected cells. Seventy-seven transcripts were detected using single-stranded ^{32}P-labeled RNA probes corresponding to *Bam*HI and *Eco*RI cloned libraries (top) of the VZV genome. (Bottom) The relative positions and direction of the 71 ORFs as described by Davison and Scott (1986a). Reprinted with permission from Reinhold *et al.* (1988).

Minson et al., 1978; Buttyan and Spear, 1981; Sandri-Goldin et al., 1983). Numerous other viruses—members of both the herpes and non-herpes families—have been shown to contain genes that could trans-activate both homologous and heterologous promoters (Green et al., 1983; Imperiale et al., 1983; Everett and Dunlop, 1984; for a review see Spector and Tevethia, 1989).

Everett and Dunlop (1984) reported that, when VZV-infected human fetal lung cells are mixed with HeLa cells previously trans-fected with HSV-1 gD- or rabbit β-globin-containing plasmids, tran-scriptional activation could be detected (Everett and Dunlop, 1984). In similar cotransfection experiments Everett (1984, 1985) demonstrated that a plasmid containing the VZV SstI-F restriction fragment acti-vated HSV-1 gD, rabbit β-globin, and ε-globin promoters. This SstI-F fragment contains what are now called ORFs 62 and 63, and the ac-tivity was subsequently localized to ORF 62 (Inchauspe et al., 1989).

Systematically, Inchauspe et al. (1989) screened the entire VZV ge-nome for restriction fragments that had trans-activating activity. Their assay used the VZV dPK promoter linked to the chloramphenicol acetyltransferase (CAT) gene. Two genes, one encoded by ORF 4 and the other by ORF 62, were proved to significantly activate expression from the dPK promoter. Figure 9 depicts a five- to 20-fold stimulation in CAT expression by these cotransfections. In a related experiment a plasmid containing a putative late VZV promoter (i.e., gpI, or p68CAT), was placed upstream from CAT and used to study trans-activation. When pG-ORF 4 and/or pORF 62 were cotransfected alone, a two- to six-fold increase in CAT activity was seen, but when they were used together, a synergistic 22-fold increase in CAT expression was observed. Stimulation was at the RNA level, as demonstrated by an increase in the steady-state message level. Inchauspe et al. (1989) also detected a VZV-encoded trans-repressor contained within a re-striction fragment that possesses ORFs 60 and 61. Cotransfection of this plasmid into cells, along with pG-ORF 4 and/or pORF 62 and the target p1tkCAT, resulted in inhibition of the trans-activation. Analy-ses of the RNA revealed no increase in the steady-state CAT RNA, implying that the trans-repression acts at the RNA level (S. Nagpal and J. M. Ostrove, unpublished observations).

The capability of VZV-encoded trans-activators to stimulate hetero-logous promoters, such as the human immunodeficiency virus long terminal repeats (Gendelman et al., 1986; Inchauspe and Ostrove, 1989b), c-myc, HSV IE175, and HSV thymidine kinase (Inchauspe and Ostrove, 1989b) was demonstrated. This activation is achieved either by VZV infection or by cotransfection with the appropriate trans-activators.

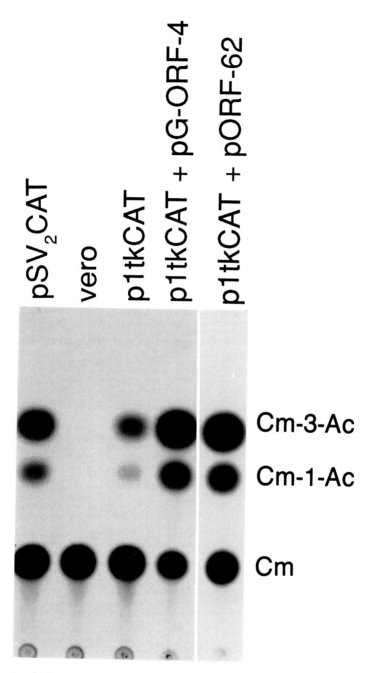

FIG. 9. CAT assay demonstrating the activation of the VZV dPK promoter (p1tkCAT) by the gene products of ORFs 4 and 62. Vero cells were either mock-transfected (vero) or transfected with pSV₂CAT (simian virus 40 promoter/enhancer CAT) or p1tkCAT alone or with VZV trans-activating genes. Cm, Cm-1-Ac, and Cm-3-Ac represent Chloramphenicol and its acetylated forms.

B. Genetics

Attempts to use genetic analyses on VZV to better understand the function of selected viral genes have been hampered by the inability to prepare a high-titered cell-free virus. Classical studies such as mutagenesis of the virus followed by selection of temperature-sensitive (ts) or host–range mutants has not been feasible. In a novel approach Felser *et al.* (1987) used VZV to complement well-characterized temperature-sensitive (ts) mutants of HSV-1 grown at their nonpermissive temperature. Several HSV-1 mutants were screened, including DNA polymerase, DNA-binding protein, gD, and immediate early proteins ICPs 27 and 4. VZV clearly complemented the growth of tsB21 mapping in the HSV-1 ICP 4 trans-activating gene. Restriction fragments mapping to VZV *Eco*A were also able to complement tsB21 in transfection assays. pGi26, a plasmid which contains a single copy of VZV ORF 62, was shown to confer the complementing activity. Felser *et al.* (1988), established Vero cell lines containing a stable integrated copy of VZV ORF 62, using cotransfection with a neomycin resistance gene and G418 selection. One cell line, FI-14, showed no basal level of expression of ORF 62 by Northern hybridization or Western blot analyses, yet supported the growth of tsB21 and d120, a deletion mutant in ICP 4.

It was noted that, on infection with HSV, the 4.3-kb ORF 62 transcript was induced to the high levels in the FI-14 cells, and the 175-kDa protein was detected by Western blotting with a rabbit antibody to a synthetic peptide composed of the predicted carboxy-terminal 12 amino acids of ORF 62. In addition, HSV-1 infections of FI-14 cells carried out in the presence of cycloheximide result in the superinduction of VZV IE175 RNA. This implies that IE175, like HSV-1 ICP 4, can autoregulate its own expression. The data were interpreted as indicating that the tegument protein of HSV-1 (VP16 or α-TIF) is interacting with the TAATGARAT DNA sequence element in the promoter of VZV gene *62* and stimulating its expression (Felser *et al.*, 1988).

DNA sequence analysis and amino acid comparison of HSV-1 ICP 4 and VZV ORF 62 indicate an overall homology of 29%, but three regions of 28 amino acids, 187 amino acids, and 442 amino acids contain 50%, 50%, and 54% amino acid homology, respectively, and many domains of nearly 100% homology in the carboxy-terminal 442 amino acids exist. These presumably represent functional domains (Davison and McGeoch, 1986; DeLuca and Schaffer, 1988). HSV-2, which contains an α-TIF-like protein, has been shown to activate IE175 protein expression (D. Alcendor and J. M. Ostrove, unpublished observations) in FI-14 cells, but pseudorabies, cytomegalovirus, equine herpesvirus, and bovine herpesvirus could not.

Further indication that a genetic system for VZV can be developed came from work by Lowe *et al.* (1987), who constructed a plasmid containing the VZV gpI promoter and hydrophobic leader sequence directing the synthesis of Epstein–Barr virus membrane gp350/220. This construct was flanked by VZV *dPK* gene sequences and cotransfected along with infectious VZV Oka DNA into MRC-5 cells. Positive recombinant plaques expressing Epstein–Barr virus gp350/220 under VZV gpI control were identified by the erythrocyte rosetting–hemadsorption property of this glycoprotein. These experiments indicate that VZV can undergo recombination and, hence, one should be able to construct site-specific mutants in VZV. Also, VZV has a potential to be used as a live virus vector for the expression of foreign genes.

Evidence for *in vivo* and *in vitro* recombination has been reported by Gelb *et al.* (1987) and by Dohner *et al.* (1988). Two distinguishable strains of VZV with known restriction endonuclease fragment-sized differences were grown together in whole human embryo cells (Flow 5000) for three passages. Analysis of the DNA from a plaque-purified virus showed that three of 58 of the isolates had one or more *BgI*I restriction fragment-length polymorphisms not found in either parent. These changes were consistent with recombination of the two parent strains and also the well-documented recombination that occurs in HSV-1 (for a review see Schaffer, 1981).

C. Genetic Relationship and Sequence Homology among VZV and Other Herpesviruses

VZV, HSV-1, and HSV-2 are members of the subfamily Alphaherpesvirinae, based on properties of the virion, host range, sites of latency, and the replication cycle (Matthews, 1982). The genetic relationship of these viruses was first studied at the DNA level by Davison and Wilkie (1983). Recombinant plasmids spanning their genomes were used in a series of cross-hybridization experiments performed under different stringencies. At low stringencies (e.g., 30–40% formamide, 37°C) cross-hybridization across approximately one-half of the genome was detected. As the stringency was increased by bringing formamide concentrations to 60%, only a small region of VZV, mapping to the short repeats, still retained detectable hybridization to HSV-1. The results of cross-hybridization studies predicted that the organization of genes within the U_L region of VZV is inverted relative to the prototypical HSV-1 genome.

It was not until the publishing of the VZV DNA sequence (Davison, 1983; Davison and Scott, 1986a) and its comparison to published sequences of HSV (McGeoch *et al.*, 1988a) or Epstein–Barr virus (Baer *et*

TABLE III

VZV GENES

Gene	Molecular weight[a]	Function	HSV homolog
1	12,103		
2	25,983		
3	19,149		UL55
4	51,540	trans-Activation, HSV-1 ICP 27 homolog	UL54
5	38,575		UL53
6	122,541	DNA replication[b]	UL52
7	28,245		UL51
8	44,816	dUTPase	UL50
9	32,845		UL49
10	46,573	Tegument protein[b]	UL48
11	91,825		UL47
12	74,269		UL46
13	34,531	Thymidylate synthetase	
14	61,350	gpV; HSV-1 gC homolog	UL44
15	44,522		UL43
16	46,087	DNA replication[b]	UL42
17	51,365	Virion protein (host shut-off)[b]	UL41
18	35,395	Small subunit ribonucleotide reductase	UL40
19	86,823	Large subunit ribonucleotide reductase	UL39
20	53,969	Virion protein[b]	UL38
21	115,774		UL37
22	306,325	Virion protein[b]	UL36
23	24,416		UL35
24	30,451	Virion protein[b]	UL34
25	17,460		UL33
26	65,692		UL32
27	38,234		UL31
28	134,041	DNA polymerase	UL30
29	132,133	Major DNA-binding protein	UL29
30	86,968		UL28
31	98,062	gpII; HSV-1 gB homolog	UL27
32	15,980		
33	66,043	Capsid protein[b]	UL26
34	65,182	Virion protein[b]	UL25
35	28,973		UL24
36	37,815	Deoxypyrimidine kinase	UL23
37	93,646	gpIII; HSV-1 gH homolog	UL22
38	60,395		UL21
39	27,078		UL20
40	154,971	Major capsid protein	UL19
41	34,387		UL18
42, 45[c]	82,752		UL15, exons 1 and 2
43	73,905		UL17
44	40,243		UL16

TABLE III

(*Continued*)

Gene	Molecular weight[a]	Function	HSV homolog
46	22,544		UL14
47	54,347		UL13
48	61,268	Exonuclease[b]	UL12
49	8907		UL11
50	48,669		UL10
51	94,370	DNA replication[b]	UL9
52	86,343	DNA replication[b]	UL8
53	37,417		UL7
54	86,776	Virion protein[b]	UL6
55	98,844	DNA replication[b]	UL5
56	27,166		UL4
57	8079		
58	25,093		UL3
59	34,375	Uracil-DNA glycosylase	UL2
60	17,616		UL1
61	50,913	trans-Repression,[d] IE110,[e] ICP 0 homolog[b]	
62, 71	139,989	trans-Activation, IE175,[e] ICP 4 homolog[b]	
63, 70	30,494	HSV-1 IE ICP 47 homolog[b]	US1
64, 69	19,868	Virion protein[b]	
65	11,436	HSV-1 tegument phosphoprotein	US9
66	43,677	Protein kinase	US2
67	39,362	gpIV; HSV-1 gI homolog[b]	US7
68	69,953	gpI; HSV-1 gE homolog[b]	US8

[a]Predicted from open reading frames (Davison and Scott, 1986a).
[b]HSV-1 homolog (McGeoch *et al.,* 1988a).
[c]Presumed spliced open reading frames.
[d]Inchauspe *et al.* (1989).
[e]IE, Immediate early.

al., 1984) that the true genetic relationships of these herpesviruses could be compared. Table III lists the VZV ORFs and describes possible functions based on homology to HSV-1 (Davison and Scott, 1986a; McGeoch *et al.,* 1988a). For example, comparison of amino acid homology between HSV-1 and VZV in the short segment of the genome (Fig. 10) clearly demonstrates that VZV genes *62–71* (VZV *RS1–3* are genes *62, 63,* and *64,* while *US1–4* are genes *65, 66, 67,* and *68*) all have homologs in HSV-1, while HSV-1 genes *US2, 4, 5, 6, 11,* and *12* lack VZV homologs (Davison and McGeoch, 1986).

Similar comparisons between VZV and Epstein–Barr virus led to

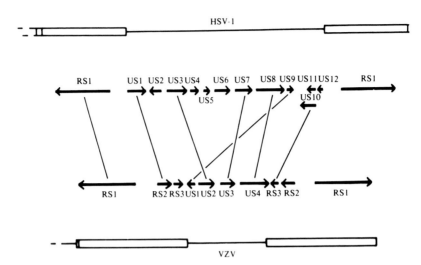

FIG. 10. Seven VZV genes (ORFs *62–69*) that map in the short segment of the genome all have homology to HSV-1 genes mapping in similar locations. *RS2* and *RS3* (ORFs 63 and 64) have moved during the evolution of these viruses from the repeat region of VZV, where they are diploid to the U$_S$ region of HSV-1. Reprinted with permission from Davison and McGeoch (1986).

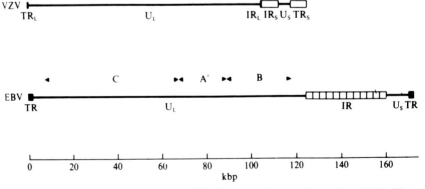

FIG. 11. Genomic organization of VZV relative to Epstein–Barr virus (EBV). Three conserved blocks of the VZV genome (A, B, and C) are represented in EBV, but in a rearranged configuration with inversion of the A segment (A'). Reprinted with permission from Davison and Taylor (1987).

the observation that a majority of VZV genes share homology, although the genome organization differs due to apparent inversions of different regions of the genome (Fig. 11). It has been observed that another 14 ORFs possibly contain conserved sequences, but due to the limitations of our knowledge of protein structure or the ability of the computer to more accurately predict homology, more detailed comparisons have yet to be performed.

Tentative functions have been ascribed to 40 of the 67 unique genes of the VZV sequence, based on their structure and relationship to other herpesvirus genomes (Davison and Scott, 1986a). The five VZV glycoproteins all have homologs in HSV (Tables II and III).

Challberg (1986) and McGeoch et al. (1988b) identified seven genes that were essential for HSV-1 DNA replication. Each of these genes has a homolog in VZV (Table IV), implying that the biochemical processes of DNA replication are similar. One major difference in their mode of DNA replication might come from the fact that HSV-1 contains two origins of DNA replication, each of which can function independently. Ori_S of HSV-1 has been mapped to the short repeat region of the genome (Stow, 1982; Mocarski and Roizman, 1982), while Ori_L was mapped to the U_L segment (Spaete and Frenkel, 1982; Weller et al., 1985).

TABLE IV

DNA REPLICATION GENES

Gene	Function	HSV-1 homolog
Essential[a]		
6	Helicase, primase complex[b]	UL52
16	Double-stranded DNA-binding protein	UL42
28	DNA polymerase	UL30
29	Single-stranded DNA-binding protein	UL29
51	Ori-binding protein,[c] ICP 8	UL9
52	Helicase, primase complex[b]	UL8
55	Helicase, primase complex[b]	UL5
Nonessential		
8	dUTPase	UL50
13	Thymidylate synthetase	None
18	Ribonucleotide reductase, small subunit	UL40
19	Ribonucleotide reductase, large subunit	UL39
48	Exonuclease	UL12
59	Uracil-DNA glycosylase[d]	UL2

[a]From Challberg (1986).
[b]From Crute et al. (1989) and Hodgman (1988).
[c]From McGeoch et al. (1988b) and Davison and Scott (1986a).
[d]From Worrad and Caradonna (1988).

Stow and Davison (1986) have mapped a single origin of replication in VZV to the short repeat segment (Ori_S). This sequence contains alternating tracts of A and T residues at the center of an almost perfect 45 base pair palindrome. VZV Ori_S shares significant DNA sequence homology with HSV-1 Ori_S and Ori_L (Stow and Davison, 1986). Plasmids containing the VZV Ori_S, when transfected into cells infected by either VZV or HSV-1, can replicate to produce many copies of high-molecular-weight species made up of tandem repeats of the input plasmid. To date, no Ori_L of VZV has been identified. This might be because it maps to a region of the genome that is extremely unstable on cloning (Straus *et al.*, 1983; Weller *et al.*, 1985; Davison and Scott, 1986a) or because VZV exists mainly as two isomers and uses a slightly different mode of DNA replication.

IX. VZV LATENCY

Studies of herpes zoster conducted in 1863 led von Barensprung to conclude that it is a disease of nervous tissue origin and that it was associated with lesions in the posterior root ganglia. This work was followed by that by Head and Campbell (1900), who studied the pathology and location of zoster lesions and were able to use this information to map areas innervated by sensory ganglia.

In 1909 von Bokay described the association of herpes zoster with varicella and it was realized that varicella could be contracted following exposure to a person with zoster. It was not until the 1950s, however, that Weller and colleagues (Weller and Stoddard, 1952; Weller, 1953; Weller *et al.*, 1958; Weller and Coons, 1954) successfully grew the viruses that caused varicella and zoster and demonstrated their identity.

Since von Bokay's (1909) observations of the apparent relationship between herpes zoster and varicella, numerous cases of varicella have been reported as a result of exposure to herpes zoster (Garland, 1943). Detailed observations on the clinical disease manifestations of varicella and zoster virus infections led Hope-Simpson to hypothesize in 1965 that VZV establishes a latent neural infection. Part of the hypothesis stated that "from every spot in the skin and mucosae of the nasopharynx, conjunctivae, bladder, etc. virus also enters the contiguous endings of the sensory nerves, whence it is transported up fibers until it arrives at the sensory ganglia, where it becomes established as a latent infection in the nuclei of the neurons. . . . In nerve and ganglion it is insulated from the now rapidly rising tide of neutralizing antibody in the circulation." Hope-Simpson also noted that seeding of the sensory ganglia can take place by the hematogenous spread of virus.

The hypothesis that the virus remains latent in the ganglia is consistent with our current knowledge regarding the state of VZV following varicella. Esiri and Tomlinson (1972) demonstrated, by electron microscopy and immunofluorescence microscopy, that VZV was latent in the trigeminal ganglia obtained from a human autopsy of a patient who died with ophthalmic zoster. Viral particles were detected in the cytoplasm of perineural cells and both the cytoplasm and the nuclei of Schwann's cells. Parts of the trigeminal ganglia demonstrated severe degeneration. Infectious virus, however, was not recovered from these tissues. In similar studies, using light and electron microscopy, virus particles were observed in the neurons and the satellite cells of a partially necrotic spinal ganglia dissected during autopsy on a person who died with herpes zoster of the abdomen.

Studies of the mechanisms by which VZV establishes and maintains a latent state have been totally hampered by the lack of a suitable animal model. In addition the recovery of VZV from ganglionic explants has not been achieved. Many attempts, including those by our own laboratory (H. A. Smith, J. M. Ostrove, and S. E. Straus, unpublished observations), have been unsuccessful. Plotkin et al. (1977) reported efforts to isolate virus by cocultivation from four to six thoracic ganglia from each of 20 adult cadavers without success. This is in contrast to HSV, which can be readily isolated from human trigeminal ganglia (Bastian et al., 1972; Baringer and Swoveland, 1973).

Our current understanding of the molecular events that occur during latency are derived mainly from the study of human trigeminal ganglia removed during autopsy. Since about 90% of adults have been exposed to varicella and approximately one-quarter of the cases of zoster occur in the face, it was assumed that, in randomly selected trigeminal ganglia, there is a high likelihood that latent VZV will be present.

Using in situ hybridization, a technique in which radiolabeled DNA or RNA is hybridized directly to the nucleic acid in tissue samples (Brahic and Haase, 1979), Hyman et al. (1983) detected VZV RNA in three of nine trigeminal ganglia removed from five individuals who lacked any signs of active zoster infection. These investigators estimated that up to 0.3% of neurons were positive for latent VZV RNA transcripts.

Gilden et al. (1983), using Southern hybridization techniques, reported the detection of VZV DNA in the sacral ganglia of an 82-year-old male who developed zoster in a sacral distribution 3 days before death. The ganglia was necrotic and possessed Cowdry type A inclusion bodies.

VZV was recovered from his spinal fluid during autopsy. Sixty copies of viral DNA per cell were detected in the sacral ganglia. In addition 13 trigeminal ganglia from nine individuals with no signs of disease were studied by Southern hybridization. Five ganglia from three males, ranging from 8 to 50 years of age, were shown to contain between 0.28 and 1 copy of VZV DNA per cell. These studies indicated that latent VZV nucleic acid could be detected in human ganglia during acute and latent infections.

Vafai et al. (1988) studied normal human trigeminal ganglia removed during autopsy for the expression of VZV-specific proteins and nucleic acids. Pieces of minced ganglia were placed in culture in the presence of [^{35}S]methionine. At daily intervals the tissue fragments were removed and lysates prepared and immunoprecipitated with human or rabbit anti-VZV serum. A 200-kDa VZV protein was precipitated with either serum after just 24 hours of incubation. Over time additional "VZV-specific" proteins could be detected by 11 days in culture; seven polypeptides of 35, 36, 45, 100, 140, 170, and 200 kDa were noted. Using antibodies that detected VZV gpI, gpII, and gpIV and the 155-kDa major capsid protein, all presumed to be expressed late in infection, no specific immunoprecipitation could be detected. Using in situ hybridization of the 12-day cultured ganglion, VZV gene 63 RNA was detected. Expression of other VZV genes apparently was not sought. The authors concluded that, during incubation of the ganglia in cell culture, no productive viral replication occurred.

Further in situ hybridization studies of human latently infected ganglia were performed by Gilden et al. (1987), who removed four thoracic ganglia from a 51-year-old male who drowned. Using ^{32}P-labeled single-stranded RNA probe homologous to VZV fragment SalI-P, whose map position overlaps genes 63 and 64, two of four ganglia from this individual showed extensive hybridization signals. In the positive tissue most of the neurons contained VZV RNA. The fact that such a high percentage of cells were positive is not consistent with the earlier work by this group and that by others (Hyman et al., 1983; Vafai et al., 1988).

The most detailed study on VZV latency in humans was performed by Croen et al. (1988), who analyzed trigeminal ganglia removed from 30 individuals, most of whom had died suddenly and were not known to be immunocompromised. Using ^{35}S-labeled single-stranded RNA probes representing almost the entire VZV genome, positive in situ hybridization results were detected in 15 of 21 individuals, ranging in age from 9 to 74 years. Of nine children ranging in age from 3 weeks to 7 months, none had positive in situ hybridization signals. This is consistent with their history of not having had varicella. Of the 15 adults with latent VZV detected in the trigeminal ganglia, all of the hybrid-

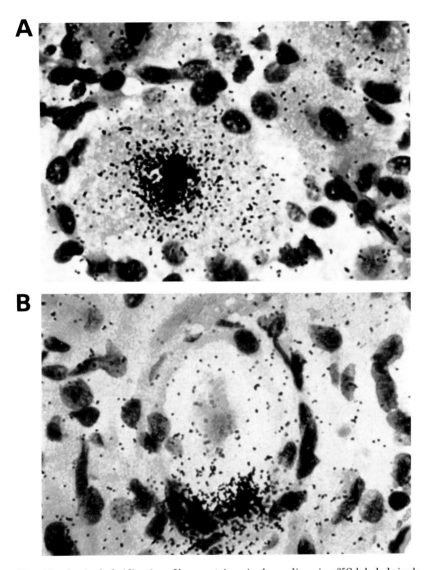

FIG. 12.   *In situ* hybridization of human trigeminal ganglia using ^{35}S-labeled single-stranded RNA probes. (A) A probe mapping to the long repeat of HSV-1 detected the HSV latency-associated transcripts in the neuronal cell. (B) A VZV probe pool detected VZV RNA in latently infected nonneuronal satellite cells. Modified from Croen *et al.* (1988).

ization signals were localized to the satellite cells surrounding the neuron within the ganglion (Fig. 12). Under no circumstance was VZV detected in the neuronal cells during latency. DNase and/or RNase treatment of the tissue section prior to hybridization proved that the signal being detected was primarily RNA.

Using discrete probes representing various regions of the genome, three regions of transcription were noted. RNA was detected with *Bam*HI-EY, *Bam*HI-J, and *Eco*RI-B. Potential transcripts contained in these ORFs include ORFs 3, 4, 5, 6, and 8, or ORFs 29, 30, 31, 32, and 62, assuming that transcription in the ganglion cell is in the same direction as that which occurs during a productive virus infection (Ostrove *et al.*, 1985; Davison and Scott, 1986a; Reinhold *et al.*, 1988). Between 0.01% and 0.15% of the nonneuronal cells exhibited latent VZV transcripts. No signals other than those found in the satellite and fibroblastlike cells were detected, even with *in situ* hybridization exposures of up to 33 days. These findings contrast sharply with those reported by others (Hyman *et al.*, 1983; Gilden *et al.*, 1987; Vafai *et al.*, 1988).

Croen *et al.* (1988) examined a greater more samples than all of the other studies combined. Controls such as Epstein–Barr virus DNA, empty plasmid, and probes transcribed in both directions from almost all regions of the VZV genome were used. Moreover, these tissue samples that were studied for latent VZV were also analyzed for latent HSV-1 using *in situ* hybridization procedures. During HSV-1 latency only a single region of the genome mapping in the long repeat is transcriptionally active. This HSV-1 latency-associated transcript is detected by *in situ* hybridization in the neuronal cells of mice (Stevens *et al.*, 1987; Spivak and Fraser, 1987), rabbits (Rock *et al.*, 1987), and humans (Croen *et al.*, 1987). Croen *et al.* (1988) detected an HSV-1 latency-associated transcript exclusively in the neuronal cells in 15 of 20 adult trigeminal ganglia, 11 of which harbored latent VZV exclusively in the nonneuronal cells.

In situ hybridization studies of the trigeminal ganglia from an 8-year-old who died during the sixth day of an active varicella infection gave Croen *et al.* (1988) the opportunity to compare latent and active VZV infections. During the active infection, in addition to non-neuronal cells, 0.6% of the neuronal cells expressed VZV RNA and all regions of the genome that were probed for were detected. Hence, active infection appears to involve both neuronal and nonneuronal cell types and, during latency, transcription is limited to nonneuronal cells. More detailed characterizations of the cell type harboring latent VZV are necessary before our understanding is complete.

This description of latent VZV transcripts present in the nonneuronal cells is consistent with models of pathogenesis, including the extreme pain and neuralgia associated with zoster. If viral reactivation requires VZV to replicate in the surrounding satellite cells, this can cause the tissue destruction and inflammation described by Shibuta *et al.* (1974). Following infection of the neuron, the virus presumably travels down many neurons within a single dermatome and results in the crop of vesicular lesions present in zoster.

Further studies of human trigeminal ganglia are needed to determine the true site(s) of latency and which genes are transcriptionally active.

X. Conclusions

The study of VZV biology and molecular biology still lags well behind that of other members of the herpesvirus family. A major advance has been in the publishing of the complete DNA sequence for both VZV and HSV. This should allow us to better understand the expression of individual VZV genes and to place them in the context of viral replication. Advances in our ability to grow the virus are needed to study temporal gene expression, viral DNA replication, and assembly. The production of a larger set of antibodies directed against the viral proteins will assist in our analysis of viral growth. Expression of VZV proteins by recombinant DNA methodology will allow us to understand the humoral and cell-mediated immune responses that develop during primary varicella, silent reinfection, or reactivation of latent virus. Studies on latency are starting to elucidate the neurotropic nature of the virus and the viral mechanisms that may be involved in maintaining latency.

Acknowledgments

I would like to thank Stephen E. Straus for critical reading of the manuscript and Rebecca Tanen and Susi Hotz for their excellent clerical support.

References

Adams, S. G., Dohner, D. E., and Gelb, L. D. (1989). *J. Med. Virol.* **29,** 38–45.
Almeida, J. D., Howatson, A. F., and Williams, M. G. (1962). *Virology* **16,** 353–355.
Andrewes, C. H. (1964). "Viruses of Vertebrates," pp. 211–212. Baillière, London.
Aragao, H. D. (1911). *Mem. Inst. Oswaldo Cruz* **3,** 309–319.
Arvin, A. M. (1987). *In* "Natural History of Varicella-Zoster Virus" (R. W. Hyman, ed.), pp. 67–130. CRC Press, Boca Raton, Florida.
Asano, Y., and Takahashi, M. (1979). *Biken J.* **22,** 81–89.
Asano, Y., and Takahashi, M. (1980). *Biken J.* **23,** 95–106.
Asano, Y., Albrecht, P., Behr, D. E., Neff, B. J., Vickers, J. H., and Rastogi, S. C. (1984). *J. Med. Virol.* **14,** 305–312.

Baer, R., Bankier, A. T., Biggins, M. D., Deininger, P. L., Farrell, P. J., Gibson, T. J., Hatfull, G., Hudson, G. S., Satchwell, S. C., Sequin, C., Tuffnell, P. S., and Barrell, B. G. (1984). *Nature (London)* **310**, 207–212.

Baringer, J. R., and Swoveland, P. (1973). *N. Engl. J. Med.* **288**, 648–650.

Bastian, F. O., Rabson, A. S., Yee, C. L., and Tratka, T. S. (1972). *Science* **178**, 306.

Becker, Y. (1988). *Virus Genes* **1**, 351–367.

Bett, W. R. (1934). "A Short History of Some Common Diseases," Ch. 1. Oxford Univ. Press, London.

Biron, K. K., Fyfe, J. A., Noblin, J. E., and Elion, G. B. (1982). *Am. J. Med.* **73**, 383–386.

Blumenthal, R. W., Rice, P. J., and Roberts, R. J. (1982). *Nucleic Acids Res.* **10**, 91–101.

Brahic, M., and Haase, A. (1979). *Proc. Natl. Acad. Sci. U.S.A.* **75**, 6125–6129.

Buttyan, R., and Spear, P. G. (1981). *J. Virol.* **37**, 459–472.

Bzik, D. J., Fox, B. A., DeLuca, N. A., and Person, S. (1984). *Virology* **133**, 301–314.

Casey, T. A., Ruyechan, W. T., Flora, M. N., Reinhold, W., Straus, S. E., and Hay, J. (1985). *J. Virol.* **31**, 172–177.

Caunt, A. E. (1963). *Lancet* **2**, 982–983.

Caunt, A. E., and Shaw, D. G. (1969). *J. Hyg.* **67**, 343–352.

Challberg, M. D. (1986). *Proc. Natl. Acad. Sci. U.S.A.* **83**, 9094–9098.

Cheng, Y. C., Tsou, T. Y., Hackstadt, T., and Mallavia, L. P. (1979). *J. Virol.* **31**, 172–177.

Christie, A. B. (1974). "Infectious Diseases: Epidemiology and Clinical Practice," 2nd Ed., pp. 278–290. Churchill-Livingstone, Edinburgh.

Cohen, G. H. (1972). *J. Virol.* **9**, 408–418.

Cole, R., and Kuttner, A. G. (1925). *J. Exp. Med.* **42**, 799–820.

Cook, M. L., and Stevens, J. G. (1968). *J. Virol.* **2**, 1458–1464.

Cook, M. L., and Stevens, J. G. (1970). *J. Ultrastruct. Res.* **32**, 334–350.

Corey, L., and Spear, P. G. (1986a). *N. Engl. J. Med.* **314**, 686–691.

Corey, L., and Spear, P. G. (1986b). *N. Engl. J. Med.* **314**, 1427–1432.

Costa, R. H., Cohen, G., Eisenberg, R., Lung, D., and Wagner, E. (1984). *J. Virol.* **49**, 287–298.

Cranage, M. P., Smith, G. L., Bell, S. E., Hart, H., Brown, C., Bankier, A. T., Tomlinson, P., Barrell, B. G., and Minson, T. C. (1988). *J. Virol.* **62**, 1416–1422.

Croen, K. D., Ostrove, J. M., Dragovic, L. J., Smialek, J. E., and Straus, S. E. (1987). *N. Engl. J. Med.* **317**, 1427–1432.

Croen, K. D., Ostrove, J. M., Dragovic, L. J., and Straus, S. E. (1988). *Proc. Natl. Acad. Sci. U.S.A.* **85**, 9773–9777.

Crute, J. J., Tsurumi, T., Zhu, L. A., Weller, S. K., Olivo, P. D., Challberg, M. D., Mocarski, E. S., and Lehman, I. R. (1989). *Proc. Natl. Acad. Sci. U.S.A.* **86**, 2186–2189.

Davison, A. J. (1983). *EMBO J.* **2**, 2203–2209.

Davison, A. J. (1984). *J. Gen. Virol.* **65**, 1969–1977.

Davison, A. J., and McGeoch, D. J. (1986). *J. Gen. Virol.* **67**, 597–611.

Davison, A. J., and Scott, J. E. (1983). *J. Gen. Virol.* **64**, 1811–1814.

Davison, A. J., and Scott, J. E. (1985). *J. Gen. Virol.* **66**, 207–220.

Davison, A. J., and Scott, J. E. (1986a). *J. Gen. Virol.* **67**, 1759–1816.

Davison, A. J., and Scott, J. E. (1986b). *J. Gen. Virol.* **67**, 2279–2286.

Davison, A. J., and Taylor, P. (1987). *J. Gen. Virol.* **68**, 1067–1079.

Davison, A. J., and Wilkie, N. M. (1983). *J. Gen. Virol.* **64**, 1927–1942.

Davison, A. J., Waters, D. J., and Edson, C. M. (1985). *J. Gen. Virol.* **66**, 2237–2242

Davison, A. J., Edson, C. M., Ellis, R. W., Forghani, B., Gilden, D. H., Grose, C., Keller, P. M., Vafai, A., Wroblewska, Z., and Yamanishi, K. (1986). *J. Virol.* **57**, 1195–1197.

Delius, H., and Clements, J. B. (1976). *J. Gen. Virol.* **33**, 125–133.

DeLuca, N. A., and Schaffer, P. A. (1988). *J. Virol.* **62**, 732–743.

de-Thé, G. (1982). *In* "Epidemiology of Epstein–Barr Virus and Associated Diseases of Man" (B. Roizman, ed.), Ch. 2. Plenum, New York.

Dobersen, M. J., Jerkofsky, M., and Green, S. (1976). *J. Virol.* **20**, 478–486.

Dohner, D. E., Adams, S. G., and Gelb, L. D. (1988). *J. Med. Virol.* **24**, 329–341.

Dumas, A. M., Geelen, J. L. M. C., Maris, W., and Van der Nordaa, J. (1980). *J. Gen. Virol.* **47**, 233–235.

Dumas, A. M. Geelen, J. L. M. C., Westrate, M. W., Wertheim, P., and Van der Nordaa, J. (1981). *J. Virol.* **39**, 390–400.

Dunkel, E. C., Siegel, M. L., Rong, B. L., and Pavan-Langston, D. (1989). *Invest. Ophthalmol. Visual Sci.* **30**, 213.

Dutia, B. M., Frame, M. C., Subak-Sharpe, J. H., Clark, W. N., and Marsden, H. S. (1986). *Nature (London)* **321**, 439–441.

Earl, P. L., Jones, E. V., and Moss, B. (1986). *Proc. Natl. Acad. Sci. U.S.A.* **83**, 3659–3663.

Ecker, J. R., and Hyman, R. W. (1982). *Proc. Natl. Acad. Sci. U.S.A.* **79**, 156–160.

Edmond, B. J., Grose, C., and Brunell, P. A. (1981). *J. Gen. Virol.* **54**, 403–407.

Edson, C. M., Hosler, B. A., Respress, R. A., Waters, D. J., and Thorley-Lawson, D. A. (1985). *J. Virol.* **56**, 333–336.

Edson, C. M., Hosler, B. A., and Waters, D. J. (1987). *Virology* **161**, 599–602.

Ellis, B. G., Furman, P. A., Fyfe, J. A., Miranda, P. D., Bauchampe, L., and Schaeffer, H. J. (1977). *Proc. Natl. Acad. Sci. U.S.A.* **74**, 5716–5720.

Ellis, R. W., Keller, P. M., Lowe, R. S., and Zivin, R. A. (1985). *J. Virol.* **53**, 81–88.

Esiri, M. M., and Tomlinson, A. H. (1972). *J. Neurol. Sci.* **15**, 35–48.

Everett, R. D. (1984). *EMBO J.* **3**, 3135–3141.

Everett, R. D. (1985). *EMBO J.* **4**, 1973–1980.

Everett, R. D., and Dunlop, M. (1984). *Nucleic Acids Res.* **12**, 5969–5978.

Felsenfeld, A. D., and Schmidt, N. J. (1975). *Infect. Immun.* **12**, 261–266.

Felsenfeld, A. D., and Schmidt, N. J. (1977). *Infect. Immun.* **15**, 807–812.

Felser, J. M., Straus, S. E., and Ostrove, J. M. (1987). *J. Virol.* **61**, 225–228.

Felser, J. M., Kinchington, P. R., Inchauspe, G., Straus, S. E., and Ostrove, J. M. (1988). *J. Virol.* **62**, 2076–2082.

Fitzgerald, M., and Shenk, T. (1981). *Cell (Cambridge, Mass.)* **24**, 251–260.

Forghani, B., Schmidt, N. J., Myoraku, C. K., and Gallo, D. (1982). *Arch. Virol.* **73**, 311–317.

Forghani, B., DuPuis, V. W., and Schmidt, N. J. (1984). *J. Virol.* **52**, 55–62.

Garland, J. (1943). *N. Engl. J. Med.* **228**, 336–337.

Gelb, L. D., and Dohner, D. (1984). *J. Invest. Dermatol.* **83**, 77–81.

Gelb, L. D., Huang, J. J., and Wellinghoff, W. J. (1980). *J. Gen. Virol.* **51**, 171–177.

Gelb, L. D., Dohner, D. E., Gershon, A. A., Steinberg, S. P., Waner, J. L., Takahashi, M., Dennchy, P. H., and Brown, A. E. (1987). *J. Infect. Dis.* **155**, 633–640.

Gelman, I. H., and Silverstein, S. (1987). *J. Virol.* **61**, 3167–3172.

Gendelman, H. E., Phelps, W., Feigenbaum, L., Ostrove, J. M., Adachi, A., Howley, P. M., Khoury, G., Ginsberg, H. S., and Martin, M. A. (1986). *Proc. Natl. Acad. Sci. U.S.A.* **83**, 9759–9763.

Gentry, G. A., Lowe, M., Alfred, G., and Nevins, K. (1988). *Proc. Natl. Acad. Sci. U.S.A.* **85**, 2658–2661.

Gershon, A. A. (1987). *Annu. Rev. Med.* **38**, 41–50.

Gershon, A. A., Cosio, L., and Brunell, P. A. (1973). *J. Gen. Virol.* **18**, 21–31.

Gershon, A. A., Steinberg, S. P., LaRussa, P., Ferrara, A., Hammerschlag, M., and Gelb, L. (1988). *J. Infect. Dis.* **158**, 132–137.

Gilden, D. H., Wroblewska, Z., Kindt, V., Warren, K. G., and Walinsky, J. S. (1978). *Arch. Virol.* **56,** 105–117.

Gilden, D. H., Shtram, Y., Friedmann, A., Wellish, M., Devlin, M., Fraser, N., and Becker, Y. (1982). *J. Gen. Virol.* **60,** 371–374.

Gilden, D. H., Vafai, A., Shtram, Y., Becker, Y., Devlin, M., and Wellish, M. (1983). *Nature (London)* **306,** 478–480.

Gilden, D. H., Rozenman, Y., Murray, R., Devlin, M., and Vafai, A. (1987). *Ann. Neurol.* **22,** 377–380.

Gilden, D. H., Devlin, M., Wellish, M., Mahalingham, R., Huff, J. C., Hayward, A., and Vafai, A. (1989). *Virus Genes* **2,** 229–305.

Gold, E. (1965). *J. Immunol.* **95,** 683–691.

Goldstein, D. J., and Weller, S. K. (1988). *J. Virol.* **62,** 2970–2977.

Goodheart, C. R., Plummer, G., and Waner, J. L. (1968). *J. Virol.* **35,** 473–475.

Graham, F., and Van der Eb, A. (1973). *Virology* **52,** 456–467.

Green, M. R., Treisman, R., and Maniatis, T. (1983). *Cell (Cambridge, Mass.)* **35,** 137–148.

Grose, C. (1980). *Virology* **101,** 1–9.

Grose, C. (1981). *J. Clin. Microbiol.* **14,** 229–231.

Grose, C. (1987). *In* "Natural History of Varicella-Zoster Virus" (R. W. Hyman, ed.), pp. 1–66. CRC Press, Boca Raton, Florida.

Grose, C., and Brunell, P. A. (1978). *Infect. Immun.* **19,** 199–203.

Grose, C., and Friedrichs, W. E. (1982). *Virology* **118,** 86–95.

Grose, C., Perrotta, D. M., Brunell, P. A., and Smith, G. C. (1979). *J. Gen. Virol.* **43,** 15–27.

Grose, C., Edmond, R. J., and Friedrichs, W. E. (1981). *Infect. Immun.* **31,** 1044–1053.

Grose, C., Edwards, D. P., Friedrichs, W. E., Weigle, K. A., and McGuire, W. L. (1983). *Infect. Immun.* **40,** 381–388.

Grose, C., Edwards, D. P., Weigle, K. A., Friedrichs, W. E., and McGuire, W. L. (1984). *Virology* **132,** 138–146.

Hackstadt, T., and Mallavia, L. P. (1978). *J. Virol.* **25,** 510–517.

Hanparian, V. V., Hilleman, M. R., and Ketlan, A. (1963). *Proc. Soc. Exp. Biol. Med.* **112,** 1040–1048.

Harbour, D. A., and Caunt, A. E. (1979). *J. Gen. Virol.* **45,** 469–477.

Hayakawa, Y., and Hyman, R. W. (1987). *Virus Res.* **8,** 25–31.

Hayakawa, Y., Torigoe, S., Shiraki, K., Yamanishi, K., and Takahashi, M. (1984). *J. Infect. Dis.* **149,** 956–963.

Hayakawa, Y., Yamamoto, T., Yamanishi, K., and Takahashi, M. (1986). *J. Gen. Virol.* **67,** 1817–1829.

Head, H., and Campbell, A. W. (1900). *Brain* **23,** 353–523.

Herrman, E. C., Jr. (1961). *Proc. Soc. Exp. Biol. Med.* **107,** 142–175.

Heuschele, W. P. (1960). *J. Am. Vet. Med. Assoc.* **136,** 256–257.

Hodgman, T. C. (1988). *Nature (London)* **333,** 22–23.

Hondo, R., and Yogo, Y. (1988). *J. Virol.* **62,** 2916–2921.

Hondo, R., Yogo, Y., Kurata, T., and Aoyama, Y. (1987). *Arch. Virol.* **93,** 1–12.

Honess, R. W., and Roizman, B. (1974). *J. Virol.* **14,** 8–19.

Honess, R. W., and Roizman, B. (1975). *Proc. Natl. Acad. Sci. U.S.A.* **72,** 1276–1280.

Hope-Simpson, R. E. (1965). *Proc. R. Soc. Med.* **58,** 9–20.

Hyman, R. W., Ecker, J. R., and Tensen, R. B. (1983). *Lancet* **2,** 814–816.

Iltis, J. P., Oakes, J. E., Hyman, R. W., and Rapp, F. (1977). *Virology* **82,** 345–352.

Imperiale, M. J., Feldman, L. T., and Nevins, J. R. (1983). *Cell (Cambridge, Mass.)* **35,** 127–136.

Inchauspe, G., and Ostrove, J. M. (1989a). *Bull. Inst. Pasteur (Paris)* **87,** 19–37.
Inchauspe, G., and Ostrove, J. M. (1989b). *Virology* **173,** 710–714.
Inchauspe, G., Nagpal, S., and Ostrove, J. M. (1989). *Virology* **173,** 700–709.
Jacobson, J. G., Leib, D. A., Goldstein, D. J., Bogard, C. L., Schaffer, P. A., Weller, S. K., and Coen, D. M. (1989). *Virology* **173,** 276–283.
Jones, F., and Grose, C. (1988). *J. Virol.* **62,** 2701–2711.
Kapsenberg, J. G. (1964). *Arch. Gesamte Virusforsch.* **15,** 67–73.
Kaufman, H. E. (1962). *Proc. Soc. Exp. Biol. Med.* **109,** 251–252.
Kaufman, H. E., Nesburn, A. B., and Maloney, E. D. (1962). *Arch. Ophthalmol. (Chicago)* **67,** 583–591.
Keis, H. M., and Gold, E. (1963). *Biochim. Biophys. Acta* **72,** 263–276.
Keller, P. M., Neff, B. J., and Ellis, R. W. (1984). *J. Virol.* **52,** 293–297.
Keller, P. M., Davison, A. J., Lowe, R. S., Bennett, C. D., and Ellis, R. W. (1986). *Virology* **152,** 181–191.
Keller, P. M., Davison, A. J., Lowe, R. S., Reimen, M. W., and Ellis, R. W. (1987). *Virology* **157,** 526–533.
Kieff, E. D., Bachenheimer, S. L., and Roizman, B. (1971). *J. Virol.* **8,** 125–132.
Killington, R. A., Yeo, L., Honess, R. W., Watson, D. H., Duncan, B. C., Halliburton, I. W., and Mumford, J. (1977). *J. Gen. Virol.* **37,** 297–310.
Kinchington, P. R., Reinhold, W. C., Casey, T. A., Straus, S. E., Hay, J., and Ruyechan, W. T. (1985). *J. Virol.* **56,** 194–200.
Kinchington, P. R., Remenick, J., Ostrove, J. M., Straus, S. E., Ruyechan, W. T., and Hay, J. (1986). *J. Virol.* **59,** 660–668.
Kinoshita, H., Hondo, R., Taguchi, F., and Yogo, Y. (1988). *J. Virol.* **62,** 1097–1100.
Kit, S. (1985). *Microbiol. Sci.* **2,** 369–375.
Kit, S., and Dubbs, D. R. (1977). *Virology* **76,** 331–340.
Kozak, M. (1984). *Nucleic Acids Res.* **12,** 857–872.
Larder, B. A., Kemp, S. D., and Darby, G. (1987). *EMBO J.* **6,** 169–175.
Leader, D. P. (1988). *Nature (London)* **333,** 308.
Leiden, J. M., Buttyan, R., and Spear, P. G. (1976). *J. Virol.* **20,** 413–424.
Lipschütz, B. (1921). *Arch. Dermatol. Syph.* **136,** 428–482.
Lopetequi, P., Matsunaga, Y., Okuno, T., Ogino, T., and Yamanishi, K. (1983). *J. Gen. Virol.* **64,** 1181–1186.
Lopetequi, P., Campo-Vera, H., and Yamanishi, K. (1985). *Microbiol. Immunol.* **29,** 569–575.
Loring, Z. A. T., and Reed, S. I. (1984). *Nature (London)* **307,** 183–185.
Lowe, R. S., Keller, P. M., Keech, B. J., Davison, A. J., Whang, Y., Morgan, A. J., Kieff, E., and Ellis, R. W. (1987). *Proc. Natl. Acad. Sci. U.S.A.* **84,** 3896–3900.
Ludwig, H., Haines, H. G., Biswal, N., and Benyesh-Melnick, M. (1972). *J. Gen. Virol.* **14,** 111–114.
Maguire, H. F., and Hyman, R. W. (1986). *Intervirology* **26,** 181–191.
Mao, J. C. H., and Robishaw, E. E. (1975). *Biochemistry* **14,** 5475.
Mar, E. C., and Huang, E. S. (1979). *Intervirology* **12,** 73–83.
Mar, E. C., Huang, Y. S., and Huang, E. S. (1978). *J. Virol.* **26,** 249–256.
Martin, J. H., Dohner, D. E., Wellinghoff, W. J., and Gelb, L. D. (1982). *J. Med. Virol.* **9,** 69–76.
Matthews, R. E. F. (1982). *Intervirology* **17,** 1–20.
Matsunaga, Y., Yamanishi, K., and Takahashi, M. (1982). *Infect. Immun.* **37,** 407–412.
May, D. C., Miller, R. L., and Rapp, F. (1977). *Intervirology* **8,** 83–91.
McGeoch, D. J. (1986). *Nucleic Acids Res.* **14,** 4281–4292.
McGeoch, D. J., and Davison, A. J. (1986). *Nucleic Acids Res.* **14,** 1765–1777.

McGeoch, D. J., Dolan, A., and Frame, M. C. (1986). *Nucleic Acids Res.* **14,** 3435–3448.
McGeoch, D. J., Darlymple, M. A., Davison, A. J., Dolan, A., Frame, M. C., McNab, D., Perry, L. J., Scott, J. E., and Taylor, P. (1988a). *J. Gen. Virol.* **69,** 1531–1574.
McGeoch, D. J., Darlymple, M. A., Dolan, A., McNab, D., Perry, L. J., Taylor, P., and Chalberg, M. D. (1988b). *J. Virol.* **62,** 444–453.
McKnight, S. L., Gavis, E. R., Kingsbury, R., and Axel, R. (1981). *Cell (Cambridge, Mass.)* **25,** 385–398.
McLauchlan, J., and Clements, J. B. (1982). *Nucleic Acids Res.* **10,** 501–512.
Melnick, J. L., Midulla, M., Wimberly, I., Barrera-Oro, J. G., and Levy, B. M. (1964). *J. Immunol.* **92,** 596–601.
Miller, R. L., and Rapp, F. (1977). *J. Gen. Virol.* **36,** 515–524.
Minson, A. C., Wildy, P., Buchan, A., and Darby, G. (1978). *Cell (Cambridge, Mass.)* **13,** 581–587.
Mishra, L., Dohner, D. E., Wellinghoff, W. J., and Gelb, L. D. (1984). *J. Virol.* **50,** 615–618.
Mocarski, E. S., and Roizman, B. (1982). *Proc. Natl. Acad. Sci. U.S.A.* **79,** 5626–5630.
Montalvo, E. A., and Grose, C. (1986). *Proc. Natl. Acad. Sci. U.S.A.* **83,** 8967–8971.
Montalvo, E. A., and Grose, C. (1987). *J. Virol.* **61,** 2877–2884.
Montalvo, E. A., Parmley, R. T., and Grose, C. (1985). *J. Virol.* **53,** 761–770.
Mori, H., Shiraki, K., Kato, T., Hayakawa, Y., Yamanishi, K., and Takahashi, M. (1988). *Intervirology* **29,** 301–310.
Myers, M. G., Duer, H. L., and Hausler, C. K. (1980). *J. Infect. Dis.* **142,** 414–420.
Myers, M. G., Stanberry, L. R., and Edmond, B. J. (1985). *J. Infect. Dis.* **151,** 106–113.
Namazue, J., Kato, T., Okuno, T., Shiraki, K., and Yamanishi, K. (1989). *Intervirology* **30,** 268–277.
Namazue, J., Campo-Vera, H., Kitamura, K., Okuno, T., and Yamanishi, N. (1985). *Virology* **143,** 252–259.
Nikas, I., McLauchlan, J., Davison, A. J., Taylor, W. R., and Clements, J. B. (1986). *Proteins* **1,** 376–384.
Oakes, J. E., Iltis, J. P., Hyman, R. W., and Rapp, F. (1977). *Virology* **82,** 353–361.
Ogino, T., Otsuka, T., and Takahashi, M. (1977). *J. Virol.* **21,** 1232–1235.
Ogino, T., Lopetequi, P., and Yamanishi, K. (1982). *Biken J.* **25,** 149–156.
Okuno, T., Yamanishi, K., Shiraki, K., and Takahashi, M. (1983). *Virology* **129,** 357–368.
Ostrove, J. M., Reinhold, W., Fan, C. H., Zorn, S., Hay, J., and Straus, S. E. (1985). *J. Virol.* **56,** 600–606.
Pahwa, S., Biron, K., Lim, W., Swenson, P., Kaplan, M. H., Sadick, N., and Pahwa, R. (1988). *JAMA, J. Am. Med. Assoc.* **260,** 2879–2882.
Pellett, P. E., Kovsovlas, K. G., Pereira, L., and Roizman, B. (1985). *J. Virol.* **53,** 243–253.
Pichini, B., Ecker, J. R., Grose, C., and Hyman, R. W. (1983). *Lancet* **2,** 1223–1225.
Plotkin, S. A., Stein, S., Snyder, M., and Immesoete, P. (1977). *Ann. Neurol.* **2,** 249.
Preston, C. M., and Cordingley, M. G. (1982). *J. Virol.* **43,** 386–394.
Provost, P. J., Keller, P. M., Banker, F. S., Keech, B. J., Klein, H. J., Lowe, R. S., Morton, D. H., Phelps, A. H., McAleer, W. J., and Ellis, R. W. (1987). *J. Virol.* **61,** 2951–2955.
Rake, G., Blank, H., Coriell, L. L., Nagler, F. P. O., and Scott, T. F. M. (1948). *J. Bacteriol.* **56,** 293–303.
Rapp, F., and Benyesh-Melnick, M. (1963). *Science* **141,** 433–434.
Rapp, F., and Vanderslice, D. (1964). *Virology* **22,** 321–330.
Reagan, R. L., Day, W. C., Moore, S., and Brueckner, A. L. (1953). *Tex. Rep. Biol. Med.* **11,** 74–78.

Reinhold, W. C., Straus, S. E., and Ostrove, J. M. (1988). *Virus Res.* **9,** 249–261.
Richards, J. C., Hyman, R. W., and Rapp, F. (1979). *J. Virol.* **32,** 812–821.
Rivers, T. M. (1926). *J. Exp. Med.* **43,** 274–287.
Rock, D. L., Nesburn, A. B., Ghiasi, H., Ong, J., Lewis, T. L., Lokensgard, J. R., and Wechsler, S. L. (1987). *J. Virol.* **61,** 3820–3826.
Ross, C. A. C., Subak-Sharpe, J., and Ferry, P. (1965). *Lancet* **2,** 708–711.
Ruska, H. (1943a). *Scientia (Milan)* **37,** 16.
Ruska, H. (1943b). *Klin. Wochenschr.* **22,** 703–704.
Ruyechan, W. T., Casey, T. A., Reinhold, W., Weir, A. C., Wellman, M., Straus, S. E., and Hay, J. (1985). *J. Gen. Virol.* **66,** 43–54.
Salahuddin, S. Z., Ablashi, D. V., Markingham, P. D., Josephs, S. F., Sturzenegger, S., Kaplan, M., Halligan, G., Biberfeld, P., Wong-Staal, F., Kramarsky, B., and Gallo, R. C. (1986). *Science* **234,** 596–601.
Salzman, N. P. (1960). *Virology* **12,** 150–152.
Sandri-Goldin, R. M., Goldin, A. L., Holland, L. E., Glorioso, J. C., and Levine, M. (1983). *Mol. Cell. Biol.* **3,** 2028–2044.
Sawyer, M. H., Ostrove, J. M., Felser, J. M., and Straus, S. E. (1986). *Virology* **149,** 1–9.
Sawyer, M. H., Inchauspe, G., Biron, K. K., Waters, D. J., Straus, S. E., and Ostrove, J. M. (1988). *J. Gen. Virol.* **69,** 2585–2593.
Schaap, G. J. P., and Huisman, J. (1968). *Arch. Gesamte Virusforsch.* **25,** 52–57.
Schaffer, P. A. (1981). *In* "The Human Herpesviruses—An Interdisciplinary Perspective" (A. J. Nahmais, W. R. Dondle, and R. S. Schinazi, eds.). Elsevier, New York.
Schmidt, N. J., Lennette, E. H., and Magoffin, R. L. (1969). *J. Gen. Virol.* **4,** 321–328.
Scott-Wilson, J. H. (1978). *Lancet* **1,** 1152.
Seidlin, M., Takiff, H. E., Smith, H. A., Hay, J., and Straus, S. E. (1984). *J. Med. Virol.* **13,** 53–61.
Shaw, V. G. (1968). "Laboratory Studies on Varicella Zoster Virus," Ph.D. thesis. Liverpool Univ., Liverpool, England.
Shemer, Y., Leventon-Kriss, S., and Sarov, I. (1980). *Virology* **106,** 133–140.
Shibuta, H., Ishikawa, T., Hondo, R., Aoyama, Y., Kurata, K., and Matumoto, M. (1974). *Arch. Gesamte Virusforsch.* **45,** 382–385.
Shipkowitz, N. L., Bower, R., Appell, R. N., Nacksen, C. W., Overby, L. R., Roderick, W. R., Schleicher, J. B., and Van Esch, A. M. (1973). *Appl. Microbiol.* **26,** 264–267.
Shiraki, K., and Hyman, R. W. (1987). *Virology* **156,** 423–426.
Shiraki, K., and Takahashi, M. (1982). *J. Gen. Virol.* **61,** 271–275.
Shiraki, K., Okuno, T., Yamanishi, K., and Takahashi, M. (1982). *J. Gen. Virol.* **61,** 255–269.
Shiraki, K., Ogino, T., Yamanishi, K., and Takahashi, M. (1983). *Biken J.* **26,** 17–23.
Shiraki, K., Ogino, T., Yamanishi, K., and Takahashi, M. (1985). *J. Gen. Virol.* **66,** 221–229.
Smith, R. F., and Smith, T. F. (1989). *J. Virol.* **63,** 450–455.
Spaete, R. R., and Frenkel, N. (1982). *Cell (Cambridge, Mass.)* **30,** 295–304.
Spector, D. J., and Tevethia, M. J. (1989). *Prog. Med. Virol.* **36,** 120–190.
Spector, D., Stonehuerner, J. G., Biron, K. K., and Averett, D. R. (1987). *Biochem. Pharmacol.* **36,** 4341–4346.
Spivak, J. G., and Fraser, N. W. (1987). *J. Virol.* **61,** 3841–3847.
Stevens, J. G., Wagner, E. K., Devi-Rao, G. B., Cook, M. L., and Feldman, L. T. (1987). *Science* **235,** 1056–1059.
Stow, N. D. (1982). *EMBO J.* **1,** 863–867.
Stow, N. D., and Davison, A. J. (1986). *J. Gen. Virol.* **67,** 1613–1623.
Straus, S. E., Aulakh, H. S., Ruyechan, W. T., Hay, J., Casey, T. A., Vande Woude, G. F., Owens, J., and Smith, H. A. (1981). *J. Virol.* **40,** 516–523.

Straus, S. E., Owens, J., Ruyechan, W. T., Takiff, H. E., Casey, T. A., Vande Woude, G. F., and Hay, J. (1982). *Proc. Natl. Acad. Sci. U.S.A.* **79**, 993–997.

Straus, S. E., Hay, J., Smith, H., and Owens, J. (1983). *J. Gen. Virol.* **64**, 1031–1041.

Straus, S. E., Reinhold, W., Smith, H. A., Ruyechan, W. T., Henderson, D. K., Blaese, R. M., and Hay, J. (1984). *N. Engl. J. Med.* **311**, 1362–1364.

Straus, S. E., Rooney, J. F., Sever, J. L., Seidlin, M., Nusinoff-Lehrman, S., and Cremer, K. (1985). *Ann. Intern. Med.* **103**, 404–419.

Straus, S. E., Ostrove, J. M., Inchauspe, G., Felser, J. M., Freifeld, A. G., Croen, K. D., and Sawyer, M. H. (1988). *Ann. Intern. Med.* **108**, 221–237.

Strobel, M., and Francke, B. (1980). *Virology* **103**, 493–501.

Svedmyr, A. (1965). *Arch. Gesamte Virusforsch.* **17**, 495–503.

Takahashi, K., Sonoda, S., Kawakami, K., Miyata, K., Oki, T., Nagata, H., Okuno, T., and Yamanishi, K. (1988). *Lancet* **1**, 1463.

Taylor, P. (1986). *Nucleic Acids Res.* **14**, 437–441.

Taylor-Robinson, D., and Caunt, A. E. (1972). *Virol. Monogr.* **12**, 1–88.

Thelander, L., and Reichard, P. (1979). *Annu. Rev. Biochem.* **48**, 133–158.

Thompson, R., Honess, R. W., Taylor, L., Morran, L., and Davison, A. J. (1987). *J. Gen. Virol.* **68**, 1449–1455.

Vaczi, L., Geder, L., Koller, M., and Jeney, E. (1963). *Acta Microbiol. Acad. Sci. Hung.* **10**, 109–115.

Vafai, A., Wroblewska, Z., Wellish, M., Green, M., and Gilden, D. (1984). *J. Virol.* **52**, 953–959.

Vafai, A., Murray, R. S., Wellish, M., Devlin, M., and Gilden, D. H. (1988). *Proc. Natl. Acad. Sci. U.S.A.* **85**, 2362–2366.

von Barensprung, F. G. F. (1863). *Ann. Chante-Krankenh. (Berlin)* **11**, 96–104.

von Bokay, J. (1909). *Wien. Klin. Wochenschr.* **22**, 1323–1326.

Walz-Cicconi, M. A., Rose, R. M., Dammin, G. J., and Weller, T. H. (1986). *Arch. Virol.* **88**, 265–277.

Watson, D. H., Russell, W. C., and Wildy, P. (1963). *Virology* **19**, 250–260.

Weller, T. H. (1953). *Proc. Soc. Exp. Biol. Med.* **83**, 340–346.

Weller, T. H., and Coons, A. H. (1954). *Proc. Soc. Exp. Biol. Med.* **86**, 789–794.

Weller, T. H., and Stoddard, M. B. (1952). *J. Immunol.* **68**, 311–319.

Weller, T. H., and Witton, H. M. (1958). *J. Exp. Med.* **108**, 869–890.

Weller, T. H., Witton, H. M., and Bell, E. J. (1958). *J. Exp. Med.* **108**, 843–863.

Weller, S. K., Spadaro, A., Schaffer, J. E., Murray, A. W., Maxam, A. M., and Schaffer, P. A. (1985). *Mol. Cell. Biol.* **5**, 930–942.

White, R. J., Simmons, L., and Wilson, R. B. (1972). *J. Am. Vet. Med. Assoc.* **161**, 690–692.

Wigdahl, B., Long, B. L., and Kinney-Thomas, E. (1986). *Virology* **152**, 384–399.

Wildy, P., Russell, W. C., and Horne, R. W. (1960). *Virology* **12**, 204–222.

Wolff, M. H. (1978). *Med. Microbiol. Immunol.* **166**, 21–28.

Worrad, D. M., and Caradonna, S. (1988). *J. Virol.* **62**, 4774–4777.

Yamanishi, K., Matsunaga, Y., Ogino, T., Takahashi, M., and Takamizawa, A. (1980). *Infect. Immun.* **28**, 536–540.

Yamanishi, K., Matsunaga, Y., Ogino, T., and Lopetequi, P. (1981). *J. Gen. Virol.* **56**, 421–430.

Yamanishi, K., Okuno, T., Shiraki, K., Takahashi, M., Kondo, T., Asano, Y., and Kurata, T. (1988). *Lancet* **1**, 1065–1067.

Zweerink, H. J., and Neff, B. J. (1981). *Infect. Immun.* **31**, 436–444.

Zweerink, H. J., Morton, D. H., Stanton, L. W., and Neff, B. J. (1981). *J. Gen. Virol.* **55**, 207–211.

ADVANCES IN VIRUS RESEARCH, VOL. 38

STRUCTURE AND FUNCTION OF THE RNA POLYMERASE OF VESICULAR STOMATITIS VIRUS

Amiya K. Banerjee and Dhrubajyoti Chattopadhyay[1]

Department of Molecular Biology
Research Institute
The Cleveland Clinic Foundation
Cleveland, Ohio 44195

I. INTRODUCTION

The RNA-dependent RNA polymerase is the key enzyme for the animal RNA viruses to transcribe and replicate their RNA genomes. Since there is no enzymatic activity of similar property in animal cells, the animal RNA viruses have evolved to code this unique enzyme within the genome RNA (e.g., picornaviruses and alphaviruses, the so-called positive-stranded RNA viruses). On the other hand, the negative-stranded RNA viruses [e.g., vesicular stomatitis virus (VSV)] and diplornaviruses (e.g., reoviruses), due to inability of their genome RNAs to directly code the enzyme, have evolved a mechanism to package the RNA polymerase during morphogenesis as structural proteins. This virion-associated RNA polymerase, in turn, plays a unique role in initiating infection by carrying out primary transcription of the genome RNA when the virus infects the host cells.

Thus, the existence of these classes of virus relies exclusively on the presence of functional RNA-dependent RNA polymerase within ma-

---

[1] Present address: Department of Biochemistry, University College of Science, Calcutta, 700019, India.

ture virions. The RNA-containing retroviruses, however, have bypassed the requirement of RNA-dependent RNA polymerase by integrating their genome RNAs into the host genome via reverse transcription, using cellular DNA-dependent RNA polymerase to produce the viral genome RNA. This novel life cycle of the retroviruses places them distinctly among the animal RNA viruses.

To understand the replication strategies of the genome RNAs of the nonretroviral animal RNA viruses, it is necessary to acquire a thorough knowledge of the structure and function of the virus-coded RNA-dependent RNA polymerase. Since the RNA-directed transcription and replication systems are found only in this class of viruses (and bacteriophages), the RNA virus RNA polymerases can be considered "alien" enzymes, a detailed study of which should provide deeper insight into the evolution of RNA viruses in general.

VSV, a prototype of negative-stranded RNA viruses, provides a model system for studying an RNA-dependent RNA polymerase which is packaged within the virion. The enzyme transcribes the genome RNA of negative polarity *in vitro* and *in vivo* into five distinct 5'-capped and 3'-polyadenylated mRNA species and a 47-base leader RNA (Banerjee, 1987a). The five mRNAs code for five structural proteins of the virus, including nucleocapsid protein (N), phosphoprotein (NS), matrix protein (M), glycoprotein (G), and the polymerase protein (L). In spite of extensive study of the VSV transcription process *in vitro* over the years, the precise functions of the virion-associated RNA polymerase still remain unclear. Only recently, using recombinant DNA technology, has understanding of the structure and the possible functions of the enzyme subunits begun. Some recent reviews have dealt in detail with the transcriptive and replicative steps in the VSV life cycle (e.g., Banerjee, 1987a,b; Emerson, 1987). Here, we describe some recent findings which have advanced our knowledge of the structure and function of the VSV RNA-dependent RNA polymerase.

II. Polypeptide Composition of VSV Transcription Complex

The transcription complex of VSV can be easily purified from virions by treatment with nonionic detergent, followed by centrifugation of the released ribonucleoprotein complex (Abraham and Banerjee, 1976a; Szilagyi and Uryvayev, 1973). Similar complexes can also be obtained from VSV-infected cells (Galet and Prevec, 1973; Toneguzzo and Ghosh, 1976). Three polypeptide components constitute the transcription complex: a 49-kDa N protein that encloses the negative-stranded genome RNA, a 241-kDa L protein, and a 29-kDa NS protein.

The latter two polypeptides are tightly associated with each other and with the template–N protein complex. A recent estimate indicates that each virion contains approximately 1200, 40, and 400 molecules of N, L, and NS proteins, respectively (Thomas *et al.,* 1985). The N protein is presumably an RNA-binding protein, which remains associated with the genome RNA template during transcription and keeps the template inaccessible to RNase action (Chanda and Banerjee, 1979; Hefti and Bishop, 1975). The L and NS proteins, on the other hand, carry out not only the transcription process, but also the posttranscriptional modifications of the newly synthesized RNA and possibly phosphorylation of the NS protein (Sánchez *et al.,* 1985).

What makes the VSV transcription complex unique compared to the other negative-stranded RNA viruses is the relative ease by which the template–N protein complex, the L protein, and the NS protein can be separated and purified free from each other (Emerson and Yu, 1975). Effective reconstitution of the transcription complex and *in vitro* RNA synthesis are achieved when the three components are mixed together. Thus, the VSV transcription complex loosely resembles the eukaryotic chromosome in its overall characteristics.

III. Some Characteristic Features of the *in Vitro* Transcription Reaction

The transcription of VSV occurs in a polar fashion starting from the 3′ end of the genome RNA (Abraham and Banerjee, 1976a; Ball and White, 1976). The first RNA synthesized is a 47-base leader RNA (Colonno and Banerjee, 1978), followed sequentially with the mRNAs coding for the N protein, NS protein, M protein, G protein, and the putative RNA polymerase L protein. The gene order of VSV, 3′ leader–N–NS–M–G–L5′, is thus established from this unique mode of transcription as well as directly by ultraviolet inactivation studies (Abraham and Banerjee, 1976b; Ball and White, 1976).

The most conclusive proof of gene order, however, came from the complete nucleotide sequence determination of the genome RNA (Colonno and Banerjee, 1978; Gallione *et al.,* 1981; McGeoch and Dolar, 1979; Rose and Gallione, 1981; Schubert *et al.,* 1984). A distinctive feature of this transcription process is that the intergenic AAAs between the leader template and the *N* gene, the intergenic dinucleotide GA and CA, and the 5′ 60 nucleotides are not transcribed and incorporated into mature mRNAs (Fig. 1). The canonical sequence CUGUU present at the beginning of each gene presumably serves as the initiation signal for mRNA synthesis. The end sequence CAU of each gene,

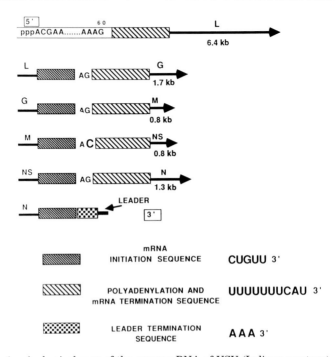

FIG. 1.   A physical map of the genome RNA of VSV (Indiana serotype).

followed by seven intergenic U stretches, probably serves as the termination and polyadenylation signal for each mRNA species.

It is interesting to note that during transcription there is a distinct pause and attenuation by the RNA polymerase at these intergenic junctions. This may be the principal reason for the establishment of a gradient of mRNA concentrations *in vitro* and *in vivo* in this order: leader RNA > N > NS > M > G > L (Iverson and Rose, 1981). The observed regulation of mRNA and protein syntheses during the replicative cycle of the virus could be mediated by this attenuation phenomenon. The precise mechanism by which the RNA polymerase (L and NS proteins) performs all of the above-mentioned transcription processes still remains unclear.

Since five genes along with the leader template are located contiguously on a linear map, coupled with the observation that transcription is polar, the most logical conclusion is that the RNA polymerase initiates RNA synthesis at the 3' end of the genome RNA. The synthesis of leader RNA and the downstream genes occur by a "stop–start" mechanism, leading to the sequential appearance of the individual mRNA species. Several experimental results point to this proposed mechanism (Emerson, 1982), although it is difficult to explain by this mechanism the proposed attenuation event occurring at the gene junctions. The polyadenylation step performed by the RNA polymerase during mRNA biosynthesis has been implicated in these attenuation phenomena (Iverson and Rose, 1981). However, the observation that negates this hypothesis is that attenuation by the RNA polymerase occurs at the leader–N gene junction without polyadenylation of the leader RNA.

An alternative transcription mechanism envisages multiple initiations by the RNA polymerase occurring simultaneously at different gene start sites, but with chain elongation of the mRNA species occurring sequentially in this order: leader RNA > N > NS > M > G > L (Testa et al., 1980). In other words, the mRNA chain completion of a transcribed gene does not occur unless the synthesis of the mRNA from its 3'-proximal gene is completed. This mechanism postulates that the physical structure of the transcription complex plays a direct role in transcription, as well as in the attenuation process.

Straightforward approaches using in vitro reconstitution reactions produced conflicting results (Emerson, 1982; Thorton et al., 1984). Thus, it remains to be conclusively shown whether the RNA polymerase indeed enters only at the 3' end of the N protein–RNA template complex and proceeds from one end to the other, resulting in the sequential synthesis of mRNA, with distinct attenuation at the gene junction. Alternatively, the polymerase binds at multiple sites and performs the same process by a mechanism involving some undetermined dynamics of the physical structure of the transcription complex.

Although the leader RNA template is an integral part of the entire transcription unit of the genome RNA, the transcribed leader RNA is uncapped and contains a 5'-triphosphate end (Colonno and Banerjee, 1978). This indicates that there is a differential mode of initiation by the RNA polymerase on the leader template compared to the rest of the gene promoter sites. Thus, the same RNA polymerase initiates the leader RNA without capping the 5' end, whereas the mRNAs are efficiently capped. Interestingly, the capping reaction in the VSV system is tightly coupled with transcription; that is, capping occurs only on the nascent mRNA chains (Abraham and Banerjee, 1976a). Thus, it seems that transcription of the leader template and the mRNA transcription units are two distinct processes with regard to their interactions with the RNA polymerase. In fact, some recent results have shown that the K_m values for ATP for syntheses of leader RNA and the mRNAs are quite different (Beckes et al., 1987). Moreover, under certain reaction conditions, leader RNA synthesis can be inhibited without affecting initiation of the N protein–mRNA chains (Thornton *et al.*, 1984; Talib and Hearst, 1983).

Another intriguing observation is that the analog of ATP, AMP-PNP, is not used by the RNA polymerase for RNA synthesis, indicating that hydrolysis of the γ-phosphate is essential for initiation of RNA chains from the leader template (Testa and Banerjee, 1979; Perrault and McLear, 1984). Whether the γ-phosphate of ATP is involved in phosphorylation of a specific protein or generates an allosteric effect on one of the constituent polypeptides of the transcription complex (Perrault and McLear, 1984) remains to be determined.

It is important to note that a switch from transcription to replication occurs precisely at the junction between the leader and the N gene, where the RNA polymerase fails to attenuate and continues transcription beyond the leader template, leading to the synthesis of the full-length complement of the positive strand. *De novo* synthesis of the N and NS proteins is essential for this replicative step during the virus life cycle (Blumberg *et al.*, 1981; Peluso and Moyer, 1983; Patton *et al.*, 1984). Curiously, under certain reaction conditions, using ribonucleoside triphosphate analogs (Chanda *et al.*, 1980, 1983) or the *polR* mutant of VSV (Perrault *et al.*, 1983), the replication reaction can occur without concurrent protein synthesis *in vitro*. The precise mechanisms by which the RNA polymerase performs this process remain elusive. Thus, the study of the structure and function of the RNA polymerase subunits L and NS is vital for understanding the steps involved in the transcription and replication reactions. In addition, a thorough study of the structure of the N protein–RNA template is important to gain insight into its role in the RNA-synthetic processes. Clearly, the template-associated N protein has to unfold transiently on a continuous basis, so that the RNA polymerase can gain contact with the template RNA. Thus, the RNA polymerase as well as the N protein perform important independent functions during transcription to successfully copy the genome RNA into distinct mRNA species in a polar fashion.

IV. L Protein

The polymerase gene (*L*) occupies more than 60% of the genome RNA. Consequently, over 90% of the spontaneous or chemically induced mutations of VSV result in the generation of *ts* (temperature-sensitive) mutants with the RNA minus phenotype and belong to complementation group I, corresponding to the *L* gene (Pringle, 1977). The large size of the L protein indicates that it is multifunctional. Curiously, attempts to demonstrate any enzymatic activity for RNA synthesis by this protein have been unsuccessful. *In vitro* reconstitution reactions have demonstrated that both the L and NS proteins are needed for transcription; however, individually, the proteins appear to be inert. Only indirect or mutational studies have indicated that the L protein is involved in RNA synthesis (Hunt and Wagner, 1974), capping and methylation (Horikami and Moyer, 1982; Hercyk *et al.*, 1988), and polyadenylation reactions (Hunt *et al.*, 1984). Recently, it was shown that the L protein alone is required only in catalytic amounts and initiates short oligonucleotide chains, indicating that the RNA polymerase activity indeed resides within it (De and Banerjee, 1985).

The perplexing phenomenon is that the L protein depends exclusively on the smaller NS protein to delineate its full RNA-synthesizing activity. This indicates that the NS protein performs some ancillary functions, possibly on the template, such that the L protein can move along the template RNA. It appears that a coordinated effort is made by the two proteins to perform the complicated transcription reactions, involving unzipping the N protein-bound genome RNA and transcription and termination of mRNA, while at the same time maintaining the structural integrity of the ribonucleoprotein template structure.

Recently, a serine-specific protein kinase activity has been shown to be associated with the L protein; it activates the NS protein function *in vitro* (Sánchez *et al.*, 1985; see Section V). However, the question still remains whether this activity indeed resides within the L polypeptide or whether a host protein kinase activity remains tightly associated with the L protein during maturation of the virion. Thus, it seems that the L protein is unique and multifunctional. A study of its primary structure is needed to gain insight into its functional role in the transcription process.

A. Primary Structure

The consensus sequences of the L gene of VSV Indiana (Schubert *et al.*, 1984) and New Jersey serotypes (Feldhaus and Lesnaw, 1988) were derived either from genomic cDNA copies or by direct dideoxy sequencing of the genome RNA. The genes for both serotypes are remarkably similar; that is, 6380 nucleotides coding for 2109 amino acids (Indiana) and 6398 nucleotides coding for the same number of amino acids (New Jersey). The estimated molecular weight of the L protein is 241,546. It is a basic protein (15% basic amino acids), but less so than the VSV M protein, which contains 17.1% basic amino acids. One interesting finding is that a high degree of mutations is detected among cDNA clones prepared from viral RNA of the same strain (i.e., Indiana serotype), indicating the high mutability of the L gene. It has been suggested that the viral polymerase itself could be a mutator protein, and the mutation found within the *L* gene could have a profound effect on the fidelity of the polymerase. This might play an important role in the rapid evolution of the virus (Schubert *et al.*, 1984).

Comparison of the amino acid sequences of the L protein of Indiana and New Jersey serotypes reveals some interesting findings. Four regions of strong similarity are found in the central region of the protein, ranging from 20 to 136 amino acids and over 90% of sequence identity (Fig. 2). The remaining portions of the L protein show about 55% similarity. The overall sequence similarity is approximately 65%. Interestingly, despite the lack of significant similarity exhibited by VSV and the L proteins of Sendai virus (Shioda *et al.*, 1986; Morgan and Rakestraw, 1986) and Newcastle disease virus (Yusoff *et al.*, 1987)— both belonging to the paramyxovirus family—two regions of considerable similarity were identified. One region included a 20-residue peptide (box A, or amino acids 530–549, Fig. 2), and the other, amino acids 670–718 (box B, Fig. 2). The degree of similarity between the Sendai virus and Newcastle disease virus amino acid sequences, allowing conservative replacements, was 51–56%. A similar degree of sequence

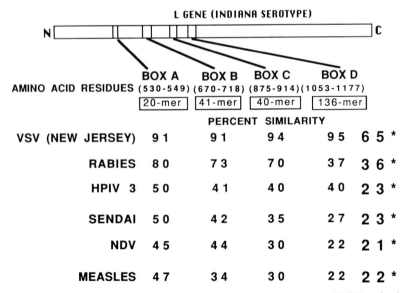

FIG. 2.   Amino acid sequence similarity between the L proteins of VSV and other rhabdoviruses and paramyxoviruses. Boxes A–D represent regions of highest similarity within the L proteins. Numbers corresponding to the asterisks represent overall sequence similarity. NDV, Newcastle disease virus; HPIV 3, Human parainfluenza virus 3.

similarity within the central region of the L protein of human para-influenza virus 3 and the VSV L protein has also been observed (Galinski *et al.*, 1988).

These results strongly suggest that these domains contain common functional sites for the L proteins of rhabdoviruses and paramyxoviruses. It is noteworthy that measles virus (another member of the paramyxovirus family) L protein, although sharing little overall amino acid sequence similarity with the VSV L protein (Blumberg *et al.*, 1988), shows appreciable similarity within the domain in box A, indicating, again, conservation of a specific functional domain in two divergently evolved proteins.

Tordo *et al.* (1988) have recently determined the complete nucleotide sequence of the genome RNA of rabies virus, a member of the rhabdovirus family. The L protein coded for by the *L* gene is a 2142-amino acid polypeptide with a calculated molecular weight of 244,206, similar to the VSV L protein. The overall amino acid identity between the L proteins of rabies virus and VSV (Indiana serotype) is 36%. Again, there are several highly conserved regions between residues 530 and 1177 (boxes A–D, Fig. 2), where the percentage of sequence identity is as high as 80%. Similarly, the rabies virus L protein and the corresponding protein in Sendai virus and Newcastle disease virus also reside in the conserved central region of the polypeptide. These observations strongly underscore that the RNA polymerase of the four unsegmented negative-stranded RNA viruses have evolved from a common ancestor. The highly conserved domains shared by these RNA virus L proteins could define the independent functional domains of these multifunctional proteins.

It is interesting to note that there is no occurrence of Gly–Asp–Asp (GDD) or Tyr–Met/Val–Asp–Asp(Y–M/V–DD) sequences flanked by highly hydrophobic regions in any of the four L proteins. These sequences are conserved in RNA-dependent RNA polymerases and reverse transcriptases from plant, animal, and bacterial viruses (Kamer and Argos, 1984). However, as noted by Galinski *et al.* (1988) and Blumberg *et al.* (1988), modified or weak GDD-like elements are present in all paramyxovirus L proteins. These observed differences in such sequences (if they are indeed ancestral RNA polymerase sequences) could relate to the profound variations in the transcription and replication strategies of the positive- and negative-stranded RNA viruses. The significance of these considerations, however, becomes equivocal, due to the virtual absence of these sequences from the L proteins of VSV and Newcastle disease virus.

Massey and Lenard (1987) have identified, within the L protein of VSV Indiana serotype, two ATP binding sequences similar to that in cellular protein kinases (i.e., Gly–X–Gly–X–X–Gly–X_{12-18}–Lys where X is any amino acid) (Kamps $et\ al.,$ 1984). These sequences are present between amino acid residues 754–778 and 1332–1351. Similar, although not identical, sequences are also present in all rhabdovirus and paramyxovirus L proteins whose amino acid sequences are known (Barik $et\ al.,$ 1990).

Recently, McClure and Perrault (1989) have found statistically significant sequence similarity between the L protein and the catalytic domains of tyrosine-specific protein kinases, such as the v-fes/v-fps kinase as well as platelet-derived growth factor receptor and its oncogene homolog, v-kit. The precise role of these sequences awaits further investigation. In any event the presence of these remnant protein kinase sequences within the L protein underscores the possible central role of phosphorylation in the VSV life cycle.

B. Functional Domains

To map the various functional domains within the L protein, it became necessary to obtain a biologically active protein using recombinant DNA technology. Schubert $et\ al.$ (1985), using a simian virus 40 (SV40) expression vector containing the entire cDNA of the L gene of VSV Indiana serotype, were successful in expressing recombinant L protein in COS cells. The expressed protein was found to be biologically active, since it efficiently complemented ts polymerase mutants of VSV at the restricted temperature. The efficiency of complementation, however, was dependent on the level of L protein expressed. Only those cells expressing low levels of the L protein were able to complement the ts defect, whereas high levels of the L protein not only greatly inhibited complementation, but also inhibited wild-type virus replication (Meir $et\ al.,$ 1987).

These results indicate that the requirement for the L protein in transcription and/or replication is catalytic, supporting the earlier observation (De and Banerjee, 1985) of a similar requirement for the L protein during $in\ vitro$ RNA synthesis. This also suggests that the unsegmented RNA viruses could have evolved in a manner such that the location of the L gene at the extreme 5' end of the genome RNA is transcribed least frequently, yielding low quantities of L protein in infected cells. Attempts to express truncated L proteins in COS cells in order to map the domains required for biological activity failed, because deletion of a few amino acids from the carboxy-terminal end of the protein resulted in complete inactivation (M. Schubert, personal communication). Thus, it seems that it would be difficult to map the functional domains of the L protein using this technique.

An alternative approach, transcription and translation of L mRNA from Sp6 transcription vectors, might have the potential to yield a biologically active protein. This method has been successfully used to obtain NS protein active *in vitro,* and considerable knowledge has been gathered regarding the functional domains of the NS protein (see Section V,B).

From the above description of the multifunctional L protein of VSV, it appears that the RNA polymerase activity and RNA capping, methylation, and polyadenylation—and possibly protein kinase—all reside within this large polypeptide. The catalytic sites of each of these functions are yet to be identified. On the basis of amino acid sequence information, it appears that they might be located in the central region of the polypeptide (boxes A–D, Fig. 2). This also indicates that these viruses have evolved from a common ancestor such that the RNA polymerase gene has conserved the functional domains appropriately positioned in the gene (i.e., at the 5′ end) for maximal replicative advantage.

The intriguing question still remains as to why the L protein alone is unable to perform RNA synthesis or other transcription-dependent enzymatic functions. The L protein relies exclusively on its interaction with the NS protein to form the RNA polymerase complex which carries out various intricate steps, leading to the synthesis of fully matured mRNAs in a sequential manner. Some of these transcriptive steps have recently been deciphered from the studies of the functional domains of the NS protein.

V. Phosphoprotein NS

Consistent with its map position on the genome RNA (i.e., near the 3′ end), the *NS* gene is transcribed and translated in infected cells in appreciable amounts. The NS protein is phosphorylated within the cell by cellular protein kinases, and a small part of the newly synthesized NS protein is packaged within the virions during maturation. By direct biochemical analyses of the NS protein, the *in vivo* phosphorylated sites (i.e., constitutive phosphorylation sites) were mapped to the aminoterminal one-third of the polypeptide between amino acid residues 35 and 106 (Hsu and Kingsbury, 1985; Bell and Prevec, 1985; Marnell and Summers, 1984). The phosphorylated NS protein remains complexed with the L protein in the virion and plays a crucial role in genome RNA transcription *in vitro* and *in vivo.*

Detailed studies of the phosphorylated states of the NS protein indicate that the protein is multiply phosphorylated, based on separation by gel electrophoresis of several different species (Clinton *et al.,* 1979).

Interestingly, some of the phosphorylated forms are resistant to phosphatase action, while others are highly sensitive (Masters and Banerjee, 1986), indicating that the NS protein exists in different conformational forms in which the phosphate groups are differentially exposed. The precise functions of these forms in transcription and replication remain unclear. However, there are several reports which suggest that differentially phosphorylated forms of the NS protein bind with varying degrees to the template and exhibit dissimilar transcription activity when complexed with the L protein (Kingsford and Emerson, 1980; Sinacore and Lucas-Lenard, 1982; Witt and Summers, 1980; Kingsbury et al., 1981). It is noteworthy that the NS protein (29,000 molecular weight) migrates slowly, as with an apparent molecular weight of 55,000 in denaturing polyacrylamide gel electrophoresis (PAGE) (Sokol et al., 1974). This anomalous migration rate of the protein is due to the high content of acidic amino acid residues (see Section V,B,3).

The in vitro reconstitution studies have established unequivocally that the NS protein is essential in the transcription process (Emerson and Yu, 1975; De and Banerjee, 1985), and its requirement in the reaction is stoichiometric or at least two to three times greater than that of the L protein. Since no other discernible enzyme activity is associated with the NS protein, it can be inferred that NS functions as a regulatory protein in the transcription process. It might have a direct role in unwinding the N protein–RNA template complex, perhaps to facilitate entry as well as movement of the L protein on the genome template (De and Banerjee, 1985). Earlier studies also indicated that the NS protein is a prerequisite for L protein binding to the N protein–RNA template complex (Mellon and Emerson, 1978). The functions of the NS protein were unclear until biologically active NS protein was obtained using recombinant DNA technology.

Another interesting observation is that the NS protein can be phosphorylated in vitro when purified transcription complex is incubated in the presence of [γ-^{32}P]ATP (Imblum and Wagner, 1974; Moyer and Summers, 1974). This observation indicated that at least one specific protein kinase was associated with the purified virions. Although most protein kinases seem to be of cellular origin, after extensive purification of the L protein from virions, a protein kinase activity was found associated with the L protein which could phosphorylate the NS protein (Sánchez et al., 1985).

Several experimental results suggested that this activity resided within the L polypeptide. The inhibition of the protein kinase activity of the L protein resulted in the total abrogation of RNA synthesis, indicating that phosphorylation of the NS protein in vitro is an essential step in obtaining an active RNA polymerase complex. Nevertheless, it remains to be conclusively proved whether the observed protein kinase activity is indeed a part of the L protein or whether a specific cellular protein kinase is tightly associated with it. Recently, D. M. Massey and J. Lenard (personal communication) have purified L protein which is 90% decreased in protein kinase activity. However, whether the residual protein kinase activity is L associated and sufficient to activate the NS protein is unknown. In any case it became cogent that phosphorylation of the NS protein is indeed an important step in the formation of an active polymerase complex for transcription of the genome RNA. Whether phosphorylation is brought about by the L protein or a specific cellular protein kinase enzyme tightly associated with the transcription complex has not been decided.

A. Primary Structure

An important observation indicating that the NS protein is indeed a unique regulatory protein came from findings that the nucleotide sequences of the NS genes of two serotypes of VSV are extremely dissimilar. A sequence relationship of only 23% is in direct contrast to the other genes, in which the sequence similarity ranges from 50% (G gene) to 70% (N gene) and 80% for the leader template (Gill and Banerjee, 1985, 1986; Rae and Elliot, 1986). In a comparison of the NS and N gene sequences of a more distant relative of the New Jersey and Indiana serotypes—Chandipura virus (Masters and Banerjee, 1987)— again, the NS gene is different from both Indiana (21%) and New Jersey serotypes (23%), while the value for the N gene is over 50% (Fig. 3). In fact, from the N and NS gene sequence comparisons it seems that Chandipura virus is equally distant from the two serotypes and roughly twice as distant from them as they are from each other. These intriguing findings clearly indicate that, in spite of being an essential component of the polymerase complex, the gene encoding the NS protein is highly mutable.

However, two observations pointed out some important commonality among the amino acid sequences of the NS protein of the different serotypes (Gill and Banerjee, 1985; Masters and Banerjee, 1987). First, overall, the protein is highly acidic, especially the amino-terminal one-third of the polypeptide spanning the constitutive phosphorylation sites. This domain contains clusters of aspartic and glutamic acid residues which contribute to the acidic nature of the polypeptide (Fig. 4). Second, the basic carboxy-terminal 22 amino acid residues in the three serotypes are most similar, with over 90% similarity between the Indiana and New Jersey serotypes and over 50% similarity between the Indiana or New Jersey serotype and Chandipura virus (Fig. 3). In addition, by a rigid computer alignment program, as many as 18

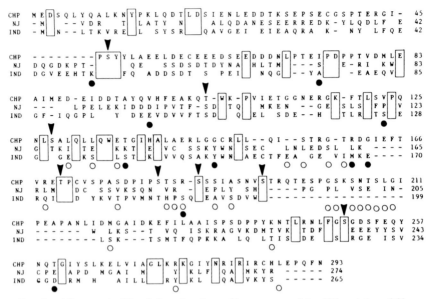

FIG. 3.   Alignment of the deduced amino acid sequences of the NS proteins of Chandipura virus (CHP); VSV New Jersey serotype (NJ), Ogden strain; and VSV Indiana serotype (IND), San Juan strain. Dashes represent gaps introduced to maximize matching amino acid residues. Spaces indicate positions where the amino acid residue is identical to that of Chandipura virus. Solid circles beneath the sequence indicate positions where the Mudd–Summers strain of the VSV Indiana serotype differs from the San Juan strain (Hudson *et al.*, 1986). Open circles beneath the sequence indicate positions where the Missouri strain of the VSV New Jersey serotype differs from the Ogden strain (Rae and Elliot, 1986). Boxed amino acids are identical in all five sequences. Arrowheads are shown above serine and threonine residues conserved in all five sequences. Data from Masters and Banerjee (1987).

potential phosphorylation sites (i.e., serine and threonine residues) are conserved between the Indiana and New Jersey serotypes (Rae and Elliot, 1986) and eight sites are conserved between the Indiana or New Jersey serotype and Chandipura viruses (unpublished observations). These observations indicated that the NS protein must have at least two domains that play some unique functional roles in the transcription processes.

B. Functional Domains

A major breakthrough in studying the functional domains within the NS protein came when biologically active NS protein was synthesized by transcription of the full-length cDNA clone of the *NS* gene *in vitro* from an Sp6 transcription vector, followed by translation of the

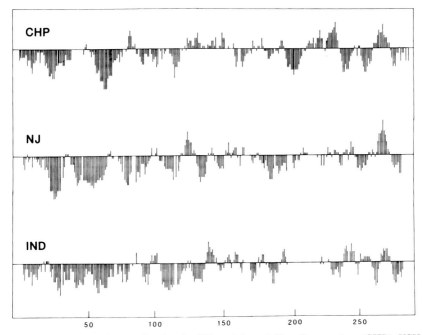

FIG. 4.   Hydropathicity plot for the NS proteins of Chandipura virus (CHP), VSV New Jersey serotype (NJ), and VSV Indiana serotype (IND). Hydrophilic regions of each amino acid sequence are shown below the axis, while hydrophobic regions are above the axis. The scale corresponds to the amino acid residue numbered from the amino terminus for the Chandipura virus NS sequence. The three sequences have been aligned as in Fig. 3. Data from Masters and Banerjee (1987).

mRNA in a cell-free extract (Gill *et al.*, 1986). The synthesized NS protein fully supported mRNA synthesis *in vitro* from N protein–RNA template complex when complexed with the L protein purified from the virion. This system provided an excellent opportunity to systematically study the various domains within the NS polypeptide by genetic manipulation of the *NS* gene in the plasmid. The result of such experiments yielded important new information regarding the functional domains of the NS protein. These are discussed in the following sections.

1. Template Binding Domain

Removal of the basic carboxy-terminal 27 amino acids from the NS protein (Fig. 5) resulted in a total loss of its capacity to bind to the N protein–RNA template complex, indicating that this domain (i.e., domain III) is the template binding region of the NS protein. In fact, the carboxy-terminal penultimate 11 basic amino acid residues (Fig. 6) are

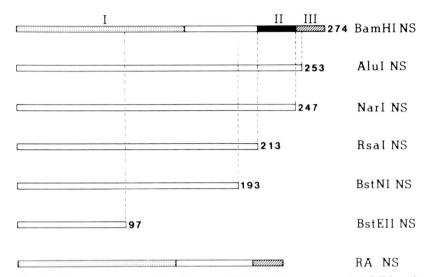

FIG. 5.   Deletion mapping of the *NS* gene (VSV New Jersey serotype). A full-length *NS* clone (856 nucleotides) was inserted into a pGEM4 vector (Gill *et al.*, 1986). For expression the vector was cut with various restriction enzymes, as shown, and the resulting fragments were transcribed and the mRNAs were translated *in vitro*. The truncated proteins were used to reconstitute transcription in the presence of purified L protein and the N protein–RNA template complex. The functional domains I–III in the NS protein are shown. The mutant clone *RA NS* was constructed by deleting domain II from the wild-type clone (Chattopadhyay and Banerjee, 1987a).

```
NJ   Q L I S K R A G V K D M T V K L T D F F G S E E E Y Y S V C P E G A
IND  T F Q P K K A G L Q D L T I S L D E F F S S R G E Y I S V G G D G R
CHP  I L A A I S P S D P P Y K N T L R N L F G S G D S F E Q Y N Q T G I
                              DOMAIN  II
NJ   A I I M G L K Y K K L F N Q A R M K Y R L
IND  A I L L G L R Y K K L F N Q A R V K Y S L
CHP  L V I A G L K R K G I Y N R I R I R C H L E P Q F N
                         DOMAIN  III
```

FIG. 6.   Amino acid sequence within domains II and III of VSV New Jersey (NJ) and Indiana (IND) serotypes and Chandipura (CHP) virus. The serine residues, including residues 213 and 247 for New Jersey serotype, within domain II that are probable phosphorylation sites are underlined. The basic amino acid residues within domain III are underlined. The amino acid residues that are conserved among all serotypes are indicated by a solid circle on top.

directly involved in binding to the template, since removal of these residues negated binding of the NS protein to the template (unpublished observations). Interestingly, however, the truncated NS protein lacking domain III can interact normally with the L protein, and the resulting complex binds to the template and effectively transcribes the template RNA. Thus, it seems that domain III is an L protein-independent template binding domain within the NS protein, whereas a separate domain within the NS protein mediates binding to the L protein, leading to transcription.

Closer examination of domain III reveals the presence of a basic amino acid stretch (NXXRXK) present in Chandipura virus and in both VSV serotypes (Fig. 6), indicating that the interaction with the template could be mediated by this basic domain. Site-directed mutagenesis experiments should provide more precise definition of the role of this domain in the transcription process.

2. L Protein Binding Domain

Removal of an additional 34 amino acid residues downstream from domain III (Fig. 5) results in total abrogation of RNA synthesis when the truncated NS protein is mixed with the L protein and the template. This indicates that domain II is directly involved in binding with the L protein to form the active RNA polymerase complex, in order to bind to the N protein–RNA template complex.

3. Ancillary Domain

The highly acidic domain I (Fig. 5), constituting the major part of the protein, seems to be an ancillary domain whose function was not immediately apparent. Domain I also associates with the L protein, since such a complex could be immunoprecipitated by antibody to the L protein (Emerson and Schubert, 1987). Coupled with the observation that the L protein functionally binds to domain II, it seems that the NS protein might have two binding sites for the L protein; one through domain II, which appears to be critical for the initiation of RNA synthesis, and the other through domain I, which might be for the elongation of RNA chains.

Since domain II appears to be the functional binding site of the NS protein, the L protein could modify the domain (e.g., by phosphorylation), so that a tight complex is formed which might facilitate its binding to the template. The sequence of domain II (Fig. 6) revealed that two serine residues at 213 and 247 are conserved in the Indiana and New Jersey serotypes of VSV. Only serine residue 247 is conserved in Chandipura NS domain II (Fig. 6). Since the protein kinase activity associated with the L protein is serine specific, these amino acid residues appear to be the most logical targets for putative phosphorylation by the L protein.

In fact, site-directed mutagenesis of the two serine residues to alanine resulted in almost complete abrogation of RNA synthesis when the mutant NS protein was mixed with purified L protein and template (Chattopadhyay and Banerjee, 1987a). The L protein-mediated binding of mutant NS protein to the template, as expected, is concomitantly decreased. These results indicated that the two serine residues within domain II play crucial roles in the binding of the NS protein to the L protein, as well as to the N protein–RNA template complex. Whether both serine residues are phosphorylated during interaction with the L protein is not yet established.

However, the interesting observation is that nascent NS protein, although presumably phosphorylated by the protein kinases present in the cell-free translation extract, is still effectively phosphorylated by the L protein. These results indicate that the serine residue(s) within domain II are somehow inaccessible to cellular protein kinases and interact specifically with the L protein. Thus, domains II and III, either independently or cooperatively, seem to be the "business" end of the NS polypeptide, which helps the L protein to bind to the template for RNA synthesis. This is underscored by the recent observation that monoclonal antibodies raised against purified NS protein seem to map predominantly at domains II and III of the polypeptide (Williams *et al.*, 1988). Similar observations were also made for the NS protein of Indiana serotype (Paul *et al.*, 1988). These observations clearly demonstrate that the structure of NS protein in solution is unique in that the observed functional domains display distinct structural configurations for binding the L protein and the template.

Probing the function of the acidic domain I of the NS protein revealed some provocative findings. From appropriately constructed Sp6 transcription vectors containing either domain I or domains II and III, it was possible to obtain two distinct polypeptides synthesized *in vitro* which correspond to domain I (21-kDa BstNI, Fig. 5) and domains II and III (10 kDa) (Chattopadhyay and Banerjee, 1987b). The 21-kDa polypeptide migrated as 50 kDa in denaturing PAGE, whereas the 10-kDa polypeptide migrated as predicted by its size. These results confirm that the acidic nature of domain I is responsible for anomalous migration in PAGE (Sokol *et al.*, 1974).

In vitro reconstitution experiments indicated that the smaller polypeptide, as predicted, binds to the template as efficiently as the native NS protein. However, no RNA synthesis occurs unless the larger polypeptide is added in trans, which restores transcription to the extent of 20%, whereas addition of a mutant NS protein lacking domain II (mutant clone *RA NS,* Fig. 5) in trans effectively activated transcription to 60% (Chattopadhyay and Banerjee, 1987b). Curiously, polyglutamic acid and tubulin (which are also acidic) were able to replace the domain I function although inefficiently (Chattopadhyay and Banerjee, 1987b).

A definitive role of domain I became apparent when a recombinant DNA was constructed that contains a full-length cDNA clone of tubulin fused to the 3'-terminal DNA segment of the *NS* clone encoding both domains II and III (Chattopadhyay and Banerjee, 1988). The resulting chimeric tubulin–NS protein efficiently supported transcription (66%) when combined with the L protein and the RNA template. This intriguing finding indicated that the acidic property of domain I is an essential feature of the NS protein which somehow plays a crucial role in the transcription process. This might also explain the apparent mutability of the *NS* gene; the gene corresponding to domain I might mutate randomly as long as the acidic nature of the region is preserved.

The role of constitutive phosphorylation sites within domain I still remains unclear. Perhaps phosphorylation of specific serine and threonine residues increases the acidity of domain I and consequently promotes transcription. Thus, it seems that the carboxy-terminal portion of the NS protein is the template binding domain and, probably in conjunction with the L protein, initiates RNA synthesis. The amino-terminal acidic domain, on the other hand, is involved in some postinitiation step, probably in RNA chain elongation. Site-directed mutagenesis of serine and threonine residues within the presumptive constitutive phosphorylation sites as well as acidic residues in domain I should reveal their roles in the transcription process.

VI. Similarity of the NS Protein with Eukaryotic Transcription Activators

The observed functional domains of the NS protein bear a remarkable resemblance to the recently discovered eukaryotic transcription activators. The most-studied yeast transcription activators, GAL4 and GCN4, contain two distinct domains: a DNA binding region and an acidic activating region, the so-called "acid blob" (Ptashne, 1988; Struhl, 1987). The latter acidic region can be replaced either by short acidic fragments of no specific primary sequence or totally unrelated acidic peptides (Ma and Ptashne, 1987; Giniger and Ptashne, 1987; Hope *et al.,* 1988).

It appears that the DNA binding region imparts specificity to the protein by binding to a specific DNA sequence upstream from the gene. This enables the acidic activator domain to interact, via an unknown mechanism, with components of the transcription complex associated with the TATA box to stimulate transcription of the gene located downstream. As observed in the NS protein, the two regions of the yeast transcription activators can also be separated and the functional units can be joined to complementary regions of unrelated activators to produce fully functional chimeric activators (Ptashne, 1988).

More recently, transcription activators with similar functional domains have been discovered in viral systems, such as adenovirus (Lillie and Green, 1989), pseudorabies virus (Feldman et al., 1982), bovine papilloma virus (Haugen et al., 1987), and human T-lymphotropic virus type I (Ballard et al., 1988). Herpesvirus contains a highly acidic structural protein, VP16, (Triezenberg et al., 1988) which does not have a DNA binding domain, but interacts with a specific cellular DNA-binding protein to activate immediate early viral genes. Recently, Sadowski et al. (1988) have synthesized a chimeric protein containing the DNA binding domain of GAL4 fused to the herpesvirus VP16. The resulting GAL4–VP16 activated the transcription of genes more efficiently than full-length GAL4 protein in mammalian cells containing a GAL4 DNA binding sequence located in proximity to a reporter gene. Similar experiments (in progress) with GAL4–NS (domain I) chimeric protein should confirm the activator function of the acidic region of the NS polypeptide.

If, indeed, the NS protein functions as a transcription activator similar to those found in yeast, plant, insects, and mammalian cells (Kakadini and Ptashne, 1988; Webster et al., 1988; Fischer et al., 1988; Ma et al., 1988), it would be the first of its kind reported in an RNA-dependent RNA transcription system. The precise mechanism by which the NS protein trans-activates the L protein, however, remains to be determined. From the data obtained thus far, it seems that domains II and III impart specificity to the NS protein by binding to the template as well as the L protein.

The acidic domain I functions as a typical acid blob and activates the L protein to move along the template. This domain must also interact with the N protein bound to the template, to transiently lift it during the course of transcription. The α-helicity of the acidic protein could be involved in the trans-activation process (Ptashne, 1988). Future studies along this line will undoubtedly shed light on the roles of these NS protein domains in the transcription process.

VII. Roles of L and NS Proteins in Transcription

From the information on the various functional domains of the VSV RNA polymerase subunits, L and NS proteins, a more definitive picture emerges with regard to the interactions of these proteins with each other and with the template, leading to transcription of the genome RNA. These can be summarized as follows.

1. The NS protein binds strongly with the N protein–RNA template complex. This binding is mediated directly by the carboxy-terminal 11 amino acids of the protein, although involvement of the entire domain III cannot be ruled out. This interaction is probably restricted to the NS protein and the N protein covering the template, and there is no evidence to suggest that the NS protein can bind directly to the genome RNA within the N protein–RNA template complex. However, there is a report that the NS protein could have some contact with the genome RNA within a region spanning the leader template (Keene *et al.*, 1981) while packaged within the nucleocapsid. Whether similar NS protein binding domains are located at distinct sites throughout the genome RNA is not known.

The question still remains as to how many NS molecules bind to the N protein–RNA template complex for effective transcription to occur. From the functional domain studies it appears that the transcription requirement for the carboxy-terminal domain of the NS protein is catalytic, whereas acidic domain I is needed in stoichiometric amounts (Chattopadhyay and Banerjee, 1988). Thus, a small amount of the NS protein could bind to the L protein and position the L protein–NS protein complex at a putative promoter site(s) on the N protein–RNA template complex to initiate transcription. Chain elongation is mediated by stoichiometric interaction of the acidic domain with the template and the L protein–NS protein complex to trans-activate the RNA polymerase. The precise mechanism by which this activation step is performed by the NS protein remains to be determined.

Clearly, these interactions must involve removal of the N protein from the RNA template to allow the L protein–NS protein complex to move along the template RNA. Acidic domain I could mimic negatively charged RNA, such that the N protein interacts preferentially with the NS protein, thus uncovering the RNA template (De and Banerjee, 1985; Hudson *et al.*, 1986). This would allow the L protein to gain contact with the template for transcription. However, a separate mechanism must exist by which the transiently removed N protein again covers the previously transcribed portion of the genome RNA. It is possible that there are separate interactions with the L and N proteins to perform the latter step.

2. Since the carboxy-terminal 27 amino acid residues are not essential for transcription, why are they conserved in the NS protein? It is important to stress again that this domain bears the most amino acid sequence similarity within the different serotypes of VSV. Perhaps domains II and III constitute the template binding domain. Domain III might be a stronger site, while domain II alone binds weakly to the template. Domain II might bind more tightly in the presence of the L protein. In fact, a chimeric protein lacking precisely domain II (mutant clone *RA NS*, Fig. 5) failed to bind to the template (Chattopadhyay and Banerjee, 1987a).

Another important involvement of domain III probably occurs during replication of the genome RNA. A series of recent observations supports this contention. The most important among the findings is that the NS protein forms an active complex with soluble N protein, making the N protein competent to interact and encapsidate the nascent RNA chains initiated from the transcription complex (Blumberg *et al.*, 1981; Peluso and Moyer, 1983; Patton *et al.*, 1984). This process allows the RNA polymerase to transcribe the genome RNA completely, without undergoing termination or reinitiation steps. Indeed, this is a necessary step for replication of the genome RNA. Recently, it has been shown that the formation of the replication-completent N protein–NS protein complex is mediated via the carboxy-terminal 27 amino acid residues (Masters and Banerjee, 1988a,b). Moreover, it has been shown that the NS protein–N protein interaction changes the configuration of the N protein such that it does not bind to nonspecific RNAs, but probably attaches to genomic or antigenomic sense VSV RNA. Thus, the conserved carboxy-terminal domain of the NS protein interacts with the N protein possibly to keep its structure in a replication-competent form.

3. Although a great deal of amino acid similarity exists among the L protein, the carboxy-terminal region of the NS protein, and the N protein of different VSV serotypes, a distinct specificity of interaction of the L and NS proteins between heterologous N protein–RNA template complexes has been observed in the two serotypes studied (De and Banerjee, 1984) (Fig. 7). Synthesis of mRNA from the N protein–RNA template complex of Indiana serotype by L and NS proteins of New Jersey serotype requires, in addition, the NS protein of Indiana serotype. In contrast, transcription of the N protein–RNA template complex of the New Jersey serotype requires specifically the L protein of the same serotype, but the NS protein of the Indiana serotype can form an active RNA polymerase complex with heterologous L protein. This specificity of interaction between the L and the NS proteins indi-

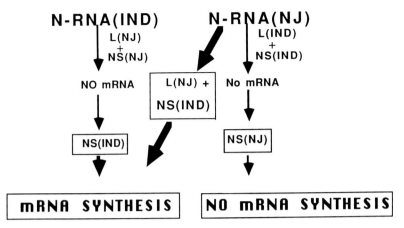

FIG. 7. Heterologous reconstitution of mRNA synthesis *in vitro* by VSV New Jersey and Indiana serotypes. The boxes contain the L and NS components from each serotype essential for mRNA synthesis from the heterologous template (see text).

cates that within the N, NS, and L proteins not only are there domains that are specific for homologous interactions, but there are other domains that regulate heterologous interactions. Further structure–function studies of these proteins should help identify the corresponding domains.

4. The L protein is undoubtedly the transcriptase component of the RNA polymerase of VSV. Alone, it fails to synthesize RNA, primarily because the configuration of the N protein–RNA template complex is unsuitable for the L protein to bind properly and perform transcription. The NS protein brings about the required change in the template structure and also activates the L protein. As discussed in Section V,B, the three identifiable domains within the NS polypeptide play crucial roles in these processes. Moreover, phosphorylation of the NS protein appears to directly regulate its binding with the template as well as with the L protein, and phosphorylation might be mediated by the L protein. It will be vital to explore the functions of the conserved domains (Fig. 2) within the L protein to gain insight into the RNA-synthesizing and -modifying functions of the domains and their roles in the transcription process.

5. The structure of the N protein–RNA template complex remains intriguing. The RNA genome contains all sequences which apparently function for signaling the RNA polymerase to initiate RNA synthesis, terminate RNA chains, and polyadenylate full-length mRNAs (Fig. 1). The template-associated N protein not only maintains the integrity of the template, but probably also plays a crucial role in its polar effect

on transcription. Thus, it is possible to predict that the microenvironment of the N protein–RNA template complex structures at the gene junctions dictates the stop–start mode of the RNA polymerase. The most important of such junctions is the leader–N gene junction, where the polymerase either stops and initiates transcription of the N gene (i.e., transcriptive mode) or continues, to form the full-length complement of the genome RNA (i.e., replicative mode).

It is tempting to speculate that the N protein, probably in association with the NS protein, plays a dynamic role in this process. In fact, as shown by Perrault *et al.* (1983), the *polR* mutants of VSV display *in vitro* an aberrant transcriptive mode which has been mapped to a point mutation in the N protein. Although no enzymatic activity is associated with the N protein, an N protein–RNA template complex-associated protein kinase activity has been demonstrated *in vitro* (Masters and Banerjee, 1986), but its function remains unclear. Thus, the precise mechanism by which the N protein plays its role in transcription and replication processes remains to be determined. *In vitro* encapsidation of VSV genomic RNA by the N protein, with subsequent reconstitution of RNA transcription by the RNA polymerase, should eventually answer many of these questions and help unravel the mechanism of genome RNA transcription in general.

ACKNOWLEDGMENTS

We thank Mark Galinski for valuable discussion and for helping to formulate the data in Fig. 2. The research in the authors' laboratory was supported by U.S. Public Health Service Grant AI 26585-02 from the National Institutes of Health.

REFERENCES

Abraham, G., and Banerjee, A. K. (1976a). *Virology* **71**, 230.
Abraham, G., and Banerjee, A. K. (1976b). *Proc. Natl. Acad. Sci. U.S.A.* **73**, 1504.
Ball, L. A., and White, C. N. (1976). *Proc. Natl. Acad. Sci. U.S.A.* **73**, 442.
Ballard, D. W., Bohnlein, E., Lowenthal, J. W., Wano, Y., Franza, B. R., and Greene, W. C. (1988). *Science* **241**, 1652.
Banerjee, A. K. (1987a). *Microbiol. Rev.* **51**, 66.
Banerjee, A. K. (1987b). *Cell (Cambridge, Mass.)* **48**, 363.
Barik, S., Rud, E. W., Luk, D., Banerjee, A. K., and Kang, C. Y. (1990). *Virology* **175**, 332.
Beckes, J. D., Haller, A. A., and Perrault, J. (1987). *J. Virol.* **61**, 3470.
Bell, J. C., and Prevec, L. (1985). *J. Virol.* **54**, 697.
Blumberg, B. M., Leppert, M., and Kolakofsky, D. (1981). *Cell (Cambridge, Mass.)* **23**, 837.
Blumberg, B. M., Crowley, J. C., Silverman, J. I., Menonna, J., Cook, S. D., and Dowling, P. C. (1988). *Virology* **164**, 487.
Chanda, P. K., and Banerjee, A. K. (1979). *Biochem. Biophys. Res. Commun.* **91**, 1337.

Chanda, P. K., Kang, C. Y., and Banerjee, A. K. (1980). *Proc. Natl. Acad. Sci. U.S.A.* **77,** 3927.

Chanda, P. K., Roy, J., and Banerjee, A. K. (1983). *Virology* **129,** 225.

Chattopadhyay, D., and Banerjee, A. K. (1987a). *Cell (Cambridge, Mass.)* **49,** 407.

Chattopadhyay, D., and Banerjee, A. K. (1987b). *Proc. Natl. Acad. Sci. U.S.A.* **84,** 8932.

Chattopadhyay, D., and Banerjee, A. (1988). *Proc. Natl. Acad. Sci. U.S.A.* **85,** 7977.

Clinton, G. M., Burge, B. W., and Huang, A. S. (1979). *Virology* **99,** 84.

Colonno, R. J., and Banerjee, A. K. (1978). *Cell (Cambridge, Mass.)* **15,** 93.

De, B. P., and Banerjee, A. K. (1984). *J. Virol.* **51,** 628.

De, B. P., and Banerjee, A. K. (1985). *Biochem. Biophys. Res. Commun.* **126,** 40.

Emerson, S. U. (1982). *Cell (Cambridge, Mass.)* **31,** 635.

Emerson, S. U. (1987). *In* "The Rhabdoviruses" (R. R. Wagner, ed.), p. 245. Plenum, New York.

Emerson, S. U., and Schubert, M. (1987). *Proc. Natl. Acad. Sci. U.S.A.* **84,** 5655.

Emerson, S. U., and Yu, Y.-H. (1975). *J. Virol.* **15,** 1348.

Feldhaus, A. L., and Lesnaw, J. A. (1988). *Virology* **163,** 359.

Feldman, L. T., Imperiale, M. F., and Nevins, J. R. (1982). *Proc. Natl. Acad. Sci. U.S.A.* **79,** 4952.

Fischer, J. A., Giniger, E., Maniatis, T., and Ptashne, M. (1988). *Nature (London)* **332,** 950.

Galet, H., and Prevec, L. (1973). *Nature (London), New Biol.* **243,** 200.

Galinski, M. S., Mink, M. A., and Pons, M. W. (1988). *Virology* **165,** 499.

Gallione, C. J., Greene, J. R., Iverson, L. E., and Rose, J. K. (1981). *J. Virol.* **39,** 529.

Gill, D. S., and Banerjee, A. K. (1985). *J. Virol.* **55,** 60.

Gill, D. S., and Banerjee, A. K. (1986). *Virology* **150,** 308.

Gill, D. S., Chattopadhyay, D., and Banerjee, A. K. (1986). *Proc. Natl. Acad. Sci. U.S.A.* **83,** 8873.

Giniger, E., and Ptashne, M. (1987). *Nature (London)* **330,** 670.

Haugen, T. H., Cripe, T. P., Ginder, G. D., Karin, M., and Turek, L. P. (1986). *EMBO J.* **6,** 145.

Hefti, E., and Bishop, D. H. L. (1975). *Biochem. Biophys. Res. Commun.* **66,** 785.

Hercyk, N., Horikami, S. M., and Moyer, S. A. (1988). *Virology* **163,** 222.

Hope, I., Subramony, M., and Struhl, K. (1988). *Nature (London)* **333,** 635.

Horikami, S. M., and Moyer, S. A. (1982). *Proc. Natl. Acad. Sci. U.S.A.* **79,** 7694.

Hsu, C.-H., and Kingsbury, D. W. (1985). *J. Biol. Chem.* **260,** 8990.

Hudson, L. D., Condra, C., and Lazzarini, R. A. (1986). *J. Gen. Virol.* **67,** 1571.

Hunt, D. M., and Wagner, R. R. (1974). *J. Virol.* **13,** 28.

Hunt, M. D., Smith, E. G., and Buckley, D. W. (1984). *J. Virol.* **52,** 515.

Imblum, R. L., and Wagner, R. R. (1974). *J. Gen. Virol.* **13,** 113.

Iverson, L. E., and Rose, J. K. (1981). *Cell (Cambridge, Mass.)* **23,** 477.

Kakadini, H., and Ptashne, M. (1988). *Cell (Cambridge, Mass.)* **52,** 161.

Kamer, G., and Argos, P. (1984). *Nucleic Acids Res.* **12,** 7269.

Kamps, M. P., Taylor, S. S., and Sefton, B. M. (1984). *Nature (London)* **310,** 589.

Keene, J. D., Thornton, B. J., and Emerson, S. U. (1981). *Proc. Natl. Acad. Sci. U.S.A.* **78,** 6191.

Kingsbury, D. W., Hsu, C.-H., and Morgan, E. M. (1981). *In* "The Replication of Negative Strand Viruses" (D. H. L. Bishop and R. W. Compans, eds.), p. 821. Elsevier/North-Holland, Amsterdam.

Kingsford, L., and Emerson, S. U. (1980). *J. Virol.* **33,** 1097.

Lillie, J. W., and Green, M. R. (1989). *Nature (London)* **338,** 39.

Ma, J., and Ptashne, M. (1987). *Cell (Cambridge, Mass.)* **52,** 179.

124   AMIYA K. BANERJEE AND DHRUBAJYOTI CHATTOPADHYAY

Ma, J., Przibilla, E., Hu, J., Bogorad, L., and Ptashne, M. (1988). *Nature (London)* **334**, 631.
Marnell, L. L., and Summers, D. F. (1984). *J. Biol. Chem.* **259**, 13518.
Massey, D. M., and Lenard, J. (1987). *J. Biol. Chem.* **262**, 8734.
Masters, P. S., and Banerjee, A. K. (1986). *Virology* **154**, 259.
Masters, P. S., and Banerjee, A. K. (1987). *Virology* **157**, 298.
Masters, P. S., and Banerjee, A. K. (1988a). *J. Virol.* **62**, 2651.
Masters, P. S., and Banerjee, A. K. (1988b). *J. Virol.* **62**, 2658.
McClure, M. A., and Perrault, J. (1989). *Virology* **172**, 391.
McGeoch, D. J., and Dolar, A. (1979). *Nucleic Acids Res.* **6**, 3199.
Meir, E., Harmison, G. G., and Schubert, M. (1987). *J. Virol.* **61**, 3133.
Mellon, M. G., and Emerson, S. U. (1978). *J. Virol.* **27**, 560.
Morgan, E. M., and Rakestraw, K. M. (1986). *Virology* **154**, 31.
Moyer, S. A., and Summers, D. F. (1974). *J. Virol.* **13**, 455.
Patton, J. R., Davis, N. L., and Wertz, G. W. (1984). *J. Virol.* **49**, 303.
Paul, P. R., Chattopadhyay, D., and Banerjee, A. K. (1988). *Virology* **166**, 350.
Peluso, R. W., and Moyer, S. A. (1983). *Proc. Natl. Acad. Sci. U.S.A.* **80**, 3198.
Perrault, J., and McLear, P. W. (1984). *J. Virol.* **51**, 635.
Perrault, J., Clinton, G. M., and McClure, M. A. (1983). *Cell (Cambridge, Mass.)* **35**, 175.
Pringle, C. R. (1977). *Comp. Virol.* **9**, 239.
Ptashne, M. (1988). *Nature (London)* **335**, 683.
Rae, B. P., and Elliot, R. M. (1986). *J. Gen. Virol.* **67**, 1351.
Rose, J. K., and Gallione, C. J. (1981). *J. Virol.* **39**, 519.
Sadowski, I., Ma, J., Triezenberg, S., and Ptashne, M. (1988). *Nature (London)* **335**, 563.
Sánchez, A., De, B. P., and Banerjee, A. K. (1985). *J. Gen. Virol.* **66**, 1025.
Schubert, M., Harmison, G. G., and Meier, E. (1984). *J. Virol.* **51**, 505.
Schubert, M., Harmison, G. G., Richardson, C. D., and Meier, E. (1985). *Proc. Natl. Acad. Sci. U.S.A.* **82**, 7984.
Shioda, T., Iwasaki, K., and Shibuta, H. (1986). *Nucleic Acids Res.* **14**, 1545.
Sinacore, M. S., and Lucas-Lenard, T. (1982). *Virology* **121**, 404.
Sokol, F., Clark, H. K., Wiktor, T. J., McFalls, M. L., Bishop, D. H. L., and Obijeski, J. F. (1974). *J. Gen. Virol.* **24**, 433.
Struhl, K. (1987). *Cell (Cambridge, Mass.)* **49**, 295.
Szilagyi, J. F., and Uryvayev, L. (1973). *J. Virol.* **11**, 279.
Talib, S., and Hearst, J. E. (1983). *Nucleic Acids Res.* **11**, 7031.
Testa, D., and Banerjee, A. K. (1979). *J. Biol. Chem.* **254**, 2053.
Testa, D., Chanda, P. K., and Banerjee, A. K. (1980). *Cell (Cambridge, Mass.)* **21**, 267.
Thomas, D., Newcomb, W. W., Brown, J. C., Wall, J. S., Hainfeld, J. F., Trus, B. L., and Steven, A. C. (1985). *J. Virol.* **54**, 598.
Thornton, G. B., De, B. P., and Banerjee, A. K. (1984). *J. Gen. Virol.* **65**, 663.
Toneguzzo, F., and Ghosh, H. P. (1976). *J. Virol.* **17**, 477.
Tordo, N., Poch, O., Ermine, A., Keith, G., and Rougeon, F. (1988). *Virology* **165**, 565.
Triezenberg, S. J., Kingsbury, R. C., and McKnight, S. L. (1988). *Genes Dev.* **2**, 730.
Webster, N., Jin, J. R., Green, S., Hollis, M., and Chambon, P. (1988). *Cell (Cambridge, Mass.)* **52**, 169.
Williams, P. M., Williamson, K. J., Emerson, S. V., and Schubert, M. (1988). *Virology* **164**, 176.
Witt, D. J., and Summers, D. F. (1980). *Virology* **107**, 34.
Yusoff, K., Millar, N. S., Chambers, P., and Emmerson, P. T. (1987). *Nucleic Acids Res.* **15**, 3961.

ADVANCES IN VIRUS RESEARCH, VOL. 38

CONTROL OF EXPRESSION AND CELL TROPISM OF HUMAN IMMUNODEFICIENCY VIRUS TYPE 1

Jerome A. Zack, Salvatore J. Arrigo, and Irvin S. Y. Chen

Division of Hematology–Oncology
Departments of Medicine and Microbiology & Immunology
University of California, Los Angeles School of Medicine
Los Angeles, California 90024

I. Introduction

The acquired immunodeficiency syndrome (AIDS) exhibits a mean clinical latency period of approximately 8 years. The slowly progressive nature of the degeneration of the immune system seen in this disease could be due in part to the rate of expression of the etiological agent, human immunodeficiency virus type 1 (HIV-1). Regulation of HIV-1 expression can be controlled both at the cellular level and by the virus itself. The various factors responsible for both the activation and the down-regulation of HIV-1 expression and the mechanism of action of these factors are the topic of this chapter. In addition, factors responsible for target cell tropism are also discussed, as these can influence virus expression.

II. Cellular Control of HIV-1 Expression

Infection of cells by HIV-1 *in vivo* is facilitated by the presence of the CD4 molecule, the receptor for the virus, on the surface of the

125

target cell. The predominant cell types infected by the virus *in vivo* are the CD4-positive T lymphocyte and the CD4-positive macrophage. These two cell types are both integral members of the human immune system; however, they exhibit vastly different functions and have different mechanisms of activation of these functions. Therefore, factors affecting the cellular control of HIV-1 expression in these cells differ and are discussed separately.

HIV expression in T lymphocytes has been linked to the proliferative state of the cell. Nondividing normal human peripheral blood T lymphocytes express the CD4 molecule and are capable of binding HIV-1; however, they do not produce HIV-1 virions. In contrast, cells stimulated by mitogen produce high levels of HIV-1 particles (McDougal *et al.*, 1985). This increased virus production is preceded by an increased proliferation of the cells. Other mitogenic stimuli [e.g., specific antigen, growth factors such as interleukin 2 and mitogenic particles of human T-cell leukemia viruses types I (HTLV-I) and II (HTLV-II)] likewise cause increased production of HIV-1 by stimulating proliferation of T cells (McDougal *et al.*, 1985; Zagury *et al.*, 1986; Zack *et al.*, 1988). The lack of expression of HIV-1 virions from unstimulated T cells does not appear to be due to an inability of the virus to enter these cells. Although studies using Southern blot technology have been unable to demonstrate HIV-1 DNA in resting lymphocytes exposed to virus (Gowda *et al.*, 1989), our laboratory has been able to demonstrate reverse transcription in resting cells by using the highly sensitive technique of polymerase chain reaction to amplify HIV-1 DNA present (Zack *et al.*, 1990). The DNA in quiescent lymphocytes is transcriptionally inactive, which could be due to at least three reasons. First, the HIV-1 DNA in these cells appears to be unintegrated, and may therefore be incapable of expressing viral mRNAs. Second, this DNA is not full length. Third, there is a paucity of transcription factors present in unstimulated T cells, and these are involved in HIV-1 expression (see below). Stimulation of resting T cells with mitogens subsequent to infection induces the expression of HIV-1 genes in these cells.

HIV-1 production by macrophage cell lines and primary macrophages can likewise be increased by treatment with growth and/or differentiation factors that increase the proliferative state of the cell (e.g., granulocyte–macrophage colony-stimulating factor and macrophage colony-stimulating factor) (Folks *et al.*, 1987; Koyanagi *et al.*, 1988). However, HIV-1 production from primary macrophages can also be increased by factors that activate the function of macrophages in the absence of inducing increased proliferation (e.g., γ-interferon) (Koyanagi *et al.*, 1988).

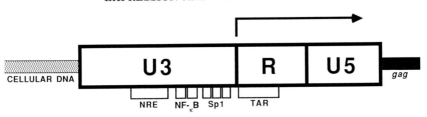

FIG. 1. The functional regions of the HIV-1 proviral long terminal repeat (LTR). The binding sites for transcription factors NF-$_\kappa$B and Sp1 are shown. The negative regulatory element (NRE), which is thought to be important in the down-regulation of virus expression by the *nef* gene product, is indicated. The trans-acting response (TAR) element and the *cis*-acting sequences required for trans-activation by *tat* are also shown. The arrow designates the site of initiation of transcription. U3, The unique 3' sequences of the viral RNAs; R, the directly repeated sequences found at both termini of the RNA molecules; U5, the unique sequences found at the 5' end of the HIV-1 RNAs. U3 and U5 sequences are duplicated during reverse transcription to form the LTR. It is not known whether integration of viral DNA into host DNA is required for the expression of viral genes.

The increased production of HIV-1 from activated T cells is linked to the presence of increased amounts of active transcription factors present in these cells. These transcription factors interact directly with the HIV-1 long terminal repeat (LTR) (Fig. 1). Activation of infected T cells increases the pool of active transcription factors known as NF-$_\kappa$B and HIVEN 86A. NF-$_\kappa$B is a DNA-binding protein originally identified in B cells that produced κ immunoglobulin light chains; it is also found in activated T cells and macrophages. This factor binds to an 11-base pair motif that is repeated twice in the enhancer region of the HIV-1 LTR (Nabel and Baltimore, 1987). HIVEN 86A is a protein that binds to the same region of the LTR (Bohnlein *et al.*, 1989) and might be related to NF-$_\kappa$B. Using LTR indicator gene constructs and gel shift analysis, Nabel and Baltimore (1987) have shown that mutations in the 11-base pair repeated sequences abolish the inducibility of LTR-directed gene expression. In addition to phorbol esters and mitogenic lectins, the peptide hormones, tumor necrosis factor α (TNF-α) (Folks *et al.*, 1989; Lowenthal *et al.*, 1989; Osborn *et al.*, 1989) and interleukin 1 (Osborn *et al.*, 1989), have been shown to increase the production of HIV-1 by acting through NF-$_\kappa$B in T cells. NF-$_\kappa$B could be induced through different pathways by TNF-α versus phorbol esters, because treatment with TNF-α does not stimulate secretion of interleukin 2 by T cells, whereas stimulation by phorbol esters does (Osborn *et al.*, 1989).

NF-$_\kappa$B regulation of HIV-1 gene expression could also be important in cells of the monocyte–macrophage lineage. NF-$_\kappa$B in this lineage appears to be associated with the differentiation state of the cell. When

promonocyte cell lines are induced to differentiate by the phorbol ester
12-O-tetradecanoylphorbol-13-acetate, NF-$_\kappa$B activity is induced.
NF-$_\kappa$B is constitutively expressed in mature monocytoid cells (Griffin *et al.*, 1989). The increased expression of HIV-1 following stimulation of
macrophages by granulocyte–macrophage colony-stimulating factor,
however, does not appear to act through the NF-$_\kappa$B binding region of the
HIV-1 LTR. In contrast, recent results from our laboratory have shown
that the region responsible for granulocyte–macrophage colony-stimulating factor activation in these cells is located slightly upstream and
partially overlaps the NF-$_\kappa$B binding regions (Koyanagi et al., 1990).

The transcription factor Sp1 binds to three guanine-rich regions in
the HIV-1 LTR just upstream from the promoter region (Jones *et al.*,
1986). Numerous other proteins present in cell extracts have been
shown to bind the HIV-1 LTR and are thought to be important in the
regulation of HIV-1 transcription (Garcia *et al.*, 1987). Initial viral
RNA transcription following infection of the cell by HIV-1 must be
dependent on cellular transcription factors, as the virion does not contain viral trans-activator proteins.

III. Stimulation of HIV-1 Expression by Other Viruses

Other viruses unrelated to the HIVs can interact either with HIV
itself or with HIV-infected cells to increase the expression of HIV RNA.
The HIV-1 LTR can be directly activated by herpesvirus-encoded trans-
acting factors, as assayed by LTR-directed expression of indicator genes
(Gendelman *et al.*, 1986; Rando *et al.*, 1987). Herpesvirus-directed induction of HIV-1 expression could be very important *in vivo,* as AIDS
patients are often coinfected with cytomegalovirus, a member of the
herpes family, or with herpes simplex viruses. The stimulation by the
herpes simplex viruses is due to genes in the immediate early region of
the virus (Ostrove *et al.*, 1987).

HTLVs can activate HIV gene expression by two distinct mechanisms. The HTLV tax (trans-activating) protein can trans-activate the
LTR of HIV-1 following deletion of certain sequences in the HIV-1
LTR known as the negative regulatory element (Siekevitz *et al.*, 1987).
The tax protein interacts with the cellular protein known as HIVEN
86A, which can bind to regions of the HIV LTR in the same area as
NF-$_\kappa$B (Bohnlein *et al.*, 1989) and could be the same or a related protein to NF-$_\kappa$B. This type of interaction may lead to increased expression of HIV-1 in cells that are coinfected with HTLV-I.

The second mechanism for HTLV activation of HIV involves direct
mitogenic stimulation of HIV-1-infected cells by both HTLV-I and

HTLV-II, which can activate division of resting human peripheral blood T cells in the absence of infection (Gazzolo and Duc Dodon, 1987; Zack *et al.*, 1988). If these cells are infected with HIV-1, this activation causes a marked increase in the production of HIV-1 (Zack *et al.*, 1988). Mitogenic induction of HIV by HTLV virions does not require coinfection of the same cell, as stimulation of mitogenic receptors on the surface of HIV-infected cells would give the same result; in fact, heat-inactivated or killed virions elicit mitogenic responses from quiescent peripheral blood lymphocytes. This type of stimulation could be provided by free HTLV particles in the serum or by contact of the HIV-1-infected cell with an HTLV-infected cell that displays the mitogenic viral product on its cell surface. Although coinfected individuals have not been shown to have an increased severity of AIDS or a quicker onset to frank AIDS, the number of individuals that are coinfected with HIV and either HTLV-I or HTLV-II is on the rise. High-risk groups for HIV and HTLV infection are overlapping, so that the number of coinfected individuals will continue to increase in the next few years. It remains to be seen, as more patients are studied, whether this coinfection will affect disease progression.

IV. VIRAL CONTROL OF HIV-1 GENE EXPRESSION

HIV-1 has evolved an extremely economic use of its 9.5 kilobases (kb) of coding sequence. It is able to encode four virion core proteins, a protease, reverse transcriptase, integrase, two envelope proteins, two trans-activators, and at least four other proteins (Fig. 2). HIV-1, as a retrovirus, is faced with the complex problem of expressing all of these proteins from a single precursor RNA. The virus must control the expression of each of these RNAs both temporally and quantitatively to ensure the appropriate levels of each protein in a variety of infected cell types. This has been accomplished through a complex pattern of RNA splicing and overlapping translational reading frames involving cis-acting regulatory elements as well as trans-acting viral proteins and response elements (Fig. 2). Several facets of this regulation have been deciphered and are discussed in detail Section IV,A–D. However, it is obvious that further levels of control over expression must exist which have yet to be discovered. The relative ratios of each singly spliced RNA, which are not yet fully characterized, must be controlled by a cis element or a trans factor which determines the relative usage of each RNA splice acceptor. The same argument can be made for the relative ratios of each doubly spliced RNA. This control might be determined solely by the sequence of the splice acceptor site or might

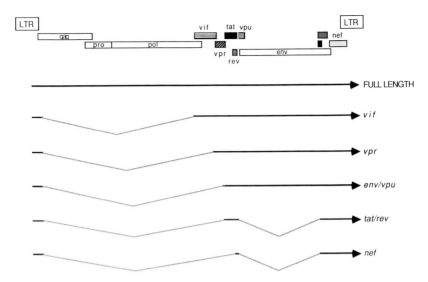

FIG. 2. The HIV-1 genome. Individual coding sequences are represented by boxes. The relative positions of these boxes illustrate open reading frames that overlap. *gag* encodes the viral core proteins, *pro* is the protease gene, *pol* encodes the polymerase of HIV-1 (reverse transcriptase), and *env* encodes the two envelope glycoproteins: gp120 and gp41. cis-acting sequences responsible for the regulation of expression are located in the long terminal repeats (see Fig. 1). RNAs postulated to encode gag, pol, vif, vpr, env, vpu, tat, rev, and nef proteins are shown.

depend on cis-acting elements, proteins, or RNA secondary structure. An additional level of control must be exerted by the virus in determining the ratio of unspliced to singly spliced RNA, perhaps dependent on cis-acting sequences such as the negative regulator of splicing found in Rous sarcoma virus (Arrigo and Beemon, 1988).

Retroviruses, as a class, derive all mRNAs from the same precursor RNA, which is identical to genomic RNA. Thus, they exhibit an unusual phenotype in that all precursor RNAs are not completely spliced to subgenomic messages; the remaining unspliced RNA is transported to the cytoplasm, retains stability in the cytoplasm, and is translated. This is extremely unusual for an RNA in a eukaryotic cell. Eukaryotic mRNAs either splice to completion or do not undergo any splicing whatsoever. Alternate splicing does occur with some eukaryotic mRNAs; however, unspliced RNA does not accumulate in the cytoplasm. Therefore, one can imagine that retroviruses have evolved a mechanism for achieving this partial splicing of RNA to maximize the coding potential of a relatively small piece of RNA.

In addition to the three basic genes found in all retroviruses—*gag,* which encodes the viral core proteins, *pol,* which encodes reverse tran-

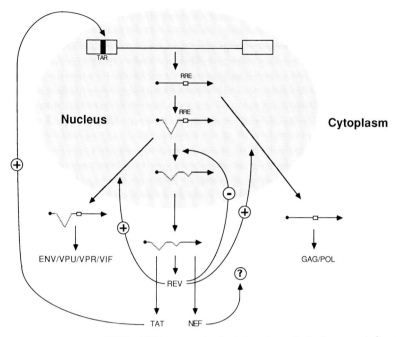

FIG. 3.   Regulation of HIV-1 RNA expression by tat and rev. A single genomic-length RNA can be either singly or doubly spliced or can remain unspliced. These RNAs are subsequently transported to the cytoplasm, where they are translated into the various HIV-1 regulatory and structural proteins. The tat protein exerts a positive effect (+) on the production of all HIV-1 RNAs expressed from the HIV-1 long terminal repeat (LTR). This effect is mediated by the trans-acting response element (TAR), which is located within the R region of the LTR, as well as within the 5′ leader sequence of all known HIV-1 RNAs. The rev protein directs cytoplasmic accumulation (+) of both full-length and singly spliced RNAs containing the *rev*-responsive element (open box). Concomitantly, the level of doubly spliced RNAs is down-regulated (−) by rev. The role of nef in this regulation is unclear.

scriptase and integrase, and *env*, which encodes the envelope glycoproteins—the HIV genome contains other genes important in the viral life cycle (Fig. 2). The functions of these genes are not fully understood; however, certain of these gene products have been implicated in the control of HIV-1 gene expression (Fig. 3). These include the *tat* gene, for trans-activator of transcription; the *rev* gene, or regulator of expression of virion proteins; and the *nef* gene, for negative factor.

A. *tat*

The *tat* gene encodes a 14-kilodalton (kDa) protein whose function is essential for HIV-1 viral replication (Dayton *et al.*, 1986; Fisher *et al.*,

1986). This protein contains a stretch of basic amino acids at its amino terminus, which appears to target the protein to the nucleus in general and, more specifically, to the nucleoli (Hauber *et al.*, 1987; Ruben *et al.*, 1989). Mutations of these basic amino acids, which lead to a loss of nuclear localization, also eliminate the trans-activating potential of tat (Ruben *et al.*, 1989). Nucleolar localization, however, is not sufficient for tat function. The nuclear localization of tat is at least partially expected, since tat is capable of trans-activating the expression of genes linked to the HIV-1 LTR (Arya *et al.*, 1985; Sodroski *et al.*, 1985a). Thus, it is not clear whether the nucleolar localization is due to a specific target or possibly nonspecific binding of negatively charged amino acids to ribosomal RNAs. The expression of numerous genes expressed from the HIV-1 LTR has been greatly augmented by tat in a wide variety of human and nonhuman cell types.

The tat protein is capable of exhibiting this effect in the absence of any other viral proteins through an element in the LTR known as the trans-activation response element (Muesing *et al.*, 1987; Rosen *et al.*, 1985). This element is located between nucleotides 19 and 42 and is thus within the 5′ leader sequence of all known HIV-1 RNAs. This location is within the R region of the LTR, and is situated at both the 5′ and 3′ ends of the proviral DNA. This positioning within both the DNA and the RNA suggests that it might exert trans-activating effects on either DNA or RNA. tat might act transcriptionally to increase the level of initiation or elongation by RNA polymerase II (Hauber *et al.*, 1987; Kao *et al.*, 1987; Peterlin *et al.*, 1986; Rice and Mathews, 1988). tat might also act posttranscriptionally to increase the cytoplasmic levels of RNA or the protein translation of LTR-driven RNAs (Feinberg *et al.*, 1986; Hauber *et al.*, 1987; Rosen *et al.*, 1986; Sodroski *et al.*, 1985b; Wright *et al.*, 1986). Since neither transcriptional nor posttranscriptional mechanisms can individually account for the increase in protein production exhibited by trans-activation with tat, a bimodal mechanism of tat action on both transcription and translation has been postulated (Cullen, 1986). Regardless of the mechanism through which tat exerts this effect, protein accumulation can be increased up to 1000-fold.

Since the *tat* gene itself is expressed under the control of the HIV-1 LTR, a positive feedback loop exists for HIV-1 LTR-driven expression. Once tat is produced, high levels of viral expression occur as a result of increasing tat protein production (Fig. 3). This type of positive feedback is theoretically ideal for a virus which may exist in a latent state, in that a switch from no expression to full expression is accomplished rapidly. Since the HIV-1 LTR is a notoriously poor promoter in most cell types, a provirus might remain latent until an outside stimulus results in the induction of tat and a subsequent cascade of rapid viral replication. In order for HIV-1-infected cells to avoid immune surveillance, an absence of viral antigens on the cell surface during latency, coupled with a rapid induction of viral replication, might be advantageous. One might imagine that tat first acts to increase HIV-1 RNA to high levels and further boosts the production of viral protein from this RNA. High levels of virus are then produced without the prior appearance of viral antigens on the cell surface, thereby eluding detection by the immune system.

B. rev

A second trans-activator of viral expression is encoded as a 19-kDa protein by the rev gene (Knight et al., 1987; Sodroski et al., 1986a). The rev protein localizes to the nucleus and the nucleoli, as does the tat protein (Felber et al., 1989). As with tat, rev seems to be directed by a stretch of basic amino acids at the amino terminus of the protein. Both the tat and rev proteins can be produced by the same doubly spliced 2-kb RNA (Muesing et al., 1987). These two proteins are encoded in overlapping reading frames; the tat initiator methionine is 5' to the rev initiation site. In the absence of rev, decreased levels of the structural proteins produced by unspliced RNAs (e.g., gag and pol) as well as by singly spliced RNAs (e.g., env) are detected (Feinberg et al., 1986; Hadzopoulou-Cladaras et al., 1989; Sodroski et al., 1986a). Indeed, within the context of a viral infection, none of the structural proteins might be produced by a rev-defective virus. The rev protein exerts this effect through the rev-responsive element, which is located within the env gene and is thus present within the 3' end of all of the RNAs encoding viral structural proteins (Arrigo et al., 1989; Felber et al., 1989; Hadzopoulou-Cladaras et al., 1989; Hammarskjold et al., 1989; Malim et al., 1989; Rosen et al., 1988). This element is removed from the doubly spliced RNAs encoding tat, rev, and nef. Thus, these RNAs would not be directly responsive to the effect of rev.

The rev protein appears to affect the nuclear-cytoplasmic distribution of HIV-1 RNAs. The cytoplasmic level of unspliced RNA is decreased in a rev mutant, and the level of tat/rev and nef RNAs are concomitantly increased (Arrigo et al., 1989). The level of env singly spliced RNA is inferred, by experiments in which env is expressed as an unspliced RNA (Emerman et al., 1989; Hammarskjold et al., 1989; Malim et al., 1988, 1989), to behave similarly to unspliced RNA. These groups have demonstrated a decrease in the accumulation of this unspliced "env" RNA in the cytoplasm of transfected cells in the absence of rev. Therefore, rev acts to increase the levels of unspliced and singly spliced RNAs, while decreasing the levels of *tat, rev,* and *nef* RNAs (Fig. 3). The mechanism by which rev accomplishes this regulation is unclear, although rev function does require the *rev*-responsive element located within the *env* sequences.

Other cis-acting elements responsible for directly or indirectly retaining RNAs in the nucleus of infected cells are thought to be present within this region (Emerman *et al,* 1989; Malim *et al.,* 1989; Rosen *et al., 1988).* rev might be necessary for the transport of unspliced and singly spliced RNAs or might interfere with the splicing of these RNAs to doubly spliced RNAs. Interestingly, the rex protein of HTLV-1 is capable of functionally replacing the rev protein of HIV-1 in allowing the appearance of HIV-1 structural proteins (Rimsky *et al.,* 1988). However, the rev protein of HIV-1 has not yet been demonstrated to functionally replace the HTLV-1 rex protein.

The final outcome of *rev* expression is the production of high levels of structural HIV-1 proteins. In the absence of RNAs capable of encoding rev, only RNAs capable of encoding tat, rev, and nef would be produced. A latent virus which is induced to produce tat would initially produce tat, rev, and nef, until the level of rev was high enough to increase the levels of RNAs encoding the structural proteins at the expense of RNA encoding tat, rev, and nef. Since tat and rev are positive and negative regulators of each other's expression, respectively, a balance of expression is accomplished (Fig. 3).

The opposing regulatory effects of these two proteins allow for the possibility of establishment of a latent proviral state. Expression of *rev* early after infection of a cell might lead to down-regulation of *tat* and, hence, a cessation of viral transcription. Without expression of *tat,* the provirus could remain latent until activated by an external stimulus. Since tat/rev can be encoded by the same mRNA, preferential expression of one of these proteins by a cellular or viral factor would lead to latency (i.e., rev) or a rapid induction of viral replication (i.e., tat).

C. nef

The *nef* gene is located at the 3′ end of the virus, just outside the envelope coding sequences, and partially overlaps the 3′ LTR. The protein product of the *nef* gene is a doublet of 25 and 27 kDa. Antibodies to the nef proteins are found in AIDS patients early after the onset of infection (Allan *et al.,* 1985). When frameshift or deletion mutations are introduced into the *nef* gene, the resulting viruses produce approximately fivefold more progeny virions that do wild-type viruses (Luciw *et al.,* 1987). These results suggest that the *nef* gene product down-regulates virus production. The *nef* gene product appears to contain both guanine nucleotide (GTP)-binding activity and GTPase activity, and is myristylated at the amino terminus of the 27-kDa protein (Guy *et al.*, 1987). The protein is associated with the cytoplasmic membrane, most likely due to the presence of the myristic acid residue. These qualities are reminiscent of characteristics of G proteins, molecules thought to be important in intracellular signal transduction.

Recent studies have shown that the *nef* gene product can act as a trans-repressor of HIV-1 gene expression by interacting with a region of the LTR known as the negative regulatory element (Ahmad and Venkatesan, 1988; Niederman *et al.*, 1989). It is thought that this activity might inhibit the trans-activation by tat, and that this activity could be responsible for HIV-1 latency. However, the inhibitory effects of nef in LTR-directed gene expression assays are low, relative to the positive effects of tat. In addition, as mentioned above, mutation of the *nef* gene in infectious virus only slightly increases virus production. The weak inhibition of HIV gene expression and the slight effect on virus production seen by *nef* mutants seem somewhat insignificant for a gene as highly conserved as *nef*. Arguments that the *nef* gene might be involved in HIV-1 latency are based on the above observations. However, other laboratories have been unable to reproduce these observations (Cullen and Greene, 1989; Hammes *et al.*, 1989; Kim *et al.*, 1989). There are probably still unidentified functions for this highly conserved gene. It appears that the appropriate assay system or cell type needed to elucidate the true function of the *nef* gene product has not yet been used.

D. *vif, vpr, and vpu*

In addition to the *tat, rev,* and *nef* genes, there are three other known genes encoded in the HIV-1 genome that are less well characterized, but that might play important roles in the virus life cycle: *vif, vpr,* and *vpu.*

1. *vif*

The *vif* gene (for virion infectivity factor) was previously known as *sor, orf*-A, -Q, or -P'. This gene is found in the 3' half of the virus overlapping the end of the *pol* gene and terminating just before the coding region for the envelope protein. It encodes as a 23-kDa protein (Kan *et al.*, 1986; Lee *et al.*, 1986; Sodroski *et al.*, 1986b) that does not appear to be required for replication of the virus or for cytopathic effects. Viruses that contain deletion mutants of the *vif* gene, however, have one-one hundredth to one-one thousandth the infectivity of wild-type viruses (Fisher *et al.*, 1987; Strebel *et al.*, 1987). The *vif* gene appears to enable cell-free virions to infect cells much more efficiently and also allows these virions to infect more efficiently by cell-to-cell contact *in vitro*.

2. *vpr*

The *vpr* gene (for virion protein R) was previously known as R. Some HIV-infected individuals produce antibodies against a 78-amino acid protein thought to be encoded by *vpr*. The *vpr* open reading frame overlaps the *vif* gene and ends just prior to the first coding exon of the *tat* gene (Wong-Staal *et al.*, 1987). To date, no function has been established for the vpr protein. However, viruses that contain mutations in this reading frame appear to induce cytopathic effects with delayed kinetics when compared with wild-type virus. When low multiplicities of infection were used, the peak of virus production from infected cells was delayed by several days when *vpu* mutants were compared with wild-type virus (Ogawa *et al.*, 1989).

3. *vpu*

The *vpu* gene (for virion protein U) encodes both 15- and 16-kDa proteins immunoreactive with AIDS patients' sera. Although all HIV-1 isolates appear to contain sequences for *vpu* in their genomes, many lack initiating methionine codons or contain a stop codon in the *vpu* open reading frame that would truncate the protein product (Cohen *et al.*, 1988; Strebel *et al.*, 1988). The product of the *vpu* gene appears to be an integral membrane protein that is phosphorylated and is not virion associated (Strebel *et al.*, 1989). Viruses containing frameshift mutations in the *vpu* open reading frame appear to express the same levels of major viral proteins as wild-type virions; however, they produce one-fifth to one-tenth the progeny virions that wild-type viruses do following infection of T lymphocytes. High levels of viral structural proteins accumulate in the cytoplasm of cells infected with *vpu* mutant viruses (Strebel *et al.*, 1988). These mutants also appear to be more cytotoxic and are able to induce the formation of multinucleated giant cells more rapidly than the wild-type virus (Terwilliger *et al.*, 1989). Therefore, the *vpu* gene product appears to function in the role of virion maturation and release. The mechanism for accomplishing this function is not yet known. The increased cytopathicity associated with *vpu* mutant viruses might be due to the intracellular accumulation of viral structural proteins.

V. HIV-1 Tropism

The life cycles of all retroviruses proceed through a number of discrete steps, which include the binding of virions to specific receptors on the surface of cells, penetration of the virus into the cells, the synthesis of viral DNA, the integration of viral DNA into host cell DNA to form the provirus, the transcription of viral mRNA, and the assembly and release of mature virions. Failure of the retrovirus to complete any of these steps results in a lack of progeny production, and the virus would then be considered nontropic for that particular target cell type. Factors affecting HIV target cell tropism are discussed in Section V,A and B.

A. The CD4 Molecule Is the Receptor for HIV

One of the chief clinical hallmarks of AIDS is the profound depletion of CD4-positive T cells seen *in vivo*. This observation first suggested that the CD4 molecule might be the receptor for the virus. Numerous lines of evidence have clearly demonstrated that the CD4 molecule is indeed the major receptor for HIV-1. Monoclonal antibodies against the CD4 molecule can block the infection of T lymphocytes by the virus (Dalgleish *et al.*, 1984; Klatzmann *et al.*, 1984). The 120-kDa envelope glycoprotein of HIV-1 can bind specifically to the CD4 molecule (McDougal *et al.*, 1986). Introduction of the human CD4 gene by transfection into CD4-negative human cells enables HIV to enter these cells (Maddon *et al.*, 1986). In addition, synthetic peptide fragments corresponding to regions of the human CD4 molecule (Lifson *et al.*, 1988), as well as soluble recombinant human CD4 (R. A. Fisher *et al.*, 1988), when placed into the culture media of CD4-positive cells *in vitro*, can efficiently block infection by HIV-1. Cells containing the CD4 molecule *in vivo* are the CD4-positive T lymphocytes and macrophages. The presence of this molecule on the surface of these cells should allow productive infection of HIV-1.

Although most strains of HIV-1 are able to infect primary T lymphocytes, various isolates have been identified that have differing tropisms for macrophages (A. G. Fisher *et al.*, 1988; Gartner *et al.*, 1986; Koyanagi *et al.*, 1987). In addition, certain isolates have differing tropisms for immortalized T-cell lines that express the CD4 molecule (Cann *et al.*, 1988; Cheng-Mayer *et al.*, 1988; Fenyo *et al.*, 1988; Sakai *et al.*, 1988). These isolates may exhibit altered kinetics of replication on these cell lines, and many do not productively infect these cells at all. The ability of HIV-1 isolates to infect a wide variety of cell lines appears to correlate with disease progression in the AIDS patients from whom these viruses were isolated (Cheng-Mayer *et al.*, 1988; Fenyo *et al.*, 1988). This has been attributed to the appearance of more aggressive viruses as the disease progresses.

An alternative interpretation is that, as a result of greater concentrations of AIDS viruses in later stages of disease, there is a greater likelihood of selecting a particular variant in culture from a larger and more heterogeneous "swarm" of HIV-1. Recent studies from our laboratory have shown that, although the normal laboratory isolates of HIV-1 that have been selected for growth in T-cell lines can infect these lines with high efficiency, the majority of fresh clinical isolates that have not been selected to grow in cell lines will not infect these cell lines, although they are fully capable of infecting primary peripheral blood T cells.

Thus, there is some type of selection process that must occur that allows certain HIV-1 isolates to infect cell lines, while others remain nontropic for transformed cells. This inability to infect transformed T-cell lines appears to be a stable genetic trait of these nontropic isolates, in that a cloned primary clinical isolate of HIV-1 (HIV-1$_{JR-CSF}$) (Koyanagi *et al.*, 1987) could not be induced to grow in any cell line tested. Superinduction of these transformed cells by phytohemagglutinin and 12-*O*-tetradecanoylphorbol-14-acetate does not induce production of this isolate. The molecular mechanism for the failure of this type of HIV-1 isolate to productively infect transformed T-cell lines has not yet been fully elucidated; however, investigation of this phenomenon is currently in progress.

Various isolates of HIV-1 have been identified that have differing tropisms for macrophages and monocytoid cell lines (A. G. Fisher *et al.*, 1988; Gartner *et al.*, 1986; Koyanagi *et al.*, 1987; Sakai *et al.*, 1988). The molecular mechanisms responsible for the ability of particular HIV-1 isolates to infect cells of the macrophage–monocytoid lineage are not fully understood. However, Cordonnier *et al.* have recently demonstrated that virions that have a single amino acid change at position 425 in the env protein lose the ability to productively infect a monocytoid cell line (U937) while they retain the ability to infect T-cell lines. This result could be due to differences in the presentation of the CD4 molecule on the surface of the two cell types, or there could be other factors that influence entry into these cells. The authors point out that this region of the glycoprotein 120 molecule might not be responsible for monocyte–macrophage tropism per se, as the wild-type virus isolate is unable to infect primary monocytes, although it is capable of infecting the monocytoid cell line. *In vivo*, HIV-1 has been identified in macrophages in the brain, lungs, lymph nodes, and Langerhans cells in the skin of AIDS patients (Baroni *et al.*, 1986; Chayt *et al.*, 1986; Koenig *et al.*, 1986; Tschachler *et al.*, 1987; Wiley *et al.*, 1986).

Infection of macrophages by HIV-1 is thought to be a very important reservoir for the virus in AIDS patients. Infected macrophages can travel throughout the body, even past the blood–brain barrier into the brain, where it is thought that they could be involved in the neuropathological symptoms seen in many AIDS patients. In addition, infected macrophages *in vitro* do not appear to develop cytopathic effects, as do infected T cells. This suggests that the infected macrophage *in vivo* might persist, harboring the virus for long periods without being killed, allowing for a greater chance of virus spread.

B. CD4-Independent Infection of Cells by HIV-1

Although the CD4 molecule has been demonstrated to be the major receptor for HIV-1, the role of this molecule as receptor in the central nervous system (CNS) is still unclear. Although the major cell type infected in the CNS appears to be of macrophage origin and would bear the CD4 molecule, there have been various reports of infection of neuronal, astrocytic, and endothelial cells in the CNS, which do not bear the CD4 molecule (Gabuzda *et al.*, 1986; Pumarola-Sune *et al.*, 1987; Wiley *et al.*, 1986). In addition, various tumor cell lines derived from the CNS can be infected by HIV-1 *in vitro* (Cheng-Meyer *et al.*, 1987; Chiodi *et al.*, 1987; Dewhurst *et al.*, 1987). Recent studies performed by Harouse *et al.* (1989) have shown that HIV-1 can infect a glioblastoma line, a medulloblastoma line, and primary human fetal neural cells; however, the investigators were unable to detect the CD4 molecule on the surface of the cells or mRNA for CD4. HIV could be rescued from these cell lines either by coculture with CD4-positive cells or by induction, using phorbol esters. In addition, the infection of CNS-derived cells by HIV-1 could not be blocked by the addition of antibodies specific to the CD4 molecule which were shown to block the infection of CD4-positive T cells.

It is not yet clear whether this CD4-independent entry of HIV-1 into cells is mediated by a specific cell surface receptor, by endocytosis, or by direct fusion of the virus with the cell membrane. In addition to cells of CNS origin, other CD4-negative cell types appear to be susceptible to infection by HIV-1. These include cells in the bowel, human rhabdomyosarcoma and osteosarcoma cell lines, and human foreskin fibroblasts (Castro *et al.*, 1989; Tateno *et al.*, 1989). The mechanism for infection of these cell types is not yet clear. However, at least in the case of the osteosarcoma cell line, the entry of HIV-1 does not appear to involve receptor-mediated endocytosis. When these cells were treated with lysosomotropic agents to block virus fusion in coated pits, no effect on infection by HIV-1 was seen (Tateno *et al.*, 1989). There appears to be a CD4-independent mechanism of HIV-1 entry into cells that bear the Fc receptor. Antibodies from some HIV-1-infected individuals or from inoculated animals could actually enhance infection of these cells by HIV-1 (Homsy *et al.*, 1989; Takedo *et al.*, 1988). Cells that bear the Fc receptor (e.g., T lymphocytes and macrophages) also bear CD4; however, soluble CD4 or antibodies against CD4 itself do not block this antibody-dependent enhancement of infection.

In addition, Homsy *et al.* have shown that this enhanced infection of macrophages could be blocked by monoclonal antibodies to the Fc receptor III, but not anti-Fc receptor I or II. Infection of CD4-positive T lymphocytes could not be blocked by this antibody, but could be inhibited by aggregated immunoglobulin G, suggesting that another receptor for immune complexes could be operative in this cell type. The establishment of a CD4-independent mechanism for HIV infection could bode ill for potential treatment regimes that involve either antibodies designed to block the CD4 binding domain of the HIV envelope glycoprotein, gp120, or the use of soluble recombinant CD4 which is intended to competitively inhibit HIV binding to CD4-positive cells.

C. HIV-1 Tropism and Regulation of Expression: What Does It All Mean?

Regulation of HIV-1 expression is extremely complex. The production of viral products and progeny virions is controlled by a myriad of factors. The type of target cell infected, its state of differentiation, and its level of transcription factors are extremely important. Macrophages and resting CD4-positive T cells could encounter the virus at the site of inoculation and subsequently travel to various parts of the body. These cells do not appear to be killed by this infection. When these infected cells are subsequently involved in an immune response to antigen, they become activated and produce progeny virions which will go on to infect fresh target cells. If the originally infected cells are not killed by HIV-1 following activation, the activation level might decrease due to immunoregulatory factors, and this may lower the production of progeny virions, due to decreased amounts of transcription factors. It is possible that these now inactive cells harbor latent HIV. Subsequent rounds of antigenic stimulation might induce further production of progeny virus. This type of modulation of HIV production, due to various rounds of antigenic stimulation, might be partially responsible for the approximate 8- to 10-year clinical latency period seen between the onset of infection with HIV and the development of symptoms.

The decrease in expression of viral products caused by deactivation of antigen-specific cells mentioned above might allow the virus to escape detection by the immune system because of the lack of expression of viral antigens on the surface of the cell. This could prolong the survival of the infected cell, because it would not be cleared by an immune response to HIV antigens. This phenomenon would tend to increase the load of infected cells *in vivo* and might have dire consequences for the infected individual.

Infection of the cells in the CNS, whether of the monocyte–macrophage lineage or of neuronal cells, could be extremely important in AIDS-related dementia. In addition, the CNS, being an immune-privileged site, might be a very important reservoir for HIV *in vivo*. Factors controlling expression of retroviruses in the CNS are not known. An additional complication of CNS infection by HIV-1 is that possible therapeutic agents might have great difficulty in crossing the blood–brain barrier and might be ineffective at eliminating this reservoir *in vivo*.

The complex interactions between the various gene products encoded by the HIV-1 genome are also extremely critical for the control of HIV-1 expression. As mentioned above, HIV-1 contains genes that can either increase or decrease expression of viral products. There is no evidence that a viral gene product can completely shut down HIV expression, thereby leading to latency, although this is a distinct possibility. Alternatively, HIV genes might not be capable of completely shutting off virus expression; however, they could be capable of causing virus expression to decrease significantly, leading to a state of persistence or chronic release of low levels of progeny virus. This type of phenomenon could also contribute to the long clinical latency period in AIDS.

It is important to distinguish between viral and clinical latency. Clinical latency is that extended period between infection of an individual with HIV-1 and the onset of symptoms. Viral latency refers to a state when a cell is infected, but no progeny virus is released. The duration of the clinical latent period in infected individuals might or might not be dependent on viral latency. Clinical latency could evolve independently of viral latency, due to the initial ability of the immune response to suppress spread of the virus by directly killing progeny virus or virally infected cells. In addition, there could be individuals infected with less cytopathic isolates of HIV-1 that eventually develop the ability to kill cells (Cheng-Mayer *et al.*, 1988; Fenyo *et al.*, 1988), therefore leading to clinical symptoms well after infection. Alternatively, viral latency could lead to the long clinical latency period due to the ability of the virus to suppress its own expression. There is some evidence to support this hypothesis *in vivo*.

A novel *in vivo* culture system for HIV-1 has recently been developed. This system involves implanting a human immune system into the SCID (severe combined immunodeficiency) mouse (McCune et al., 1988). Due to the genetic malformation of the immune system in this mouse, it cannot reject transplanted fetal xenogeneic tissue; therefore, if fetal human hematopoietic tissue is implanted in the mouse, a human immune system can develop. The resulting chimeric animal is known as a SCID-hu mouse. This mouse can be infected with HIV-1 if the virus is injected into a human fetal thymic implant (Namikawa et al., 1988). Only human cells are infected by the virus, as murine cells do not have the appropriate receptor. When tissue sections from an infected human thymus derived from SCID-hu mice were subjected to in situ hybridization for HIV-1 mRNA and stained with antibodies to HIV-1 proteins, cells were found that express viral messages, but do not express viral envelope antigens on their surface (Namikawa et al., 1988).

This result indicates that the cells are indeed making a viral message, but are probably not producing viral particles, suggesting that a viral gene product could be responsible for the lack of virion production. This would favor an argument that HIV can regulate its own latency. It is not clear whether it is more advantageous for HIV-1 to follow a productive or latent course of infection; however, it is probable that the ability to become latent could be an advantage in vivo, in that a virus might escape the immune response or might kill the host less quickly, therefore allowing it to persist for a longer period in the host.

The transfer to another host of a cell infected with a latent virus might be accomplished through sexual intercourse. This virus could subsequently be activated and perpetuate replication. However, the virus does not appear to favor a latent infection in vitro. Here, conditions are maintained that encourage cell proliferation, and no exogenous host-derived factors are present to suppress a productive infection. It is conceivable that latent infection could be induced in vivo due to a host-derived influence such as an immune response, the state of proliferation of the cell, or other factors in the microenvironment of the infected cell. These influences could alter expression of a particular viral gene (e.g., nef, tat, or rev), resulting in a decrease in virion production. If the infected cell ceases to replicate and becomes quiescent, the lack of cellular transcription factors could nonspecifically shut down virus expression entirely.

Further studies must be performed to dissect the complex interplay between cellular and viral controlling factors and their interactions that regulate HIV-1 expression. Elucidation of these complex mechanisms could lead to effective clinical intervention for this disease and help to combat the various diseases caused by the other human retroviruses.

VI. Questions to Be Addressed

There are many unanswered questions regarding factors that affect the HIV-1 life cycle and how these factors relate to disease. Among these are:

- What genes are involved in controlling HIV-1 cell tropism?
- What role does cellular tropism of HIV for tumor T-cell lines and macrophages *in vitro* play in the pathogenesis of HIV *in vivo*?
- Does a latent state for HIV exist in which provirus is present, but no virions are produced? If so, under what environmental circumstances does it occur, and what viral genes are involved in bringing about the latent state?
- How can reactivation from a latent state be prevented *in vivo*?
- Under what stimuli can the latent state of HIV be reactivated, and what viral genes are involved?

References

Ahmad, N., and Venkatesan, S. (1988). *Science* **241,**1481–1485.

Allan, J. S., Coligan, J. E., Lee, T.-H., McLane, M. F., Kanki, P. J., Groopman, J. E., and Essex, M. A. (1985). *Science* **230,** 810.

Arrigo, S. J., and Beemon, K. (1988). *Mol. Cell. Biol.* **8,** 4858–4867.

Arrigo, S. J., Weitsman, S., Rosenblatt, J. D., and Chen, I. S. Y. (1989). *J. Virol.* **63,** 4875–4881.

Arya, S. K., Chan, G., Josephs, S. J., and Wong-Staal, F. (1985). *Science* **229,** 69–73.

Baroni, C. D., Pezzella, F., Mirolo, M., Ruco, L. P., and Rosse, G. B. (1986). *Histopathology* **10,** 5.

Bohnlein, E., Siekevitz, M., Ballard, D. W., Lowenthal, J. W., Rimsky, L., Bogerd, H., Hoffman, J., Wano, Y., Franza, B. R., and Greene, W. C. (1989). *J. Virol.* **63,** 1578–1586.

Cann, A. J., Koyanagi, Y., and Chen, I. S. Y. (1988). *In* "Control of Human Retrovirus Gene Expression" (R. Franza, B. Cullen, and F. Wong-Staal, eds.), pp. 127–133. Cold Spring Harbor Lab., Cold Spring Harbor, New York.

Castro, B., Cheng-Mayer, C., Evans, L. A., and Levy, J. A. (1989). *AIDS* **882,** 517–528.

Chayt, K. J., Harper, M. E., Marselle, L. M., Lewin, E. B., Rose, R. M., Oleske, J. M., Epstein, L. G., Wong-Staal, F., and Gallo, R. C. (1986). *JAMA, J. Am. Med. Assoc.* **256,** 2356–2359.

Cheng-Meyer, C., Rutka, J. T., Rosenblum, M. L., McHugh, T., Stites, D. P., and Levy, J. A. (1987). *Proc. Natl. Acad. Sci. U.S.A.* **84,** 3526–3530.

Cheng-Mayer, C., Seto, D., Tateno, M., and Levy, J. A. (1988). *Science* **240,** 80–82.

Chiodi, F., Fuerstenberg, S., Gidlund, M., Asjo, B., and Fenyo, E. M. (1987). *J. Virol.* **61,** 1244–1247.

Cohen, E. A., Terwilliger, E. F., Sodroski, J. G., and Haseltine, W. A. (1988). *Nature (London)* **334,** 532–534.

Cullen, B. R. (1986). *Cell (Cambridge, Mass.)* **46,** 973–982.

Cullen, B. R., and Greene, W. C. (1989). *Cell (Cambridge, Mass.)* **58,** 423–426.

Dalgleish, A. G., Beverley, P. C. L., Clapham, P. R., Crawford, D. H., Greaves, M. F., and Weiss, R. A. (1984). *Nature (London)* 312, 763–767.

Dayton, A. I., Sodroski, J. G., Rosen, C. A., Goh, W. C., and Haseltine, W. A. (1986). *Cell (Cambridge, Mass.)* 44, 941–947.

Dewhurst, S., Sakai, K., Bresser, J., Stevenson, M., Evinger-Hodges, M. J., and Volsky, D. J. (1987). *J. Virol.* 61, 3774–3782.

Emerman, M., Vazeux, R., and Peden, K. (1989). *Cell (Cambridge, Mass.)* 57, 1155–1165.

Feinberg, M. B., Jarrett, R. F., Aldovini, A., Gallo, R. C., and Wong-Staal, F. (1986). *Cell (Cambridge, Mass.)* 46, 807–817.

Felber, B. K., Margarita, H. C., Cladaras, C., Copeland, T., and Pavlakis, G. N. (1989). *Proc. Natl. Acad. Sci. U.S.A.* 86, 1495–1499.

Fenyo, E. M., Morfeldt-Manson, L., Chiodi, F., Lind, B., von Gegerfelt, A., Albert, J., Olausson, E., and Asjo, B. (1988). *J. Virol.* 62, 4414–4419.

Fisher, A. G., Aldovini, A., Debouk, C., Gallo, R. C., and Wong-Staal, F. (1986). *Nature (London)* 320, 367–371.

Fisher, A. G., Ensoli, B., Ivanoff, L., Chamberlain, M., Petteway, S., Ratner, L. Gallo, R. C., and Wong-Staal, F. (1987). *Science* 237, 888–893.

Fisher, A. G., Ensoli, B., Looney, D., Rose, A., Gallo, R. C., Saag, M. S., Shaw, G. M., Hahn, B. H., and Wong-Staal, F. (1988). *Nature (London)* 334, 444–447.

Fisher, R. A., Bertonis, J. M., Meier, W., Johnson, V. A., Costopoulos, D. S., Liu, T., Tizard, R., Walker, B. D., Hirsh, M. S., Schooley, R. T., and Flavell, R. A. (1988). *Nature (London)* 331, 76–78.

Folks, T. M., Justement, J., Kinter, A., Dinarello, C. A., and Fauci, A. S. (1987). *Science* 238, 800–802.

Folks, T. M., Clouse, K. A., Justement, J., Rabson, A., Duh, E., Kehrl, J. H., and Fauci, A. S. (1989). *Proc. Natl. Acad. Sci. U.S.A.* 86, 2365–2368.

Gabuzda, D. H., Ho, D. D., de la Monte, S. M., Hirsh, M. S., Rota, T. R., and Sobel, R. A. (1986). *Ann. Neurol.* 20, 289–295.

Garcia, J. A., Wu, F. K., Mitsuyasu, R. T., and Gaynorm, R. B. (1987). *EMBO J.* 6, 3761–3770.

Gartner, S., Markovits, P., Markovitz, D. M., Kaplan, M. H., Gallo, R. C., and Popovic, M. (1986). *Science* 233, 215–219.

Gazzolo, L., and Duc Dodon, M. (1987). *Nature (London)* 326, 714–717.

Gendelman, H. E., Phelps, W., Feigenbaum, L., Ostrove, J. M., Adachi, A., Howley, P. M., Khoury, G., Ginsberg, H. S., and Martin, M. A. (1986). *Proc. Natl. Acad. Sci. U.S.A.* 83, 9759–9763.

Gowda, S. D., Stein, B. S., Mohagheghpour, N., Benike, C. J., and Engleman, E. G. (1989). *J. Immunol.* 142, 773–780.

Griffin, G. E., Leung, K., Folks, T. M., Kunkel, S., and Nabel, G. J. (1989). *Nature (London)* 339, 70–73.

Guy, B., Kieny, M. P., Riviere, Y., Le Peuch, C., Dott, K., Girard, M., Montagnier, L., and Lecocq, J.-P. (1987). *Nature (London)* 330, 226–229.

Hadzopoulou-Cladaras, M., Felber, B. K., Cladaras, C., Athanassopoulos, A., Tse, A., and Pavlakis, G. N. (1989). *J. Virol.* 63, 1265–1274.

Hammarskjold, M.-L., Heimer, J., Hammarskjold, B., Sangwan, I., Albert, L., and Rekosh, D. (1989). *J. Virol.* 63, 1959–1966.

Hammes, S. R., Dixon, E. P., Malim, M. H., Cullen, B. R., and Greene, W. C. (1989). *Proc. Natl. Acad. Sci. U.S.A.* 86, 9549–9553.

Harouse, J. M., Kunsch, C., Hartle, H. T., Laughlin, M. A., Hoxie, J. A., Wigdahl, B., and Gonzales-Scarano, F. (1989). *J. Virol.* 63, 2527–2533.

Hauber, J., Perkins, A., Heimer, E. P., and Cullen, B. R. (1987). *Proc. Natl. Acad. Sci. U.S.A.* 84, 6364–6368.

Homsy, J., Meyer, M., Tateno, M., Clarkson, S., and Levy, J. A. (1989). *Science* **244,** 1357–1360.

Jones, K. A., Kadonaga, J. T., Luciw, P. A., and Tjian, R. (1986). *Science* **232,** 755–759.

Kan, N. C., Franchini, G., Wong-Staal, F., DuBois, G. C., Robey, W. G., Lautenberger, J. A., and Papas, T. S. (1986). *Science* **231,** 1553–1555.

Kao, S.-Y., Calman, A. F., Luciw, P. A., and Peterlin, B. M. (1987). *Nature (London)* **330,** 489–493.

Kim, S., Ikeuchi, K., Byrn, R., Groopman, J., and Baltimore, D. (1989). *Proc. Natl. Acad. Sci. U.S.A.* **86,** 9544–9548.

Klatzmann, D., Champagne, E., Chamaret, S., Gruest, J., Guetard, D., Hercend, T., Gluckman, J.-C., and Montagnier, L. (1984). *Nature (London)* **312,** 767–768.

Knight, D. M., Flomerfelt, F. A., and Ghrayeb, J. (1987). *Science* **236,** 837–840.

Koenig, S., Gendelman, H. E., Orenstein, J. M., Dal Canto, M. C., Pezeshkpour, G. H., Yungbluth, M., Janotta, F., Aksamit, A., Martin, M. A., and Fauci, A. S. (1986). *Science* **233,** 1089–1093.

Koyanagi, Y., Miles, S., Mitsuyasu, R. T., Merrill, J. E., Vinters, H. V., and Chen, I. S. Y. (1987). *Science* **236,** 819–822.

Koyanagi, Y., O'Brien, W. A., Zhao, J. Q., Golde, D. W., Gasson, J. C., and Chen, I. S. Y. (1988). *Science* **241,** 1673–1675.

Koyanagi, Y., O'Brien, W. A., Yip, M. T., Zhao, J.-Q., Peterlin, B. M., Wachsman, W., and Chen, I. S. Y. (1990). Submitted for publication.

Lee, T.-H., Coligan, J. E., Allan, J. S., McLane, M. F., Groopman, J. E., and Essex, M. (1986). *Science* **231,** 1546–1549.

Lifson, J. D., Hwang, K. M., Nara, P. L., Fraser, B., Progett, M., Dunlop, N. M., and Eiden, L. E. (1988). *Science* **241,** 712–716.

Lowenthal, J. W., Ballard, D. W., Bohnlein, E., and Greene, W. C. (1989). *Proc. Natl. Acad. Sci. U.S.A.* **86,** 2331–2335.

Luciw, P. A., Cheng-Mayer, C., and Levy, J. A. (1987). *Proc. Natl. Acad. Sci. U.S.A.* **84,** 1434–1438.

Maddon, P. J., Dalgleish, A. G., McDougal, J. S., Clapham, P. R., Weiss, R. A., and Axel, R. (1986). *Cell (Cambridge, Mass.)* **47,** 333–348.

Malim, M. H., Hauber, J., Fenrick, R., and Cullen, B. R. (1988). *Nature (London)* **335,** 181–183.

Malim, M. H., Hauber, J., Le, S.-Y., Maizel, J. V., and Cullen, B. R. (1989). *Nature (London)* **338,** 254–257.

McCune, J. M., Namikawa, R., Kaneshima, H., Shultz, L. D., Lieberman, M., and Weissman, I. L. (1988). *Science* **241,** 1632–1639.

McDougal, J. S., Mawle, A., Cort, S. P., Nicholson, J. K. A., Cross, G. D., Scheppler-Campbell, J. A., Hicks, D., and Sligh, J. (1985). *J. Immunol.* **135,** 3151–3162.

McDougal, J. S., Kennedy, M. S., Sligh, J. M., Cort, S. P., Mawle, A., and Nicholson, J. K. A. (1986). *Science* **231,** 382–385.

Muesing, M. A., Smith, D. H., and Capon, D. J. (1987). *Cell (Cambridge, Mass.)* **48,** 691–701.

Nabel, G., and Baltimore, D. (1987). *Nature (London)* **326,** 711–714.

Namikawa, R., Kaneshima, H., Lieberman, M., Weissman, I. L., and McCune, J. M. (1988). *Science* **242,** 1684–1686.

Niederman, T. M. J., Thielan, B. J., and Ratner, L. (1989). *Proc. Natl. Acad. Sci. U.S.A.* **86,** 1128–1132.

Ogawa, K., Shibata, R., Kiyomasu, T., Higuchi, I., Kishida, Y., Ishimoto, A., and Adachi, A. (1989). *J. Virol.* **63,** 4110–4114.

Osborn, L., Kunkel, S., and Nabel, G. J. (1989). *Proc. Natl. Acad. Sci. U.S.A.* **86,** 2336–2340.

Ostrove, J. M., Leonard, J., Weck, K. E., Rabson, A. B., and Gendelman, H. E. (1987). *J. Virol.* **61**, 3726–3732.

Peterlin, M., Luciw, P., Barr, P., and Walker, M. (1986). *Proc. Natl. Acad. Sci. U.S.A.* **83**, 9734–9738.

Pumarola-Sune, T., Navia, B. A., Cordon-Cardo, C., Cho, E.-S., and Price, R. W. (1987). *Ann. Neurol.* **21**, 490–496.

Rando, R. F., Pellett, P. E., Luciw, P. A., Bohan, C. A., and Srinivasan, A. (1987). *Oncogene* **1**, 13–18.

Rice, A. P., and Mathews, M. B. (1988). *Nature (London)* **322**, 551–553.

Rimsky, L., Hauber, J., Dukovich, M., Malim, M., Langlois, A., Cullen, B., and Greene, W. (1988). *Nature (London)* **335**, 738–740.

Rosen, C. A., Sodroski, J. G., and Haseltine, W. A. (1985). *Cell (Cambridge, Mass.)* **41**, 813–823.

Rosen, C. A., Sodroski, J. G., Goh, W. C., Dayton, A., Lippke, J., and Haseltine, W. A. (1986). *Nature (London)* **319**, 555–559.

Rosen, C. A., Terwilliger, E. F., Dayton, A. I., Sodroski, J. G., and Haseltine, W. A. (1988). *Proc. Natl. Acad. Sci. U.S.A.* **85**, 2071–2075.

Ruben, S., Perkins, A., Purcell, R., Joung, K., Sia, R., Burghoff, R., Haseltine, W. A., and Rosen, C. A. (1989). *J. Virol.* **63**, 1–8.

Sakai, K., Dewhurst, S., Ma, X., and Volsky, D. J. (1988). *J. Virol.* **62**, 4078–4085.

Siekevitz, M., Josephs, S. F., Dukovich, M., Peffer, N., Wong-Staal, F., and Greene, W. C. (1987). *Science* **238**, 1575–1578.

Sodroski, J., Patarca, R., Rosen, C., Wong-Staal, F., and Haseltine, W. (1985a). *Science* **229**, 74–77.

Sodroski, J., Rosen, C., Wong-Staal, F., Salahuddin, S. Z., Popovic, M., Arya, S., Gallo, R. C., and Haseltine, W. A. (1985b). *Science* **227**, 171–173.

Sodroski, J., Goh, W. C., Rosen, C., Dayton, A., Terwilliger, E, and Haseltine, W. (1986a). *Nature (London)* **321**, 412–417.

Sodroski, J., Goh, W. C., Rosen, C., Tartar, A., Portetelle, D., Burny, A., and Haseltine, W. (1986b). *Science* **231**, 1549–1553.

Strebel, K., Daugherty, D., Clouse, K., Cohen, D., Folks, T., and Martin, M. A. (1987). *Nature (London)* **328**, 728–731.

Strebel, K., Klimkait, T., and Martin, M. A. (1988). *Science* **241**, 1221–1223.

Strebel, K., Klimkait, T., Maldarelli, F., and Martin, M. A. (1989). *J. Virol.* **63**, 3784–3791.

Takeda, A., Tuazon, C. U., and Ennis, F. A. (1988). *Science* **242**, 580–583.

Tateno, M., Gonzalez-Scarano, F., and Levy, J. A. (1989). *Proc. Natl. Acad. Sci. U.S.A.* **86**, 4287–4290.

Terwilliger, E. F., Cohen, E. A., Lu, Y., Sodroski, J. G., and Haseltine, W. A. (1989). *Proc. Natl. Acad. Sci. U.S.A.* **86**, 5163–5167.

Tschachler, E., Groh, V., Popovic, M., Mann, D. L., Konrad, K., Safai, B., Eron, L., diMarzo Veronese, F., Wolff, K., and Stingl, G. (1987). *J. Invest. Dermatol.* **88**, 233–237.

Wiley, C. A., Schrier, R. D., Nelson, J. A., Lampert, P. W., and Oldstone, M. B. A. (1986). *Proc. Natl. Acad. Sci. U.S.A.* **83**, 7089–7093.

Wong-Staal, F., Chanda, P. K., and Ghrayeb, J. (1987). *AIDS Res. Hum. Retroviruses* **3**, 33–39.

Wright, C., Felber, B., Paskalis, H., and Pavlakis, G. (1986). *Science* **234**, 988–992.

Zack, J. A., Cann, A. J., Lugo, J. P., and Chen, I. S. Y. (1988). *Science* **240**, 1026–1029.

Zack, J. A., Arrigo, S. J., Weitsman, S. R., Go, A. S., Haislip, A., and Chen, I. S. Y. (1990). *Cell (Cambridge, MA)* **61**, in press.

Zagury, D., Bernard, J., Leonard, R., Cheynier, R., Feldman, M., Sarin, P. S., and Gallo, R. C. (1986). *Science* **231**, 850–853.

ADVANCES IN VIRUS RESEARCH, VOL. 38

INTERFERON-INDUCED PROTEINS AND THE ANTIVIRAL STATE

Peter Staeheli

Institute for Immunology and Virology
University of Zürich
Zürich CH-8028, Switzerland

I. INTRODUCTION

Cells infected with virus are stimulated to produce and secrete interferons (IFNs). IFNs, in turn, act on other cells by binding to specific cell surface receptors, thereby inducing a complex pattern of physiological changes, including establishment of an antiviral state (for a review see Pestka *et al.*, 1987; De Maeyer and De Maeyer-Guignard, 1988). All activities of IFNs are thought to be mediated by IFN-regulated cellular proteins. Some new proteins are synthesized after treatment of cells with IFNs and the levels of many other proteins are increased. A large number of cDNAs corresponding to IFN-induced mRNAs have been isolated and their sequences have been determined. An exciting field of IFN research is identification of the biological activities of IFN-induced proteins and establishment of their roles in the antiviral state.

The initial events following the binding of IFN molecules to specific cell surface receptors are poorly understood. Biochemical analysis suggests that IFN-α and -β molecules make use of the same receptor, whereas IFN-γ binds to a distinct receptor (Aguet and Morgensen, 1983). The primary structures of the human and mouse IFN-γ receptors have been elucidated (Aguet *et al.*, 1988; Hemmi *et al.*, 1989),

147

whereas that of the IFN-α/β receptor is still unknown. The intracellular signaling events induced by the interactions of the IFN receptors with their ligands are completely unknown. It was recently shown that treatment of human cells with IFN-α rapidly activated a latent DNA-binding protein, designated E factor, apparently present in the cytoplasm of untreated cells (Dale *et al.*, 1989). Once activated, E factor is believed to rapidly migrate into the nucleus and to stimulate, in concert with additional factors, the transcription of IFN-responsive genes. Activated E factor was shown to bind to a highly conserved 14-base pair DNA sequence. This so-called IFN-responsive element is present in the 5'-regulatory regions of most IFN-responsive human and mouse genes studied (Porter *et al.*, 1988; Reich *et al.*, 1987; Cohen *et al.*, 1988; Hug *et al.*, 1988; Rutherford *et al.*, 1988; Shirayoshi *et al.*, 1988; Reid *et al.*, 1989; Wathelet *et al.*, 1987, 1988).

Although the physiological functions of the majority of the IFN-induced proteins remain obscure, we are beginning to understand the roles of a few of them in the antiviral state. In Section II of this chapter, I review recent work on the best-characterized IFN-induced proteins. The focus is on IFN-induced proteins with assigned functions, namely, protein kinase P1, 2-5A synthetase, mouse and human Mx proteins, indolamine 2,3-dioxygenase, and a few other IFN-induced proteins. The Mx proteins are discussed in greatest detail, because the Mx system is under investigation in our laboratory.

The conclusion of Section II of this chapter is that individual IFN-induced proteins have distinct biochemical activities that lead to discrete physiological changes in IFN-treated cells. For example, the IFN-induced protein kinase P1 can inhibit the multiplication of many different viruses by reducing the translation rates of viral mRNAs. In contrast, the antiviral activities of 2-5A synthetases and Mx proteins show a high degree of specificity for particular classes of viruses. A large body of data supports the view that activation of the 2-5A synthetase pathway could induce an antiviral state with high specificity toward picornaviruses. The 2-5A synthetase pathway might also be responsible, in part, for the IFN-mediated inhibition of vaccinia virus. The IFN-dependent resistance of mice to influenza viruses is due to Mx protein. Only mice able to synthesize Mx protein, but not mice with defective *Mx* genes, show a high degree of resistance to experimental infections with influenza viruses. Genetically engineered cells that synthesized Mx protein in a constitutive manner acquired selective resistance to influenza viruses, directly demonstrating the antiinfluenza virus potential of the mouse Mx protein. One of the IFN-induced human proteins with homology to the mouse Mx protein not only inhibited influenza viruses when expressed in transfected mouse cells, but also interfered with the multiplication of vesicular stomatitis virus (VSV), a rhabdovirus. Some other IFN-induced proteins have interesting biochemical properties, but their physiological functions—in particular, their roles in the IFN-induced antiviral state—have not been resolved. The IFN-induced enzyme indolamine 2,3-dioxygenase might not contribute to the antiviral state. Rather, its physiological role could be the inhibition of intracellular parasites.

A wide range of different RNA and DNA viruses are sensitive to the antiviral actions of IFNs. Since the multiplication strategies of animal viruses are extremely diverse, it seems logical that IFNs inhibit different viruses by more than one mechanism. Many laboratories are presently engaged in precisely defining the IFN-sensitive multiplication steps of different viruses. In Section III of this chapter, I discuss recent work on the molecular mechanisms of IFN action toward many DNA and RNA viruses. This chapter should complement and update a comprehensive article on this subject by Samuel (1988). The conclusion is that IFN-induced viral growth restrictions are indeed extremely diverse. For example, in the case of the simian virus 40, IFN appears to block an early step of the virus replication cycle, probably the uncoating of the virions. In many instances, early transcription of the viral genomes was identified as the main target of IFN action. In other cases the stability of viral mRNAs was reduced in IFN-treated cells, or the translation of viral mRNAs was impaired. Finally, in the case of cells chronically infected with retroviruses, IFN seems to interfere with the maturation of viral particles. In some systems IFN might inhibit more than one viral multiplication step. It is conceivable that inhibition at multiple stages of the virus replication cycle is most effective, because the resulting effects might not simply be additive, but synergistic.

II. INTERFERON-INDUCED PROTEINS AND THEIR RECOGNIZED FUNCTIONS

The best-characterized IFN-induced proteins of human and mouse cells are listed in Tables I and II, respectively. Some of these proteins are induced by all types of IFNs, whereas others are induced preferentially either by IFN-α and IFN-β or by IFN-γ. Among the IFN-induced proteins we find some well-known proteins (e.g., β_2-microglobulin or the major histocompatibility antigens), but most IFN-induced proteins are novel.

Tables I and II are most likely incomplete and biased in favor of abundant gene products found in cell lines commonly used by cell

TABLE I

IFN-Induced Proteins and mRNAs in Human Cells

Designation	Alternative name	Protein size (kDa)	Enzymatic activity	Biological functions recognized	Location in cell	Induction by IFNs	Chromosomal location	cDNA cloned	Reference
2-5A synthetase	—	100	Synthetase	None	Cytoplasm	α, β, γ	11(?)	No	Benech et al. (1985a,b)
	—	69	Synthetase	None	Cytoplasm or nucleus	α, β, γ	11(?)	No	Williams et al. (1986)
	cDNA E18	46	Synthetase	None	Cytoplasm	α, β, γ	12	Yes	Hovanessian et al. (1987)
	cDNA E16	40	Synthetase	Picornavirus inhibition	Cytoplasm	α, β, γ	12	Yes	Chebath et al. (1987b)
P1 kinase	P68, DAI, dsI, P1/eIF-2α kinase	68	Kinase	Preferential inhibition of viral mRNA translation	Cytoplasm	α, β, γ	?	No	Kitajewski et al. (1986), Galabru and Hovanessian (1987)
MxA	p78, IFI-78K	76	?	Influenza virus and VSV inhibition	Cytoplasm	α, β >> γ	21	Yes	Staeheli and Haller (1985), Horisberger et al. (1988), Pavlovic et al. (1990)
MxB	—	73	?	None	Cytoplasm	α, β >>> γ	21	Yes	Aebi et al. (1989)
GBP-1	67K GBP	67	Guanylate binding	None	Cytoplasm	α, β, γ	?	Yes	Cheng et al. (1983, 1985), Cheng et al. (1986b)
GBP-2	—	—	—	None	?	α, β, γ	?	Yes	Cheng (1990)
ISG56	56-kDa protein, pIF-2, mRNA 561, IFI-56K, C56	56	?	None	Cytoplasm	α, β >>> γ	10	Yes	Chebath et al. (1983), Larner et al. (1984), Wathelet et al. (1986, 1987), Cheng et al. (1986c), Kusari et al. (1987)

ISG54	pIF-1	—	—	None	—	$\alpha, \beta >> \gamma$	?	Yes	Levy et al. (1986)
6-26	Thymosin β4	5	—	None	Secreted	α	?	Yes	McMahon et al. (1986)
6-16	10Q	12	?	None	?	$\alpha, \beta >> \gamma$	?	Yes	Friedman et al. (1984)
1-8 Family	cDNA 1-8	—	—	—	—	α, β, γ	?	Yes	Friedman et al. (1984), Reid et al. (1989)
	cDNA 9-27	—	—	—	—	α, β, γ	?	Yes	
ISG15	15-kDa protein, ubiquitin cross-reactive protein	15	?	None	Cytoplasm	$\alpha, \beta >> \gamma$	?	Yes	Reich et al. (1987), Haas et al. (1987), Knight et al. (1988)
17-kDa protein	20-kDa protein	17	?	Cessation of cell growth	Surface	α, β	?	No	Knight et al. (1985), Hillman et al. (1987)
IP-10	Platelet factor 4-related protein	7	—	Mediation of inflammatory response	Secreted	$\gamma >> \alpha, \beta$	4	Yes	Luster et al. (1985), Luster et al. (1987), Kaplan et al. (1987)
IP-30	—	25	?	None	Lysosomal	$\gamma >> \alpha, \beta$	?	Yes	Luster et al. (1988)
Indolamine 2,3-dioxygenase	—	42	Tryptophan degradation	Inhibition of *Toxoplasma* and *Chlamydia*	?	$\gamma >> \alpha, \beta$	?	No	Pfefferkorn (1984), Rubin et al. (1988)
Complement factor B	—	—	—	Via immune system	Secreted	α, β, γ	6	Yes	Strunk et al. (1985)
MHC[a] class I antigens	—	—	—	Via immune system	Surface	α, β, γ	6	Yes	Rosa et al. (1986)
β2-Microglobulin	—	—	—	Via immune system	Surface	α, β, γ	15	Yes	Wallach et al. (1982)
MHC[a] class II antigens	—	—	—	Via immune system	Surface	$\gamma >> \alpha, \beta$	6	Yes	Rosa et al. (1986)

[a] MHC, major histocompatibility complex.

TABLE II
IFN-Induced Proteins and mRNAs in Murine Cells

Designation	Alternative name	Protein size (kDa)	Enzymatic activity	Biological functions recognized	Location in cell	Induction by IFNs	Chromosomal location	cDNA cloned	Reference
2-5A synthetase	—	100	Synthetase	None	Cytoplasm	α, β, γ	?	No	Lengyel (1982)
	—	40	Synthetase	None	Nuclear	α, β, γ	?	No	St. Laurent et al. (1983)
	Human E18-related	42	Synthetase	None	?	α, β, γ	?	Yes	Ichii et al. (1986), Flenniken et al. (1988)
P1 kinase	P1/eIF-2α kinase	67	Kinase	Preferential inhibition of viral mRNA translation	—	α, β, γ	?	No	Berry et al. (1985)
Mx1	Mx protein	72	?	Selective inhibition of influenza viruses	Nuclear	α, β >> γ	16	Yes	Horisberger et al. (1983), Staeheli et al. (1986b,c), Reeves et al. (1988)
Mx2	—	—	—	None	?	α, β >> γ	16	Yes	Staeheli and Sutcliffe (1988)
GBP-1	GBP-65K	65	Guanylate binding	None	?	α, β, γ	3	Yes	Staeheli et al. (1984b), Prochazka et al. (1985)
202	—	56	?	—	?	α, β	1	Yes	Samanta et al. (1984), Engel et al. (1985)
203 Family	—	—	—	—	—	α, β	1	Yes	Engel et al. (1988)
204	202 Related	72	?	None	?	α, β	1	Yes	Opdenakker et al. (1989)
1–8	Human 1–8 related	—	—	—	—	α, β	?	Yes	Flenniken et al. (1988)
IRF-1	—	36	DNA binding	IFN gene regulation	Nuclear	α, β	?	Yes	Miyamoto et al. (1988)
IRF-2	—	38	DNA binding	IFN gene regulation	Nuclear	α, β	?	Yes	Harada et al. (1989)
MHCa class I	—	—	—	Via immune system	Surface	α, β, γ	17	Yes	Rosa et al. (1986)
MHCa class II	—	—	—	Via immune system	Surface	γ >> α, β	17	Yes	Rosa et al. (1986)

aMHC, Major histocompatibility complex.

biologists. Presumably, many additional IFN-induced mRNAs and proteins occurring at low abundance have not yet been identified. Also, if some IFN-induced proteins were to exist exclusively in some highly specialized cell types, they almost certainly have escaped identification. It is unknown which of the 12 abundant IFN-α-induced proteins and which of the 28 abundant IFN-γ-induced proteins of human fibroblasts that were identified on two-dimensional gels (Beresini *et al.*, 1988) are included. Not included in Tables I and II are proteins whose syntheses are only marginally induced in IFN-treated cells, such as metallothionein (Friedman *et al.*, 1984), tumor necrosis factor receptor (Ruggiero *et al.*, 1986), intercellular adhesion molecule 1 (Dustin *et al.*, 1986), interleukin-2 receptor (Holter *et al.*, 1986), and fibronectin (Cofano *et al.*, 1984).

The primary structures of most IFN-induced proteins shown in Table I are known. The most notable exceptions are the P1 kinase and the large 2-5A synthetases; cDNA clones corresponding to these proteins have not been described. In some cases the sequence information is incomplete, because only partial cDNA clones were obtained. IFN-induced proteins have heterogeneous structures, and no common sequence motifs have been recognized. IFN-induced proteins are also heterogeneous with respect to intracellular localization. A mouse and a rat Mx protein and some forms of 2-5A synthetase accumulate in the nucleus of IFN-treated cells, whereas other IFN-induced proteins are located in the cytoplasm. Still other IFN-induced proteins are secreted or inserted into cell membranes. Some IFN-induced proteins constitute small protein families. These might be generated from alternatively spliced mRNAs, as in the case of the 40- and 46-kDa forms of human 2-5A synthetase, or might represent the products of closely related but distinct genes, as in the cases of the *Mx, GBP,* and *1–8* genes or in the case of the mouse *202* and *203* gene complexes. Some of the genes encoding IFN-induced proteins have been mapped genetically. A large number of chromosomes were found to carry IFN-responsive genes; only closely related genes occurred as gene clusters.

Tables I and II show that the biochemical properties of most IFN-induced proteins are still poorly characterized. Only three of these proteins have recognized enzymatic activities: 2-5A synthetase can synthesize 2′-5′-linked oligomers of adenosine, P1 kinase can phosphorylate itself and the α subunit of eukaryotic initiation factor 2 (eIF-2), and indolamine 2,3-dioxygenase can degrade tryptophan. Some IFN-induced proteins, designated GBPs, can bind to guanine nucleotides, whereas double-stranded (ds) RNA-binding activity was demonstrated for P1 kinase and 2-5A synthetase. A few IFN-induced proteins show sequence homology to proteins with known functions, suggesting similar biochemical properties. The IFN-induced protein ISG15 is homologous to ubiquitin, a protein implicated in the regulation of protein turnover. IFN-γ-induced IP-10 is related to platelet factor 4 and β-thromboglobulin, two chemotactic proteins released by platelets on degranulation that play a role in inflammation and wound healing. Finally, Mx proteins are related to Vps1, a yeast protein that plays a role in the complex process of intracellular protein sorting.

Tables I and II illustrate that the physiological functions of the majority of the IFN-induced proteins have not been recognized. A few of these proteins were shown to have intrinsic antiviral properties: namely, 2-5A synthetase, P1 kinase, and certain forms of Mx proteins. Indolamine 2,3-dioxygenase appears to be responsible for the IFN-γ-induced inhibition of intracellular parasites. A poorly characterized IFN-induced 17-kDa protein seems to cause cessation of cell growth. Finally, the increased expression in IFN-treated cells of β_2-microglobulin, histocompatibility antigens, and certain complement factors can cause a stimulation of the host immune responses against viruses, parasites, and malignant cells.

In the following section IFN-induced proteins with recognized physiological or biochemical functions are discussed in more detail.

A. 2-5A Synthetase

A few of the IFN-induced proteins are enzymes exhibiting 2-5A synthetase activity. These enzymes catalyze the conversion of ATP into 2′-5′A oligomers. The activity of 2-5A synthetases is dependent on dsRNA. Although 2-5A synthetases accumulate in IFN-treated cells, these enzymes remain inactive until activation by dsRNA. dsRNA is usually not present in eukaryotic cells, but is produced in significant quantities in cells infected with certain viruses. Thus, 2′-5′A oligomers, which activate a latent ribonuclease, designated RNase L, are believed to be present exclusively in virus-infected cells. [For a comprehensive review of the biochemical properties of 2-5A synthetases and of the 2′-5′A oligonucleotide-dependent RNase L pathway, refer to recent articles by Pestka et al. (1987) and De Maeyer and De Maeyer-Guignard (1988).] It is worth mentioning that 2′-5′A oligomers, which are required for the activation of RNase L, have low metabolic stability, thus permitting tight control of the RNA degradation process. A localized activation of the 2-5A synthetase/RNase L pathway was proposed (Baglioni and Nilsen, 1983). In infected cells, 2-5A synthetase is believed to recognize and bind to double-stranded replicative complexes of RNA viruses. 2′-5′A oligomers would thus be synthesized at the site of viral RNA synthesis. If so, RNase L would be activated in a localized way, and viral RNA could be preferentially cleaved.

cDNA clones corresponding to two alternatively spliced mRNAs originating from a single human gene (Merlin *et al.*, 1983; Benech *et al.*, 1985a,b; Saunders *et al.*, 1985) located on chromosome 12 (Williams *et al.*, 1986) have been cloned and sequenced. They encode proteins of 40 and 46 kDa, respectively, both of which accumulate in the cytoplasm of IFN-treated cells (Chebath *et al.*, 1987b). Both forms of 2-5A synthetase were produced in *Escherichia coli*, and it was demonstrated that the 40- and 46-kDa proteins both exhibit dsRNA-dependent 2-5A synthetase activity (Mory *et al.*, 1989). Chebath *et al.* (1987a) provided convincing proof that the 40-kDa form of the human 2-5A synthetase has antiviral activity. When the corresponding cDNA was expressed in Chinese hamster ovary cells under the control of a constitutive promoter, the multiplication of Mengo virus was strongly inhibited, whereas the multiplication of unrelated viruses (e.g., VSV and herpes simplex virus) was not affected. Similarly, expression of the 40-kDa form of 2-5A synthetase in human and mouse cells conferred resistance to encephalomyocarditis virus (EMC), but not to VSV or herpes simplex virus (Rysiecki *et al.*, 1989; Coccia *et al.*, 1988), suggesting a beneficial role of this protein in IFN-mediated cellular resistance to picornaviruses. It is unknown whether the 46-kDa form of 2-5A synthetase can interfere with picornaviruses replication.

It was proposed that the 2-5A synthetase/RNase L system might play a role in the cellular defense against vaccinia virus (see Section III). But the question of whether transfected cells expressing 2-5A synthetase in a constitutive manner might also have acquired resistance to vaccinia virus has not been addressed.

Large IFN-induced 2-5A synthetases were identified with the help of specific monoclonal antibodies (Hovanessian *et al.*, 1987). In human cells these induced proteins are 69 and 100 kDa. The 100-kDa form of 2-5A synthetase was found to be tightly associated with microsomes, whereas the 69-kDa form was present in both the cytoplasm and the nucleus of IFN-treated cells (Hovanessian *et al.*, 1988b). Since the mRNAs coding for these proteins have not been cloned, the relationship of the different 2-5A synthetases remains unclear. Antibodies raised to a peptide common to the 40- and 46-kDa forms of 2-5A synthetase also detected the 69- and 100-kDa forms of the enzyme (Chebath *et al.*, 1987a), indicating the presence of at least one common structural motif in the different proteins. The overall homology between these proteins might not be very high, however, since DNA probes derived from the 40-kDa form of 2-5A synthetase failed to unambiguously identify the putative mRNAs encoding the larger forms of 2-5A synthetase (Benech *et al.*, 1985a). These findings indicate that the large 2-5A synthetases are encoded by one or more genes distinct from the well-characterized 2-5A synthetase gene on chromosome 12. Enhanced 2-5A synthetase activity was detected in mouse or human hybrid cell lines harboring human chromosome 11, but not chromosome 12 (Shulman *et al.*, 1984). It will be interesting to learn whether the genes, which code for the large 2-5A synthetases, indeed map to human chromosome 11.

It is unclear whether the different forms of 2-5A synthetase have distinct physiological functions. It was noted that the activation requirements of the large 2-5A synthetases are different from the small ones, in that less dsRNA was required for activation (Ilson *et al.*, 1986).

The 2-5A synthetase system of the mouse is less well characterized. As in human cells, both large and small forms of the enzyme were detected (St. Laurent *et al.*, 1983). A mouse cDNA related to the human 40- or 46-kDa 2-5A synthetase gene has been identified (Ichii *et al.*, 1986; Flenniken *et al.*, 1988). The sequences of this 42-kDa form of mouse 2-5A synthetase and the human 46-kDa enzyme are about 70% identical. It is not clear whether the cloned mouse 2-5A synthetase represents the previously described nuclear 40-kDa form of the 2-5A synthetase (St. Laurent *et al.*, 1983).

B. P1 Kinase

A second IFN-induced dsRNA-dependent enzyme is P1 kinase. The P1 kinase is a single polypeptide of 68 kDa in human cells and 65 kDa in mouse cells. The human and mouse P1 kinases have been purified, and monoclonal antibodies were produced (Berry *et al.*, 1985; Penn and Williams, 1985; Laurent *et al.*, 1985; Kostura and Mathews, 1989). cDNAs coding for polypeptides recognized by monoclonal antibodies to P1 kinase have recently been identified (Hovanessian *et al.*, 1988a), but the deduced structure of the human P1 kinase has not been published.

Significant constitutive levels of P1 kinase are present in noninduced cells, and these levels increase severalfold after exposing the cells to IFN-α, -β, or -γ. The activity of P1 kinase is highly dependent on activating factors, the best known of which is dsRNA (Galabru and Hovanessian, 1987). P1 kinase can also be activated by polyanions, such as heparin, dextran sulfate, or poly(L-glutamine) (Hovanessian and Galabru, 1987). The human P1 kinase appears to possess high- and low-affinity binding sites for dsRNA; the high-affinity binding site might be responsible for P1 kinase activation (Galabru *et al.*, 1989). Once activated by dsRNA, P1 kinase phosphorylates itself. The phosphorylated P1 kinase can then catalyze the phosphorylation of certain exogenous substrates. *In vivo*, the P1 kinase seems to selectively phosphorylate itself and the α subunit of the protein synthesis initiation factor elF-2. Phosphorylation of elF-2 on its α subunit leads to the inactivation of this factor (Jagus *et al.*, 1981), and, as a consequence, to inhibition of protein synthesis.

Although P1 kinase is synthesized at increased rates in IFN-treated cells, the enzyme remains inactive until activation by dsRNA or other factors. A localized activation of the P1 kinase by dsRNA is conceivable and might be similar to the scenario discussed above for a localized 2-5A synthetase activation. P1 kinase might bind to dsRNAs that accumulate in the cell compartments of active virus replication, elF-2α might get phosphorylated at these sites, and the translation of viral mRNAs might be blocked efficiently without severely affecting the overall host cell protein synthesis.

Several viruses have developed unique strategies to overcome the host cell restrictions associated with P1 kinase. The strategies include the specific inactivation of P1 kinase by viral factors, the specific degradation of P1 kinase, and the masking of viral dsRNA to prevent P1 kinase activation. In the case of adenovirus, P1 kinase is inactivated by virus-encoded small RNAs, designated VAI RNAs. The growth of a virus mutant that is unable to produce VAI RNA was strongly inhibited by IFN, whereas wild-type virus was not affected (Kitajewski *et al.*, 1986). It was shown that VAI RNA can prevent activation of P1 kinase by dsRNA *in vitro*. Irreversible inactivation of P1 kinase by VAI RNA was demonstrated using purified enzymes (Galabru *et al.*, 1989). The binding affinity of VAI RNA to P1 kinase was lower than that of synthetic dsRNA, but, in spite of its lower binding affinity, VAI RNA could not be displaced by synthetic dsRNA or reovirus dsRNA.

In reovirus the viral σ3 protein interferes with the activation of P1 kinase by dsRNA (Imani and Jacobs, 1988). However, this inhibition was overcome by an excess of dsRNA, suggesting that reovirus σ3 protein strongly binds to dsRNA and thereby prevents P1 kinase activation. In poliovirus-infected cells, activated P1 kinase was shown to be degraded at an increased rate (Black *et al.*, 1989). The viral proteins (or virus-induced host cell factors) involved in this specific P1 kinase degradation process are unknown. In influenza virus-infected cells, a virus-encoded product was postulated which blocks the autophosphorylation of P1 kinase (Katze *et al.*, 1988). Interaction of this inhibitor with P1 kinase appears to be irreversible, reminiscent of adenovirus VAI RNA-mediated inactivation of P1 kinase. Cells infected with vaccinia virus were shown to contain a specific kinase inhibitory factor. This factor is a poorly characterized protein most likely coded for by the virus genome (Rice and Kerr, 1984; Whitaker-Dowling and Youngner, 1984).

Several lines of evidence indicate that activated P1 kinase indeed plays a role in the antiviral state toward VSV, EMC, and some other viruses. First, the experiments described above showed that adenovirus mutant strains lacking the gene that codes for VAI RNA, but not wild-type adenovirus, was susceptible to IFN treatment of the host cell (Kitajewski et al., 1986). Second, VSV and EMC, which usually grow poorly in IFN-treated cells, were found to multiply to high titers in IFN-treated cells coinfected with vaccinia virus. The above-mentioned vaccinia virus-encoded P1 kinase inhibitor (alone or together with additional vaccinia virus-encoded functions) can apparently relieve some of the IFN-induced growth restrictions toward VSV and EMC. These results thus indirectly demonstrate pronounced inhibitory effects of the P1 kinase toward VSV and EMC (Whitaker-Dowling and Youngner, 1983, 1986; Paez and Esteban, 1984a).

C. Mx Proteins

The discovery of the mouse influenza virus resistance locus Mx more than 25 years ago (Lindenmann, 1962) and the stages of research toward the understanding of this system, including identification of the Mx protein, have been reviewed elsewhere (Haller, 1981; Staeheli and Haller, 1987). Here, a condensed description of mammalian and nonmammalian Mx proteins is given, and the biological activities of these proteins are discussed.

1. Mouse Mx Protein

The mouse Mx protein (also designated Mx1 protein) is an IFN-induced 72-kDa protein accumulating in the nuclei of many different cell types of influenza virus-resistant Mx$^+$ mice, but not in cells from influenza virus-susceptible Mx$^-$ mice (Horisberger et al., 1983; Dreiding et al., 1985). Antibodies to Mx protein were produced (Staeheli et al., 1985; Horisberger and Hochkeppel, 1985) and Mx cDNAs were cloned and sequenced (Staeheli et al., 1986b). Mx protein consists of 631 amino acids, more than 30% of which are charged. The positively and negatively charged amino acids occur at comparable frequencies. At the carboxy terminus the Mx protein contains many arginine and lysine residues, which are essential for translocation of the Mx protein into the nucleus (Noteborn et al., 1987). By immunostaining, the distribution of Mx protein in the nuclei was nonuniform and appeared as a granular or punctate pattern (Dreiding et al., 1985). Large quantities of Mx protein were produced in Escherichia coli (Schein and Noteborn,

1988) and in insect cells infected with recombinant baculovirus (R. Krug, personal communication; H. Arnheiter, personal communication). Purification of the Mx protein under nondenaturing conditions was difficult, mainly due to the strong tendency of the Mx protein to form insoluble aggregates. Furthermore, since no *in vitro* activity of the Mx protein has been recognized, the purification of biologically active Mx protein cannot be monitored.

The *Mx1* gene (the gene encoding the Mx protein) was mapped to the distal part of mouse chromosome 16 (Staeheli *et al.*, 1986c) in the vicinity of the protooncogene *ets-2* (Reeves *et al.*, 1988). This region of mouse chromosome 16 is homologous to the part of human chromosome 21 that, when trisomic, results in Down's syndrome. The *Mx1* gene consists of 14 exons spread over more than 55 kilobase pairs of DNA (Hug *et al.*, 1988). An additional 5'-noncoding exon consisting of 72 base pairs is used by different mouse strains at greatly variable frequencies (Hug *et al.*, 1988). Unusual features of the *Mx1* gene are that its first exon consists of only 27 base pairs and that this exon is separated from the second exon by an intron that is larger than 30 kilobase pairs.

Synthesis of the Mx protein is under tight transcriptional control by IFN-α/β. In cultured cells grown in absence of IFN, neither the Mx protein nor *Mx* mRNAs were detectable. Treatment of the cells with IFN-α/β led to a rapid but transient induction of the *Mx* gene, as shown by nuclear run-off experiments (Staeheli *et al.*, 1986a). Significant levels of *Mx* mRNA were detectable by 90 minutes postinduction. By 8 hours postinduction *Mx* mRNA was abundant in IFN-treated mouse embryo cells, reaching levels corresponding to approximately 0.1% of the polyadenylated RNAs, or about 1000 *Mx* mRNA molecules per cell (Staeheli *et al.*, 1986a). The cloned *Mx1* promoter responded similarly well to IFN when it was introduced into mouse L cells by transfection (Hug *et al.*, 1988). High levels of Mx protein accumulated in the nucleus of cultured cells from influenza virus-resistant Mx^+ mice overnight induction with IFN-α/β (Dreiding *et al.*, 1985). In these cells IFN-γ did not induce significant amounts of the Mx protein (Staeheli *et al.*, 1984a; Dreiding *et al.*, 1985).

The transfected *Mx1* promoter and the endogenous *Mx1* gene both responded to some activating factors present in cells infected with either Newcastle disease virus or influenza virus (Hug *et al.*, 1988; Bazzigher *et al.*, 1990). Activation of the *Mx1* gene (and other IFN-responsive genes) by virus was apparently not mediated by IFN. The physiological significance of this IFN-independent induction of IFN-responsive genes is not quite clear. In influenza virus virus-induced Mx protein seemed to arrive too late in the infected cells to confer more than minimal protection (Bazzigher *et al.*, 1990). This might be because the

Mx protein interferes with influenza virus multiplication at an early
step (see Section III).

In healthy Mx^+ mice under normal laboratory conditions, Mx protein is not present in significant quantities in most cells of the animal.
A few Mx protein-expressing cells were consistently found in the liver
and other organs, most likely representing sessile macrophages and
endothelial cells (Jülke and Haller, 1990). Furthermore, macrophages
freshly explanted from the peritoneal cavity of healthy Mx^+ mice
contained high levels of the Mx protein (Dreiding et al., 1985). These
results suggest either that macrophages can respond more efficiently
to low levels of endogenous IFN or that these cells frequently encounter viruses and other agents capable of activating the Mx gene.

Using double-immunofluorescence techniques, Mx protein synthesis
was studied in the course of influenza virus infections (Jülke and
Haller, 1990). A hepatotropic influenza virus strain was used that
grows in hepatocytes and causes fulminant generalized liver necrosis
and death in Mx^- mice. In Mx^+ mice this virus produced only small
necrotic foci, which were self-limiting. The noninfected cells surrounding the viral lesions contained the Mx protein. The interpretation of
this result was that infected hepatocytes produced and secreted IFN,
which, in turn, induced Mx protein synthesis and influenza virus resistance in the surrounding cells.

Influenza virus-susceptible Mx^- mice lack the ability to synthesize
the Mx protein. Southern blot analysis of genomic DNA allowed three
distinct Mx restriction fragment-length polymorphism (RFLP) types to
be defined among the 41 inbred mouse strains that were tested
(Staeheli et al., 1988). Mx RFLP type 1 was found in strain A2G and
two other influenza virus-resistant mouse strains. Mx RFLP type 2
was found in strain BALB/c and most other strains, all of which were
influenza virus susceptible. Mx RFLP type 3 was found in strain
CBA/J and some other influenza virus-susceptible mouse strains. Sequence analysis of cloned cDNAs from representative strains of the
three Mx RFLP types (A2G, BALB/c, and CBA/J) revealed that Mx^-
mice have mutations in the coding regions of their Mx1 genes. A deletion which includes the coding exons 9–11 of the Mx1 gene was found
in BALB/c and in the other RFLP type 2 strains (Staeheli et al., 1988).
The deletion removes 424 bases of coding sequence of Mx1 mRNA,
destroys its open reading frame, and leads to premature termination of
its translation. As a result IFN-treated cells from these strains synthesize a shorter Mx1 mRNA, with greatly decreased metabolic stability.
The Mx^- phenotype of the Mx RFLP type 3 strain CBA/J is due to a
point mutation that converts the lysine codon AAA at position 389 to a
TAA termination codon (Staeheli et al., 1988). This mutation abolishes

synthesis of the mature Mx protein and also results in a decreased metabolic stability of *Mx1* mRNA.

To assess the biological activity of the Mx protein in transfected cells, *Mx1* cDNA was cloned behind the simian virus 40 early promoter and introduced into influenza virus-susceptible Mx⁻ mouse 3T3 cells (Staeheli *et al.*, 1986b). Originally, no stable cell lines were obtained with this cDNA construct. Therefore, a single cell analysis based on immunofluorescence techniques was performed. It was possible to show that the Mx protein conferred to transfected cells a high degree of resistance to influenza virus, but not to VSV (Staeheli *et al.*, 1986b). Using the promoter of the mouse 3-hydroxy-3-methylglutaryl-co-enzyme A reductase gene to drive expression of *Mx1* cDNA in trans-fected cells, stable 3T3 and L cell lines were eventually obtained that expressed high levels of Mx protein in a constitutive manner (Pavlovic *et al.*, 1990). These cell lines survived a challenge with high doses of influenza virus, but were destroyed by VSV and other viruses, includ-ing picornaviruses and herpes simplex virus. The influenza viral titers at different times after infection of the Mx protein-expressing cells were much lower (less than 1%) than those of control cells not express-ing the Mx protein. All other viruses tested (i.e., VSV, two picor-naviruses, two togaviruses, and herpes simplex virus) grew to compara-ble titers in cells expressing or lacking the Mx protein (Pavlovic *et al.*, 1990).

An alternative approach toward assessing the biological activity of the Mx protein was to neutralize the IFN-induced influenza virus re-sistance by the microinjection of monoclonal antibodies to the Mx pro-tein (Arnheiter and Haller, 1988). The antibodies were microinjected into the cytoplasm of mouse Mx⁺ cells, before the cells were treated with IFN and infected with virus. Cells microinjected with antibody 2C12, but not with two other monoclonal antibodies to Mx protein, failed to develop resistance to influenza virus. IFN-induced resistance to VSV was not affected by any of these antibodies. None of the anti-bodies seemed to interfere with the nuclear accumulation of the Mx protein, suggesting that antibody 2C12 binds to a region of the Mx protein necessary for antiviral function (Arnheiter and Haller, 1988). The epitope on mouse Mx1 protein recognized by 2C12 has not been mapped; it most likely includes conserved sequences, because 2C12 also binds to the mouse Mx2 protein (T. Zürcher and P. Staeheli, un-published observations), all rat Mx proteins (Meier *et al.*, 1988), human MxA protein (Staeheli and Haller, 1985; Aebi *et al.*, 1989; Weitz *et al.*, 1989), and Mx proteins of goats and hamsters (Mortier and Haller, 1987).

It is unclear whether the nuclear location of the mouse Mx1 protein

is necessary for its function toward influenza virus. Noteborn *et al.*
(1987) found that the Mx1 protein did not always accumulate efficiently in the nuclei of transfected cells and that transfected cells with
the Mx1 protein predominantly in the cytoplasm still acquired resistance to influenza virus. However, when the nuclear translocation
signal (located within the last 31 carboxy-terminal amino acids) was
removed, the antiviral activity of Mx1 protein was destroyed (Noteborn *et al.*, 1987). A similar result was obtained when the carboxy-
terminal 34 amino acids of the mouse Mx1 protein were replaced by
the 34 carboxy-terminal amino acids of the mouse Mx2 protein. The
two sequences are 65% identical, but the Mx2 sequence contains no
nuclear translocation signal. The hybrid protein accumulated to high
levels in the cytoplasm of permanently transfected 3T3 cells, and these
cells remained fully susceptible to influenza virus (T. Zürcher and P.
Staeheli, unpublished observations). These results clearly establish
that the carboxy terminus of the Mx1 protein is necessary for its function. A detailed point mutation analysis is necessary to distinguish
between the possibility that this part of Mx1 protein is critical simply
because it harbors the nuclear translocation signal and the possibility
that this region of Mx1 protein harbors an additional domain required
for function against influenza virus.

A transgene consisting of about 2 kilobases (kb) of the mouse *Mx1*
promoter, the coding sequence of *Mx1* cDNA, and 3'-noncoding sequences from the mouse β-globin gene was recently introduced into
Mx⁻ mice, and it was demonstrated that these animals acquire the
ability to synthesize the Mx1 protein when exposed to IFN or virus (H.
Arnheiter, personal communication). Transgenic mice expressing significant amounts of the Mx1 protein survived infection with pathogenic influenza virus. These results confirm earlier conclusions from
work with *Mx*-congenic mice that influenza virus resistance in mice is
due to the Mx1 protein.

2. Mouse Mx2 Protein

The haploid mouse genome contains two *Mx* genes, rather than one.
Southern blot analysis first suggested the presence of some *Mx*-related
sequences, and cDNA cloning experiments eventually confirmed the
existance of a related but distinct *Mx* gene, designated *Mx2* (Staeheli
and Sutcliffe, 1988). Two *Mx* genes also exist in humans and probably
in most other mammals (see below).

In all inbred mouse strains tested the *Mx2* gene was crippled by
mutations. As a consequence mouse Mx2 protein is not present in
commonly used mouse strains and cell lines, and *Mx2* mRNAs of the
mouse strains BALB/c and CBA/J have greatly decreased metabolic

stabilities (Staeheli and Sutcliffe, 1988). The sequence of *Mx2* mRNA, as deduced from cloned cDNAs, was more than 90% identical to *Mx1* mRNA in the region corresponding to the amino-terminal halves of the encoded polypeptides. The sequences corresponding to the carboxy-terminal halves of the Mx1 protein and the putative Mx2 protein are about 60% identical (Staeheli and Sutcliffe, 1988).

A computer-assisted alignment of the *Mx1* and *Mx2* cDNA sequences allowed the identification of a hypothetical supernumerary nucleotide in the *Mx2* mRNA of strains BALB/c and CBA/J that destroys the long open reading frame (Staeheli and Sutcliffe, 1988). Recently, this extra nucleotide has been removed from the cloned *Mx2* cDNA by site-directed mutagenesis. This engineered polypeptide presumably has the same sequence as the putative mouse Mx2 protein, consists of 655 amino acids, and is about 74 kDa in size. When expressed in transfected 3T3 cells, the repaired *Mx2* cDNA gave rise to a cytoplasmic protein recognized by polyclonal antisera to the mouse Mx1 protein and by monoclonal antibody 2C12 (T. Zürcher and P. Staeheli, unpublished observations). The putative mouse Mx2 protein is about 90% identical to the rat Mx2 and Mx3 proteins (E. Meier and H. Arnheiter, personal communication), both of which accumulate in the cytoplasm of IFN-treated rat cells (see below).

The *Mx2* gene maps to mouse chromosome 16 and is closely linked to *Mx1* (Staeheli and Sutcliffe, 1988). Regulation of the *Mx1* and *Mx2* genes by IFN is similar (Staeheli and Sutcliffe, 1988).

As mentioned above, strains BALB/c and CBA/J (and presumably most other *Mx* RFLP types 2 and 3 strains) carry an *Mx2* gene that is nonfunctional due to a frameshift mutation. The three *Mx* RFLP type 1 strains tested failed to express *Mx2* mRNA on induction by IFN, indicating that the *Mx2* promoter might be defective in these strains (Staeheli and Sutcliffe, 1988). It is unknown whether wild mice possess functional *Mx2* genes.

Since Mx2 protein is not found in commonly used inbred mouse strains or cell lines, its potential antiviral function could not be assessed. Mouse cell lines expressing a repaired *Mx2* cDNA construct were recently established in our laboratory. With the help of these transfected cell lines, a functional study of mouse Mx2 protein should now be feasible.

3. Human MxA Protein

Human MxA protein is a 76-kDa protein found in IFN-treated human fibroblasts, peripheral blood lymphocytes, and some established cell lines. It was originally discovered with the help of anti-mouse Mx protein monoclonal antibody 2C12 (Staeheli and Haller,

1985). MxA protein accumulates in the cytoplasm of IFN-treated cells. By immunostaining, the distribution of MxA protein throughout the cytoplasm of IFN-treated fibroblasts (or mouse cells transfected with *MxA* cDNA) was nonuniform and appeared as a granular or punctate pattern (Staeheli and Haller, 1985; Pavlovic *et al.*, 1990). It is unclear whether this staining pattern is artifactual or might indicate an association of the MxA protein to some vesicles or Golgi compartments.

Horisberger and Hochkeppel (1987) purified the MxA protein, which they called p78, prepared monoclonal antibodies, and eventually isolated corresponding cDNA clones, with the help of synthetic oligonucleotides designed to fit the amino-terminal sequence of p78 (Horisberger *et al.*, 1988). Using mouse *Mx1* cDNA as a hybridization probe, *MxA* cDNA clones were also isolated by Aebi *et al.* (1989). MxA protein (as deduced from the cloned cDNAs) consists of 662 amino acids and is 75.5 kDa in size. MxA protein is 67% and 77% identical to the mouse Mx1 protein and the hypothetical Mx2 protein, respectively. The similarity between these proteins is most pronounced in their amino-terminal regions. Like mouse Mx1 and Mx2 proteins, the human MxA protein is rich in charged amino acids (Aebi *et al.*, 1989). Weitz *et al.* (1989) purified the MxA protein from IFN-treated human foreskin fibroblasts by immunoaffinity chromatography, using antibody 2C12. The amino-terminal sequence of the purified protein was as predicted from the cloned cDNA, except for the initiator, methionine, which was absent from mature MxA protein.

The *MxA* gene maps to human chromosome 21 (Horisberger *et al.*, 1988). Details of the *MxA* gene organization are unknown.

The MxA protein was not detectable in cells cultured in the absence of IFN, but it was abundantly present in human fibroblasts or lymphocytes treated for 18 hours with IFN-α (Staeheli and Haller, 1985; Horisberger and Hochkeppel, 1987). Cells treated with IFN-γ contained low levels of the MxA protein. Induction of the *MxA* gene by IFNs was rapid and occurred in cells with blocked protein synthesis (Goetschy *et al.*, 1989; Aebi *et al.*, 1989). Treatment of human fibroblasts with interleukin-1α, interleukin-1β, or tumor necrosis factor induced the *MxA* gene (Goetschy *et al.*, 1989). The possibility that *MxA* induction by these cytokines was via IFN has not been excluded.

The lymphocytes of all blood donors tested to date responded to treatment with IFN-α by synthesizing *MxA* mRNA or the MxA protein, suggesting that, unlike in the Mx system of the mouse, nonfunctional *MxA* alleles might not exist in the human population (von Wussow *et al.*, 1989; J. Fäh, P. Staeheli, and O. Haller, unpublished observations). However, some established cell lines, including HeLa,

HEp-2, K562, and Daudi, failed to synthesize the MxA protein or *MxA* mRNA when treated with IFN (Staeheli and Haller, 1985; Horisberger and Hochkeppel, 1987; J. Fäh, F. Pitossi, J. Pavlovic, and P. Staeheli, unpublished observations). The nature of the defect in these cell lines is unknown.

Cell lines lacking the ability to synthesize the MxA protein when treated with IFN failed to resist infection with influenza virus (J. Pavlovic, F. Pitossi, and P. Staeheli, unpublished observations). This suggests, but does not prove, that the human MxA protein plays a role in cellular defense against influenza virus. To directly assess the antiviral potential of the MxA protein, appropriate *MxA* cDNA constructs were transfected into mouse 3T3 cells, and permanent cell lines were established that express high levels of human MxA protein in a constitutive manner (Pavlovic *et al.*, 1990). As in the natural situation the MxA protein accumulated in the cytoplasm of the transfected mouse cells and appeared granular or punctate when visualized by immunostaining techniques. MxA protein-expressing mouse 3T3 cells exhibited significant resistance to influenza virus. At different times after infection with influenza virus, the viral titers in the supernatants of MxA protein-expressing cells were about one one-hundredth that in the control cells. Immunofluorescence analysis of such cultures revealed that a small portion of the infected cells failed to acquire the resistance phenotype, although these cells contained detectable amounts of the MxA protein. It thus seems that high levels of the MxA protein are necessary for efficient antiviral protection of transfected mouse cells (Pavlovic *et al.*, 1990).

Surprisingly, MxA protein-expressing mouse 3T3 cells were also resistant to VSV. At different times after infection, the VSV titers in the supernatants of MxA protein-expressing cells were about one one-hundredth that in the control cells. All other viruses tested (i.e., two picornaviruses, two togaviruses, and a herpes simplex virus) grew to comparable titers in cells either expressing or lacking the MxA protein (Pavlovic *et al.*, 1990). More work is required to assess the full spectrum of the antiviral activity of the MxA protein. Nevertheless, these data clearly indicate that the human MxA protein has a broader antiviral activity than the mouse Mx1 protein, comprising orthomyxoviruses and rhabdoviruses. It will be interesting to learn whether the replication of rabies virus, a rhabdovirus causing severe disease in humans, is inhibited in cells expressing the human MxA protein.

The mode of action of the MxA protein toward influenza virus and VSV has not been elucidated. The human MxA protein presumably blocks a cytoplasmic multiplication step of influenza virus, VSV, and possibly other viruses. The modes of action of the cytoplasmic human

MxA protein and the nuclear mouse Mx1 protein presumably are of a distinct nature. The view of distinct modes of action of human MxA and mouse Mx1 proteins is supported by recent studies by Weitz *et al.* (1989) which showed that the cytoplasmic microinjection of antibody 2C12 cannot alter the virus susceptibility of IFN-treated human cells, although this antibody could efficiently neutralize the IFN-induced antiviral state to influenza virus in mouse or rat cells (Arnheiter and Haller, 1988). This result might indicate that some functionally important regions of the nuclear rodent Mx1 proteins are recognized by antibody 2C12, but that the corresponding sequences of human MxA protein are not located in a region important for MxA function.

Whatever the antiviral mechanisms of Mx proteins might be, it is of interest to determine to what extent, if any, transgenic expression of the human MxA protein can complement the genetic defect of Mx⁻ mice. From the experiments with transfected 3T3 cells, one expects that transgenic Mx⁻ mice expressing the MxA protein are influenza virus resistant.

4. Human MxB Protein

The MxB protein is an IFN-induced 73-kDa protein whose sequence is about 63% identical to that of the MxA protein. In IFN-treated fibroblasts or lymphocytes, the MxB protein and its corresponding mRNA are about 3–5 fold less abundant than the MxA protein and *MxA* mRNA. The MxB protein accumulates in the cytoplasm of IFN-treated human fibroblasts (Aebi *et al.*, 1989).

MxB mRNA was originally cloned by virtue of its homology to mouse *Mx1* cDNA (Aebi *et al.*, 1989). The cDNA sequence predicted a protein of 82 kDa, but antibodies raised against a fusion protein between β-galactosidase and a polypeptide encoded by *MxB* mRNA revealed that the major form of natural MxB protein is smaller—namely, about 73 kDa. Since the methionine codon at the beginning of the long open reading frame of *MxB* mRNA is located in a unfavorable context for translation initiation, it was hypothesized that a downstream AUG codon might serve as start site for translation (Aebi *et al.*, 1989). The similarity of the MxB protein to mouse Mx1 and Mx2 proteins is 56% and 62%, respectively. The MxB protein is thus more distantly related to the mouse Mx proteins than is the MxA protein. The highest degree of sequence identity between the MxB proteins and the other Mx proteins is found near their amino termini.

Like *MxA*, the *MxB* gene also maps to human chromosome 21 (Huber *et al.*, 1990). Details of the *MxB* gene organization are unknown.

The MxB protein and *MxB* mRNA were not detectable in cells

cultured in the absence of IFN, but they became readily detectable after treatment with IFN-α (Aebi *et al.*, 1989). Cells treated with IFN-γ contained low levels of *MxB* mRNA. Induction of the *MxB* gene by IFNs was rapid and occurred even in cells with blocked protein synthesis (Aebi *et al.*, 1989). The lymphocytes of all blood donors tested and all cell lines analyzed responded to IFN by accumulating *MxB* mRNA or the MxB protein, with the exception of an HeLa cell line (J. Fäh, F. Pitossi, J. Pavlovic, and P. Staeheli, unpublished observations).

No antiviral or other biological activity of the MxB protein has been recognized. *MxB* cDNA constructs were transfected into mouse 3T3 cells, and permanent cell lines expressing the MxB protein in a constitutive manner were established as described above for the MxA protein (Pavlovic *et al.*, 1990). Transfected cells that expressed the MxB protein remained susceptible to infection with a large number of different viruses, including influenza virus and VSV.

5. Rat Mx1 Protein

The rat Mx1 protein is an IFN-induced 72-kDa protein closely related to the mouse Mx1 protein. The rat Mx1 protein is recognized by all monoclonal antibodies raised against the mouse Mx1 protein. It is also recognized by a serum raised against a peptide corresponding to the 16 carboxy-terminal amino acids of the mouse Mx1 protein. The Mx1 proteins of rats and mice both accumulate in the nuclei of IFN-treated cells (Meier *et al.*, 1988). cDNA clones corresponding to the rat Mx1 protein were identified by virtue of their cross-hybridization to mouse *Mx1* cDNA (Meier *et al.*, 1988). The sequence of the rat Mx1 protein is about 90% identical to that of the mouse Mx1 protein (E. Meier and H. Arnheiter, personal communication).

The *Mx1* gene maps to rat chromosome 11, which is syntenic to mouse chromosome 16 and human chromosome 21 (C. Szpirer and O. Haller, personal communication). Details about the gene organization are not known. *Mx1* expression is under the control of IFN-α and -β, much like the mouse *Mx* genes (Meier *et al.*, 1988). IFN-γ is a poor inducer of the rat *Mx1* gene.

Cells of all 26 inbred rat strains analyzed synthesized the Mx1 protein on induction with IFN-α (Meier *et al.*, 1988). Wild rats were not analyzed. Since most inbred rat strains were derived from only a few founder stocks, the significance of this result is unclear. It might indicate that, in contrast to the situation in mice, *Mx1* mutations are rare or nonexistent in rats.

The IFN-induced antiviral state of rat cells toward influenza virus could be neutralized by the microinjection of monoclonal antibody

2C12 (Arnheiter and Haller, 1988), indicating that one or more rat Mx proteins have the potential to inhibit influenza virus. Indeed, transient cDNA expression experiments showed that the rat Mx1 protein can block the multiplication of influenza virus (E. Meier and H. Arnheiter, personal communication).

6. Rat Mx2 and Mx3 Proteins

The rat Mx2 and Mx3 proteins are IFN-induced proteins of about 75 kDa. These proteins are recognized by monoclonal antibody 2C12, but not by some other monoclonal antibodies to the mouse Mx1 protein. The Mx2 and Mx3 proteins are both cytoplasmic (Meier et al., 1988). cDNA clones encoding these Mx proteins were isolated (Meier et al., 1988). Sequence analysis revealed that the Mx2 and Mx3 proteins are almost identical, differing at only a few amino acid positions (E. Meier and H. Arnheiter, personal communication). The rat Mx2 and Mx3 proteins are about 90% identical to the mouse Mx2 protein and about 65% identical to the mouse Mx1 protein.

All Mx-related sequences of the rat genome mapped to rat chromosome 11 (C. Szpirer and O. Haller, personal communication), indicating that the genes which code for the different rat Mx proteins are all located on this chromosome. Meier et al. (1988) isolated three distinct classes of rat Mx cDNAs, each appearing to code for one of the three rat Mx proteins (E. Meier and H. Arnheiter, personal communication). This suggests that rats have three distinct Mx genes. In contrast, mice and humans have only two Mx genes (see above). The possibility that the cells used for the cDNA cloning experiments were derived from a heterozygous rat has not been excluded. It is therefore possible that Mx2 and Mx3 cDNAs originated from allelic variants of a single gene, rather than from distinct genes with highly conserved sequences.

Rat strains with nonfunctional alleles of Mx2 or Mx3 genes have not been identified. No data on the antiviral potentials of the Mx2 and Mx3 proteins are available.

7. Mx-Related Genes of Mammals and Fish

IFN-induced proteins that cross-react with monoclonal antibodies to mouse or human Mx proteins were found not only in rat cells, but also in hamster and goat cells (Mortier and Haller, 1987) and in cells from bovines and other mammals (Horisberger, 1988), indicating a rather strict conservation of some Mx protein domains. Interestingly, all of these cross-reactive proteins accumulated predominantly in the cytoplasm of IFN-treated cells. It thus appears that mice and rats are exceptional in that they possess nuclear and cytoplasmic Mx proteins. When "Zoo-blots" prepared from mammalian and nonmammalian

DNAs were hybridized to radiolabeled probes derived from the murine *Mx1* gene, hybridization signals were obtained with all DNAs tested, although the intensities of the signals varied among species (R. Grob and O. Haller, personal communication).

Recently, a fish gene was identified that is remarkably similar to the mouse *Mx1* gene (Staeheli *et al.*, 1989). The cloned fish DNA contained blocks of sequences related to mouse *Mx1* exons 3–8, which appear to represent the exons of a bona fide fish gene, because they are separated by intron sequences flanked by consensus splice acceptor and donor sites. The cloned putative coding sequence of the fish *Mx* gene is 67% identical to the mouse Mx1 protein sequence (Staeheli *et al.*, 1989). Injection of synthetic dsRNA or virus into the peritoneal cavity of fish resulted in five- to 10-fold elevated levels of two liver mRNAs about 2.0–2.5 kb in length that hybridized to the cloned fish *Mx* DNA (Staeheli *et al.*, 1989), suggesting that the fish *Mx* genes are under control by IFN similar to that of the mammalian *Mx* genes. The complete sequences of the two putative fish Mx proteins are still unknown. Nothing is known about their intracellular location.

An important question is whether the Mx proteins of nonmammalian species such as fish also play a role in the cellular defense against viruses. A variety of fish viruses and viruslike agents have been recognized, most of which are associated with pathological conditions (McAllister, 1979). Fish might need efficient defense mechanisms to cope with all kind of viruses in their environment; one such defense mechanism might involve the Mx proteins. This idea is particularly intriguing in view of some new results (discussed above) that the human MxA protein can inhibit multiplication of the rhabdovirus VSV. Rhabdoviruses are among the most important viral pathogens of fish (McAllister, 1979).

8. Yeast Vps1

Rothman *et al.* (1990) recognized a remarkably high sequence similarity between yeast Vps1 and the mouse Mx1 protein. Over their entire lengths the two proteins have 27% identical amino acid sequences, and 38% of the Mx1 sequence is similar to the Vps1 sequence when conservative amino acid replacements are included in the comparison. The Vps1 protein is related, to similar degrees, to mouse Mx2, human MxA, human MxB, and fish Mx proteins. Sequence comparisons revealed that two blocks of sequences, located in the aminoterminal halves of the Mx and Vps1 proteins, are particularly well conserved in all of these proteins (Fig. 1). A first block of 41 amino acids, which is almost invariable in all vertebrate Mx proteins characterized to date, has a well-conserved counterpart in the Vps1 protein.

```
LPAIAVIGDQSSGKSSVLEALSG-VALPRGSGIVTRCPLVLKL   Mx1  (36-77)
LPAIAVIGDQSSGKSSVLEALSG-VALPRGSGIVTRCPLVLKL   Mx2  (63-104)
LPAIAVIGDQSSGKSSVLEALSG-VALPRGSGIVTRCPLVLKL   MxA  (70-111)
LPAIAVIGDQSSGKSSVLEALSG-VALPRGSGIVTRCPLVLKL   MxB  (118-159)
LPAIAVIGDQSSGKSSVLEALSG-VALPRGSGIVTRCPLELKM   fish Mx
*********************************** **          common to mouse, man & fish
 **  *  * **********    *   **** ***** ** *     common to all
LPQINVVGSQSSGKSSVLENIVGRDFLPRGTGIVTRRPLVLQL   yeast Vps1 (29-71)
    GXXXXGK                                    GTP-binding consensus


GISDKLISLDVSSPNVPDLTLIDLPGITRVAVGNQPADIGRQIKRLIKTYIQKQETINLVVVPSNVDIATTEALSMAQEVDPEGDRTIGVLTKPDLVDRG   Mx1  (122-221)
GISDKLISLDVSSPNVPDLTLIDLPGITRVAVGNQPADIGRQIKRLIKTYIQKQETINLVVVPSNVDIATTEALSMAQEVDPEGDRTIGILTKPDLVDRG   Mx2  (149-248)
GISHELITLEISSRDVPDLTLIDLPGITRVAVGNQPADIGYKIKTLIKKYIQRQETISLVVVPSNVDIATTEALSMAQEVDPEGDRTIGILTKPDLVDKG   MxA  (156-255)
GISHELITLEISSRDVPDLTLIDLPGITRVAVGNDQPRDIGLQIKALIKKYIQRQQTINLVVVPCNVDIATTEALSMAHEVDPEGDRTIGILTKPDLMDRG   MxB  (203-302)
GISDDLISLEIASPDVPDLTLIDLPGIARVAVKGQPENIGDQIKRLIQKFIKRQETISLVVVPCNVDIATTEALKMAQEVDPDGERTLGILTKPDLVDKG   fish Mx
***  **  *   ***** ****** **** ** ** ** **   * ** ***** ********** ** **** * ** * ***** * *            common to mouse, man & fish
*** *  * *  ** **** * ** ** * ** *   * * * * *  * * *   *  *  * **** * ** * *** ** * *                 common to all
GISSVPINLRIYSPHVLTLTLVDLPGLTKVPVGDQPPDIERQIKDMLLKYISKPNAIILSVNAANTDLANSDGLKLAREVDPEGTRTIGVLTKVDLMDQG   yeast Vps1 (156-255)
      DXXG                                                                                TKXD        GTP-binding consensus
```

FIG. 1. Comparison of amino acid sequences of vertebrate Mx proteins and yeast protein Vps1. Shown are two regions of high sequence similarity between the mouse Mx1 protein, hypothetical mouse Mx2 protein, human MxA protein, human MxB protein, putative fish Mx protein, and yeast Vps1. The numbers identify the protein fragments that were included in the comparison. The asterisks mark the presence of identical amino acids at corresponding positions.

The corresponding Vps1 sequence is about 65% identical to that of the vertebrate Mx proteins. A second block of 100 amino acids, which exhibits a sequence conservation of 67% among the mouse, human, and fish Mx proteins, is also well conserved in the Vps1 protein: 44 of the 67 invariable amino acids of the Mx protein are found in the Vps1 sequence.

The physiological role of the Vps1 protein in yeasts is poorly understood. Vps1 is a constitutively expressed 79-kDa protein localized in the cytoplasm of yeast cells (Rothman *et al.*, 1990). Immunostaining revealed a punctate cytoplasmic labeling pattern, indicating an association of Vps1 with some vesicles. Null mutations in the *VPS1* gene are not lethal, but yeast cells with a defective *VPS1* gene fail to grow at elevated temperatures (Rothman *et al.*, 1990). The product of the yeast *VPS1* gene is required for the correct sorting of proteins to the yeast lysosomelike vacuole and for the normal organization of intracellular membranes (Rothman and Stevens, 1986). The mechanisms by which the Vps1 protein performs these cellular functions have not been resolved.

How does the VPS1 system of yeasts relate to the vertebrate Mx system? Could the vertebrate Mx proteins be involved in the complex process of intracellular protein sorting, and could such cellular activities of the Mx proteins result in antiviral protection? One might hypothesize that the Mx proteins are involved in the down-regulation of certain transport processes across biological membranes, thereby controlling, for example, endocytosis, secretion, and transport to or from the cell nucleus. In IFN-treated cells the flux through certain of these transport systems might be reduced, due to the action of Mx proteins. It is conceivable that reduced transport capacities would not greatly interfere with the metabolism of growth-arrested cells, but would greatly compromise the needs of a fast-growing virus.

9. Mx Proteins Could Have GTP Binding Activity

Sequence comparison suggested that the amino-terminal halves of the Mx and Vps1 proteins are likely to perform similar biochemical functions and that highly conserved sequences are probably essential for this activity. On careful inspection of the highly conserved protein domains, we discovered (Pavlovic et al., 1990) that all Mx protein sequences and the Vps1 sequence contain the three consensus sequence elements typically found in GTP-binding proteins (Dever et al., 1987). These consensus elements—GXXXXGK, DXXG, and TKXD—occur in the expected order (Fig. 1), and the spacing between these elements in the Mx proteins is similar to that in bona fide GTP-binding proteins. These findings suggest that the Mx proteins are GTP-binding proteins and that this putative biochemical activity might be essential for the biological activities of the Mx proteins.

D. Guanylate-Binding Proteins

1. Human GBP-1

GBP-1 is an IFN-induced 67-kDa protein of human fibroblasts that strongly binds to either agarose-immobilized or free guanine nucleotides (Cheng et al., 1983, 1985). GBP-1 binds to guanosine mono-, di-, and triphosphates, but not to those of adenosine or cytosine. The protein was not detected in fibroblasts that were cultured in the absence of IFN, but treatment of the cells with IFN-α, -β, or -γ induced the synthesis of large quantities of GBP-1. GBP-1 was purified from IFN-treated fibroblasts, and polyclonal rabbit antisera were produced (Cheng et al., 1985). Immunostaining experiments showed that GBP-1 is confined to the cytoplasm of IFN-treated cells. cDNA clones encoding GBP-1 were isolated by screening a λgt11 expression library with the specific antiserum (Cheng et al., 1986b). cDNA sequence analysis revealed that GBP-1 consists of 592 amino acids and that it contains sequences related to, but distinct from, the GTP-binding sites of other proteins with guanylate-binding properties (Y.-S. E. Cheng, personal communication).

GBP-1 is readily detected in the lymphocytes of patients treated with IFN-α or in lymphocytes treated with IFN-α or -γ in vitro. The cells of all blood donors tested responded to IFN treatment by synthesizing GBP-1

(Cheng *et al.*, 1988). In contrast, some established cell lines failed to synthesize GBP-1 when treated with IFN-α or -γ (Cheng *et al.*, 1986a). The physiological role of human GBP-1—in particular, its role in the IFN-induced antiviral state—is unknown.

2. Human GBP-2

When a λgt11 expression library was screened with a polyclonal serum raised against purified guanine nucleotide-binding proteins, a cDNA clone was isolated whose restriction map differed greatly from that of GBP-1 cDNA. Northern blot analysis indicated that this clone was derived from a distinct mRNA, designated GBP-2. cDNA sequence analysis revealed that GBP-1 and -2 mRNAs originated from closely related, but distinct, genes (Y.-S. E. Cheng, personal communication). The putative GBP-2 protein is about 75% identical to GBP-1 and has a similar size (Y.-S. E. Cheng, personal communication).

3. Mouse GBP-1

Mouse GBP-1 is the predominant 65-kDa species of a family of guanylate-binding proteins found in IFN-treated mouse cells (Cheng *et al.*, 1983; Staeheli *et al.*, 1983, 1984b). It is synthesized in mouse fibroblasts treated with either IFN-α/β or IFN-γ and in the macrophages or spleen cells of mice treated with IFN-inducing agents. Some mouse strains failed to respond to IFNs by synthesizing GBP-1; the ability to synthesize GBP-1 was inherited as a single autosomal dominant trait, designated *Gbp-1*. Mice carrying the *Gbp-1ᵃ* allele can synthesize GBP-1, whereas mice homozygous for *Gbp-1ᵇ* cannot (Staeheli *et al.*, 1984b). The *Gbp-1* locus was mapped to mouse chromosome 3, close to the minor histocompatibility locus *H-23* (Prochazka *et al.*, 1985).

Using the human GBP-2 cDNA as a hybridization probe, a mouse cDNA clone was isolated that most likely codes for GBP-1. On Northern blots the cloned cDNA detected an IFN-induced 3-kb mRNA of *Gbp-1ᵃ* cells that was absent from IFN-treated *Gbp-1ᵇ* cells (P. Staeheli and Y.-S. E. Cheng, unpublished observations). The structure of the mouse GBP-1 protein has not been determined.

Mice lacking the ability to synthesize GBP-1 are healthy, and both alleles of *Gbp-1* were found in outbred lines of wild mice (Prochazka *et al.*, 1985). No increased virus susceptibility of *Gbp-1ᵇ* mice was noticed. Thus, unlike the case of the Mx system, the availability of the mutant GBP-1 mouse strain has not helped to recognize the physiological role of this protein.

E. ISG15

An abundant 15-kDa protein, ISG15, was found in human cells treated with IFN-α or -β (Korant et al., 1984). This protein was purified, most of its sequence was determined, and a corresponding cDNA clone was then isolated and characterized (Blomstrom et al., 1986). It was found that the 15-kDa protein is synthesized from a 17-kDa precursor by carboxy-terminal processing (Reich et al., 1987; Knight et al., 1988). The mature 15-kDa protein consists of 156 amino acids and is found in the cytoplasm of IFN-treated cells.

Haas et al. (1987) noted a significant sequence homology between the ISG15 protein and ubiquitin. Their polyclonal antisera to ubiquitin detected a cross-reactive protein of 15 kDa that was absent from untreated control cells, but abundantly present in human or mouse fibroblasts treated with IFN-β. A comparison of the gel mobility and the immunoreactivity of an authentic purified sample of ISG15 protein and the human ubiquitin cross-reactive protein revealed the identity of the two proteins (Haas et al., 1987). The ISG15 protein is composed of two domains, each bearing a regular pattern of homology to ubiquitin. Within the amino-terminal domain 23 of the 80 amino acids are conserved (29% identity); the homology is about 40% if conservative amino acid replacements are taken into account. Within the carboxy-terminal domain 28 of the 76 amino acids are conserved (37% identity); the homology increases to about 60% if conservative amino acid replacements are included (Haas et al., 1987).

It is unknown whether the ISG15 protein has an ubiquitinlike activity. Authentic ubiquitin has no recognized enzymatic activity, and the physiological functions of this protein are extremely complex. Within the cytoplasm ubiquitin is ligated to a wide array of soluble and membrane-bound target proteins. Ubiquitin ligations might serve regulatory roles, but in other cases ubiquitin ligation is an obligatory step for the multienzyme energy-dependent proteolytic pathway apparently responsible for the degradation of various short-lived and abnormal proteins (see references cited by Haas et al., 1987). It is conceivable that the ISG15 protein serves similar functions in IFN-treated cells, thereby restricting the replication of certain viruses. It will be interesting to learn whether expression of the ISG15 protein in transfected eukaryotic cells results in antiviral protection.

F. 17-kDa Protein

An IFN-induced 17-kDa protein (originally referred to as the 20-kDa protein) with a powerful growth inhibition potential has been described. This protein was found only in cells whose growth could be inhibited by IFN-β (Knight *et al.*, 1985). The 17-kDa protein was shown to be localized on the exterior of cell membranes, and it was partially purified from the membranes of IFN-β-treated Daudi cells (Hillman *et al.*, 1987). When the purified 17-kDa protein was added to the culture medium of Daudi, Namalwa, or HeLa cells, it caused a cessation of cell growth. The same membrane fraction from untreated control Daudi cells caused no growth inhibition. The mechanism of the 17-kDa protein-mediated growth inhibition is unknown. It was specu- lated that the purified 17-kDa protein might insert itself into the plasma membrane of the target cells in the same location it occupied originally in the IFN-treated cells, thereby causing a cessation of cell growth (Hillman *et al.*, 1987). The 17-kDa protein might cause, di- rectly or indirectly, a reduction in the binding of essential growth factors (Hillman *et al.*, 1987). The structure of the 17-kDa protein is unknown. cDNA clones coding for this protein have not been identified.

G. IP-10

In the human lymphoma cell line U937 treated with IFN-γ, IP-10 mRNA was induced more than 30-fold (Luster *et al.*, 1985). The gene coding for IP-10 was mapped to human chromosome 4 (Luster *et al.*, 1987). The IP-10 cDNA sequence predicted a 12-kDa protein, which contains a 21-amino acid signal peptide typically found in the precur- sors of secreted proteins. The secreted portion of this protein would consist of 98 amino acids with a size of 10 kDa (Luster *et al.*, 1985). However, polyclonal monospecific antisera raised against recombinant IP-10 protein detected a secreted protein of 6–7 kDa, not the predicted 10-kDa polypeptide, suggesting additional posttranslational processing (Luster and Ravetch, 1987). Amino-terminal sequence analysis of the secreted 6- to 7-kDa form of IP-10 confirmed that the signal peptide was absent from the mature protein and that the amino-terminal sequence of IP-10 was exactly as predicted from the analysis of the cloned IP-10 cDNA. Thus, the IP-10 precursor protein appeared to undergo extensive carboxy-terminal processing, resulting in the carboxy-trimmed mature 6- to 7-kDa form of the IP-10 (Luster and Ravetch, 1987).

The IP-10 sequence is about 35% identical to platelet factor 4 and β-thromboglobulin over a region of 62 amino acids (Luster *et al.*, 1985). The latter proteins are secreted by platelets on degranulation and are believed to be important mediators in inflammatory and wound-heal- ing responses. The similarities of IP-10, platelet factor 4, and β-throm- boglobulin suggest that IP-10 might also be involved in the inflammatory response. Direct proof that IP-10 performs such functions is difficult to obtain. Kaplan *et al.* (1987) studied the expression of IP-10 after evoking a delayed-type response to a purified derivative of tuberculin. They noted the presence of the IP-10 protein in dermal macrophages and endothelial cells. Intense IP-10-specific immunostaining of the basal layer of the epidermal keratinocytes was prominent at 2 days, and by 1 week the entire epidermis was staining. IP-10 expression was also observed after injection of IFN-γ and in the natural lesions of tuberculoid leprosy and cutaneous leishmaniasis. Kaplan *et al.* (1987) proposed that the local production of IFN-γ by T cells of the dermal infiltrate induces synthesis of the IP-10 protein, which, in turn, might recruit a broad spectrum of cells involved in the inflammatory response. From this point of view, IP-10 might be regarded as a "secondary cytokine" (Kaplan *et al.*, 1987).

H. Indolamine 2,3-Dioxygenase

Human epithelial or fibroblast cells treated with IFN-γ degrade tryptophan at an increased rate due to greatly elevated levels of indolamine 2,3-dioxygenase (IDO) activity (Pfefferkorn, 1984; Pfefferkorn *et al.*, 1986b; Byrne *et al.*, 1986a); the enzyme IDO degrades tryptophan to N-formylkynurenine. Antibodies to IDO were produced (Rubin *et al.*, 1988). These polyclonal rabbit sera neutralized IDO activity of IFN-γ-treated cells and precipitated a single 42-kDa protein. The isolation of IDO cDNA clones has not been reported. However, as pointed out by Rubin *et al.* (1988), a previously characterized cDNA clone derived from an IFN-γ-induced 2.2-kb mRNA (Caplen and Gupta, 1988) is likely to code for IDO.

The IDO activity in IFN-γ-treated cells has been implicated in the inhibition of intracellular parasites such as *Toxoplasma gondii* and *Chlamydia psittaci* (Pfefferkorn, 1984; Byrne *et al.*, 1986b; Pfefferkorn *et al.*, 1986a; Carlin *et al.*, 1989). It is believed that the inhibition is due to IDO-mediated tryptophan starvation of the parasites. It was demonstrated that the antitoxoplasma activity induced in host cells by IFN-γ was strongly dependent on the tryptophan concentration in the medium (Pfefferkorn, 1984). Progressively higher minimal inhibitory concentrations of IFN-γ were observed as the tryptophan concentration in the culture medium was increased. However, evidence has been presented that degradation of tryptophan is not the only mechanism by which IFN-γ inhibits the replication of intracellular parasites (de la Maza *et al.*, 1985; Turco and Winkler, 1986). Proof that IDO can inhibit the replication of certain intracellular parasites requires experiments similar to those discussed above for 2-5A synthetase and the Mx protein. Cell lines expressing the cloned IDO cDNA under the control of a foreign promoter should exhibit increased resistance to infections with parasites.

III. Interferon-Induced Inhibition of Virus Replication

Our present knowledge about how IFNs inhibit the multiplication of a large number of different viruses is summarized in Table III. For several reasons it is often difficult to unambigously determine the relevant viral multiplication steps blocked in IFN-treated cells. First, properties of the host cell might largely determine the mode of IFN action. For example, IFN has dramatically different effects on the multiplication of influenza virus in mouse cells differing in the Mx locus (Staeheli and Haller, 1987). Cells from mice with a functional

TABLE III

Virus Multiplication Steps Inhibited in IFN-Treated Cells

Virus	Principal site of inhibition	IFN-induced protein responsible
DNA viruses		
Simian virus 40	Some early step, probably virus uncoating	Unknown
Herpes simplex virus	Trans-activation of immediate early genes by VP16, release of mature infectious virions	Unknown
Vaccinia virus	Maintenance of viral RNA pools	2-5A synthetase
Adenovirus	Translation of viral mRNAs	P1 kinase
Hepatitis B virus	Virion assembly(?)	Unknown
RNA viruses		
Vesicular stomatitis virus	Accumulation of primary viral mRNAs	Human MxA protein and unknown others
	Translation of viral mRNAs	P1 kinase
Mengo virus, encephalomyocarditis virus	Maintenance of viral mRNA pools, translation of viral mRNAs	2-5A synthetase P1 kinase and unknown others
Reovirus	Translation of viral mRNAs	P1 kinase and unknown others
Influenza virus	Early step: accumulation or translation of primary viral mRNAs	Mouse Mx1 and human MxA proteins, respectively
Retroviruses	Before intregration of proviral DNA, budding of particles	Unknown

Mx system synthesize the Mx protein in response to IFN treatment, and they become resistant to a large number of viruses, including influenza virus. In contrast, cells from mice with nonfunctional *Mx* genes lack the ability to produce the Mx protein. When treated with IFN, such cells remain almost completely susceptible to influenza virus. It is well known that established cell lines often differ with respect to the accumulation of IFN-induced proteins. Therefore, conflicting results about the mode of action of IFN toward certain viruses might often be attributed to cell line differences. If a virus is strongly inhibited by two IFN-induced proteins, each acting by a unique mechanism, the modes of inhibition might appear to be completely different, depending on whether a cell line is used which lacks either the first or the second IFN-induced protein.

Second, the choice of the type of IFN might greatly determine the conclusions from studies about the mode of IFN action. IFN-α/β and IFN-γ induce only partially overlapping sets of IFN-induced proteins (Beresini *et al.*, 1988). Consequently, the antiviral effects induced by the different IFNs might also differ considerably. Ulker and Samuel (1985) presented evidence that the molecular mechanisms of VSV inhibition by IFN-γ in human amnion U cells are fundamentally different from those of IFN-α.

Further difficulties arise because IFN-mediated inhibition of virus multiplication often occurs at many different levels. Defining the principal virus multiplication step inhibited by IFN is usually straightforward, but defining additional steps of less pronounced inhibition can be difficult. One should bear in mind that even minor effects at multiple sites might eventually lead to drastic overall effects.

A. DNA Viruses

1. Simian Virus 40

The results of several independent studies permit the conclusion that the principal site of IFN action against simian virus 40 (SV40), a member of the Papovaviridae, is the uncoating process of the infecting virions. It is not known which of the IFN-induced proteins is responsible for this effect.

Work by Oxman and Levin (1971) established that the levels of early SV40 RNAs and viral T antigen polypeptides were decreased greatly in cells treated with IFN before infection. It was then noted that in permissive monkey cells appreciable protection against SV40-induced cytopathic effects was achieved only when treatment with IFN preceded SV40 infection by at least 10 hours. IFN treatment after infection was ineffective (Mozes and Defendi, 1979), demonstrating that IFN affected an early step of the SV40 multiplication cycle and that later steps were probably not the principal sites of IFN action. Metz *et al.* (1976) found that attachment and penetration of SV40 virions were not affected significantly by IFN treatment of the host cells. Yamamoto *et al.* (1975) observed that the degree of protection by IFN decreased with increasing multiplicity of infection and that viral macromolecule synthesis was much less inhibited by IFN when monkey cells were infected with the naked DNA, rather than with intact virions. Together, these data suggested either that uncoating of the virions was blocked in IFN-treated cells or that the viral DNA was degraded immediately after uncoating.

Yakobson *et al.* (1977) found that the treatment of monkey cells with IFN after virus infection resulted in a moderate (33–50%) inhibition of viral mRNA translation, indicating that late stages of SV40 multiplication are also sensitive to IFN. However, no such effects on SV40 mRNA translation were observed in a similar study by Daher and Samuel (1982).

Brennan and Stark (1983) reinvestigated the mode of IFN action toward SV40. They compared the IFN-induced antiviral states of monkey CV-1 cells and SV40-transformed COS cells that constitutively express T antigen. Multiplication of SV40 was strongly inhibited in IFN-treated CV-1 cells, but not in IFN-treated COS cells, suggesting that T antigen might play a crucial role in overcoming the IFN-induced cellular resistance to SV40. IFN treatment efficiently blocked the growth of VSV in both cell lines, demonstrating that the COS cells were otherwise responsive to IFN. To further investigate the role of T antigen, a series of elegant experiments was performed with a temperature-sensitive mutant of SV40. Mutant virus *tsA58* encodes a T antigen which is inactive at 41°C, but which regains its activity at 32°C. At 41°C transcription of early viral genes is observed, but replication of the mutant virus does not occur, due to the lack of functional T antigen.

Brennan and Stark (1983) first infected CV-1 cells at 41°C with *tsA58* to permit accumulation of the mutant T antigen. They then treated the cells with IFN at 41°C for 24 hours before shifting the temperature down to 32°C and superinfecting the cells with a second SV40 virus. Under these conditions replication of the second virus was not blocked significantly, demonstrating that an antiviral state toward SV40 could not be maintained in the presence of T antigen, reminiscent of the situation in IFN-treated COS cells.

The authors concluded from these results that IFN pretreatment must inhibit SV40 multiplication by blocking the onset of early viral transcription. They proposed that, in cells pretreated with IFN, a virion component might not be removed from the DNA of the infecting particle, thus preventing its transcription. Viral DNA replication (which is T antigen dependent) is not blocked in IFN-treated cells. Thus, if T antigen is provided in trans, viral DNA replication is initiated, despite incomplete uncoating. Since the newly synthesized viral DNA is presumably free of coat components, transcription from these templates can proceed unhindered in IFN-treated cells. If such a mechanism were operative, one would expect that even traces of T antigen in an infected cell were sufficient to overcome the restrictions of the IFN system toward SV40. Indeed, the effectiveness of IFN toward SV40 is greatly reduced when high doses of virus are used (Yamamoto *et al.*, 1975). A high multiplicity of infection might increase the chance that the uncoating in IFN-treated cells of at least one infecting virion is possible, which, in turn, would be transcribed and would direct the synthesis of T antigen.

2. Herpes Simplex Virus

The principal stage of herpes simplex virus (HSV) multiplication blocked by IFN appears to be an early step, possibly an event between uncoating of the infecting virion and the onset of immediate early viral gene transcription. In IFN-treated cells VP16, a structural protein of HSV, might fail to trans-activate viral gene transcription. Inhibition of some late step in virus morphogenesis in IFN-treated cells has also been described. There are no indications as to which of the IFN-induced proteins might be responsible for HSV inhibition.

Lipp and Brandner (1980) documented a nearly complete impairment of HSV-1-specific protein and DNA syntheses in infected primary African green monkey kidney cells pretreated with the IFN-β-inducing agent poly(I) : poly(C). Similarly, in human diploid fibroblasts or in a mouse cell line treated with homologous IFNs before infection with HSV-1, the synthesis of all classes of viral proteins was inhibited (Lipp and Brandner, 1985). When macrophages were infected with HSV-1, viral DNA polymerase activity was low in cells pretreated with IFN as compared to infected cells not treated with IFN (Domke *et al.*, 1985; Domke-Opitz *et al.*, 1986). A strong inhibition of early (α) protein synthesis was observed, whereas the concentration of mRNA encoding the α protein ICP 4 was moderately reduced (Straub *et al.*, 1986). These results indicated that IFN acts early during the HSV multiplication cycle.

Analysis of viral DNA in the nuclei of permissive cells early after infection revealed that the uptake of virions, their transport to the nucleus, and the stability of viral DNA were not altered in IFN-treated cells (Mittnacht *et al.*, 1988). Nuclear run-off experiments revealed a significant reduction of immediate early gene transcription in IFN-treated cells (Mittnacht *et al.*, 1988; Oberman and Panet, 1988). When nuclear run-off RNA from IFN-treated and control cells was hybridized to short probes from either the 5' or the 3' end of the immediately early gene *ICP 4*, a comparable reduction of the signal intensity was observed, suggesting that the onset of transcription, rather than elongation of the RNA chains, was blocked in IFN-treated cells (Mittnacht *et al.*, 1988).

When ICP 4 promoter-controlled marker genes were transfected into fibroblast cells, transcription from the ICP 4 promoter was not influenced by IFN (LaMarco and McKnight, 1990; H. Jacobsen, personal communication). However, when these transfected cells were infected with HSV-1, strong trans-activation of the marker genes was observed only in the absence of IFN (De Stasio and Taylor, 1989). Treatment of the transfected cells with IFN-α or -γ prior to infection with HSV abolished trans-activation of the marker gene, but had no effect on the transcription of a cellular gene. Experiments by LaMarco and McKnight (1990) suggested that the trans-activating HSV protein VP16 fails to function properly in IFN-treated cells. Cotransfection of a VP16 construct and an ICP 4 promoter-controlled marker gene resulted in strong trans-activation, and this VP16-mediated trans-activation was abolished in cells treated with IFN. Based on the observation that the ICP 4 promoter and the promoters of IFN-induced genes contain related sequences, LaMarco and McKnight (1990) proposed that some induced transcription factors of IFN-treated cells might bind to the ICP 4 promoter, thereby blocking the enhancer effect of VP16.

Muñoz and Carrasco (1984) investigated the effects of IFN on HSV-1 multiplication in HeLa cells. They found no strong inhibition of viral protein synthesis and no inhibition of the formation of new virus particles from cells treated with IFN-α, as assessed by electron microscopy and biochemical analysis. However, the infectivity of virions released from IFN-treated cells was reduced. This effect of IFN on virus morphogenesis was observed in cells treated with IFN-α, but not in cells treated with IFN-γ. In IFN-γ-treated cells HSV-1 multiplication was blocked at an early stage between virus entry and the transcription of immediate early genes (Feduchi *et al.*, 1989). Altinkilic and Brandner (1988) recently undertook a reinvestigation of the mode of action of IFN-α toward HSV. They failed to confirm the conclusion of Muñoz and Carrasco (1984) that IFN-α-treated cells release noninfectious particles. Using the same virus strains and HeLa cells lines, they observed, rather, that virus-specific proteins failed to accumulate to normal levels.

Chatterjee *et al.* (1984) found similar levels of viral DNA synthesized in both IFN-α-treated and untreated human fibroblasts, but the total number of HSV-1 particles released from IFN-treated cells was reduced greatly (Chatterjee *et al.*, 1985). The synthesis of two specific viral glycoproteins (i.e., D and B) was found to be drastically reduced or delayed in IFN-treated human or monkey cells (Chatterjee *et al.*, 1985; Chatterjee and Whitley, 1989). These data suggest that IFN can also block a late stage in HSV morphogenesis.

3. Vaccinia Virus

Inhibition of vaccinia virus multiplication appears to result from the decreased stability of some early viral mRNAs in IFN-treated cells. This RNA degradation might result from activation of the dsRNA-dependent 2-5A synthetase pathway.

Metz and Esteban (1972) demonstrated that viral protein synthesis was strongly inhibited in IFN-treated L cells infected with vaccinia virus, whereas total viral RNA accumulation was not much reduced. This inhibition correlated with an activation of the 2-5A synthetase pathway: Degradation of viral and cellular RNAs was observed (Esteban *et al.*, 1984). Vaccinia virus replication was relatively resistant to IFN in other lines of L cells which produced 2-5A oligonucleotides late in the infection. Vaccinia virus-encoded ATPase and phosphatase activities were discovered (Paez and Esteban, 1984b; Rice *et al.*, 1984), which presumably degrade 2-5A oligonucleotides and thereby interfere with the IFN-induced antiviral state.

Reduced levels of vaccinia virus early mRNAs were observed in IFN-treated chick cells (Bialy and Colby, 1972). Jungwirth *et al.* (1972) concluded from their experiments with chick cells that IFN-mediated inhibition of vaccinia virus protein synthesis was probably due to the nonefficient translation of early virus-specific mRNAs, rather than the inhibition of RNA synthesis. Pulse-labeling experiments indeed showed (Grün *et al.*, 1987) that early vaccinia virus mRNAs were synthesized, but that these mRNAs failed to accumulate to normal levels in IFN-treated cells. When extracted from IFN-treated cells, vaccinia virus early mRNAs were largely degraded, suggesting that the 2-5A synthetase pathway was activated (Grün *et al.*, 1987).

4. Adenovirus

Human adenoviruses are generally more resistant to the antiviral action of IFN than are other animal viruses. This resistance appears to be caused by the presence in infected cells of virus-encoded small RNA molecules, which prevent activation of the IFN-induced P1 protein kinase by dsRNA.

Work with virus mutants led to the discovery that IFN resistance of adenovirus was dependent on VAI gene function (Schneider *et al.*, 1985). VAI RNA, an approximately 160-nucleotide-long highly structured RNA coded for by the virus genome, is found in large amounts at late stages of infections with wild-type adenovirus. The deletion mutant *dl331* that fails to produce VAI RNA grew poorly in human 293 cells that possess a functional P1 kinase system, but grew nearly as well as wild-type virus in human GM2767A cells that lack detectable P1 kinase activity. Inhibition of P1 kinase by adenovirus VAI RNA was also demonstrated *in vitro* with extracts of IFN-treated cells and purified VAI RNA (Kitajewski *et al.*, 1986).

5. *Hepatitis B Virus*

Treatment with IFN-α efficiently decreased hepatitis B virus (HBV) particles in patient sera (Lai *et al.*, 1987). Since *in vitro* culture systems for HBV are not available, analysis of the mechanism by which IFN blocks HBV multiplication is difficult. Available data suggest that IFN might interfere with HBV core particle assembly.

Transient or stable transfection of HBV DNA into human hepatoma cells can yield replicative intermediates and mature virus. When Hayashi and Koike (1989) studied the effect of IFN on HBV in this system, they found reduced amounts of viral DNA associated with intracellular HBV core particles. IFN had no effect on viral RNA and viral protein syntheses. These data therefore indicated that IFN might inhibit some step in the genome RNA-primed assembly of HBV core particles. Apparently, this step was not rate limiting in the culture system used, because IFN treatment had no effect on the accumulation of HBV particles secreted into the culture medium (Hayashi and Koike, 1989).

B. *RNA Viruses*

1. *Vesicular Stomatitis Virus*

Multiplication of VSV is extremely sensitive to the antiviral action of IFNs. Treatment of human fibroblasts or other cell types with low doses of IFN-α can reduce VSV yields to one one-millionth of their original values. Sensitivity to IFN-γ is also quite high, making VSV a suitable virus for assaying the biological activity of IFNs. Despite these dramatic effects of IFNs on VSV titers, biochemical analysis of the mode of VSV inhibition in IFN-treated cells has proved to be difficult, and the precise nature of the IFN-induced inhibitory activities is still obscure. Presumably, induced inhibitory activities are operative at multiple stages of the VSV multiplication cycle, leading to a pronounced synergistic effect. Depending on the cell line and the type of IFN studied, IFN was variably found to interfere with uptake of VSV particles, transcription of the VSV genome, and translation of the VSV mRNAs. The activity of the dsRNA-dependent P1 kinase is responsible, at least in part, for the observed inhibitory effects on VSV mRNA translation. The mechanism by which the human MxA protein (see Section II) interferes with VSV multiplication has not been resolved.

Treatment of human, mouse, or chick cells with homologous IFN-α or -β can lead to reduced rates of pinocytosis. The uptake of certain medium components or VSV virions was reduced to one-quarter to one-half of its original level in IFN-treated cells (Wilcox et al., 1983; Whitaker-Dowling et al., 1983), suggesting that VSV and presumably other viruses that penetrate via pinocytotic vesicles might encounter a first IFN-induced barrier as they penetrate the host cell. Some cell lines commonly used for IFN research do not show this phenotype (Wilcox et al., 1983; Masters and Samuel, 1983a; Belkowski and Sen, 1987); these cell lines, rather, permit equally efficient VSV adsorption and penetration, irrespective of IFN treatment.

Early work demonstrated that the primary transcription of VSV was reduced to about 20% in IFN-α-treated human fibroblasts (Manders et al., 1972). Similar observations were made with other cell types. Marcus and Sekellick (1978) found that the primary transcription of VSV was reduced to about one-fourth in IFN-α-treated Vero monkey cells. In human amnion U cells treated with IFN-γ, primary viral RNA synthesis was reduced to about one-quarter (Ulker and Samuel, 1985). Interestingly, treatment of amnion U cells with IFN-α instead of IFN-γ did not affect the primary transcription of VSV (Masters and Samuel, 1983b). Belakowski and Sen (1987) measured the accumulation of primary VSV transcripts in human and mouse cells treated with IFN-α. They found about 90% reduced levels of primary VSV transcripts in infected cells pretreated with IFN. Together, these data strongly suggest that VSV primary transcription is indeed a major site of IFN action.

Different mechanisms can be envisaged: VSV virion uncoating might be incomplete in IFN-treated cells, as in the case of SV40 (see above), virion RNAs might become degraded more rapidly in IFN-treated cells, or VSV polymerase activity might be inhibited by an IFN-induced protein. Recently, RNase activity was found to be associated with purified VSV nucleocapsids prepared from IFN-treated cells. This RNase was not associated with nucleocapsids prepared from infected control cells, suggesting that specific VSV RNA degradation

184   PETER STAEHELI

might play a role in IFN-induced cellular resistance toward this virus (Kumar *et al.*, 1988a). Further experiments are required to clarify whether this RNase activity can account for the inhibition of VSV primary transcription observed in IFN-treated cells. It will also be interesting to know whether this newly discovered RNase activity is related to the previously described RNase L activity, which is held responsible for the impaired mRNA transcription by reovirus cores isolated from IFN-treated cells (Lengyel *et al.*, 1980). The human MxA protein, recently shown to interfere with VSV multiplication when transfected into mouse 3T3 cells (Pavlovic *et al.*, 1990), could inhibit an early virus replication step.

The translation of VSV mRNAs is also inhibited in IFN-treated cells. This was shown in several independent studies, including a more recent report by Masters and Samuel (1983b). Simultaneous measuring of viral RNA and protein pools permitted these authors to estimate the average reduction in the rate of *in vivo* translation of VSV mRNAs. The calculated values for IFN-α-treated amnion U cells indicated that viral protein synthesis proceeded at about 10% of the unimpaired rate in control cells. Further evidence that VSV mRNA translation is inhibited in IFN-treated cells comes from double-infection experiments with VSV and vaccinia virus. The latter virus synthesizes a protein factor, designated specific kinase inhibitory factor (SKIF), which inhibits activation of the P1 kinase by dsRNA. Mouse L cells treated with IFN-α/β were resistant to VSV, but resistance to VSV was lost when the cells were superinfected with vaccinia virus (Thacore and Youngner, 1973; Whitaker-Dowling and Youngner, 1983; Paez and Esteban, 1984a). Apparently, SKIF inhibits the activity of the P1 kinase present in IFN-treated cells, thus permitting VSV mRNAs to be translated and VSV proteins to accumulate in sufficient quantities so that infectious virions can be assembled and released. These results establish a key role of the dsRNA-dependent P1 kinase in IFN-induced cellular resistance toward VSV.

Additional IFN-induced factors probably also play a role in reducing the efficiency of VSV mRNA translation. De Ferra and Baglioni (1981) found that VSV mRNAs were undermethylated in IFN-treated cells. Specifically, a high proportion of VSV mRNAs from IFN-treated cells had nonmethylated cap structures, and these RNAs were translated inefficiently *in vitro*, suggesting that an IFN-induced function selectively interfered with viral mRNA cap synthesis. Sahni and Samuel (1986) postulated that VSV mRNAs might have some intrinsic structural determinants recognized by an IFN-induced cellular factor in a rather sequence-specific manner. In this model factor-bound VSV mRNA would exhibit a reduced translation efficiency. To test this hypothesis, Sahni and Samuel (1986) transiently transfected COS cells with a VSV cDNA construct that permitted the expression of the VSV *G* gene from an SV40 promoter. Indeed, they found that IFN-treated and control COS cells contained similar concentrations of VSV mRNAs, but that VSV G protein synthesis was reduced to one-fifth in IFN-treated cells. Since no control transfections with a cellular cDNA were performed, it is still possible that the observed inhibition of VSV mRNA translation in IFN-treated COS cells was not virus specific. There is evidence that the P1 kinase can become activated during transient COS cell transfection experiments (Kaufman and Murtha, 1987), presumably by dsRNA resulting from illegitimate bidirectional transcription of circular plasmid DNA. Thus, additional experiments are required to firmly establish the existence of unique structural elements in VSV mRNAs that would permit the protein synthesis machinery of an IFN-treated cell to discriminate between cellular and viral mRNAs.

2. Mengo Virus and Encephalomyocarditis Virus

Studies of the IFN action toward picornaviruses were mainly performed with Mengo virus and EMC virus. The principal stage of picornavirus multiplication affected in IFN-treated cells is viral protein synthesis. Inhibition of viral protein synthesis is, at least in part, a result of greatly increased degradation of viral and cellular RNA in IFN-treated cells infected with picornaviruses. The 2-5A synthetase/RNase L system appears to play a key role in this degradation process. The dsRNA-dependent P1 kinase might also contribute to picornavirus inhibition by lowering the efficiency of viral mRNA translation in IFN-treated cells.

Early studies established that the principal inhibitory action of IFN against Mengo virus is exerted at a step in the multiplication cycle which immediately follows uncoating of the virion, but precedes *de novo* viral RNA synthesis. Viral protein synthesis was largely absent in IFN-treated cells, apparently due to the inhibition of polysome assembly (Levy and Carter, 1968). Conceivably, these effects resulted, at least in part, from increased degradation of the viral and cellular RNAs (Levy and Carter, 1968; Wreschner *et al.*, 1981; Watling *et al.*, 1985). The results of several investigations demonstrated that activation of the dsRNA-dependent 2-5A synthetase/RNase L system can severely inhibit picornavirus multiplication (Lengyel, 1982; Baglioni and Nilsen, 1983). First, it could be demonstrated that high levels of 2-5A oligonucleotides indeed occurred in IFN-treated cells after infection with EMC (Williams *et al.*, 1979) and that ribosomal RNA was getting degraded in such cells (Wreschner *et al.*, 1981).

Second, a chemically synthesized analog of natural 2-5A
oligonucleotides could inhibit the activation of the 2-5A-dependent
RNase L in IFN-treated mouse L cells by EMC virus, inhibit ribosomal
RNA degradation, and partially restore EMC RNA synthesis and virus
yields (Watling *et al.*, 1985). Third, when a cDNA encoding the human
40-kDa form of 2-5A synthetase was constitutively expressed in transfected Chinese hamster ovary cells, strong inhibition of Mengo virus,
but not VSV or HSV-2 multiplication, was observed (Chebath *et al.*,
1987a). Constitutive expression of this cDNA in mouse (Coccia *et al.*,
1988) or human cells (Rysiecki *et al.*, 1989) also reduced EMC virus
yields. These experiments thus provided direct proof that the 2-5A
synthetase/RNase L pathway plays a role in picornavirus inhibition.

The other dsRNA-dependent enzyme, P1 kinase, also appears to play
a beneficial role in IFN-mediated cellular resistance to picornaviruses.
When IFN-treated mouse L cells were coinfected with vaccinia virus
and EMC virus, the EMC virus yields were increased up to 1000-fold
compared to infections with EMC virus alone (Whitaker-Dowling and
Youngner, 1986). This result suggests that inhibition of P1 kinase
activity by a vaccinia virus-encoded factor permitted EMC virus to
grow to fairly high titers, despite activation of the 2-5A syn-
thetase/RNase L pathway.

The conclusion that both IFN-induced dsRNA-dependent enzyme
pathways are involved in picornavirus inhibition is compatible with
the results of several investigations of cell clones with altered 2-5A
synthetase expression patterns (Kumar *et al.*, 1987, 1988b; Lewis,
1988). In some of these clones, IFN-induced EMC protection was observed in the absence of a functional 2-5A synthetase/RNase L pathway. In a few clones IFN was shown to induce an antiviral state
against picornaviruses, although neither RNase L nor the P1 kinase
were activated, indicating that additional unidentified IFN-induced
proteins might further contribute to picornavirus inhibition.

3. Reovirus

The inhibitory effects of IFN on the multiplication of human reovirus serotype 3 (Dearing strain) and serotype 1 (Lang strain) have
been studied by several laboratories. Clearly, the principal stage of
IFN-mediated inhibition is the translation of early reovirus mRNAs.
Inhibition of protein synthesis is largely due to activation of the IFN-
induced dsRNA-dependent P1 kinase in cells infected with reovirus.

Wiebe and Joklik (1975) demonstrated that reovirus adsorption to
mouse L cells and conversion of parental virions to subviral particles
were resistant to the action of IFN. The effect of IFN on the accumulation and translation of early reoviral mRNAs was studied by using a

temperature-sensitive reovirus mutant. At the nonpermissive temperature accumulation of the viral mRNAs was only slightly reduced in IFN-treated cells, whereas the synthesis of early reovirus proteins was reduced dramatically (Wiebe and Joklik, 1975). This result suggested that transcription by the virion-associated polymerase was not affected significantly by IFN. Rather, it indicated that the translation of primary viral transcripts was blocked in IFN-treated cells. This inhibition appeared to be quite selective for reovirus mRNAs. Translation of cellular (Wiebe and Joklik, 1975) or coinfecting SV40 virus mRNAs was not adversely affected (Daher and Samuel, 1982).

The precise mechanism of reovirus polypeptide synthesis inhibition is not known. It was shown that the treatment of host cells with IFN inhibited reovirus cap methylation (Sen et al., 1977), increased the degradation rate of reovirus mRNAs (Nilsen et al., 1982; Baglioni et al., 1984), and altered components required for translation of reovirus mRNAs (Samuel, 1979; Nilsen et al., 1982; Samuel et al., 1984). When reovirus mRNAs lacking methylated cap structures were incubated with extracts from IFN-treated cells, they were less efficiently methylated by these extracts as compared to extracts from cells not pretreated with IFN (Sen et al., 1975). However, in vivo cap methylation of reovirus mRNAs was not diminished in IFN-treated cells under conditions of IFN treatment that reduced reovirus yields by more than 98% (Desrosiers and Lengyel, 1979), suggesting that the impairment of reovirus cap methylation is not the principal site of IFN action in vivo. When RNA extracted from infected cells was analyzed by the Northern blotting technique, full-sized reovirus mRNAs were detected in untreated cells, but not in IFN-treated cells (Nilsen et al., 1982), suggesting enhanced RNA degradation in cells treated with IFN. The concentration of 2-5A oligonucleotides in reovirus-infected cells increased to significant levels and an RNase activity became detectable in such cells which cleaved ribosomal RNA (Nilsen et al., 1982), suggesting that the 2-5A synthetase/RNase L pathway was activated.

Results of several other studies suggested that reovirus mRNA translation might be impaired mainly as a consequence of increased phosphorylation of protein synthesis initiation factor eIF-2 by activated dsRNA-dependent P1 kinase. Phosphorylation of eIF-2 was indeed increased in IFN-treated cells infected with reovirus serotype 3 (Samuel et al., 1984), and the kinetics of both induction and decay of P1 kinase correlated with the induction and decay of the antiviral state against reovirus (Samuel and Knutson, 1982). Furthermore, when single reovirus cDNAs were transiently expressed in transfected COS monkey cells, IFN treatment caused a strong inhibition of reovirus protein synthesis in such cells, whereas the reoviral mRNA concentrations were not decreased significantly (George and Samuel, 1988). It is likely that the observed inhibition was caused by an activation of the IFN-induced P1 kinase. This might indicate that reovirus mRNAs have intrinsic properties that cause the activation of the P1 kinase. Alternatively, the P1 kinase might have been activated by dsRNA molecules or other inducing factors apparently present in cells transiently transfected with plasmid DNAs of either viral or cellular origin (Kaufman and Murtha, 1987).

Imani and Jacobs (1988) observed that the σ3 protein of reovirus serotype 1 (Lang strain) can efficiently block the activation of P1 kinase by dsRNA. It was shown that this inhibition could be overcome by an excess of dsRNA, suggesting that the σ3 protein binds to dsRNA and thereby prevents activation of P1 kinase. The serotype 1 strain of reovirus is considerably more resistant to IFN than serotype 3 strains. Imani and Jacobs (1988) found that the serotype 1 strain also contained more kinase inhibitory activity, suggesting that P1 kinase indeed plays a key role in the IFN-mediated cellular resistance to reovirus.

Jacobs *et al.* (1988) demonstrated that reovirus mRNA preparations which were extensively depleted of contaminating dsRNA molecules failed to activate the P1 kinase *in vitro.* Surprisingly, however, such reovirus mRNAs were still translated at poor efficiency in extracts from IFN-treated cells. These results indicate that additional IFN-induced activities could restrict the translation of reovirus mRNAs and that P1 kinase might not be the only IFN-induced protein blocking reovirus multiplication.

4. Influenza Virus

Influenza virus occupies a special place in IFN history. IFN was discovered and first defined as an agent capable of inducing an antiviral state against influenza virus in chick cells (Isaacs and Lindenmann, 1957). Analysis of the precise mode of influenza virus inhibition in IFN-treated cells has proved to be difficult, and different researchers have drawn conflicting conclusions from their experiments. It is clear that IFN blocks the influenza virus multiplication cycle at an early stage, shortly before, during, or immediately after primary transcription of the viral genome. In mouse cells this inhibitory effect is mainly due to the IFN-induced nuclear Mx protein. The homologous human MxA protein also interferes with the multiplication of influenza viruses, although this protein accumulates in the cytoplasm of IFN-treated cells. This suggests different modes of action of the human and mouse Mx proteins.

Bean and Simpson (1973) studied the effect of IFN on the early transcription of influenza virus in chick cells. They infected the cells with ^{32}P-labeled virus, extracted the cellular RNAs at different times after infection, allowed annealing of any complementary RNAs in the sample, and then monitored the conversion of the labeled virion RNA into a ribonuclease-resistant (double-stranded) form. With this technique it was possible to demonstrate that influenza viral transcription reached significant levels in control chick cells 60–180 minutes postinfection, whereas IFN pretreatment of chick cells abolished the accumulation of significant amounts of influenza virus-specific transcripts (Bean and Simpson, 1973).

Using similar methods, Repik et al. (1974) also found greatly reduced concentrations of influenza virus transcripts in IFN-treated chick cells. However, since comparably low concentrations of influenza virus transcripts were also measured in cells treated with the protein synthesis inhibitor cycloheximide, this result was thought to indicate that IFN-mediated inhibition of influenza virus did not operate at the level of primary transcription, but at an intermediate step between primary and secondary transcriptions, such as viral protein synthesis (Repik et al., 1974). Clearly, the sensitivity of this assay was too low to permit any firm conclusions.

Haller et al. (1980) discovered that expression of the genetically determined influenza virus resistance phenotype of the mouse was regulated by IFN. The product of the resistance gene, the Mx protein, was later shown to render cells selectively resistant to infections with influenza virus (Staeheli et al., 1986b; see Section II). Studying the IFN-mediated inhibition of influenza virus multiplication in cells from congenic mice either possessing (Mx$^+$) or lacking (Mx$^-$) a functional Mx system offered the unique opportunity to separately evaluate the contributions of the Mx protein and additional IFN-induced proteins. Using explanted macrophages of Mx-bearing animals, Horisberger et al. (1980) were able to demonstrate that IFN failed to inhibit influenza virus adsorption and penetration. It was further shown that influenza virus uncoating presumably was not a principal target of IFN action in Mx-bearing mouse cells. When virus uncoating was measured as a function of radiolabeled input virion RNA becoming accessible to RNase digestion in permeabilized cells, the kinetics of viral uncoating were found to be essentially the same in Mx$^+$ and Mx$^-$ cells, and this process was not influenced by IFN treatment in either cell type (Meyer and Horisberger, 1984). However, virus protein synthesis was abolished almost completely in IFN-treated Mx$^+$ cells, whereas viral protein synthesis was only marginally reduced in Mx$^-$ cells. Hence, the target of IFN action in Mx$^+$ mouse cells was probably after virion uncoating and before viral protein synthesis. It was further possible to conclude from these studies that IFN-induced proteins other than Mx protein did not contribute significantly to the IFN-mediated resistance of mouse cells against influenza virus.

Meyer and Horisberger (1984) and Krug *et al.* (1985) independently studied the effect of IFN on influenza virus primary transcription in Mx$^+$ mouse cells. If cells are infected with influenza virus in the presence of protein synthesis inhibitors such as cycloheximide or anisomycin, virus-specific primary transcription is permitted, but not replication or amplification of the viral genome. Since the synthesis of primary transcripts is catalyzed by virion-associated enzymes, such transcripts accumulate to fairly high levels in cycloheximide-treated cells after their infection with influenza virus. Conflicting conclusions were reached with respect to the step of influenza virus multiplication blocked in IFN-treated Mx$^+$ cells. Meyer and Horisberger (1984) detected comparable amounts of primary influenza virus transcripts in IFN-treated Mx$^+$ and Mx$^-$ macrophages, and these transcripts were shown to function as mRNAs *in vitro*. However, *in vivo* translation of these mRNAs was blocked in IFN-treated Mx$^+$ cells. Meyer and Horisberger (1984) therefore concluded that the principal stage of influenza virus inhibition by Mx protein was after the synthesis of primary transcripts, presumably at the level of mRNA translation.

In contrast, Krug *et al.* (1985) detected primary influenza virus transcripts only in IFN-treated Mx$^-$ cells. They failed to detect significant levels of influenza virus-specific transcripts in IFN-treated Mx$^+$ embryo cells. When the rate of influenza viral mRNA synthesis was measured in permeabilized cells, influenza virus-specific transcription was observed in IFN-treated Mx$^-$ cells, but not IFN-treated Mx$^+$ cells. Krug *et al.* (1985) concluded from these results that the Mx protein most likely blocked the onset of influenza virus transcription. It is difficult to find any reasonable explanation for the apparent discrepancies between the results of the two groups. Possibly, the use of different virus strains and cell cultures was critical. The recently established transfected 3T3 cell lines expressing high levels of the mouse Mx protein in a constitutive manner (Pavlovic *et al.*, 1990) should help to definitively define the site of Mx protein action.

Most strains of influenza virus grow to high titers in bovine cells, and the treatment of such cells with human IFN-α resulted in a dramatic reduction of virus yields (Horisberger, 1988). Ransohoff *et al.* (1985) studied the molecular basis of this inhibition. No effect of IFN treatment was observed on the nuclear accumulation of virion RNAs by 2 hours postinfection, suggesting that penetration and uncoating of influenza virus were not blocked to significant degrees in IFN-treated bovine cells. However, when primary transcription was measured as a function of polyadenylated influenza viral RNAs present in the cytoplasm of cycloheximide-treated cells by 7 hours postinfection, a strong inhibitory effect of IFN was observed (Ransohoff et al., 1985). The failure of IFN-treated bovine cells to accumulate influenza viral primary transcripts under these conditions indicates either that viral transcription was directly inhibited or that viral transcripts showed increased lability in IFN-treated cells. Similarly, no polyadenylated primary influenza viral RNAs were found in IFN-treated rat cells by 5 hours postinfection (Meier et al., 1988). Mechanistic studies on the antiinfluenza activity of IFN in human cells have not been published. Recent experiments in our laboratory demonstrated that one of the Mx-related human proteins, MxA, is active against influenza virus. Mouse 3T3 cells transfected with appropriate cDNA constructs accumulated the MxA protein in the cytoplasm and thereby acquired resistance to influenza virus (Pavlovic et al., 1990).

5. Retroviruses

Production of infectious retrovirus by chronically infected cell lines is greatly reduced when such cells are treated with IFNs. IFN treatment can also dramatically reduce the virus production of cells newly infected with exogenous retrovirus. IFN-mediated inhibition appears to operate at more than one level. An early step of the retrovirus multiplication cycle, presumably proviral DNA formation, and a late step, presumably the release of infectious viral particles, are inhibited in IFN-treated cells. It is unclear which of the IFN-induced proteins might cause these effects.

The multiplication of Rous sarcoma virus in chick embryo cells was strongly inhibited when the cells were treated with IFN prior to de novo infection with virus (Strube et al., 1982). Greatly reduced levels of virus-specific proteins and RNAs were measured 3 days after infection of the IFN-treated cells, and the level of virus-specific DNA was about one-twentieth that in control cells not treated with IFN (Strube and Jungwirth, 1985; Zens et al., 1989). These results suggested that IFN blocks Rous sarcoma virus replication in chick cells during or before viral DNA synthesis. The Kirsten strain of murine sarcoma virus failed to transform mouse 3T3 cells, provided these cells were treated with IFN before infection with the virus (Morris and Burke, 1979; Avery et al., 1980). Most of the IFN-pretreated cells contained no detectable proviral DNA. Since IFN treatments before infection and up to 4 hours postinfection were equally effective, it seemed that IFN did not act by inhibiting adsorption, penetration, or uncoating of the virus (Avery et al., 1980). More likely, the synthesis or integration of proviral DNA was blocked in IFN-treated cells.

Huleihel and Aboud (1983) studied the effect of IFN on the multiplication of murine sarcoma virus in rat kidney cells. At various times after infection, DNA was extracted from the cell cytoplasm and the nucleus, and the appearance of unintegrated and integrated proviral DNA was monitored. IFN was found to delay viral DNA synthesis, but the amount of viral DNA eventually formed in IFN-treated cells was normal. The transport of viral DNA into the nucleus was strongly delayed in IFN-treated cells, and this DNA failed to integrate into the host genome (Huleihel and Aboud, 1983), suggesting that IFN might inhibit proviral DNA integration by blocking its supercoiling.

In contrast to the sarcoma viruses just discussed, the early multiplication steps of Friend murine leukemia viruses were not inhibited by the IFN treatment of mouse 3T3 cells (Riggin and Pitha, 1982). The synthesis of proviral DNA and its subsequent integration into the cell genome proceeded unaffected by IFN, indicating that the reduced virus production was due to inhibition of a later step.

IFN can transiently reduce the yields of infectious retroviruses of chronically infected cells. Sen and Sarkar (1980) found that IFN treatment greatly reduced the concentration of mouse mammary tumor virus produced by an established cell line, without significantly influencing the rates of viral RNA and protein syntheses. Electron microscopy of thin sections from the control and IFN-treated cells revealed no differences in any subcellular viral structures; however, IFN-treated cells showed an increased number of virus particles on the cell surface (Sen and Sarkar, 1980). These results suggested a block at a late stage of virus morphogenesis in IFN-treated cells. Similarly, IFN treatment of mouse 3T3 cells chronically infected with Moloney murine leukemia virus inhibited virus release by about 95%. This inhibition was accompanied by an accumulation of intracellular viral structures which appeared to lack the RNA genome (Aboud and Hassan, 1983), suggesting a defect in virus assembly in the IFN-treated cells. IFN-treated cells were found to produce noninfectious virus particles lacking the viral glycoprotein gp71 (Bilello et al., 1982). When treated with IFN for 18 hours, chick cells chronically infected with Rous sarcoma virus (RSV) strains of receptor subgroups A–C produced one-twentieth to one onehundredth the infectious virus (Zens et al., 1989). Viral RNA and protein syntheses were not affected, suggesting an IFN-mediated inhibition of virus maturation. IFN treatment of human or monkey cells chronically infected with Mason–Pfizer monkey virus reduced the release of infectious particles by more than 97%, whereas the number of intracytoplasmic particles remained normal (Chatterjee and Hunter, 1987). Hence, IFN apparently also affected a late step in the life cycle of this retrovirus.

Kornbluth *et al.* (1989) recently studied the inhibitory effect of IFN on the multiplication of human immunodeficiency virus (HIV) in cultured human macrophages. In cultures treated with IFN-α, -β, or -γ, little or no infectious HIV or p24 core antigen was released into the supernatants, no virions were seen by electron microscopy, no viral RNA or DNA was detectable in the cell lysates, and no cytopathology occurred. Inhibition was also pronounced when macrophages were treated with IFNs 3 days after HIV infection (Kornbluth *et al.*, 1989), suggesting a block in the virus life cycle beyond the early events of virus binding, penetration, and uncoating. Further experiments are necessary to distinguish between the possibilities of an IFN-mediated inhibition of proviral DNA synthesis or integration and inhibition of a later step in viral morphogenesis.

IV. CONCLUSIONS

Studies of the action of IFN against different viruses have clearly shown that the IFN-induced inhibitory mechanisms are multifaceted. This result is not expected, considering that IFN-treated cells contain a large number of induced proteins believed to mediate all IFN-induced effects.

A picture is slowly emerging which shows that individual IFN-induced proteins have distinct biochemical activities that lead to discrete physiological changes in IFN-treated cells. Consequently, a given IFN-induced protein might dramatically inhibit the multiplication of one group of viruses and at the same time have no detectable effects on a second group of viruses. Examples of IFN-induced proteins with restricted antiviral properties are 2-5A synthetase, which inhibits picornaviruses and possibly vaccinia virus, and the Mx proteins, which inhibit influenza viruses and, in the case of the human MxA protein, the rhabdovirus VSV. Other IFN-induced proteins could regulate some basic cellular functions (e.g., protein synthesis), thereby affecting the multiplication of a large panel of unrelated viruses. An example of an IFN-induced protein of this type is the protein kinase P1. It is reasonable to assume that the IFN-induced antiviral state should be considered the sum of antiviral activities of a large number of IFN-induced proteins acting in concert.

It is important to note that at least some of the IFN-induced proteins confer antiviral protection when constitutively expressed in permanently transfected cell lines. This indicates that the constitutive expression of individual IFN-induced proteins in transfected cells and screening of the resulting cell lines for acquired antiviral properties might be reasonable experimental approaches, which should permit us to define the potential antiviral and other activities of a large number of IFN-induced proteins with still unknown functions.

ACKNOWLEDGMENTS

I thank H. Arnheiter, Y.-S. E. Cheng, A. Hovanessian, H. Jacobsen, R. Krug, S. McKnight, E. Meier, T. Stevens, and B. Williams, who contributed to this chapter by communicating unpublished results, and M. Aebi, M. Aguet, O. Haller, J. Lindenmann, and J. Pavlovic for helpful discussions and comments on the manuscript.

REFERENCES

Aboud, M., and Hassan, Y. (1983). *J. Virol.* **45**, 489–495.
Aebi, M., Fäh, J., Hurt, N., Samuel, C. E., Thomis, D., Bazzigher, L., Pavlovic, J., Haller, O., and Staeheli, P. (1989). *Mol. Cell. Biol.*, **9**, 5062–5072.
Aguet, M., and Mogensen, K. E. (1983). *Interferon* **5**, 1–22.
Aguet, M., Dembic, Z., and Merlin, G. (1988). *Cell (Cambridge, Mass.)* **55**, 273–280.
Altinkilic, B., and Brandner, G. (1988). *J. Gen. Virol.* **69**, 3107–3112.
Arnheiter, H., and Haller, O. (1988). *EMBO J.* **7**, 1315–1320.
Avery, R. J., Norton, J. D., Jones, J. S., Burke, D. C., and Morris, A. G. (1980). *Nature (London)* **288**, 93–95.
Baglioni, C., and Nilsen, T. W. (1983). *Interferon* **5**, 23–42.
Baglioni, C., de Benedetti, A., and Williams, G. J. (1984). *J. Virol.* **52**, 865–871.
Bazzigher, L., Haller, O., and Staeheli, P. (1990). Submitted for publication.
Bean, W. J., Jr., and Simpson, R. W. (1973). *Virology* **56**, 646–651.
Belkowski, L. S., and Sen, G. C. (1987). *J. Virol.* **61**, 653–660.
Benech, P., Merlin, G., Revel, M., and Chebath, J. (1985a). *Nucleic Acids Res.* **13**, 1267–1281.
Benech, P., Mory, Y., Revel, M., and Chebath, J. (1985b). *EMBO J.* **4**, 2249–2256.
Beresini, M. H., Lempert, M. J., and Epstein, L. B. (1988). *J. Immunol.* **140**, 485–493.
Berry, M. J., Knutson, G. S., Lasky, S. R., Munemitsu, S. M., and Samuel, C. E. (1985). *J. Biol. Chem.* **260**, 11240–11247.
Bialy, H. S., and Colby, C. (1972). *J. Virol.* **9**, 286–289.
Bilello, J. A., Wivel, N. A., and Pitha, P. M. (1982). *J. Virol.* **43**, 213–222.
Black, T. L., Safer, B., Hovanessian, A., and Katze, M. G. (1989). *J. Virol.* **63**, 2244–2251.
Blomstrom, D. C., Fahey, D., Kutny, R., Korant, B. D., and Knight, E. (1986). *J. Biol. Chem.* **261**, 8811–8816.
Brennan, M. B., and Stark, G. R. (1983). *Cell (Cambridge, Mass.)* **33**, 811–816.
Byrne, G. I., Lehmann, L. K., Kirschbaum, J. G., Borden, E. C., Lee, C. M., and Brown, R. R. (1986a). *J. Interferon Res.* **6**, 389–396.
Byrne, G. I., Lehmann, L. K., and Landry, G. J. (1986b). *Infect. Immun.* **53**, 347–351.
Caplen, H. S., and Gupta, S. L. (1988). *J. Biol. Chem.* **255**, 8390–8393.
Carlin, J. M., Borden, E. C., and Byrne, G. I. (1989). *J. Interferon Res.* **9**, 329–337.
Chatterjee, S., and Hunter, E. (1987). *Virology* **157**, 548–551.
Chatterjee, S., and Whitley, R. J. (1989). *Virus Res.* **12**, 33–42.
Chatterjee, S., Lakeman, A. D., Whitley, R. J., and Hunter, E. (1984). *Virus Res.* **1**, 81–87.

Chatterjee, S., Hunter, E., and Whitley, R. (1985). *J. Virol.* **56**, 419–425.
Chebath, J., Merlin, G., Metz, R., Benech, P., and Revel, M. (1983). *Nucleic Acids Res.* **11**, 1213–1226.
Chebath, J., Benech, P., Revel, M., and Vigneron, M. (1987a). *Nature (London)* **330**, 587–588.
Chebath, J., Benech, P., Hovanessian, A., Galabru, J., and Revel, M. (1987b). *J. Biol. Chem.* **262**, 3852–3857.
Cheng, Y.-S. E. (1990). Manuscript in preparation.
Cheng, Y.-S. E., Colonno, R. J., and Yin, F. H. (1983). *J. Biol. Chem.* **258**, 7746–7750.
Cheng, Y.-S. E., Becker-Manley, M. F., Chow, T. P., and Horan, D. C. (1985). *J. Biol. Chem.* **260**, 15834–15839.
Cheng, Y.-S. E., Becker-Manley, M. F., and Rosenblum, M. G. (1986a). In "The Biology of the Interferon System 1985." (W. E. Stewart and H. Schellekens, eds.), pp. 135–138. Elsevier, Amsterdam.
Cheng, Y.-S. E., Nguyen, T. D., and Becker-Manley, M. (1986b). In "The Biology of the Interferon System 1985." (W. E. Stewart and H. Schellekens, eds.), pp. 139–142. Elsevier, Amsterdam.
Cheng, Y.-S. E., Becker-Manley, M. F., Nguyen, T. D., Degrado, W. F., and Jonak, G. J. (1986c). *J. Interferon Res.* **6**, 417–427.
Cheng, Y.-S. E., Becker-Manley, M. F., Rucker, R. G., and Borden, C. (1988). *J. Interferon Res.* **8**, 385–391.
Coccia, E. M., Romeo, G., Affabris, E., Battistini, A., Fiorucci, G., Marziali, G., Orsatti, R., Albertini, R., Chebath, J., and Rossi, G. B. (1988). *Abstr. Annu. Meet. Int. Soc. Interferon Res.*
Cofano, F., Comoglio, P. M., Landolfo, S., and Tarone, G. (1984). *J. Immunol.* **133**, 3102–3106.
Cohen, B., Peretz, D., Vaiman, D., Benech, P., and Chebath, J. (1988). *EMBO J.* **7**, 1411–1419.
Daher, K. A., and Samuel, C. E. (1982). *Virology* **117**, 379–390.
Dale, T. C., Ali Imam, A. M., Kerr, I. M., and Stark, G. R. (1989). *Proc. Natl. Acad. Sci. U.S.A.* **86**, 1203–1207.
De Ferra, F., and Baglioni, C. (1981). *Virology* **112**, 426–435.
de la Maza, L. M., Peterson, E. M., Fennie, C. W., and Czarniecki, C. W. (1985). *J. Immunol.* **135**, 4198–4200.
De Maeyer, E. M., and De Maeyer-Guignard, J. (1988). "Interferons and Other Regulatory Cytokines." Wiley (Interscience), New York.
Desrosiers, R. C., and Lengyel, P. (1979). *Biochim. Biophys. Acta* **562**, 471–480.
De Stasio, P. R., and Taylor, M. W. (1989). *Biochem. Biophys. Res. Commun.* **159**, 439–444.
Dever, T. E., Glynias, M. J., and Merrick, W. C. (1987). *Proc. Natl. Acad. Sci. U.S.A.* **84**, 1814–1818.
Domke, I., Straub, P., Jacobson, H., Kirchner, H., and Panet, A. (1985). *J. Gen. Virol.* **66**, 2231–2236.
Domke-Opitz, I., Straub, P., and Kirchner, H. (1986). *J. Virol.* **60**, 37–42.
Dreiding, P., Staeheli, P., and Haller, O. (1985). *Virology* **140**, 192–196.
Dustin, M. L., Rothlein, R., Bhan, A. K., Dinarello, C. A., and Springer, T. A. (1986). *J. Immunol.* **137**, 245–254.
Engel, D. A., Samanta, H., Brawner, M. E., and Lengyel, P. (1985). *Virology* **142**, 389–397.
Engel, D. A., Snoddy, J., Toniato, E., and Lengyel, P. (1988). *Virology* **166**, 24–29.
Esteban, M., Benavente, J., and Paez, E. (1984). *Virology* **134**, 40–51.
Feduchi, E., Alonso, M. A., and Carrasco, J. (1989). *J. Virol.* **63**, 1354–1359.

Flenniken, A. M., Galabru, J., Rutherford, M. N., Hovanessian, A. G., and Williams, B. R. G. (1988). *J. Virol.* **62**, 3077–3083.

Friedman, R. L., Manly, S. P., McMahon, M., Kerr, I. M., and Stark, G. R. (1984). *Cell (Cambridge, Mass.)* **38**, 745–755.

Galabru, J., and Hovanessian, A. G. (1987) *J. Biol. Chem.* **262**, 15538–15544.

Galabru, J., Katze, M. G., Robert, N., and Hovanessian, A. G. (1989). *Eur. J. Biochem.* **178**, 581–589.

George, C. X., and Samuel, C. E. (1988). *Virology* **166**, 573–582.

Goetschy, J.-F., Zeller, H., Content, J., and Horisberger, M. A. (1989). *J. Virol.* **63**, 2616–2622.

Grün, J., Kroon, E., Zöller, B., Krempien, U., and Jungwirth, C. (1987). *Virology* **158**, 28–33.

Haas, A. L., Ahrens, P., Bright, P. M., and Ankel, H. (1987). *J. Biol. Chem.* **262**, 11315–11323.

Haller, O. (1981). *Curr. Top. Microbiol. Immunol.* **92**, 25–52.

Haller, O., Arnheiter, H., Gresser, I., and Lindenmann, J. (1980). *Nature (London)* **283**, 660–662.

Haller, O., Acklin, M., and Staeheli, P. (1987). *J. Interferon Res.* **7**, 647–656.

Harada, H., Fujita, T., Miyamoto, M., Kimura, Y., Maruyama, M., Furia, A., Miyata, T., and Taniguchi, T. (1989). *Cell (Cambridge, Mass.)* **58**, 729–739.

Hayashi, Y., and Koike, K. (1989). *J. Virol.* **63**, 2936–2940.

Hemmi, S., Peghini, P., Metzler, M., Merlin, G., Dembic, Z., and Aguet, M. (1989). *Proc. Natl. Acad. Sci. U.S.A.,* **86**, 9901–9905.

Hillman, M. C., Jr., Knight, E., Jr., and Blomstrom, D. C. (1987). *Biochem. Biophys. Res. Commun.* **148**, 140–147.

Holter, W., Grunow, R., Stockinger, H., and Knapp, W. (1986). *J. Immunol.* **136**, 2171–2175.

Horisberger, M. A. (1988). *Virology* **162**, 181–186.

Horisberger, M. A., and Hochkeppel, H. K. (1985). *J. Biol. Chem.* **260**, 1730–1733.

Horisberger, M. A., and Hochkeppel, H. K. (1987). *J. Interferon Res.* **7**, 331–343.

Horisberger, M. A., Haller, O., and Arnheiter, H. (1980). *J. Gen. Virol.* **50**, 205–210.

Horisberger, M. A., Staeheli, P., and Haller, O. (1983). *Proc. Natl. Acad. Sci. U.S.A.* **80**, 1910–1914.

Horisberger, M. A., Wathelet, M., Szpirer, J., Szpirer, C., Islam, Q., Levan, G., Huez, G., and Content, J. (1988). *Somatic Cell Mol. Genet.* **14**, 123–131.

Hovanessian, A. G., and Galabru, J. (1987). *Eur. J. Biochem.* **167**, 467–473.

Hovanessian, A. G., Laurent, A. G., Chebath, J., Galabru, J., Robert, N., and Svab, J. (1987). *EMBO J.* **6**, 1273–1280.

Hovanessian, A. G., Galabru, J., Thomas, N., Shon, B., Robert, N., Svab, J., Brown, R. E., Kerr, I., Williams, B. R. G., and Meurs, E. (1988a). *Abstr. Annu. Meet. Int. Soc. Interferon Res.*

Hovanessian, A. G., Svab, J., Marié, I., Robert, N., Chamaret, S., and Laurent, A. G. (1988b). *J. Biol. Chem.* **263**, 4945–4949.

Huber, P., Aebi, M., Grob, R., Pravtcheva, D., Ruddle, F. H., and Haller, O. (1990). Manuscript in preparation.

Hug, H., Costas, M., Staeheli, P., Aebi, M., and Weissmann, C. (1988). *Mol. Cell. Biol.* **8**, 3065–3079.

Huleihel, M., and Aboud, M. (1983). *J. Virol.* **48**, 120–126.

Ichii, Y., Fukunaga, R., Shiojiri, S., and Sokawa, Y. (1986). *Nucleic Acids Res.* **14**, 1011–1017.

Ilson, D. H., Torrence, P. F., and Vilcek, J. (1986). *J. Interferon Res.* **6**, 5–12.

Imani, F., and Jacobs, B. L. (1988). *Proc. Natl. Acad. Sci. U.S.A.* **85**, 7887–7891.

Isaacs, A., and Lindenmann, J. (1957). *Proc. R. Soc. London, Ser. B* **147**, 258–267.
Jacobs, B. L., Miyamoto, N. G., and Samuel, C. E. (1988). *J. Interferon Res.* **8**, 617–631.
Jagus, R., Anderson, W. F., and Safer, B. (1981). *Prog. Nucleic Acid Res. Mol. Biol.* **25**, 127–185.
Jülke, P., and Haller, O. (1990). Manuscript in preparation.
Jungwirth, C., Horak, I., Bodo, G., Lindner, J., and Schultze, B. (1972). *Virology* **48**, 5Cohn, Z. A. (1987). *J. Exp. Med.* **166**, 1098–1108.
Katze, M. G., Tomita, J., Black, T., Krug, R. M., Safer, B., and Hovanessian, A. G. (1988). *J. Virol.* **62**, 3710–3717.
Kaufman, R. J., and Murtha, P. (1987). *Mol. Cell. Biol.* **7**, 1568–1571.
Kitajewski, J., Schneider, R. J., Safer, B., Munemitsu, S. M., Samuel, C. E., Thimmappaya, B., and Shenk, T. (1986). *Cell (Cambridge, Mass.)* **45**, 195–200.
Knight, E., Jr., Fahey, D., and Blomstrom, D. C. (1985). *J. Interferon Res.* **5**, 305–313.
Knight, E., Jr., Fahey, D., Cordova, B., Hillman, M., Kutny, R., Reich, N., and Blomstrom, D. (1988). *J. Biol. Chem.* **263**, 4520–4522.
Korant, B. D., Blomstrom, D. C., Jonak, G. J., and Knight, E. (1984). *J. Biol. Chem.* **259**, 14835–14839.
Kornbluth, R. S., Oh, P. S., Munis, J. R., Cleveland, P. H., and Richman, D. D. (1989). *J. Exp. Med.* **169**, 1137–1151.
Kostura, M., and Mathews, M. B. (1989). *Mol. Cell. Biol.* **9**, 1576–1586.
Krug, R. M., Shaw, M., Broni, B., Shapiro, G., and Haller, O. (1985). *J. Virol.* **56**, 201–206.
Kumar, R., Tiwari, R. K., Kusari, J., and Sen, G. C. (1987). *J. Virol.* **61**, 2727–2732.
Kumar, R., Chattopadhyay, D., Banerjee, A. K., and Sen, G. C. (1988a). *J. Virol.* **62**, 641–643.
Kumar, R., Choubey, D., Lengyel, P., and Sen, G. C. (1988b). *J. Virol.* **62**, 3175–3181.
Kusari, J., Szabo, P., Grzeschik, K.-H., and Sen, G. C. (1987). *J. Interferon Res.* **7**, 53–59.
Lai, C.-L., Lok, A. S.-F., Lin, H.-J., Wu, P.-C., Yeoh, E.-K., and Yeung, C.-Y. (1987). *Lancet* **2**, 877–880.
LaMarco, K. L., and McKnight, S. L. (1990). Submitted for publication.
Larner, A. C., Jonak, G., Cheng, Y.-S. E., Korant, B., Knight, E., and Darnell, J. E. (1984). *Proc. Natl. Acad. Sci. U.S.A.* **81**, 6733–6737.
Laurent, A. G., Krust, B., Galabru, J., Svab, J., and Hovanessian, A. G. (1985). *Proc. Natl. Acad. Sci. U.S.A.* **82**, 4341–4345.
Lengyel, P. (1982). *Annu. Rev. Biochem.* **51**, 251–282.
Lengyel, P., Desrosiers, R., Broeze, R., Slattery, E., Taira, H., Dougherty, J., Samanta, H., Pichon, J., Farrell, P., Ratner, L., and Sen, G. C. (1980). *In* "Microbiology—1980" (D. Schlessinger, ed.), pp. 219–226. Am. Soc. Microbiol., Washington, D.C.
Levy, H. B., and Carter, W. A. (1968). *J. Mol. Biol.* **31**, 561–577.
Levy, D., Larner, A., Chaudhuri, A., Babiss, L. E., and Darnell, J. E. (1986). *Proc. Natl. Acad. Sci. U.S.A.* **83**, 8929–8933.
Lewis, J. A. (1988). *Virology* **162**, 118–127.
Lindenmann, J. (1962). *Virology* **16**, 203–204.
Lipp, M., and Brandner, G. (1980). *J. Gen. Virol.* **47**, 97–111.
Lipp, M., and Brandner, G. (1985). *In* "The Biology of the Interferon System 1984" (H. Kirchner and H. Schellekens, eds.), pp. 355–360. Elsevier, Amsterdam.
Luster, A. D., and Ravetch, J. V. (1987). *J. Exp. Med.* **166**, 1084–1097.
Luster, A. D., Unkeless, J. C., and Ravetch, J. V. (1985). *Nature (London)* **315**, 672–676.
Luster, A. D., Jhanwar, S. C., Chaganti, R. S. K., Kersey, J. H., and Ravetch, J. V. (1987). *Proc. Natl. Acad. Sci. U.S.A.* **84**, 2868–2871.
Luster, A. D., Weinshank, R. L., Feinman, R., and Ravetch, J. V. (1988). *J. Biol. Chem.* **263**, 12036–12043.

Manders, E. K., Tilles, J. G., and Huang, A. S. (1972). *Virology* **49**, 573–581.
Marcus, P. I., and Sekellick, M. J. (1978). *J. Gen. Virol.* **38**, 391–408.
Masters, P. S., and Samuel, C. E. (1983a). *J. Biol. Chem.* **258**, 12019–12025.
Masters, P. S., and Samuel, C. E. (1983b). *J. Biol. Chem.* **258**, 12026–12033.
McAllister, P. E. (1979). *Compr. Virol.* **14**, 401–470.
McMahon, M., Stark, G. R., and Kerr, I. M. (1986). *J. Virol.* **57**, 362–366.
Meier, E., Fäh, J., Grob, M. S., End, R., Staeheli, P., and Haller, O. (1988). *J. Virol.* **62**, 2386–2393.
Merlin, G., Chebath, J., Benech, P., Metz, R., and Revel, M. (1983). *Proc. Natl. Acad. Sci. U.S.A.* **80**, 1904–4908.
Metz, D. H., and Esteban, M. (1972). *Nature (London)* **238**, 385–388.
Metz, D. H., Levin, M. J., and Oxman, M. N. (1976). *J. Gen. Virol.* **32**, 227–240.
Meyer, T., and Horisberger, M. A. (1984). *J. Virol.* **49**, 709–716.
Mittnacht, S., Straub, P., Kirchner, H., and Jacobsen, H. (1988). *Virology* **164**, 201–210.
Miyamoto, M., Fujita, T., Kimura, Y., Maruyama, M., Harada, H., Sudo, Y., Miyata, T., and Taniguchi, T. (1988). *Cell (Cambridge, Mass.)* **54**, 903–913.
Morris, A. G., and Burke, D. C. (1979). *J. Gen. Virol.* **43**, 173–181.
Mortier, C., and Haller, O. (1987). In "The Biology of the Interferon System 1986." (K. Cantell and H. Schellekens, eds.), pp. 79–84. Nijhoff, Dordrecht, The Netherlands.
Mory, Y., Vaks, B., and Chebath, J. (1989). *J. Interferon Res.* **9**, 295–304.
Mozes, L. W., and Defendi, V. (1979). *Virology* **93**, 558–568.
Muñoz, A., and Carrasco, L. (1984). *J. Gen. Virol.* **65**, 1069–1078.
Nilsen, T. W., Maroney, P. A., and Baglioni, C. (1982). *J. Virol.* **42**, 1039–1045.
Noteborn, M., Arnheiter, H., Richter-Mann, L., Browning, H., and Weissmann, C. (1987). *J. Interferon Res.* **7**, 657–669.
Oberman, F., and Panet, A. (1988). *J. Gen. Virol.* **69**, 1167–1177.
Opdenakker, G., Snoody, J., Choubey, D., Toniato, E., Pravtcheva, D. D., Seldin, M. F., Ruddle, F. H., and Lengyel, P. (1989). *Virology* **171**, 568–578.
Oxman, M. N., and Levin, M. J. (1971). *Proc. Natl. Acad. Sci. U.S.A.* **68**, 299–302.
Paez, E., and Esteban, M. (1984a). *Virology* **134**, 12–28.
Paez, E., and Esteban, M. (1984b). *Virology* **134**, 29–39.
Pavlovic, J., Zürcher, T., Haller, O., and Staeheli, P. (1990). Submitted for publication.
Penn, L. J. Z., and Williams, B. R. G. (1985). *Proc. Natl. Acad. Sci. U.S.A.* **82**, 4959–4963.
Pestka, S., Langer, J. A., Zoon, K. C., and Samuel, C. E. (1987). *Annu. Rev. Biochem.* **56**, 727–777.
Pfefferkorn, E. R. (1984). *Proc. Natl. Acad. Sci. U.S.A.* **81**, 908–912.
Pfefferkorn, E. R., Eckel, M., and Rebhun, S. (1986a). *Mol. Biochem. Parasitol.* **20**, 215–224.
Pfefferkorn, E. R., Rebhun, S., and Eckel, M. (1986b). *J. Interferon Res.* **6**, 267–279.
Porter, A. C. G., Chernajovsky, Y., Dale, T. C., Gilbert, C. S., Stark, G. R., and Kerr, I. M. (1988). *EMBO J.* **7**, 85–92.
Prochazka, M., Staeheli, P., Holmes, R. S., and Haller, O. (1985). *Virology* **145**, 273–279.
Ransohoff, R. M., Maroney, P. A., Nayak, D. P., Chambers, T. M., and Nilsen, T. W. (1985). *J. Virol.* **56**, 1049–1052.
Reeves, R. H., O'Hara, B. F., Pavan, W. J., Gearhart, J. D., and Haller, O. (1988). *J. Virol.* **62**, 4372–4375.
Reich, N., Evans, B., Levy, D., Fahey, D., Knight, E., Jr., and Darnell, J. E. (1987). *Proc. Natl. Acad. Sci. U.S.A.* **84**, 6394–6398.
Reid, L. E., Brasnett, A. H., Gilbert, C. S., Porter, A. C. G., Gewert, D. R., Stark, G. R., and Kerr, I. M. (1989). *Proc. Natl. Acad. Sci. U.S.A.* **86**, 840–844.

Repik, P., Flamand, A., and Bishop, D. H. L. (1974). *J. Virol.* **14**, 1169–1178.

Rice, A. P., and Kerr, I. M. (1984). *J. Virol.* **50**, 229–236.

Rice, A. P., Roberts, W. K., and Kerr, I. M. (1984). *J. Virol.* **50**, 229–236.

Riggin, C. H., and Pitha, P. M. (1982). *Virology* **118**, 202–213.

Rosa, R. M., Cochet, M. M., and Fellous, M. (1986). *Interferon* **7**, 47–87.

Rothman, J. H., and Stevens, T. H. (1986). *Cell (Cambridge, Mass.)* **47**, 1041–1051.

Rothman, J. H., Raymond, C. K., Stevens, T. H., Gilbert, T., and O'Hara, P. J. (1990). Submitted for publication.

Rubin, B. Y., Anderson, S. L., Hellermann, G. R., Richardson, N. K., Lunn, R. M., and Valinsky, J. E. (1988). *J. Interferon Res.* **8**, 691–702.

Ruggiero, V., Tavernier, J., Fiers, W., and Baglioni, C. (1986). *J. Immunol.* **136**, 2445–2450.

Rutherford, M. N., Hannigan, G. E., and Williams, B. R. G. (1988). *EMBO J.* **7**, 751–759.

Rysiecki, G., Gewert, D. R., and Williams, B. R. G. (1989). *J. Interferon Res.*, **9**, 649–657.

Sahni, G., and Samuel, C. E. (1986). *J. Biol. Chem.* **261**, 16764–16768.

Samanta, H., Pravtcheva, D. D., Ruddle, F. H., and Lengyel, P. (1984). *J. Interferon Res.* **4**, 295–300.

Samuel, C. E. (1979). *Proc. Natl. Acad. Sci. U.S.A.* **76**, 600–604.

Samuel, C. E. (1988). *Prog. Nucleic Acid Res. Mol. Biol.* **35**, 27–72.

Samuel, C. E., and Knutson, G. S. (1982). *J. Biol. Chem.* **257**, 11796–11801.

Samuel, C. E., Duncan, R., Knutson, G. S., and Hershey, J. W. B. (1984). *J. Biol. Chem.* **259**, 13451–13457.

Saunders, M. E., Gewert, D. R., Tugwell, M. E., McMahon, M., and Williams, B. R. G. (1985). *EMBO J.* **4**, 1761–1768.

Schein, C. H., and Noteborn, M. H. M. (1988). *Biotechnology* **6**, 291–294.

Schneider, R. J., Safer, B., Munemitsu, S. M., Samuel, C. E., and Shenk, T. (1985). *Proc. Natl. Acad. Sci. U.S.A.* **82**, 4321–4325.

Sen, G. C., and Sarkar, N. H. (1980). *Virology* **102**, 431–443.

Sen, G. C., Lebleu, B., Brown, G. E., Rebello, M. A., Furuichi, Y., Morgan, M., Shatkin, A. J., and Lengyel, P. (1975). *Biochem. Biophys. Res. Commun.* **65**, 427–434.

Sen, G. C., Shaila, S., Lebleu, B., Brown, G. E., Desrosiers, R. C., and Lengyel, P. (1977). *J. Virol.* **21**, 69–83.

Shirayoshi, Y., Burke, P. A., Appella, E., and Ozato, K. (1988). *Proc. Natl. Acad. Sci. U.S.A.* **85**, 5884–5888.

Shulman, L. M., Barker, P. E., Hart, J. T., Messer-Peters, P. G., and Ruddle, F. H. (1984). *Somatic Cell Mol. Genet.* **10**, 247–257.

Staeheli, P., and Haller, O. (1985). *Mol. Cell. Biol.* **5**, 2150–2153.

Staeheli, P., and Haller, O. (1987). *Interferon* **8**, 1–23.

Staeheli, P., and Sutcliffe, J. G. (1988). *Mol. Cell. Biol.* **8**, 4524–4528.

Staeheli, P., Colonno, R., and Cheng, Y.-S. E. (1983). *J. Virol.* **47**, 563–567.

Staeheli, P., Horisberger, M. A., and Haller, O. (1984a). *Virology* **132**, 456–461.

Staeheli, P., Prochazka, M., Steigmeier, P. A., and Haller, O. (1984b). *Virology* **137**, 135–142.

Staeheli, P., Dreiding, P., Haller, O., and Lindenmann, J. (1985). *J. Biol. Chem.* **260**, 1821–1825.

Staeheli, P., Danielson, P., Haller, O., and Sutcliffe, J. G. (1986a). *Mol. Cell. Biol.* **6**, 4770–4774.

Staeheli, P., Haller, O., Boll, W., Lindenmann, J., and Weissmann, C. (1986b). *Cell (Cambridge, Mass.)* **44**, 147–158.

Staeheli, P., Pravtcheva, D., Lundin, L.-G., Acklin, M., Ruddle, F., Lindenmann, J., and Haller, O. (1986c). *J. Virol.* **58**, 967–969.

Staeheli, P., Grob, R., Meier, E., Sutcliffe, J. G., and Haller, O. (1988). *Mol. Cell. Biol.* **8**, 4518–4523.

Staeheli, P., Yu, Y.-X., Grob, R., and Haller, O. (1989). *Mol. Cell. Biol.* **9**, 3117–3121.

St. Laurent, G., Yoshie, O., Floyd-Smith, G., Samanta, H., Sehgal, P. B., and Lengyel, P. (1983). *Cell (Cambridge, Mass.)* **33**, 95–102.

Straub, P., Domke, I., Kirchner, H., Jacobsen, H., and Panet, A. (1986). *Virology* **150**, 411–418.

Strube, W., and Jungwirth, C. (1985). *Eur. J. Cell Biol.* **39**, 232–235.

Strube, W., Strube, M., Kroath, H., Jungwirth, C., Bodo, G., and Graf, T. (1982). *J. Interferon Res.* **2**, 37–49.

Strunk, R. C., Cole, F. S., Perlmutter, D. H., and Colten, H. R. (1985). *J. Biol. Chem.* **260**, 15280–15285.

Thacore, H. R., and Youngner, J. S. (1973). *Virology* **56**, 505–511.

Turco, J., and Winkler, H. H. (1986). *Infect. Immun.* **53**, 38–46.

Ulker, N., and Samuel, C. E. (1985). *J. Biol. Chem.* **260**, 4324–4330.

von Wussow, P., Jakschies, D., Hochkeppel, H., Horisberger, M. A., Hartung, K., and Deicher, H. (1989). *Arthritis Rheum.* **32**, 914–918.

Wallach, D., Fellous, M., and Revel, M. (1982). *Nature (London)* **299**, 834–836.

Wathelet, M., Moutschen, S., Defilippi, P., Cravador, A., Collet, M., Huez, G., and Content, J. (1986). *Eur. J. Biochem.* **155**, 11–17.

Wathelet, M. G., Clauss, I. M., Nols, C. B., Content, J., and Huez, G. A. (1987). *Eur. J. Biochem.* **169**, 313–321.

Wathelet, M. G., Clauss, I. M., Content, J., and Huez, G. A. (1988). *Eur. J. Biochem.* **174**, 323–329.

Watling, D., Serafinowska, H. T., Reese, C. B., and Kerr, I. M. (1985). *EMBO J.* **4**, 431–436.

Weitz, G., Bekisz, J., Zoon, K., and Arnheiter, H. (1989). *J. Interferon Res.*, **9**, 679–689.

Whitaker-Dowling, P., and Youngner, J. S. (1983). *Virology* **131**, 128–136.

Whitaker-Dowling, P., and Youngner, J. S. (1984). *Virology* **137**, 171–181.

Whitaker-Dowling, P., and Youngner, J. S. (1986). *Virology* **152**, 50–57.

Whitaker-Dowling, P. A., Wilcox, D. K., Widnell, C. C., and Youngner, J. S. (1983). *Proc. Natl. Acad. Sci. U.S.A.* **80**, 1083–1086.

Wiebe, M. E., and Joklik, W. K. (1975). *Virology* **66**, 229–240.

Wilcox, D. K., Whitaker-Dowling, P. A., Youngner, J. S., and Widnell, C. C. (1983). *Mol. Cell. Biol.* **3**, 1533–1536.

Williams, B. R. G., Golgher, R. R., Brown, R. E., Gilbert, C. S., and Kerr, I. M. (1979). *Nature (London)* **282**, 582–586.

Williams, B. R. G., Saunders, M. E., and Willard, H. F. (1986). *Somatic Cell Mol. Genet.* **12**, 403–408.

Wreschner, D. H., James, T. C., Silverman, R. H., and Kerr, I. M. (1981). *Nucleic Acids Res.* **9**, 1571–1581.

Yakobson, E., Prives, C., Hartman, J. R., Winocour, E., and Revel, M. (1977). *Cell (Cambridge, Mass.)* **12**, 73–81.

Yamamoto, K., Yamaguchi, N., and Oda, K. (1975). *Virology* **68**, 58–70.

Zens, W., Degen, H.-J., Barnekow, A., Gelderblom, H., and Jungwirth, C. (1989). *Virology* **171**, 535–542.

EXPRESSION OF A PLANT VIRUS-CODED TRANSPORT FUNCTION BY DIFFERENT VIRAL GENOMES

Joseph G. Atabekov and Mikhail E. Taliansky

Department of Virology and
Moscow State University
Moscow 119899
U. S. S. R.

I. INTRODUCTION

The systemic spread of plant virus infection proceeds in two ways: (1) slow cell-to-cell (i.e., short-distance) transport within the parenchyma and (2) rapid migration over a long distance via the conducting tissues (i.e., long-distance transport) (for a review see Matthews, 1981; Hull, 1989). The systemic spread of infection includes at least the following four transport events: (1) a number of consecutive short-distance transport occur within the epidermal and parenchymal tissues of the initially infected leaf (i.e., the first transport barrier should be overcome); (2) the virus genome should be transferred from the infected parenchyma into the conducting tissue, which signifies the transition from short- to long-distance transport (second transport barrier); (3) the long-distance transport of infection occurs (third transport barrier); and (4) the viral infection migrates from the conducting system back to the parenchymal cells (fourth transport barrier) with subsequent cell-to-cell spreading over the secondarily infected parenchymal tissues.

Until recently, the systemic spread of plant virus infection (i.e., transport of the viral genetic material) was commonly assumed to be a genetically passive process (i.e., not a virally encoded one). The infectious material that accumulated in the primary infected cells was believed to migrate passively to the surrounding healthy cells through the plasmodesmata.

During the early 1980s evidence accumulated in support of the theory that the transport of infection is under the control of a plant virus genome. This means that a particular virus-specific function—namely, the transport function (TF) coded by the virus genome—is essential for movement of the viral genetic material over the infected plant. A particular virally encoded protein(s), the transport protein(s) (TP), is responsible for the expression of TF. In other words, the cell-to-cell transport of the genomes of at least some plant viruses (possibly all of them) is coded by the virus genome itself. Different aspects of the plant virus encoded transport phenomenon have been reviewed by Atabekov and Dorokhov (1984).

In recent years our knowledge of virus-coded TF has markedly increased:

1. The possibility of comparing the putative genes coding for the TF (transport genes) in viruses differing in genomic structure arose. Such analysis was of special interest, as the genomic RNAs of many viruses have been sequenced. Comparative analysis of these sequences revealed homology in the primary structure among the nonstructural proteins encoded by viruses belonging to quite different taxonomic groups. In practically all plus-sense RNA plant viruses the conserved consensus sequences in genes for the putative viral RNA polymerases were found. The putative genes for proteases of the viruses whose RNA is translated into a polyprotein contain homologous sequences as well (for a review see Goldbach, 1986).

On the other hand, comparison of nucleotide sequences of putative transport proteins of different plant viruses revealed significant variability in their structure. No homology was found between the putative viral TPs of some taxonomically unrelated viruses. The structural diversity between the TPs of plant viruses probably reflects their host specifity, but this fact also points to the possible existence of several mechanisms for the cell-to-cell transport of a virus infection. Nevertheless, it should be assumed that the number of such mechanisms is limited. For the sake of simplicity, we assumed it possible to divide some plant viruses into tentative transport groups on the basis of the sequence similarity between their TPs (even when it was not pronounced).

2. Virus infection requires the expression of two genomes: that of

the virus and that of the plant. Experimental evidence now available supports the theory that TF belongs to this type of viral function (i.e., it is coded by the virus and the host as well).

In this chapter we summarize current knowledge on the subject of TF expression by viruses belonging to different taxonomic groups, with special emphasis on the two aspects noted above. Plant virus-encoded transport has recently been reviewed by Meshi and Okada (1987) and by Hull (1989).

II. Comparative Analysis of Plant Virus Genome Organization and Transport Function Expression

A. Essential and Nonessential Elements of the Viral Genome

Numerous monopartite and multipartite plant plus-sense RNA viruses have been classified into two large supergroups (Goldbach, 1987) based on the interviral sequence homologies between plant and animal RNA viruses: the "Sindbis-like" viruses (e.g., tobamo-, bromo-, cucumo-, tobra-, and ilarviruses) and "picornalike" viruses (e.g., poty-, como-, and nepoviruses). The most distinguishing feature is that genomic RNAs of the first supergroup act functionally monocistronic, even when they contain more than one gene. Only the 5'-proximal open reading frame (ORF) can be translated directly; all of the 5'-distal ORFs are translationally silent and expressed through the subgenomic RNAs. On the other hand, RNAs of viruses belonging to the second supergroup are continually translated into a single large polyprotein from which the functional proteins are generated by proteolytic processing. These two modes of translation are referred to below as "the 5'-proximal translation with other gene subgenomization" and "continuous translation with polyprotein processing," respectively.

The genome of any virus is a combination of trans-acting coding sequences (i.e., genes) and noncoding cis-acting regulatory sequences. The data available allow the estimation of the minimal number of functions (genes) essential or nonessential for the replication of the viral RNA genome (in different viruses they can occur in different combinations). The following genes can be considered essential: (1) a block of genes (usually not less than two) encodes components of RNA-dependent RNA polymerase, ensuring synthesis of virus genomic and subgenomic RNAs; and (2) viruses whose RNA is translated into a polyprotein should have a gene(s) for processing a protease(s) to generate RNA-replicating and other virus-specific proteins. The rest of the genes can be regarded as nonessential for viral genome replication in virus-infected cells, although they could express some important functions (e.g., symptom development, virus genome cell-to-cell transport, and mature particle formation). (a) The gene(s) responsible for TF can be regarded as nonessential, since the viruses deficient in transport can nevertheless normally replicate and produce virus progeny in primarily infected cells or in isolated protoplasts. (b) One or more genes coding for a viral coat protein(s) are also nonessential, since, in the absence of an active coat protein, RNAs of many multipartite and monopartite viruses can replicate to form infectious daughter RNAs which spread from cell to cell (e.g., Jockusch, 1968; Bruening, 1977; Nassuth et al., 1981; Goldbach et al., 1980; Robinson et al., 1980; French and Ahlquist, 1987). The only exception is associated with members of the ilarvirus group. In this case the coat protein per se or its mRNA should be present in the inoculum to activate the genome replication (Bol et al., 1971). (c) The gene encoding a so-called helper component protein required for aphid transmission of the virus (see, e.g., Thornbury et al., 1985) should be considered nonessential as well.

Arrangement of Essential and Nonessential Genes in a Plus-Sense Genomic RNA

The ribosome-scanning model of the translation initiation process has been accepted for most eukaryotic mRNAs (for a review see Kozak, 1986a,b). This model suggests that a 40 S ribosome primarily binds to the 5' end of the mRNA and then moves along the nontranslated sequence until it approaches the first AUG codon. If this codon occurs in the optimal sequence context, the ribosome stops and initiates translation. However, the initiation of translation at AUGs downstream from the first AUG, based on the phenomenon of leaky scanning and subsequent reinitiation, has been proposed (Kozak, 1986a,b). This mechanism can explain the translation of eukaryotic bicistronic mRNAs.

It seems reasonable to assume that plus-sense RNAs of plant viruses function in translation as typical eukaryotic messengers; that is, only the 5'-proximal ORFs are available for the ribosomes. The initiation of translation of internal genes of eukaryotic viral RNAs is possible in principle (e.g., Shin et al., 1987; Herman, 1987; Jang et al., 1988); however, this type of translation has never been proved to occur in plant viruses. Several general principles regulating the arrangement of essential and nonessential genes within the monopartite and multipartite genome can be proposed for the plus-sense RNA viruses with different translational strategies.

a. Viruses Which Use the Strategy of the 5'-Proximal Translation and Other Gene Subgenomization. (1) The essential genes should be either localized 5' proximally within a monopartite genome or autonomized within a separate monocistronic genomic RNA in a multipartite genome. It is obvious that RNA replicase components should be produced on the parental genomic RNA by direct translation. Consequently, the polymerase gene(s) should not be positioned 5' distally (i.e., expressed through the subgenomic RNA), since the polymerase itself is needed for subgenomic RNA synthesis.

There is a good possibility for the direct translation of two 5'-proximal essential genes tandomly positioned: (a) The first (5'-proximal) gene of the RNA-replicating enzyme could be followed by a suppressible "leaky" termination codon, ensuring the read-through translation of a gene downstream and in frame (for a review see Goldbach, 1986). This possibility can be realized in both monopartite [tobacco mosaic virus (TMV)] and multipartite [tobacco rattle virus (TRV)] genomes. (b) Initiation of translation of both of the two overlapping 5'-proximal genes can occur; according to the leaky ribosome scanning model each of the AUGs can initiate the translation (Kozak, 1986a,b). (c) The mechanism of ribosomal frameshifting might be considered an alternative way to express the two tandomly positioned (or overlapping) 5'-proximal essential genes if these genes lie in different translational reading frames.

(2) The nonessential genes, dispensable for virus replication (e.g., transport and coat protein genes), should not have the 5'-proximal position if they are located within one genomic RNA together with the replicase gene(s).

(3) However, a nonessential gene could have the 5'-proximal localization if it is disposed within a separate RNA component of a multipartite genome [together with another nonessential gene(s)]. It is obvious that such an RNA component would be dispensable for RNA replication. In other words, when a nonessential gene occupies the 5'-proximal position within the genomic RNA component of the multipartite genome, it points to the fact that all of the downstream genes would be in the nonessential category as well.

b. Viruses Which Use the Strategy of Continuous Translation with Polyprotein Processing. Arrangement and localization of essential and nonessential genes within the genomic RNA do not demonstrate visual limits in this case. Thus, the coat protein gene(s) can have the 3'- or 5'-proximal localization in potyviruses and picornaviruses, respectively. The same seems to be true for the transport genes which can be localized 5' proximally within the continually translatable multipartite or monopartite genome (e.g., the tentative transport genes of nepo-, como-, and potyviruses) (see Section II,B,1).

B. Division of Plant Viruses into Tentative Transport Groups

As mentioned in Section I, it seems reasonable to divide plant viruses into several transport groups on the basis of the similarity between their putative TPs. This division is made independently of the character of the viral genome (RNA or DNA), its structure (monopartite or multipartite), and its mode of translational expression. It is obvious that the viruses coding for the TPs with no homology were divided into different groups. We are aware that this approach is rather speculative, particularly in cases when the viruses coding for the TPs with a low level of similarity were tentatively included in one group. We hope that our transport groups reflect the real differences between the molecular mechanisms of the TF coded by different viral genomes; however, we are aware that future studies could bring changes in this classification. Independently of us, a similar (but not identical) division has been offered by Hull (1989).

1. First Transport Group: Tobamo-, Tobra-, Caulimo-, Nepo-, Como-, and Potyviruses

In this section the TPs and the putative virally encoded TPs of the viruses belonging to six quite different taxonomic groups are considered. Rodlike tobamoviruses and tobraviruses are single-stranded (ss) RNA viruses possessing a capped monopartite and bipartite genome, respectively (Fig. 1). Both are members of the Sindbis-like supergroup of viruses, in terms of Goldbach (1987). Spherical nepoviruses and comoviruses have a bipartite genome (Fig. 1) consisting of two ssRNAs carrying the a 5′-terminal genome-linked protein (VPg) at the 5′ end and 3′ polyadenylated. The monopartite genome of filamentous potyviruses consists of one ssRNA (Fig. 1) containing a VPg and a poly(A) tract. The three latter groups belong to the picornalike supergroup of viruses (Goldbach, 1987). Finally, caulimoviruses contain a circular double-stranded DNA molecule. Although structural and taxonomic differences are so obvious, we consider it possible to unite tobamo-, tobra-, and caulimoviruses in discussing the problem of virus transport. It seems logical, since the putative transport proteins of these viruses possess a rather high level of homology, which might reflect their functional similarity. Some of the poty-, como-, and nepoviruses coded for polypeptides not as similar to the TP of tobamoviruses might be tentatively included in the same group as possible members.

    a. TF of Tobamoviruses. The genomic RNA of TMV (molecular weight 2.1×10^6) encodes three nonstructural proteins (126, 183, and 30 kDa) and the coat protein (Fig. 1) (Beachy *et al.*, 1976); Bruening *et al.*, 1976; Goelet and Karn, 1982; Goelet *et al.*, 1982; Ohno *et al.*, 1984).

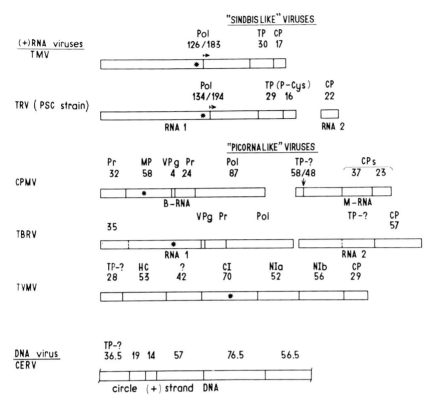

FIG. 1. Tentative transport group I of plant viruses. Arrangement of genes in the genomes of tobacco mosaic virus, TMV (Hirth and Richards, 1981; Goldbach, 1987); tobacco rattle virus, TRV (Cornelissen *et al.*, 1986; Goldbach, 1987); cowpea mosaic virus, CPMV (Lomonossoff and Shanks, 1983; Van Wezenbeek *et al.*, 1983); tomato black ring virus, TBRV (Meyer *et al.*, 1986; Goldbach, 1987); tobacco vein mottling virus, TVMV (Domier *et al.*, 1987); and carnation etched ring virus, CERV (Hull *et al.*, 1986). The numbers correspond to the molecular mass of the proteins in kilodaltons. Pol, Virus-specific RNA polymerase component(s); TP, transport protein; CP, coat protein; P-Cys, cystein-rich protein; Pr, protease; MP, membrane-binding protein; VPg, 5′-terminal genome-linked protein; HC, helper component; CI, cylindrical inclusion protein; NIa, NIb, nuclear inclusion proteins a and b; ORF, open reading frame; →, leaky termination codon; ↓, additional initiation codon. The dashed line in TBRV RNA 2 indicates that, to date, the proteins have not been identified.

Expression of two genes (i.e., the 30-kDa protein and the coat protein) requires the synthesis of at least two subgenomic RNAs, 3′ coterminal with genomic RNA (for a review see Hirth and Richards, 1981).

It has been proposed that the 30-kDa protein plays a transport role in TMV infection (Atabekov and Morozov, 1979). The 30-kDa protein has been shown *in vivo* in TMV-infected leaf tissue (Joshi *et al.*, 1983) and protoplasts (Watanabe *et al.*, 1984; Ooshika *et al.*, 1984).

The temperature-sensitive (*ts*) TMV mutant Ls1 derived from a Japanese temperature-resistant (*tr*) tomato strain L by Nishiguchi *et al.*, (1978, 1980) played a key role in the study of virally encoded TF. Its temperature sensitivity manifests itself as a block to the cell-to-cell transport of infection at nonpermissive temperatures (32–33°C). Under such conditions Ls1 replicates only in primary infected cells or in isolated protoplasts.

Leonard and Zaitlin (1982) found a small difference between the peptide maps of the 30-kDa proteins coded for by Ls1 and L (wild type): the replacement of a proline by a serine codon at the 153 position of the polypeptide (Ohno *et al.*, 1983). However, the possibility remained that an additional mutation(s) occurred in another gene(s) of the Ls1 genome which might have been responsible for the ts transport of Ls1. Finally, Meshi *et al.* (1987) proved that without a doubt the substitution from a proline to a serine in the 30-kDa protein due to the one-base substitution in the 30-kDa gene is responsible for the ts TF of TMV Ls1. Using the *in vitro* expression systems that allow production of infectious TMV RNA from cloned full-length DNA copies, they investigated the function of the 30-kDa protein by a reverse-genetics approach. Introduction of a single point mutation into the 30-kDa gene of the parental TMV-L resulted in phenotype ts in transport (Meshi *et al.*, 1987). The authors proved experimentally that the ts behavior of Ls1 is due to a certain amino acid exchange in the 30-kDa protein.

In a separate series of experiments they constructed different types of frameshift and deletion mutants carrying the mutations at different positions of the 30-kDa gene (or lacking most of this gene) Meshi *et al.* (1987) proved that the 30-kDa protein is involved in TF and therefore can be regarded as the TP. It was found also that this gene is not essential for virus replication in protoplasts (Meshi *et al.*, 1987). Thus, Ls1 can be regarded as a classical model system for TF analysis.

A nitrous acid-induced TMV mutant, Ni2519, was also reported to be ts in TF (Jockusch, 1968). Phenotypically, the *ts* mutation in the *Ni2519* genome reflects the blockage of the virus assembly and TF inhibition. The virus assembly ts phenotype was shown not to result from the ts behavior of any virus-specific protein; rather, the virion RNA molecule itself is temperature-sensitive due to a cis-acting mutation (Taliansky *et al.*, 1982a,b; Kaplan *et al.*, 1982). On the other hand, the ts TF of Ni2519 could be complemented by a helper virus (TMV vulgare) (Taliansky *et al.*, 1982a). Zimmern and Hunter (1983) revealed an arginine-to-glycine transition in the 30-kDa gene of Ni2519, compared with its parental strain A 14. This transition, which was located at position 5332 in Ni2519 RNA, close to the assembly origin, could have a dual effect: altering the conformation of the assembly origin (which is responsible for *ts* assembly if Ni2519 RNA), resulting in the temperature sensitivity of the 30-kDa protein (TP).

This conclusion is consistent in that the revertant virus with a regained wild-type sequence at position 5332 was temperature resistant in both functions (Zimmern and Hunter, 1983). These results strongly suggested that the replacement of a glycine for an arginine in residue 144 (Ni2519) as well as the replacement of a proline by serine at position 153 (Ls1) of the 30-kDa protein resulted in the temperature sensitivity of TF.

Beachy and colleagues cloned cDNA encoding the 30-kDa gene of TMV (Oliver *et al.*, 1986) and utilized plant-transforming technology to directly define the function of the 30-kDa gene of TMV (Deom *et al.*, 1987). Transgenic tobacco plants expressed the 30-kDa protein mRNA and accumulated the 30-kDa protein. This approach allowed Deom *et al.* (1987) to study the function of the 30-kDa gene in the absence of the expression of other viral genes. The authors showed that expression of the 30-kDa gene in transgenic plants complements Ls1 transport at the temperatures nonpermissive for Ls1 movement in normal tobacco. These results conclusively demonstrate that the 30-kDa protein [TP, or M protein, as it was designated by Palukaitis and Zaitlin (1986)] potentiates movement of the virus from the infected to the adjacent healthy cells.

A complete nucleotide sequence for the transport genes has been determined, and the amino acid sequences for the 30-kDa proteins (TPs) of five tobamoviruses have been predicted: common strains of TMV vulgare (Goelet *et al.*, 1982) and OM (Meshi *et al.*, 1982a), a tomato strain of TMV-L (Takamatsu *et al.*, 1983), sunn-hemp mosaic virus (or cowpea strain of TMV) (Meshi *et al.*, 1982b), and cucumber green mottle mosaic virus (Meshi *et al.*, 1983). Analysis of the amino acid sequences of the TPs of tobamoviruses revealed some highly conserved regions in their structures (Ohno *et al.*, 1983; Saito *et al.*, 1988). It is important that amino acid substitutions in the TPs of TMV mutants Ls1 and Ni2519, ts in TF, occur in one of these conserved regions. In the case of the Ls1 mutant, in the 153 position serine replaces proline, conservedly present in this position in the rest of the tobamoviruses (Ohno *et al.*, 1983). This substitution, as was shown, decreases the rigidity of the Ls1 TP molecule, which might account for its conformational instability at nonpermissive temperatures (Ohno *et al.*, 1983). Another *ts* mutant, Ni2519, mentioned above, has the replacement arginine for glycine in the 144 position (Zimmern and Hunter, 1983). In spite of these data, it should be mentioned that, in general, the TP (30-kDa protein) structure of tobamoviruses is less conserved in both size and sequence than other TMV-encoded nonstructural proteins (Ohno *et al.*, 1984).

The time course of 30-kDa protein synthesis was examined in TMV-infected protoplasts (Watanabe *et al.*, 1984) and in intact leaves (Joshi *et al.*, 1983). The 30-kDa protein and its subgenomic RNA are transiently synthesized at an early stage (i.e., 2–9 hours postinoculation) in TMV-infected protoplasts (Watanabe *et al.*, 1984), that is, before any significant number of virions have been formed. Other virally encoded proteins were synthesized continually; that is, the expression of the 30-kDa gene is controlled by a mechanism different from that controlling the other genes. The results of the studies by Watanabe *et al.* (1986) suggest that the 126- and 183-kDa proteins are involved in the production and regulation of subgenomic RNA for the 30-kDa protein: mutations in the 126- or 183-kDa gene of the attenuated strain of TMV cause a reduction in the synthesis of the 30-kDa protein and its mRNA, but do not affect the synthesis of other viral proteins and RNAs in protoplasts. In intact tissue a maximum amount of the 30-kDa protein was accumulated by about 1 day postinoculation (Joshi *et al.*, 1983) and survived for at least 2 days in the systemic host (Moser *et al.*, 1988).

It is interesting that the 30-kDa protein associated with the nuclear fraction after being synthesized in protoplasts (Watanabe *et al.*, 1986). This fact allowed the authors to suggest that the virally encoded TP might be involved in the regulation of host cell transcription. It should be noted, however, that the other TMV-encoded proteins are also capable of forming complexes with the host chromatin (Van Telgen *et al.*, 1985a–c), but the functional significance of this phenomenon remains obscure. Localization of the 30-kDa protein in the nuclear fraction in protoplasts did not allow the determination of its intracellular localization within intact tissue (i.e., under conditions of normal TF expression), since protoplasts lack plasmodesmata and cell walls and therefore cannot have intercellular communication.

Tomenius *et al.* (1987) reported that the 30-kDa protein was accumulated inside the plasmodesmata, but not in the nuclei or in any other part of the cell. It could be found in the plasmodesmata, at least after a 16-hour infection, with a maximum after a 24-hour infection, and after that there was apparently a decrease in the amount of 30-kDa protein present in the plasmodesmata (Tomenius *et al.*, 1987).

Consistent with these results, Moser *et al.* (1988) showed that the 30-kDa TMV protein appeared transiently in a crude membrane fraction, but was accumulated more stably in the cell walls during a TMV infection of the systemic host (*Nicotiana tabacum* L.cv. Samsun) In cv.

Samsun NN, the hypersensitive host of TMV, the amount of 30-kDa protein detected in the cell wall fraction decreased sharply as soon as necrosis was visible. This drop in the amount of 30-kDa protein is most easily interpreted as an effect of the hypersensitive reaction and might explain why TMV infection becomes localized in the leaves of cv. Samsun NN (Moser *et al.*, 1988).

Tomenius *et al.* (1987) suggested that the 30-kDa protein might be accumulated in plasmodesmata at a particularly early stage of infection, before the viral genome passes through them. These observations are in agreement with the suggestion by Atabekov and Dorokhov (1984) that TF is an early function, which is eventually switched off. Transport of the viral RNA into adjoining cells at a certain stage of virus replication might compete with virion maturation: the "flux" of the newly synthesized viral RNA changes its direction and is switched from the transport to the assembly of viral particles that probably do not participate in transport (Atabekov and Dorokhov, 1984).

It is probable that the TMV 30-kDa protein operates particularly in the plasmodesmata, thus modifying them to allow cell-to-cell transport of the infectious entity. However, it remains unknown what happens with plasmodesmata, modified by TP, and whether TP is a purely structural protein or an enzymatically active one and, if so, what the nature of this activity is (see Section IV).

b. TF of Tobraviruses. The bipartite genome of TRV (Fig. 1) consists of two genomic RNA species: RNA 1, with a molecular weight of about 2.4×10^6, and RNA 2, with a molecular weight varying between isolates, ranging from 0.6 to 1.4×10^6 (for a review see Harrison and Robinson, 1986). RNA 1 of TRV can replicate, spread from cell to cell, and cause symptoms in the absence of RNA 2 (Harrison and Robinson, 1981); that is, it should contain the transport gene and the gene(s) of RNA polymerase. The nucleotide sequences reported for RNAs 1 and 2 of different TRV strains (Bergh *et al.*, 1985; Cornelissen *et al.*, 1986; Boccara *et al.*, 1986; Angenent *et al.*, 1986; Hamilton *et al.*, 1986, 1987) and comparison of the *in vitro* TRV RNA translation data (Mayo *et al.*, 1976; Robinson *et al.*, 1983; Hughes *et al.*, 1986) allowed evaluation of the genomic organization of TRV (Fig. 1). The 5'-proximal ORF of TRV RNA 1 codes for two polypeptides (134- and 194-kDa proteins), the putative components of the viral RNA polymerase. The larger protein is a read-through translation product. The next downstream (internal) ORF codes for the 29-kDa protein and, finally, the 3'-proximal ORF codes for the 16-kDa cysteine-rich protein. Both genes are expressed through subgenomic RNA synthesis. It was proposed (Boccara *et al.*, 1986; Hamilton *et al.*, 1986) that TMV and TRV have a common ancestor and that the TRV genome can be considered a bipartite

212 JOSEPH G. ATABEKOV AND MIKHAIL E. TALIANSKY

variant of TMV (Goldbach, 1987). This proposal was based on similar morphology, gene arrangement, translational strategy, and homology between 29/30-kDa and 194/183-kDa proteins in TRV and TMV, respectively.

Significant sequence homology between the 29-kDa protein of TRV and the 30-kDa TP of TMV suggests that both function as TPs. Homology between the 30-kDa TP of TMV and the 29-kDa protein of TRV extends over almost all of the protein molecules, especially in the middle portion of the proteins (Boccara et al., 1986; Cornelissen et al., 1986). As mentioned above, in this region of the 30-kDa protein of TMV, the conserved sequences of the TPs of all tobamoviruses are localized. It has been shown recently (Ziegler-Graf et al., quoted by Godefroy-Colburn et al., 1989) that a frameshift mutant of TRV in the 29-kDa gene was deficient in TF which could be complemented by TMV. These studies implicated the 29-kDa protein gene of TRV in the cell-to-cell transport of virus infection, which is in agreement with the fact that RNA 1 of TRV, containing this gene, is capable of systemic spreading over the infected plant in the absence of RNA 2. The pronounced homology between the TPs of TMV and TRV suggests that the molecular mechanisms of the transport function expression are similar for tobamoviruses and tobraviruses. The function of the 16-kDa gene is obscure.

c. TF of Caulimoviruses. Cauliflower mosaic virus (CaMV), a member of the caulimovirus group, contains a double-stranded circular DNA molecule, with neither strand covalently linked (for a review see Mason et al., 1987). CaMV uses a retroviral strategy for replication (Hohn et al., 1985). A model for CaMV replication has been proposed in which a full-length CaMV RNA transcript is used as a template for CaMV DNA synthesis by reverse transcriptase (Hull and Covey, 1985; Pfeiffer and Hohn, 1983). The positive-stranded DNA of CaMV comprises six major ORFs. Functions have been assigned to some of these putative gene products: aphid transmission factor (ORF 2) (Woolston et al., 1983), coat protein precursor (ORF 4) (Daubert et al., 1982), putative reverse transcriptase (ORF 5) (Toh et al., 1983; Takatsuji et al., 1986), and main inclusion body matrix protein (ORF 6) (Covey and Hull, 1981). Comparison of CaMV DNA nucleotide sequences with that of carnation etched ring virus (CERV (Fig. 1) and figwort mosaic virus, other members of the caulimovirus group, showed that these viruses have a similar genome organization (Hull et al., 1986; Richins et al., 1987).

Taking into account the differences in genomic structure and expression among tobamo-, tobra-, and caulimoviruses, it was unexpected that a significant sequence homology existed between putative ORF 1 products of CaMV, CERV, and figwort mosaic virus and the TPs of TMV (30-kDa protein) and TRV (29-kDa protein) (Hull et al., 1986; Richins et al., 1987). The divergence of caulimovirus ORF 1 products from the TMV TP is not greater than that between the TPs of different TMV strains. The similarity between caulimovirus ORF 1 products and the TMV 30-kDa protein occurs in a central region of the polypeptides (Hull and Covey, 1985; Hull et al., 1986; Richins et al., 1987). ORF 1 products of caulimoviruses were suggested to be good candidates for the role of TP (Hull and Covey, 1985; Hull et al., 1986; Martinez-Izquierdo et al., 1987). Therefore, there is a question about the possibility of a similarity in mechanisms of cell-to-cell transport of RNA and DNA genomes of plant viruses, although it should not be excluded that the spread of caulimovirus infection from cell to cell could occur through RNA transcripts.

The probable participation of the ORF 1 product in the control of cell-to-cell transport of CaMV infection is supported by the presence of this protein in the cell wall-enriched fraction of CaMV-infected leaves observed at late-stage infection (Albrecht et al., 1988). These results are in agreement with the data obtained by Linstead et al. (1988), who demonstrated that the CaMV ORF 1 product is located around modified plasmodesmata between mesophyll cells and the ends of phloem parenchymal cells in small vascular elements. Sections of CaMV-infected turnip tissue showed modifications of the plasmodesmata between mesophyll cells, including an increased diameter of the intersymplastic channel and the appearance of a fine granular tubular structure within a cell wall extension (Linstead et al., 1988). These modifications were similar to those reported for CaMV-infected Chinese cabbage leaves (Conti et al., 1972; Bassi et al., 1974), dahlia mosaic caulimovirus-infected zinnia leaves (Kitajima and Lauritis, 1969), and CERV-infected Saponaria vaccaria and Dyanthus caryophyllus tissues (Lawson and Hearon, 1974).

Linstead et al. (1988) suggest that the ORF 1 product is involved in the cell-to-cell movement of caulimoviruses, turning into a structural component of plasmodesmata, although it cannot be excluded that the putative TP of these viruses possesses some enzymatic activity. In this connection it is interesting that Martinez-Izquierdo et al. (1987) found similarity in the amino acid sequence of the CaMV ORF 1 product and the ATP binding site of some kinases (see Section III,A,4). Hull et al. (1986) observed a significant similarity in the amino acid sequences of the CERV ORF 1 product, plastocyanin, and the fourth intron of the yeast mitochondrial apocytochrome b gene; homology between the latter and the TMV TP was reported earlier by Zimmern (1983). The significance of such a similarity to nonviral proteins remains unknown.

    d. TF of Nepoviruses. The genome of nepoviruses (Fig. 1) comprises two RNA species with molecular weights of approximately 2.8×10^6 (RNA 1) and $1.3–2.4 \times 10^6$ (RNA 2), both essential for the virus to infect plants (for a review see Murant, 1981). Both RNAs contain a VPg at their 5' end and poly(A) at their 3' end. RNA 1 (but not RNA 2) of tomato black ring nepovirus (TBRV) is able to replicate independently in inoculated protoplasts (Robinson *et al.*, 1980), but cannot spread from cell to cell (D. J. Robinson, quoted by Meyer *et al.*, 1986; Randles *et al.*, 1977). It has been suggested, therefore (Meyer *et al.*, 1986), that TF is encoded in RNA 2 of TBRV, whereas RNA 1 codes for the protein(s) involved in RNA replication.

    As with picorna-, poty-, and comoviruses, the functionally active nepovirus-specific proteins should be released by the proteolytic cleavage of the polyprotein precursor, the product of continuous translation of viral RNAs (Fritsch *et al.*, 1980). RNA 1 of nepoviruses codes for a protease which cleaves the RNA 1 and RNA 2 translation products (for a review see Wellink and van Kammen, 1988). RNA 2 of nepoviruses codes for at least two stable virus-specific proteins; (1) the coat protein located in the carboxy-terminal one-third of the polyprotein molecule and (2) the protein corresponding to the amino-terminal larger part of the polyprotein (Meyer *et al.*, 1986), which should be regarded as the most probable candidate for TF expression. The amino acid sequences of certain regions of the tentative TP coded by TBRV RNA 2 slightly resembled parts of the 30-kDa TP of tobamoviruses and the tentative TP of cowpea mosaic comovirus (Meyer *et al.*, 1986).

    e. TF of Comoviruses. There is a similarity in the genomic organization (Fig. 1) and expression of comoviruses and nepoviruses. Cowpea mosaic virus (CPMV), a member of the comoviruses, has a bipartite positive RNA genome with molecular weights of 2.2×10^6 (bottom component, or B-RNA) and 1.2×10^6 (middle component, or M-RNA). Both RNAs have a VPg and a poly(A) at their 5' and 3' ends, respectively (reviewed by Goldbach and van Kammen, 1985). Similarly, nepovirus RNA 1, the B-RNA of CPMV, is capable of independent replication in isolated protoplasts, but does not spread from cell to cell in the leaves of inoculated plants (Goldbach *et al.*, 1980; Rezelman *et al.*, 1982). Like nepovirus RNA 2, the M-RNA is not capable of independent replication and replicates only in the presence of B-RNA (Goldbach *et al.*, 1980). The restriction of B-RNA multiplication to inoculated cells is due to localization of the transport-controlling gene in M-RNA.

The mode of translation is similar in comoviruses and nepoviruses (i.e., continuous translation of RNA with subsequent polyprotein processing) (Goldbach and van Kammen, 1985; Wellink *et al.*, 1987). Upon translation, M-RNA gives the 105-kDa polyprotein from the carboxy-terminal part from which two coat proteins of CPMV are derived (Van Wezenbeek *et al.*, 1983; Wellink *et al.*, 1986, 1987). The amino-terminal part of the 105-kDa polyprotein corresponds to the nonstructural 48- or 58-kDa protein which has been proposed to play the role of TP (Rezelman *et al.*, 1982). The presence of a 48-kDa protein in the membrane fraction of infected leaves (Wellink *et al.*, 1987) is consistent with this proposition.

It was recently shown by J. van Lent (personal communication) that the 48-kDa protein is associated with tubular structures protruding from the cell wall, which were found only in infected cells and were thought to be involved in transport. Similar results were obtained with red clover mottle comovirus (RCMV) (G. P. Lomonossoff and M. Shanks, personal communication). The 48- and 58-kDa proteins show limited homology to both the 30-kDa TP of tobamoviruses and the tentative TP coded by RNA 2 of TBRV (Meyer *et al.*, 1986). Wellink *et al.* (1987) reported that considerable amounts of the 48-kDa protein were excreted into the culture medium by CPMV-infected protoplasts. It is interesting to consider whether the excretion into the medium in protoplast culture can be considered a common feature of viral TPs.

f. TF of Potyviruses. The potyviral monopartite genome (Fig. 1) consists of approximately 10-kilobase (kb) ssRNA with VPg covalently linked to the 5' end and the 3'-terminal poly(A) tract. The nucleotide sequences of tobacco etch virus (TEV) (Allison *et al.*, 1986) and tobacco vein mottling virus (TVMV) (Domier *et al.*, 1986) have been determined. It is clear that the typical potyviral genome contains a single ORF coding for a polyprotein precursor, which is then processed (for a review see Hiebert and Dougherty, 1988). The potyviral genome is believed to code for at least seven virus-specific proteins which are located within the polyprotein precursor: the coat protein, helper component involved in aphid transmission, VPg, cylindrical inclusion (CI) protein, two nuclear inclusion proteins (NIa and NIb), and the putative protein encoded in the 5' termini of potyviral RNAs (28 and 34 kDa for TVMV and TEV, respectively) (for a review see Hiebert and Dougherty, 1988).

It appeared that the 28-kDa polypeptide of TVMV (but not the 34-kDa polypeptide of TEV) is, to some degree, homologous to the 30-kDa TP of TMV cowpea strain and to the 29-kDa TP of TRV (Domier *et al.*, 1987). Therefore, this protein can be regarded as a possible candidate for TF expression. However, the lack of homologous relationships between the 28-kDa protein of TVMV and the 34-kDa protein of TEV makes such an assignment more tentative (Domier *et al.*, 1987).

Andrews and Shalla (1974) and Forster *et al.* (1988) suggested the possibility that the CI proteins can play a role in the cell-to-cell movement of potyviruses. This suggestion was based on observations (Murant and Roberts, 1971; Lawson and Hearon, 1971; Lawson *et al.*, 1971; Andrews and Shalla, 1974) that the CI proteins are closely associated with plasmodesmata. It should be noted that some homology has been found between the CI proteins of TVMV and TEV and the 58-kDa B-RNA-coded protein of CPMV, the 37-kDa (2C) protein of encephalomyocarditis virus, the 42-kDa protein of beet necrotic yellow vein furovirus, and the 25- to 26-kDa proteins of two potexviruses (Domier *et al.*, 1987; Forster *et al.*, 1988; Skryabin *et al.*, 1988). All of these proteins contain a nucleoside triphosphate (NTP)-binding motif located in a similar position. The possibility of CI participation (as well as the participation of other viral NTP-binding proteins) in TF has not been ruled out.

2. Second Transport Group: Tricornaviruses (Bromo-, Cucumo-, and Ilarviruses) and Dianthoviruses

In this section we discuss data concerning TF encoded by ilarviruses (including alfalfa mosaic virus, AlMV), bromoviruses, and cucumoviruses, since the members of these groups share a number of important features of genomic organization. The basic similarity is that all of these viruses have tripartite ssRNA genomes (Fig. 2). RNAs 1 and 2 are monocistronic and code for two large polypeptides, the putative components of RNA polymerase. RNA 3 contains two nonoverlapping genes (Fig. 2): the 5'-proximal gene coding for the 32- to 35-kDa (3a) protein, a putative TP, and the 3'-proximal coat protein gene which is expressed through the subgenomic RNA (RNA 4) (for a review see Van Vloten-Doting and Jaspars, 1977; Hull and Maule, 1985; Van Vloten-Doting, 1985).

The transport genes mapping to genomic RNA 3 of tricornaviruses follow, first of all, from the fact that their RNAs 1 and 2 were sufficient to induce viral RNA synthesis in protoplasts in the absence of RNA 3 (Kiberstis *et al.*, 1981; Nassuth *et al.*, 1981; Nassuth and Bol, 1983). In contrast, the plants inoculated with a mixture of RNAs 1 and 2 from the bromo-, cucumo-, or ilarviruses contained no detectable viral RNA and showed no symptoms (Peden and Symons, 1973; Lot *et al.*, 1974; Nassuth and Bol, 1983; Kiberstis *et al.*, 1981). It is generally assumed that, for viruses with the tripartite genome, the 3a (32- to 35-kDa) proteins encoded by RNA 3 are involved in cell-to-cell transport (Van Vloten-Doting, 1985, 1987). Several virus isolates, particularly

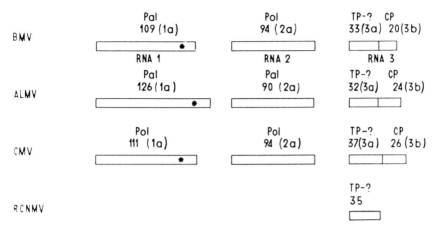

FIG. 2. Tentative transport group II of plant (+) RNA viruses. Arrangement of genes in the genomes of brome mosaic virus (BMV), alfalfa mosaic virus (ALMV), cucumber mosaic virus (CMV), and red clover necrotic mosaic virus (RCNMV) RNA 2 (Goldbach, 1986, 1987; Symons, 1985; Rezaian et al., 1985; Lommel et al., 1988). For definitions of abbreviations, see the legend to Fig. 1.

the ts mutant of cowpea chlorotic mottle bromoviruses (Dawson et al., 1975) lacking functional coat protein, are able to spread from cell to cell. Therefore, virus-specific transport function is unlikely to be provided by the coat protein itself.

The situation is rather complicated in the case of AIMV, taking into account the ability of its coat protein to activate viral genome replication. Nevertheless, it is obvious that in this case as well the products of RNAs 1 and 2 are involved in viral RNA replication and that the RNA 3-encoded product is necessary for the systemic spread of infection in infected plants. Huisman et al. (1986) identified several ts mutants of AIMV with a so-called non-ts phenotype in cowpea and tobacco protoplasts which were able to replicate in isolated protoplasts, but not in leaf tissues at nonpermissive temperatures.

It was suggested by Huisman et al. (1986) that these mutants were ts in TF, due to the mutation in the 3a gene and that the 32-kDa protein can be considered functionally equivalent to the 30-kDa TMV TP. However, two mutants identified by Huisman et al. (1986) as ts in TF originally were characterized as the mutants in RNA 2 (Van Vloten-Doting et al., 1980). These results might suggest that an AIMV RNA 2-encoded function is involved in cell-to-cell transport as well. However, it is possible that these mutants have accumulated additional mutations since their original characterization (Huisman et al., 1986).

Some indirect evidence supporting the theory that the 3a protein of AIMV plays a role as TP comes from analysis of the kinetics of accumulation of the three nonstructural proteins of AIMV in tobacco plants (Berna et al., 1986). It was found that the 32-kDa protein reached its maximum concentration in the membrane fraction sooner and disappeared more rapidly than in the other nonstructural proteins. The authors suggested that either the 32-kDa protein is synthesized transiently, as in the case of the 30-kDa TMV TP, or it decays faster than the other two. The 32-kDa protein of AIMV has been found to accumulate in the cell wall fraction of tobacco leaves infected with AIMV (Godefroy-Colburn et al., 1986, 1989). This protein was visualized immunocytochemically in the middle lamella of the walls of parenchymal or epidermal cells just reached by the infection front and in which viral multiplication has just begun. The 32-kDa protein was not found later when AIMV had accumulated to high levels in infected walls (Stussi-Garaud et al., 1987).

These findings also support the concept that the 32-kDa protein is involved in the spread of viral infection from cell to cell; that is, it is the TP of AIMV. The data obtained by MacKenzie and Tremane (1988) somewhat contradict these results. They located the 3a protein of cucumber mosaic cucumovirus (CMV) (equivalent to the 32-kDa protein of AIMV) in the nucleoli of cells recently infected with virus (24–30 hours postinoculation) as well as in cells from systemically infected leaf tissue.

Despite some differences, the tricornaviruses AIMV, BMV, and CMV basically have similar genomic organizations and code for functionally similar nonstructural proteins. It is important that amino acid sequences of the 3a proteins of these viruses also exhibit a certain sequence homology (for a review see Goldbach, 1986). The most extensive homology is found between the brome mosaic virus (BMV) and CMV proteins (34%) and the least is found between the AIMV and CMV proteins (14%) (Murthy, 1983; Savithri and Murthy, 1983). In addition to sequence homology, significant similarities in the predicted charge distributions, hydrophobicity profiles, and α-helix, β-sheet, and β-turn propensities occur throughout the 3a proteins of BMV, CMV, and AIMV, confirming an ancestral relationship and possible functional similarity among these proteins (Davies and Symons, 1988).

Members of the dianthovirus group contain a bipartite ssRNA genome encapsidated in isometric particles. It was clear from pseudorecombination experiments (Osman and Buck, 1987) that RNA 1 encoded the viral replicase and the coat protein (Morris-Krsinich et al., 1983), whereas ability to invade the plants systemically was dependent primarily on RNA 2, suggesting that the transport gene was located there. This suggestion is in agreement with the finding by Osman and Buck (1987), Pappu and Hiruki (1988), and Paje-Manalo and Lommel (1989) that RNA 1 can replicate independently of RNA 2 in isolated protoplasts, whereas the systemic spread of infection in plants requires both RNAs 1 and 2. The complete nucleotide sequence of red clover necrotic mosaic dianthovirus (RCNMV) RNA 2 has been determined by Lommel et al. (1988). RNA 2 of RCNMV is 1448 nucleotides in length, with a 5′-terminal cap but no 3′-terminal poly(A) tail. The only ORF has been identified in RNA 2, which can encode a polypeptide of 35 kDa, observed in vivo (Lommel et al., 1988). A certain similarity between the RCNMV 35-kDa protein and BMV 3a protein was shown by Lommel et al. (1988). Consequently, this 35-kDa protein of RCNMV might be involved in the TF expression of this virus. Thus, the putative TP (35 kDa) of RCNMV might be tentatively included in transport group II.

3. Third Transport Group: Potex-, Hordei-, and Furoviruses

It might seem rather strange that in discussing TF expression we unite such different groups of viruses: potexviruses, which have a monopartite genome and filamentous particles, beet necrotic yellow vein furovirus (BNYVV); and barley stripe mosaic hordeivirus (BSMV). The latter two are rod-shaped viruses with multipartite genomes. However, the sequence data showed a certain similarity of the gene organization and homology between the putative proteins coded by them, which possibly reflects, in particular, the similarity in TF coding by these three types of genomes.

a. TF of Potexviruses. Potato virus X (PVX) is a member of the potexviruses. The filamentous virions of PVX contain ss 5′-capped and 3′-polyadenylated plus-sense RNA (molecular weight 2.1×10^6) (Purcifull and Edwardson, 1981; Morozov et al., 1981). The complete nucleotide sequence of the PVX genome, reported by Kraev et al. (1988), Skryabin et al. (1988), and Huisman et al. (1988), contains five relatively long ORFs. ORF 1, nearest to the 5′ terminus, coded for a polypeptide with a molecular weight of about 165,000. Two regions of this protein display sequence similarities to the putative RNA-replicating proteins of different plant viruses. Downstream to the 165-kDa gene the triple block of ORFs similar to those in BSMV and BNYVV genomes (see below) was revealed (Fig. 3). The putative 25-kDa polypeptide coded for by ORF 2 is clearly related to the presumptive NTP binding domain of the 165-kDa protein. Consequently, two NTP-binding proteins are encoded by PVX RNA.

ORF 3, coding for a putative 12-kDa protein, overlaps ORF 2 at its 5′

terminus and ORF 4 at its 3' terminus, thus forming a triple block of overlapping genes (Fig. 3). The predicted translation product of ORF 4 is a small protein of about 8 kDa. Both the 12- and 8-kDa proteins of PVX as well as analogous pairs of proteins of BSMV (14 and 17 kDa) and BNYVV (13 and 15 kDa) (Fig. 3) have essential features of membrane-bound proteins (Skryabin *et al.*, 1988). ORF 4 is followed by the PVX coat protein gene at the 3' end of genomic RNA. Comparison of nucleotide sequences of the PVX genome with those of other potexviruses—white clover mosaic virus (Forster *et al.*, 1988), potato aucuba mosaic virus (Bundin *et al.*, 1986), narcissus mosaic virus (Zuidema *et al.*, 1989), and lily virus X (J. F. Bol, personal communication)—shows that different potexviruses have similar genome organization.

It was demonstrated by Taliansky *et al.* (1982c) that the cell-to-cell movement of TMV mutant Ls1 (ts in TF) can be complemented by PVX in mixed infected plants. Consequently, the PVX genome is capable of coding for a protein(s) which is a functional analog of the 30-kDa TMV TP. However, the proteins structurally homologous to the 30-kDa TMV protein are not encoded by the genome of PVX and by the genomes of three other viruses (i.e., BSMV, BNYVV, and narcissus mosaic virus) which contain the triple blocks of the overlapping ORFs mentioned above. Conversely, the genomes of viruses belonging to transport groups I and II do not contain this triple block of genes. One

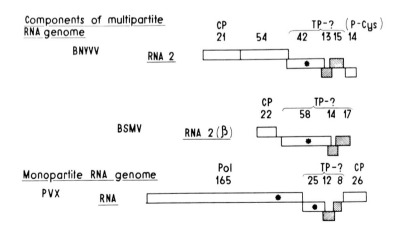

FIG. 3. Tentative transport group III of plant (+) RNA viruses. Arrangement of genes in the genomes of potato virus X (PVX) RNA, barley stripe mosaic virus (BSMV) RNA 2 (β) (Gustafson and Armour, 1986), and beet necrotic yellow vein virus (BNYVV) RNA 2 (Bouzoubaa *et al.*, 1986). For definitions of abbreviations, see the legend to Fig. 1. The genes coding for the membrane-bound hydrophobic proteins are shaded.

can speculate that one or both of the putative membrane-bound proteins coded by PVX are involved in virus-specific TF. The participation of the 25-kDa NTP-binding protein in TF expression is more questionable, although it cannot be ruled out. It is noteworthy that the TPs of transport groups I and II show no homology to any of the proteins coded by animal viruses. This observation probably reflects the fact that animal viruses do not need the TF at all. On the other hand, the NTP-binding motif has been revealed in various proteins coded by both animal and plant viruses. Therefore, it seems probable that, at least in the case of transport groups I and II, the RNA-replicating NTP-binding proteins are not involved in TF expression. In the case of potexviruses (group III), the NTP-binding motif was doubled in two virus-specific proteins: the replicase coded by the 5'-proximal ORF 1 and the 25-kDa protein coded by an internal ORF 2 (Kraev et al., 1988; Skryabin et al., 1988). The function of the 25-kDa PVX protein as well as of the NTP-binding proteins doubled in the genomes of furoviruses (the 42-kDa protein of BNYVV RNA 2) and hordeiviruses (the 58-kDa protein of BSMV RNA 2) (Gorbalenya et al., 1989) is obscure.

b. TF of BSMV, a Hordeivirus. BSMV is a rod-shaped virus with a tripartite RNA genome, but at the strain level virion RNA can be composed of four RNA segments. These RNA species have been designated RNA 1 (or α), RNA 2a (or β), RNA 2b (or γ), RNA 3 (or γ), and RNA 4 (or λ) in accordance with decreasing length (Gustafson et al., 1982, 1987). Atabekov and Novikov (1989) suggested that BSMV RNAs should be designated RNAs 1, 2, 3_{II}, 3_{III}, and 4, respectively. The typical tripartite (orthotripartite) strains contain RNAs 1, 2, and 3_{III}. In pseudobipartite strains the second (RNA 2) and third (RNA 3_{II}) are of similar size, and the pseudoquadripartite strains contain an additional virion RNA 4 nonessential for BSMV genome replication (for a review see Atabekov and Novikov, 1989). For the sake of simplicity only the orthotripartite genome is considered below.

RNA 1 codes for a component of RNA replicase and is monocistronic. RNA 3_{III} contains two cistrons: the 5'-proximal cistron, which encodes the second component of RNA replicase, and the 3'-proximal gene, which is expressed in the subgenomic RNA which directs the synthesis of the 17-kDa cysteine-rich protein (see Atabekov and Novikov, 1989). Based on the general assumptions formulated above (Section II,A,1), one can suggest that the 17-kDa protein would not be essential for BSMV genome replication. The function of this polypeptide is unknown. Probably, TF expression is controlled by the protein(s) coded by RNA 2 of BSMV. RNA 2 (Fig. 3) contains the coat protein gene in the 5'-proximal position and in three additional ORFs (Gustafson and Armour, 1986). In accordance with the general assumptions mentioned

222 JOSEPH G. ATABEKOV AND MIKHAIL E. TALIANSKY

above, this fact suggests that all of the genes in RNA 2 should be nonessential for BSMV RNA replication; that is, RNA 2 might contain the transport gene(s). ORF 2, which codes for the putative 58-kDa NTP-binding protein, is located downstream from the coat protein gene of BSMV RNA 2. ORF 3 codes for the 14-kDa protein and overlaps ORF 2 and ORF 4, which code for the 17-kDa protein (Fig. 3). As mentioned above, the similar triple block of genes has been revealed in the potexvirus genome and in BNYVV, furovirus, and RNA 2 (Fig. 3).

It is important to note that the cell-to-cell systemic spread of TMV can be complemented by BSMV in doubly infected plants (Hamilton and Dodds, 1970; Dodds and Hamilton, 1972). In particular, the complementation of TF of the TMV *ts* mutant Ls1 (ts in TF) by BSMV was demonstrated by Malyshenko *et al.* (1987). Thus, both BSMV and PVX should produce a transport protein(s) which is functionally analogous to the 30-kDa TMV TP.

As for the potexviruses (see above), the 14-kDa and/or the 17-kDa membrane-bound proteins of BSMV should be considered the most probable candidates for the role of TPs. This is in agreement with the results of I. T. D. Petty and A. O. Jackson (quoted by Hull, 1989), showing that the TF of BSMV is associated with RNA 2.

c. TF of BNYVV, a Furovirus. BNYVV is a multicomponent soilborne fungus-transmitted RNA virus with rigid rod-shaped virions (Steven *et al.*, 1981). Most BNYVV isolates contain four ss 5′-capped and 3′-polyadenylated RNA molecules. RNAs 1 and 2 have lengths of about 7100 and 4600 nucleotides, respectively, whereas RNAs 3 and 4 often differ in length, depending on the isolate used (Richards *et al.*, 1985; Bouzoubaa *et al.*, 1986, 1987). It is important that BNYVV isolates maintained in *Chenopodium quinoa* wild type often lack RNA 3 and/or RNA 4 (Richards *et al.*, 1985; Kuszala *et al.*, 1986; Burgemeister *et al.*, 1986); that is, these two smallest RNA segments are not essential for virus replication. Consequently, the genome of BNYVV can be considered bipartite, at least when propagated in *C. quinoa*.

The complete nucleotide sequences of BNYVV RNAs 1 (Bouzoubaa *et al.*, 1987) and 2 (Bouzoubaa *et al.*, 1986) have been reported. RNA 1 has one long ORF encoding the 237-kDa protein thought to be involved in viral RNA replication (Bouzoubaa *et al.*, 1987). Unfortunately, to our knowledge, there is no information on the capability of RNA 1 of BNYVV to replicate alone (i.e., without RNA 2) in the infected cells. BNYVV RNA 1 is monocistronic and obviously does not code for TF. Therefore, the TF should be encoded by one or more of the ORFs in RNA 2. There are six ORFs in the sequence of RNA 2 (Fig. 3). The 5′-proximal coat protein gene is followed by the ORF coding for the 54-kDa putative protein of unknown function. A block of three ORFs corresponds to three tentative proteins (42, 13, and 15 kDa) that are similar to the respective proteins encoded by the RNAs of potexviruses and BSMV (see above). Finally, the 3'-proximal ORF which codes for the tentative 14-kDa cysteine-rich protein.

One might assume that the 54- and 14-kDa BNYVV-coded proteins are not involved in TF, although this assumption is questionable. The basis for this assumption is the fact that only the block of three ORFs coding for the NTP-binding protein and two hydrophobic short polypeptides is constantly repeated in genomes of hordeiviruses (RNA 2), potexviruses, and furoviruses (RNA 2) (see Fig. 3). Thus, one might regard one or more of these proteins (probably the membrane-bound ones) as the candidate(s) for the role of TPs of the three listed taxonomic groups of plant viruses. None of these membrane-bound proteins possesses any similarity to the putative TPs of plant viruses that we included in transport groups I and II. Thus, hordei-, furo-, and potexviruses might be accommodated in the tentative group III.

4. Miscellaneous Viruses

This tentative group is proposed to separate those viruses for which information concerning their TPs is obscure or absent. In several of them no transport genes could be accurately recognized in a viral genome, although their complete nucleotide sequence was known, for example, carnation mottle virus (Guilley et al., 1985), maize chlorotic mottle virus (Nutter et al., 1989), turnip yellow mosaic virus (Morch et al., 1988), and southern bean mosaic virus (Wu et al., 1987). So far as we know, no data concerning the TPs of closterovirus, tombusvirus, tobacco necrosis virus, and pea enation mosaic virus are available, nor are data available for the plant minus-sense ssRNA viruses and plant reoviruses.

Geminiviruses have circular small (2.5–3.0 kb) ssDNA genomes packaged in twin (geminate) quasiisometric particles. The majority of geminiviruses can be divided into two subgroups: geminiviruses whose genomes are divided between two circular DNA molecules (i.e., bipartite genome) and those whose genomes consist of one DNA molecule (i.e., monopartite genome) (for a review see Davies, 1987). The first subgroup of geminiviruses includes cassava latent virus (CLV), bean golden mosaic virus (BGMV), and tomato golden mosaic virus (TGMV), all of which are whitefly-transmitted and infect dicotyledonous plants. The nucleotide sequences of CLV (Stanley and Gay, 1983), TGMV (Hamilton et al., 1984), and BGMV (Howarth et al., 1985) have been determined. Comparison of CLV, BGMV and TGMV sequences revealed six ORFs conserved in approximate size and position arrangement, suggesting common or similar functions and possibly a common origin for these viruses (Davies, 1987). Four of these ORFs, including the putative gene of the coat protein, reside on DNA 1 (A component). The two remaining ORFs are located on DNA 2 (B component).

It has been conclusively demonstrated that both DNAs are required for the infectivity of CLV, TGMV, and BGMV (Stanley and Gay, 1983). On the other hand, CLV DNA protoplast experiments (Townsend *et al.*, 1985) showed that DNA 2 was not required for replication. In agreement with this result, Rogers *et al.* (1986) and Sunter *et al.* (1987) demonstrated that all of the virally encoded functions necessary for DNA replication are carried on DNA 1 (or the A component) of TGMV. It might be inferred that DNA 1 is autonomous in single cells, but DNA 2 (the B component) provides a cell-to-cell spread (i.e., transport) function, possibly expressing one or both of the 33.7-kDa (BL 1) or 29.3 kDa (BR 1) products. Thus, one or both of these proteins can be regarded as the putative TPs of bipartite geminiviruses; insertion and deletion mutagenesis in both genes in DNA 2 of CMV and TGMV indicate that they are both required for the successful establishment of infection (Brough *et al.*, 1988; Etessami *et al.*, 1988). However, the mutants of TGMV in the AL 2 ORF also show phenotypes typical for the defective TF [DNA replication (+) and infectivity in plants (−)]. Thus, it can be suggested that one of the proteins coded by DNA 1 (the A component) AL 2 might also be involved in the process of cell-to-cell movement (Elmer *et al.*, 1988; D. M. Bisaro, personal communication). Unfortunately, at present we do not have at our disposal any data on the similarity of amino acid sequences of these proteins nor on the tentative TPs of other plant viruses.

It is obvious that the coat protein itself is not required for the cell-to-cell spread of CLV and TGMV, since deletion coat protein mutants of these viruses are still infectious in plants giving normal systemic symptoms (Stanley and Townsend, 1986; Gardiner *et al.*, 1988). However, rather surprisingly, it was found that the extensive deletions across the CLV coat protein gene limited the spread of the genomic component in plants, although infectivity was restored to these latent mutants by the introduction of a DNA fragment of a suitable size (even if it was foreign) to replace the deleted sequences (Etessami *et al.*, 1989). These data indicate that, at least in the case of CLV, there is a relatively specific size requirement in order for DNA 1 to spread (Etessami *et al.*, 1989).

Unlike CLV, TGMV, and BGMV, the monopartite geminiviruses (maize streak virus, wheat dwarf virus, and beet curly top virus) are transmitted by leafhoppers and are not readily mechanically transmitted (Stanley *et al.*, 1986). The sequence data on these viruses were all accommodated in one circle bearing a superficial resemblance to DNA 1 of the bipartite geminiviruses. DNA 2, coded for by putative transport protein(s) of bipartite viruses, appeared to be missing. Data on possible candidates for the TPs of monopartite geminiviruses are absent. One might suggest that geminiviruses, which are strictly limited to phloem (e.g., beet curly top virus) (Thomas and Mink, 1979) do not require TF at all.

5. Do the Phloem-Limited Viruses Have a Transport Gene?

The restriction of so-called phloem-limited plant viruses (e.g., luteoviruses and some geminiviruses) to phloem tissues has been regarded as a particular example of TF blockage in the virus-infected plant (Atabekov and Dorokhov, 1984). This is in agreement with the fact that isolated mesophyll protoplasts can be infected with different phloem-limited viruses (Kubo and Takanami, 1979; Barker and Harrison, 1982). Barker and Harrison (1986) suggested that companion cells are the main sites of potato leaf roll luteovirus (PLRV) replication. For the phloem-associated viruses the companion cells can be regarded as target cells primarily infected by vectors. After replication in these cells, the viral genetic material seemingly moves into the sieve tubes, which could serve predominantly as a route for transporting virus over the plant (Barker and Harrison, 1986). It is not clear whether this process requires the expression of a viral gene(s) (i.e., whether phloem-limited viruses require some transport protein for movement along a conductive system). On the other hand, a particular feature of the phloem-limited viruses is their inability to transport their genome from the phloem into parenchymal cells.

This can be considered failure of a transport event, since movement of the phloem-limited viruses into the parenchyma can be complemented by an appropriate helper virus: (1) Carr and Kim (1983) reported that the geminivirus BGMV moves into the parenchyma in plants doubly infected with BGMV and TMV; (2) Barker (1987) suggested that PLRV can be transported into the parenchyma in the presence of potato Y potyvirus; and (3) Atabekov et al. (1984) suggested that PLRV moves into parenchymal cells as a result of TF complementation by PVX or TMV. It can be assumed that PLRV is deficient in TF which can be complemented by different foreign viruses belonging to unrelated groups (for a review see Hull, 1989). The complete nucleotide sequences of the genomic RNAs of barley yellow dwarf luteovirus (Miller et al., 1988), beet western yellow virus (Veidt et al., 1988), and PLRV (Mayo et al., 1989; van der Wilk et al., 1989) have been reported. No evidence for a transport function was found in the sequences of these luteovirus RNAs; indeed, the viruses are largely confined to phloem and might therefore lack one (M. A. Mayo, personal communication).

III. Host Dependence and Interviral Phenomenological Exchangeability of Transport Function

A. Host Dependence of Virus-Coded Transport

1. Virus–Host Compatibility Needed for TF Expression

The systemic spread of a virus requires the expression of two genomes: that of a virus and that of a host. A given plant virus can express its TF only in a strictly definite plant species. However, frequently, a virus can replicate in inoculated cells or in isolated protoplasts of the resistant plants, but it is unable to express TF in the total plant tissues. As a result the virus remains located in primarily infected cells and replicates only there; the plant looks immune (i.e., "nonhost") because of the blockage of TF.

The term "subliminal symptomless infection" was offered to designate virus–host interactions when no visible symptoms develop and extremely little infective virus progeny can be recovered from the inoculated leaves (Cheo, 1970; Cheo and Gerard, 1971). At least in the cases studied, TF could not be expressed in the subliminally infected hosts (Sulzinski and Zaitlin, 1982). The best example of this phenomenon is provided by the TMV mutant Ls1, ts in TF: Replication of Ls1 at 32–35°C is restricted to the initially infected cells (Nishiguchi et al., 1978, 1980).

It can be assumed that the establishment of subliminal infection in a given plant species is also due to its nonfunctional virus-coded TPs. Apparently, the 30-kDa TMV TP is nonfunctional in cowpea and cotton as is the Ls1 TP at 33°C in the normal host plant tobacco. In general, it can be proposed that virus-coded TF is host specific (i.e., host dependent), since the viral TPs are capable of operating in only some strictly definite plant species or even in some varieties. Obviously, some kind of functional compatibility must occur between the virus-coded TPs and some components of the host for the systemic spread of infection to develop.

2. Host Resistance Breaking Mutations in the Transport Gene

The resistance of tomatoes to the TMV-L strain conferred by the Tm-2 gene is due to the blockage of TF in these tomato lines (Motoyoshi and Oshima, 1975, 1977; Stobbs and MacNeill, 1980; Taliansky *et al.*, 1982d). Meshi *et al.* (1989) showed that the amino acid changes occurring in the 30-kDa protein (Cys-68 to Phe, Glu-133 to Lys, or Glu-52 to Lys) overcome the Tm-2 resistance. The mutants are able to spread over Tm-2 tomatoes, unlike their parent strain L. The authors proposed that the 30-kDa TP might interact with a putative host factor(s) possibly localized at the plasmodesmata and that TPs coded by different viruses might have structural features which enable them to interact with a putative host factor(s) of various host plants. Alternatively, a resistance host factor might solely inhibit function of the 30-kDa TP, preventing TF (Meshi *et al.*, 1989).

3. Host-Dependent Suppression of ts TF

Mushegyan *et al.* (1990) demonstrated that the phenotypic expression of *ts* mutations in the TF of TMV mutants Ni2519 and Ls1 was host dependent. These mutants appear to be ts in TF when multiplying in *N. tabacum* (see above). Both mutants spread systemically in another host, *Amaranthus caudatus* L., at both 24°C and 33°C. Consequently, the mutational *ts* defect in TF could be suppressed in *A. caudatus*, enabling the virus to spread normally in this plant at the temperature nonpermissive for TF in *N. tabacum*. Moreover, Ls1 can serve as a helper virus at 33°C, complementing in *A. caudatus* the spreading of cucumber green mottle mosaic virus and red clover mosaic virus, the viruses which infect this plant subliminally (Mushegyan *et al.*, 1990). It was suggested by the authors that the TPs of Ls1 and Ni2519 can be functionally stabilized or recognized in *A. caudatus* by a putative factor(s) provided by the host.

4. Role of cAMP in TF

The role of cAMP as the secondary messenger in animal, fungal, and bacterial cells is well documented (for a review see Flockart and Corbin, 1982). cAMP dependence was also shown in plants for (1) antiviral factor formation (Rosenberg *et al.*, 1982), (2) metabolism of starch (Tu, 1977), and (3) some morphogenetic processes (Chopra and Sharma, 1985).

Mushegyan *et al.* (1986) showed that nicotinic acid (known to reduce cAMP levels in certain animal cells) blocked the systemic spread and accumulation of TMV in tobacco leaf disks, producing no significant changes in the TMV level in tobacco protoplasts. Nicotinic acid is likely to block the transport of the infection, rather than TMV replication. A decrease in the virus-specific transport seems to be based on the reduction of the intracellular cAMP level as the inhibiting effect of nicotinic acid could be overcome by exogenous dibutyryl-cAMP. In accordance with this, dibutyryl-cAMP and papaverine (increasing the cAMP level in animal cells) were shown to rescue the *ts* mutant Ls1 transport from cell to cell at nonpermissive temperatures, although no stimulation of its replication occurred in protoplasts. These results imply that cAMP is involved in TMV TF expression.

cAMP is known to participate in different processes in eukaryotic cells by activating cAMP-dependent protein kinases that catalyze the phosphorylation of different cellular polypeptides (for a review see Severin and Kochetkova, 1985). Therefore, it is of importance to mention that the 30-kDa TP of TMV was reported to be phosphorylated (D. Zimmern, personal communication, quoted by Hull, 1989). One can speculate that a putative host factor (mentioned above) is a specifically acting (i.e., recognizing the 30-kDa protein) cAMP-dependent protein kinase. Assuming that only a phosphorylated 30-kDa protein is involved in TF, one can speculate what the roles of cAMP and of the hypothetic host factor are (cAMP-dependent protein kinase?) in the process of virus cell-to-cell movement. It is worth mentioning that Saito *et al.* (1988) found, in the central part of the TMV 30-kDa protein, a motif similar to the triphosphate binding sites of several nucleotide-binding proteins, including some protein kinases. Previously, Martinez-Izquierdo *et al.* (1987) also observed in ORF 1 (the putative transport gene of CaMV) product sequences similar to sites involved in the ATP binding of some kinases and sites involved in the Mg^{a+}-ATP binding of other ATP-requiring proteins.

These peculiarities in the structure of TPs allow speculation as to their possible involvement in TF expression, either by providing energy or by modifying cellular structures (e.g., plasmodesmata) via kinase activity (Martinez-Izquierdo *et al.*, 1987), which can be activated by host-directed cAMP-dependent phosphorylation.

5. Tissue Specificity of the Action of Viral Transport Proteins

It might be assumed that virus-specific TPs can not only be host specific, but in some virus–host combinations be tissue specific as well; that is, they are functional only in certain tissues of the host plant. There are several examples of the systemic spread of a virus in one type of tissue and of its inability to penetrate into another type of tissue of the same plant. The virions of *Chloris* striate mosaic geminivirus (CSMV) can be detected in all tissues of *Chloris gayana* leaves except the epidermis (Francki *et al.*, 1979). It is possible that the TP of CSMV is nonfunctional in epidermis. In agreement with this possibility, CSMV is not mechanically transmissible. It is not excluded that the ability of a plant virus to express its TF in the epidermis is obligatory for the mechanical transmissibility of plant viruses.

Replication of many viruses is restricted to phloem cells, which probably reflects their inability to express TF in nonconducting plant tissues (see Section II,B,5).

Another phenomenon that possibly represents a transport event is the seed transmission of viruses. Numerous viruses cannot be transmitted through seed, which may reflect, at least in some cases, their inability to express the TF essential for penetration of the virus genome into generative cells.

6. Possible Role of Some Virus-Specific Factors Other Than the TPs in TF

Apparently some minimum threshold amount of the viral RNA and the TP should be reached in the infected cell to ensure subsequent cell-to-cell transport to the healthy cells. In other words, there should be a dependence between the synthesis of viral genomic and subgenomic viral RNAs and the rate of cell-to-cell transport of infection. Consequently, the defect in RNA replication would manifest itself as a decline in TF expression.

Hall and co-workers (Dreher et al., 1989) revealed the relationship between the replication of the genomic BMV RNA 3 and the ability of the virus to infect plants systemically. Using infectious RNA transcripts of wild-type RNAs 1 and 2 and RNA 3 variants with mutations in the 3' tRNA-like region, the authors found that the systemic spread of BMV was prevented by the decreased replication of RNA 3. It is likely that inadequate amounts of RNA 3 gene products were established in cells to permit effective systemic spread. Using different combinations of infectious in vitro RNA transcripts from two bromovirus (BMV and CCMV) cDNA clones, Allison et al. (1988) examined the role of RNA 3 products. Wild-type combinations of BMV and CCMV transcripts induced normal systemic infection in their selective hosts (barley and cowpea, respectively). Surprisingly, the heterologous combinations (e.g., BMV RNAs 1 and 2 and RNA 3 from CCMV) were infectious only in protoplasts. The hybrid viruses in which RNA 3 was exchanged between BMV and CCMV were capable of restricted cell-to-cell spread within the local lesion in C. hybridum, but failed to spread systemically over the natural host of either virus.

It could be suggested that the 32-kDa TP coded by RNA 3 requires a certain host factor(s) compatible for the respective BMV or CCMV in the systemically infectible host. However, the following facts might be important in understanding these results: (1) The level of heterologous CCMV RNA 3 replication (even in protoplasts) was considerably lowered in combination with BMV RNAs 1 and 2 compared with homologous combination (i.e., the accumulation of CCMV RNA 3 was "barely detectable"). (2) The replication of BMV RNA 3 in combination with CCMV RNAs 1 and 2 was significant, but accumulation of the latter was suppressed in this combination. The possibility has not excluded that the inability of both BMV and CCMV hybrids derived by RNA 3 exchange to spread over the nonhost plants (i.e., barley or cowpea) was due to the lowered (in comparison to the parent virus) level of RNA accumulation.

An alternative explanation suggests that RNA 3 substitution is not sufficient to change the host range of the two bromoviruses, since virus-specific factors other than the TPs [e.g., the protein(s) encoded by RNA 1 and/or RNA 2] are involved directly in cell-to-cell movement.

The data by Schoelz and Shepherd (1988) with recombinant genomes of caulimovirus (CaMV) from different strains of the virus confirmed the theory that more than a single locus of the CaMV genome might be involved in the control of the viral systemic spread. One locus was mapped to ORF 6 and coded for an inclusion body protein, and another locus was localized within ORFs 1–5. Taking into account the data mentioned above (see Section II,B,1,c) on the role of the ORF 1 product as a putative TP, one can assume that this protein might interact with the product of ORF 6 or that both of these proteins should be combined in order to express TF.

The role of the viral coat protein gene in TF expression by different viruses has been discussed (for a review see Atabekov and Dorokhov, 1984). Different TMV strains, defective or ts in the coat protein gene, are known to spread from cell to cell in infected leaves, although their long-distance transport is severely damaged. Tobravirus (TRV) RNA 1 coded for the TP, but lacking the coat protein gene, is also capable of cell-to-cell spreading (see above). In accordance with this, Takamatsu et al. (1987) showed that a coatless TMV mutant, lacking most of the coat protein gene, is defective only in long-distance movement, but its cell-to-cell transport is not damaged. Thus, one might suggest that in most cases the coat protein gene is not essential for short-distance cell-to-cell transport.

On the other hand, Wellink and van Kammen (1989) reported that the comovirus (CPMV) infection can be restricted to the initially inoculated cells by the introduction of mutations not only in the 48- or 58-kDa TP gene, but also in the coat protein genes. The results strongly indicate that, during normal infection, coat proteins of CPMV play an essential role in cell-to-cell movement. The authors suggest that in this case the TP is realized in the form of viral particles. These findings seem to conflict with the results of the studies by Malyshenko et al. (1988), who reported that it is possible to complement the transport of B-RNA (with neither the TP gene nor coat protein genes) of another comovirus, red clover mosaic virus, with a tobamovirus helper (see below). Wellink and van Kammen (1989) and Goldbach *et al.* (1989) suggested that CPMV is transported as particles, not as naked RNA, whereas the tobamovirus could transport viral genetic material in another form, lacking viral coat proteins.

However, it seems that an alternative suggestion could also be proposed. It has not been excluded that the CPMV genome is transported in the form of viral ribonucleoprotein (vRNP), described by Dorokhov *et al.* (1984) for TMV and PVX. vRNP was shown to contain several proteins and the coat protein as a major structural element, probably protecting RNA. In the case of TMV coat protein mutants, the systemic spread (especially the long-distance transport) is considerably suppressed. This is in agreement with the fact that absence of the coat protein from vRNP makes it unstable. This effect might be even more important in the case of multipartite viruses such as CPMV.

To clarify the relationship between the ability to form virus particles and long-distance movement, T. Saito and Y. Okada (personal communication) constructed a TMV mutant that could produce a normal coat protein with full activity, but could not form the virus particles, due to multiple mutations introduced in the assembly origin region (so as to retain the coded 30-kDa protein sequence). The progeny mutant RNA and coat protein were normally accumulated only in the inoculated leaves. In systemic (noninoculated) leaves accumulation of viral RNA and protein decreased to less than one-tenth compared to the wild type. Thus, it can be suggested that long-distance transport is realized with the help of viral particles or, alternatively, that the origin of assembly plays an essential role in the formation of a stable (i.e., containing coat protein) vRNP special transport form, the structure and function of which were described earlier (for a review see Atabekov and Dorokhov, 1984).

B. Transport Function Complementation between Related and Unrelated Viruses

1. Phenomenological Nonspecificity of Viral TF on Complementation

The transport-blocking mechanism operating in a resistant plant (e.g., in a subliminally infected plant) does not prevent superinfection of this plant by other viruses or their systemic spread. Under conditions of mixed infection, the spreading virus (i.e., helper) might provide the TP to the transport-deficient (i.e., dependent) virus, facilitating its movement over the resistant plant. Consequently, this type of resistance to the virus can be overcome; that is, the nonhost plant can be infected by this virus in the presence of another (helper) virus which can spread normally. In a model experiment the mutation in the transport gene of TMV *ts* mutant Ls1 was complemented by TMV tr strain L used as a helper virus (Taliansky *et al.*, 1982c). In the presence of helper, the mutant, (ts in TF) spreads systemically at a temperature which is nonpermissive for the ts virus alone. As mentioned in Section III,A,2, resistance of Tm-2 tomato plants to TMV-L is due to the blockage of TF, which can be overcome by the complementation of transport by the unrelated helper virus (i.e., PVX) (Taliansky *et al.*, 1982d). Thus, in the first complementation system the defective (ts) 30-kDa TP of Ls1 was substituted by the functional (tr) 30-kDa TP of TMV-L, whereas in the second example the 30-kDa TP of TMV-L nonfunctional in Tm-2 tomatoes was substituted by putative PVX-coded TP, which was effective in this host. As a result the nonhost plant whose resistance to a virus was due to the blockage of TF was systemically infected in both cases.

When members of the same (i.e., tobamovirus) group (e.g., TMV, sunn-hemp mosaic virus, cucumber green mottle mosaic virus, and tobamovirus from orchids) were used in different combinations and in different plants, they were capable of complementing one another's systemic spread. The transport of the dependent virus was relatively efficient in the nonhost plant and depended totally on the presence of the related helper tobamovirus (Malyshenko *et al.*, 1990).

Table I and Fig. 4 summarize numerous examples of the transport-deficient virus complementation by different unrelated viruses. Systemic spread of phloem-limited viruses (transport-deficient in parenchyma) can be complemented by different foreign helper viruses. Geminivirus BGMV was shown to spread systemically in beans coinfected with TMV (bean strain) (Carr and Kim, 1983). Complementation of the spread of luteovirus PLRV into mesophyll cells was effected by PVX (Atabekov *et al.*, 1984) and potato virus Y (Barker, 1987). Barker (1989) examined the interaction between PLRV and beet western yellow luteovirus with a range of 19 sap-transmissible viruses belonging to 10 different groups. The accumulation of PLRV increased up to 10-fold in plants doubly infected with each of several potyviruses, narcissus mosaic potexvirus, carrot mottle virus, or various tobraviruses (see Table I). Fuentes and Hamilton (1988) demonstrated that protoplasts of a nonpermissive host (*Phaseolus vulgaris*) were readily susceptible to infection by southern bean mosaic virus (SBMV).

In the mixed infection of bean leaves (SBMV and sunn-hemp mosaic virus, tobamovirus helper), SBMV spread within the inoculated leaf. On the other hand, the accumulation of PLRV was not substantially

TABLE I

POSSIBLE COMPLEMENTATION OF THE SYSTEMIC SPREAD
BETWEEN UNRELATED PLANT VIRUSES[a]

Tentative transport group	Helper virus Taxonomic group	Name	Nonspreading (helper-dependent virus) Name	Taxonomic group[c]	Reference
I	Tobamovirus	TMV[b]	RCMV	Comovirus (I)	Malyshenko et al. (1988)
		SHMV[b]	RCMV	Comovirus (I)	Malyshenko et al. (1988)
		TMV[b]	BMV	Bromovirus (II)	Taliansky et al. (1982d) Malyshenko et al. (1988)
		SHMV[b]	BMV	Bromovirus (II)	Taliansky et al. (1982d)
		TMV[b]	BSMV	Hordeivirus (III)	Malyshenko et al. (1990)
		TMV[b]	PLRV	Luteovirus (phloem limited)	Atabekov et al. (1984)
		SHMV[b]	BGMV	Geminivirus (phloem limited)	Carr and Kim (1983)
		SHMV[b]	SBMV	Sobemovirus (?)	Fuentes and Hamilton (1988)
	Tobravirus	TRV	TMV-Ls1	Tobamovirus (I) (33°C)	Malyshenko et al. (1990)
		TRV	PLRV	Luteovirus (phloem limited)	Barker (1989)
		PEBV	PLRV	Luteovirus (phloem limited)	Barker (1989)
		PEBV	BWYV	Luteovirus (phloem limited)	Barker (1989)
	Comovirus	RCMV	TMV-Ls1	Tobamovirus (I) (33°C)	Malyshenko et al. (1990)
	Nepovirus	ArMV	RCMV	Comovirus (I)	Malyshenko et al. (1990)
		ArMV	TMV	Tobamovirus (I)	Malyshenko et al. (1990)
	Potyvirus	PVY[b]	PVX	Potexvirus (III)	Unpublished observations

(continued)

TABLE I

(*Continued*)

Tentative transport group	Helper virus		Nonspreading (helper-dependent virus)		Reference
	Taxonomic group	Name	Name	Taxonomic group[c]	
		PVY[b]	PLRV	Luteovirus	Barker (1987, 1989)
		PVV	PLRV	Luteovirus (phloem limited)	Barker (1989)
		PVV	BWYV	Luteovirus (phloem limited)	Barker (1989)
		PVA	PLRV	Luteovirus	Barker (1989)
II	Cucumovirus	CMV	RCMV	Comovirus (I)	Malyshenko et al. (1990)
	Bromovirus	BMV[b]	TMV	Tobamovirus (I)	Hamilton and Nichols (1977)
		BMV[b]	PVX	Potexvirus (III)	Malyshenko et al. (1990)
	Ilarvirus	ALMV	BMV	Bromovirus (II)	Malyshenko et al. (1990)
III	Potexvirus	PVX[b]	TMV	Tobamovirus (I)	Taliansky et al. (1982d)
		PVX[b]	RCMV	Comovirus (I)	Malyshenko et al. (1990)
		PVX[b]	PLRV	Luteovirus (phloem limited)	Atabekov et al. (1984)
		NMV	PLRV	Luteovirus (phloem limited)	Barker (1989)
	Hordeivirus	BSMV	TMV	Tobamovirus (I)	Dodds and Hamilton (1972), Taliansky et al. (1982d)
		BSMV[b]	PVX	Potexvirus (III)	Malyshenko et al. (1990)
Unclassified	Carlavirus	PVM[b]	PVX	Potexvirus (III)	Unpublished observations

(*continued*)

TABLE I

(Continued)

Tentative transport group	Helper virus		Nonspreading (helper-dependent virus)		Reference
	Taxonomic group	Name	Name	Taxonomic group[c]	
	Unclassifed virus	CMotV	PLRV	Luteovirus (phloem limited)	Barker (1989)
		CMotV	BWYV	Luteovirus (phloem limited)	Barker (1989)
	Unclassified virus	CLSV	CGMMV	Tobamovirus (I)	Weber et al. (1985)

[a] ALMV, Alfalfa mosaic virus; ArMV, arabis mosaic virus; BGMV, bean golden mosaic virus; BMV, brome mosaic virus; BSMV, barley stripe mosaic virus; BWYV, beet western yellow virus; CGMMV, cucumber green mottle mosaic virus; CLSV, cucumber leaf spot virus; CMotV, carrot mottle virus; CMV, cucumber mosaic virus; NMV, narcissus mosaic virus; PEBV, pea early-browning virus; PLRV, potato leaf roll virus; PVA, potato virus A; PVM, potato virus M; PVV, potato virus V; PVX, potato virus X; PVY, potato virus Y; RCMV, red clover mosaic virus; SBMV, southern bean mosaic virus; SHMV, sunn-hemp mosaic virus; TMV, tobacco mosaic virus; TRV, tobacco rattle virus.

[b] Virus combinations for which the complementation in TF is proved directly.

[c] The tentative transport group of the dependent virus is shown in parentheses.

affected by coinfection with either of two nepoviruses, cucumber mosaic cucumovirus, pea enation mosaic virus, or parsnip yellow fleck virus (Barker, 1989). In our experiments such combinations of viruses also occurred when no complementation was observed. In particular, we failed to determine complementation between TRV (tobravirus) and RCMV (comovirus) in tobacco plants, BSMV (hordeivirus) and TRV in wheat, TRV and BSMV in tobacco, and BSMV and AIMV (ilarvirus) in wheat (Malyshenko et al., 1990). However, these negative results are difficult to interpret; therefore, only positive results are discussed below.

On the whole, the TF complementation seems to be phenomenologically rather nonspecific, although certain exceptions are, nevertheless, possible. One can suggest at least two explanations for this phenomenon.

1. It is possible that nonhomologous TPs coded by unrelated viruses are functionally similar. An example of such structurally unrelated,

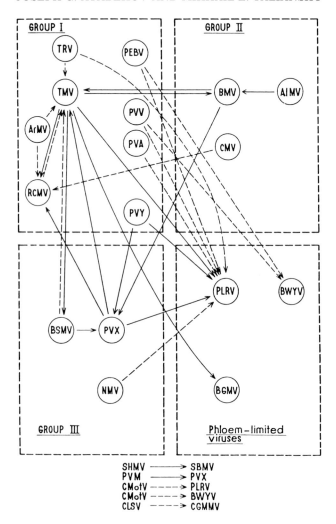

SHMV ⟶ SBMV
PVM ⟶ PVX
CMotV ----→ PLRV
CMotV ----→ BWYV
CLSV ----→ CGMMV

Fig. 4.   Summarized data on complementation between unrelated viruses belonging to different transport groups. For the abbreviations of viruses see footnote a of Table I. The arrows indicate the helper-dependent virus. The solid lines correspond to the virus combinations when the complementation of TF was supported by direct evidence. The dashed lines were used when the accumulation of the dependent viruses was increased, but no direct evidence of TF complementation was provided. Virus combinations that have at least one unclassified virus are presented at the bottom.

but functionally similar, proteins is provided by ilarviruses, the coat protein of which has, in addition to its structural role, an early function in the replication cycle (Bol et al., 1971; Alblas and Bol, 1978). Members of the ilarvirus group (including AIMV) can activate one another's genome with their respective coat proteins (Van Vloten-Doting, 1975; Gonsalves and Fulton, 1977), although these proteins might have no homology at the nucleotide or amino acid level (Brederode et al., 1980; Cornelissen et al., 1984).

It is obvious from Fig. 4 and Table I that taxonomically and structurally unrelated viruses can complement one another's TF in the same plant species. For example, N. tabacum can serve as a host for the complementation of (1) TMV (Ls1 at 33°C) TF by potexvirus (PVX) and by tobravirus (TRV); (2) RCMV comovirus TF by TMV, cucumovirus (cucumber mosaic virus), and potexvirus (PVX); (3) BSMV hordeivirus TF by TMV; and (4) TRV TF by TMV. Analogously, Cucumis sativus can serve as a host for the complementation of RCMV comovirus and TMV tobamovirus by arabis mosaic nepovirus (Malyshenko et al., 1990). On the other hand, complementation of TF of a given virus can occur in quite different plant species (even in monocotyledons and dicotyledons). Thus, the transport of TMV can be complemented in tobacco, vigna, cucumber, tomato (Tm-2 cell line), and barley plants (Malyshenko et al., 1990; Taliansky et al., 1982d; Hamilton and Dodds, 1970).

Apparently, the viral TPs should find in the infected plant a certain host-encoded factor(s) to express the TF. However, it does not seem likely that different TPs of diverse viruses belonging to different transport groups need different host factors in each of numerous plant species. For the sake of simplicity we assume that the TPs of different viruses use for TF expression a rather limited number of host factors which are similar and available, being widely distributed among various plant species.

2. On the other hand, it cannot be ruled out that the TPs belonging to different transport groups (or even the TPs of different viruses within the same groups) use different molecular mechanisms for TF expression. If this is true, the TP of a foreign helper virus should promote transport over its systemic host not only of its own genome, but also that of an unrelated (helper-dependent) virus. The presence of functionally active TP(s), coded by a helper virus, in plant cells seems to be a prerequisite for the complementation of helper-dependent virus TF. However, is this a sufficient condition?

This question can probably be answered by experiments with transgenic plants expressing transport genes of plant viruses. Beachy and co-workers showed that the tobacco plants transgenic in the TMV (common strain) 30-kDa protein gene could be systemically infected with TMV ts mutant Ls1 at the nonpermissive temperature (Deom *et al.*, 1987) and with unrelated phloem-limited luteovirus (PLRV), but inoculation of these plants with BMV, maize dwarf mosaic potyvirus, or CaMV did not result in any infection (R. N. Beachy, personal communication, quoted by Hull, 1989).

At first, these results seem to contradict those mentioned above on the phenomenological nonspecifity of TF complementation. However, to solve this dilemma, one might assume that, at least in certain virus–host combinations, the presence of an active TP per se is a necessary but insufficient condition to ensure the systemic spread of a dependent virus in nonhost plants. One can speculate that other virus-specific products should be adapted to the plant (host factor) and/or to the TP(s). We can imagine, for example, that after modification of the plasmodesmata by the TP the viral genomic RNA(s) of only this virus should be present in the cell (possibly, in the form of vRNP, as described by Dorokhov *et al.*, 1983, 1984) to act in concert with the TP. The possible role of vRNP might be to trigger the opening of the TP-modified plasmodesmata, after which these "gates" could be used nonspecifically by foreign viruses.

2. Lack of Complementation between the Viroid and Viruses

Viroids are small ss circular and linear RNAs that do not code for proteins and replicate using a host-encoded enzyme system (for a review see Riesner and Gross, 1985). Viroids are effectively transported in infected plant cells (Palukaitis, 1987), perhaps using entirely host-controlled or viroid-induced mechanisms.

It has been shown that under conditions of mixed infection the potato spindle tuber viroid does not complement transport of the TMV Ls1 mutant (ts in TF) at nonpermissive temperatures. The latter can be due principally to different mechanisms of virus and viroid transport in plants (Kondakova *et al.*, 1989).

IV. CONCLUDING REMARKS AND SUGGESTIONS

A. General Conclusions

This chapter summarizes the current status of our knowledge of plant virus TF by viruses belonging to different taxonomic groups and of the role of a host in TF expression.

1. Comparison of nucleotide sequences of transport genes and de-

duced amino acid sequences of putative transport proteins of different plant viruses allow division of the viruses into several transport groups.

2. Current experiments on transport function complementation suggest that in many cases the TPs coded by unrelated plant viruses are functionally interchangeable: The viruses belonging to different transport groups can complement one another.

3. Virus transport requires expression of two genomes: that of a virus and that of a host. Apparently, the viral TPs should find in the infected plant a certain host-encoded factor(s) in order to express the TF.

B. Hypothetical Elements of Transport Function

Although the molecular mechanisms of TF remain obscure, it seems advisable to postulate several more or less well-founded assumptions based on current experimental data and some analogies. Although we are aware that future experimentation will result in changes in these assumptions, at least in part they can be experimentally proved at this time.

1. Threshold Concentration of TP

In order for the viral genome to leave the infected cell, the TP (and the RNA coding for it) should be accumulated in this cell in concentrations exceeding a certain threshold. This barrier level might vary in different tissues and in different plants; for example, it could differ for short-distance and long-distance transport. Therefore, TF obviously depends on the effectiveness of viral RNA synthesis (see Section III,A,6). This dependence is especially pronounced for the multipartite viruses.

2. Compartmentalization of TF

All of the elements involved in TF expression probably should be localized at or close to the plasmodesmata. It is not quite clear what the fate of the TPs is after they perform TF. Probably, they are eliminated from the plasmodesmata and eventually degraded (Godefroy-Colburn et al., 1989).

3. Tropism of TP for Plasmodesmata

The TP should be transported within the cytoplasm of the infected cell and should reach the plasmodesmata. In other words, the virally encoded TPs should have an affinity for certain structural components of the plasmodesmata of the susceptible plant (or acquire such an affinity by posttranslational modification). The efficiency of tropism of the TPs to the plasmodesmata can vary in different virus–host combinations, influencing the level of the threshold TP concentration mentioned above.

4. Compatibility between the TP and a Hypothetical Host Factor(s)

It has been assumed (see Sections III,A,2 and 3 and III,B,1) that the TPs bind to a putative host factor(s), probably the protein(s). At least some of the host factors should be localized in the plasmodesmata. Consequently, the TPs should have structural features which enable them to recognize host factors of different host plants. Therefore, the tropism of the TP for the plasmodesmata depends on the affinity of the TP to host factors, the structural components of the plasmodesmata. The level of affinity can vary for different TP–host plant combinations, controlling thereafter the host range of a virus. Apparently, the compatibility between the viral TP and host factors exists in a systemically infectible plant. On the other hand, a TP produced by a heterologous virus in the subliminally infected plant is thought to be ineffective due to its inability to bind host factors and/or to modify the plasmodesmata of the nonhost plant.

The following well-known facts seem important in answering this question. First, a given plant species (e.g., *C. amaranticolor*) can be infected by dozens of different viruses conceivably coding for structurally unrelated TPs. Does this mean that each of these viruses finds in *C. amaranticolor* its own particular host factor? Second, a given plant virus can infect numerous plant species (e.g., cucumber mosaic virus infects hundreds of species from dozens of different families, including monocotyledons and dicotyledons). Does this mean that in each of these plant species a given viral TP finds structurally identical host factors?

For the sake of simplicity, we assumed (see Section III,B,1) that the number of different host factors is rather limited and that the putative host factor proteins would be widely distributed and rather conserved among the plants. This assumption seems to contradict the results by Meshi *et al.* (1989), who showed that two (or even one) amino acid substitutions in the 30-kDa TP (Glu-133 to Lys and Glu-52 to Lys) overcome the resistance of Tm-2 tomatoes to TMV-L. This obviously supports the theory of a high specificity for the TP–host factor interaction, although these substitutions should be regarded as fundamental and might strongly influence the conformation of the 30-kDa TP, making it nonfunctional.

Therefore, accepting the assumption mentioned above, it seems desirable to discuss an additional speculation which would account for the differences in the activity of a given TP in different (sometimes closely related) plant species. This speculation proposes that one of the host factors is involved in the posttranslational modification of the TP, thus modulating its activity. If the 30-kDa TMV TP is phosphorylated (D. Zimmern, quoted by Hull, 1989), the protein kinase (host encoded and possibly virus induced) might variously phosphorylate the 30-kDa proteins coded for by different TMV mutants. Analogously, the differences in phosphorylation of a given TP in different plant species might control the efficiency of TF expression and movement of this virus in these plants. The hypothetical posttranslational modification of TPs might serve as an additional mechanism which allows a limited number of host factors to regulate the function of the TPs in numerous plant species.

5. vRNP as a Possible Additional Element of TF

The modification of plasmodesmata by TPs is possibly a necessary but insufficient event in TF expression. We assume that additional factors are required for the movement of infection. It can be imagined that the role of such an additional element involved in TF is played by the particles of vRNP which were suggested earlier to serve as a transport form of virus infection. The interaction of vRNPs with the TP-modified plasmodesmata might trigger some reactions, resulting in the "opening of the gates" (see Section III,B,1). If this is the case, at least two virus-specific factors participate in TF: namely, the TP and vRNP.

6. Is the TF Energy Dependent?

After or even before binding to a host factor the TP might induce some enzymatic reaction(s). It is important to recognize whether TF is an energy-dependent process which requires the hydrolysis of NTPs and NTPase activity. In considering analogies, it is worth discussing the phenomenon of messenger ribonucleoprotein (mRNP) transport from the nucleous to the cytoplasm through the nuclear pores. mRNPs are too large to leave the nucleus by passive diffusion and viruses cannot move passively through the normal (i.e., unmodified) plasmodesmata. The transport of mRNP through the nuclear pore is an energy-dependent process which is mediated by an NTPase within the nuclear envelope. Nuclear envelopes also contain protein kinase and the transport stimulatory protein, the effect of which is enhanced by cAMP-dependent phosphorylation. Thus, cAMP-dependent protein kinase is involved in the control of the NTPase activity required for nucleocytoplasmic transport of mRNPs (for a review see Schroder et al., 1987).

Data discussed in Section III,A,4 imply that cAMP is involved somehow and that the protein phosphorylation event might contribute to the process of TF expression: (1) Exogenous cAMP rescued the TMV mutant ts in TF; (2) the 30-kDa TMV TP is phosphorylated; and (3) sequence homology exists between the TPs of some plant viruses and protein kinases. Together, these facts make rather attractive the theory of cAMP-dependent protein kinase contribution to TF. The protein kinase might be either virus encoded or (more probably) host encoded, but virus induced. It has not been ruled out that cAMP or/and TP might function as activators of protein kinase (or that TP itself acts as a protein kinase). By analogy with the nucleocytoplasmic transport of mRNPs, it seems conceivable that the phosphorylation (or autophosphorylation) of certain proteins in the virus-infected cell might be necessary for the subsequent activation of NTPase activity (if TF is energy dependent). Alternatively, the possibility has not been excluded that TP activates NTPase by direct binding. Finally, cAMP might operate as a transcriptional regulator, increasing in the plant the amount of the host factor(s) necessary for TF expression.

ACKNOWLEDGMENTS

We are most grateful to H. Barker, D. M. Bisaro, T. Godefroy-Colburn, R. Goldbach, T. C. Hall, R. I. Hamilton, R. Hull, G. Lebeurier, S. A. Lommel, G. P. Lomonossoff, M. A. Mayo, T. Meshi, A. van Kammen, J. Stanley, and J. Wellink for providing us with reprints, preprints, and other materials containing unpublished data. We thank S. Y. Morozov and A. R. Mushegyan for helpful discussions. Thanks are due to M. I. Goldstein for help in translating this text.

REFERENCES

Alblas, F., and Bol, J. F. (1978). *J. Gen. Virol.* **46**, 653–656.
Albrecht, H., Geldreich, A., Menissier de Murcia, J., Kirchherr, D., Mesnard, J.-M., and Lebeurier, G. (1988). *Virology* **163**, 503–508.
Allison, R. F., Johnston, R. E., and Dougherty, W. G. (1986). *Virology* **154**, 9–20.
Allison, R. F., Janda, M., and Ahlquist, P. (1988). *J. Virol.* **62**, 3581–3588.
Andrews, J. H., and Shalla, T. A. (1974). *Phytopathology* **64**, 1234–1243.
Angenent, G. C., Linthorst, H. J. M., van Belkum, A. F., Cornelissen, B. J. C., and Bol, J. F. (1986). *Nucleic Acids Res.* **14**, 4673–4682.
Atabekov, J. G., and Dorokhov, Y. L. (1984). *Adv. Virus Res.* **29**, 313–364.
Atabekov, J. G., and Morozov, S. Y. (1979). *Adv. Virus Res.* **25**, 1–91.
Atabekov, J. G., and Novikov, V. K. (1989). *CMI/AAB Descriptions of Plant Viruses*, in press.
Atabekov, J. G., Taliansky, M. E., Drampyan, A. H., Kaplan, I. B., and Turka, I. E. (1984). *Biol. Nauki (Moscow)* **10**, 28–31 (in Russian).
Barker, H. (1987). *J. Gen. Virol.* **68**, 1223–1227.
Barker, H. (1989). *Ann. Appl. Biol.*, in press.
Barker, H., and Harrison, B. D. (1982). *Plant Cell Rep.* **1**, 247–249.
Barker, H., and Harrison, B. D. (1986). *Ann. Appl. Biol.* **109**, 595–604.

Bassi, M., Favali, M. A., and Conti, G. G. (1974). *Virology* **47**, 694–700.
Beachy, R. N., Zaitlin, M., Bruening, G., and Israel, H. W. (1976). *Virology* **73**, 498–507.
Bergh, S. T., Koziel, M. G., Huang, S. C., Thomas, R. A., Guilley, D. P., and Siegel, A. (1985). *Nucleic Acids Res.* **13**, 8507–8518.
Berna, A., Briand, J. P., Stussi-Garaud, C., and Godefroy-Colburn, T. (1986). *J. Gen. Virol.* **67**, 1135–1147.
Boccara, M., Hamilton, W. D. O., and Baulcombe, D. C. (1986). *EMBO J.* **5**, 223–229.
Bol, J. F., Van Vloten-Doting, L., and Jaspars, E. M. J. (1971). *Virology* **46**, 73–85.
Bouzoubaa, S., Zigler, V., Beck, D., Guilley, H., Richards, K., and Jonard, G. (1986). *J. Gen. Virol.* **67**, 1689–1700.
Bouzoubaa, S., Quillet, L., Guilley, H., Jonard, G., and Richards, K. (1987). *J. Gen. Virol.* **68**, 615–626.
Brederode, F. T., Koper-Zwarthoff, E. C., and Bol, J. F. (1980). *Nucleic Acids Res.* **8**, 2213–2223.
Brough, C. L., Hayes, R. J., Morgan, A. J., Coutts, R. H. A., and Buck, K. W. (1988). *J. Gen. Virol.* **69**, 503–514.
Bruening, G. (1977). *Compr. Virol.* **11**, 55–141.
Bruening, G., Beachy, R. N., Scalla, R., and Zaitlin, M. (1976). *Virol.* **71**, 498–517.
Bundin, V. S., Vishnyakova, O. A., Zakharyev, V. M., Morozov, S. Y., Atabekov, J. G., and Skryabin, K. G. (1986). *Dokl. Akad. Nauk SSSR* **290**, 728–733 (in Russian).
Burgemeister, W., Koenig, R., Weich, H., Sebald, W., and Leseman, D. E. (1986). *J. Gen. Virol.* **61**, 1–14.
Carr, R. J., and Kim, K. S. (1983). *J. Gen. Virol.* **64**, 2489–2492.
Cheo, P. C. (1970). *Phytopathology* **60**, 41–46.
Cheo, P. C., and Gerard, J. S. (1971). *Phytopathology* **61**, 1010–1012.
Chopra, R. N., and Sharma, P. (1985). *J. Plant Physiol.* **117**, 293–296.
Conti, G. G., Vegetti, G., Bassi, M., and Favali, M. (1972). *Virology* **47**, 694–700.
Cornelissen, B. J. C., Jansen, H., Zuidema, D., and Bol, J. F. (1984). *Nucleic Acids Res.* **12**, 2427–2437.
Cornelissen, B. J. C., Linthorst, H. J. V., Brederode, F. T., and Bol, J. F. (1986). *Nucleic Acids Res.* **14**, 2157–2169.
Covey, S. N., and Hull, R. (1981). *Virology* **111**, 463–474.
Daubert, S., Richins, R., Shepherd, R. J., and Gardner, R. C. (1982). *Virology* **122**, 444–449.
Davies, J. W. (1987). *Microbiol. Sci.* **4**, 18–23.
Davies, C., and Symons, R. H. (1988). *Virology* **165**, 216–224.
Dawson, J. R. O., Motoyoshi, F., Watts, J. W., and Bancroft, J. B. (1975). *J. Gen. Virol.* **29**, 99–107.
Deom, K. M., Oliver, M. J., and Beachy, R. N. (1987). *Science* **237**, 389–394.
Dodds, J. A., and Hamilton, R. I. (1972). *Virology* **50**, 404–411.
Domier, L. L., Franklin, K. M., Shahabuddin, M., Hellmann, G. M., Overmeyer, J. H., Hiremath, S. T., Siaw, M. F. E., Lomonossoff, G. P., Shaw, J. G., and Rhoads, R. E. (1986). *Nucleic Acids Res.* **14**, 5417–5430.
Dorokhov, Y. L., Alexandrova, N. M., Miroshnichenko, N. A., and Atabekov, J. G. (1983). *Virology* **127**, 237–252.
Dorokhov, Y. L., Alexandrova, N. M., Miroshnichenko, N. A., and Atabekov, J. G. (1984). *Virology* **135**, 395–405.
Dreher, T. W., Rao, A. L. N., and Hall, T. C. (1989). *J. Mol. Biol.*, in press.
Elmer, J. S., Brand, L., Sunter, G., Gardiner, W. E., Bisaro, D. M., and Rogers, S. G. (1988). *Nucleic Acids Res.* **16**, 7043–7060.
Etessami, P., Callis, R., Ellowood, S., and Stanley, J. (1988). *Nucleic Acids Res.* **16**, 4811–4829.

Etessami, P., Watts, J., and Stanley, J. (1989). *J. Gen. Virol.* **70**, 277–289.

Flockart, D. A., and Corbin, J. D. (1982). *CRC Crit. Rev. Biochem.* **12**, 133–186.

Forster, R. L. S., Bevan, M. W., Harbison, S. A., and Gardner, R. C. (1988). *Nucleic Acids Res.* **16**, 291–303.

Francki, R. I. B., Hatta, T., Grylls, N. E., and Grivell, C. J. (1979). *Ann. Appl. Biol.* **91**, 51.

French, R., and Ahlquist, P. (1987). *J. Virol.* **61**, 1457–1465.

Fritsch, C., Mayo, M., and Murant, A. F. (1980). *J. Gen. Virol.* **46**, 381–389.

Fuentes, L., and Hamilton, R. I. (1988). *Abstr. Int. Congr. Plant Pathol. 5th.*

Gardiner, W. E., Sunter, G., Brand, L., Elmer, J. S., Rogers, S. G., and Bisaro, D. M. (1988). *EMBO J.* **7**, 899–904.

Godefroy-Colburn, T., Gadey, M. J., Berna, A., and Stussi-Garaud, C. (1986). *J. Gen. Virol.* **67**, 2233–2239.

Godefroy-Colburn, T., Schoumacher, F., Erny, C., Berna, O., Moser, M.-J., Gagey, C., and Stussi-Garaud, C. (1989). Submitted for publication.

Goelet, P., and Karn, J. (1982). *J. Mol. Biol.* **154**, 541–550.

Goelet, P., Lomonossoff, G. P., Butler, P. J. G., Akam, M. E., Gait, M. J., and Karn, J. (1982). *Proc. Natl. Acad. Sci. U.S.A.* **79**, 5818–5822.

Goldbach, R. (1986). *Annu. Rev. Phytopathol.* **24**, 289–310.

Goldbach, R. (1987). *Microbiol. Sci.* **4**, 197–202.

Goldbach, R., and van Kammen, A. (1985). In "Molecular Plant Virology" (J. W. Davies, ed.), Vol. 2, pp. 83–120. CRC Press, Boca Raton, Florida.

Goldbach, R., Rezelman, G., and van Kammen, A. (1980). *Nature (London)* **286**, 297–300.

Goldbach, R., Eggen, R., de Jager, C., van Kammen, A., van Lent, J., Rezelman, G., and Wellink, J. (1989). Submitted for publication.

Gonsalves, D., and Fulton, R. W. (1977). *Virology* **81**, 398–407.

Gorbalenya, A. E., Blinov, V. M., Donchenko, A. P., and Koonin, E. V. (1989). *J. Mol. Evol.* **28**, 256–268.

Guilley, H., Carrington, J. C., Balazc, E., Jonard, G., Richards, K., and Morris, T. J. (1985). *Nucleic Acids Res.* **13**, 6663–6677.

Gustafson, G. D., and Armour, S. L. (1986). *Nucleic Acids Res.* **14**, 3895–3909.

Gustafson, G. D., Milner, I. I., McFarland, J. E., Pedersen, K., Larkins, B. A., and Jackson, A. O. (1982). *Virology* **120**, 182–193.

Gustafson, G. D., Hunter, B., Hanau, R., Armour, S. L., Jackson, A. O. (1987). *Virology* **158**, 394–406.

Hamilton, R. I., and Dodds, J. A. (1970). *Virology* **42**, 266–268.

Hamilton, R. I., and Nichols, C. (1977). *Phytopathology* **67**, 484–489.

Hamilton, W. D. O., Stein, V. E., Coutts, R. H. A., and Buck, K. N. (1984). *EMBO J.* **3**, 2197–2205.

Hamilton, W. D. O., Boccara, M., and Balcombe, D. C. (1986). *Abstr. EMBO Workshop Mol. Plant Virol.* p. 16.

Hamilton, W. D. O., Boccara, M., Robinson, D. J., and Baulcombe, D. C. (1987). *J. Gen. Virol.* **68**, 2563–2575.

Harrison, B. D., and Robinson, D. J. (1981). In "Handbook of Plant Virus Infections and Comparative Diagnosis" (E. Kurstak, ed.), pp. 516–540. Elsevier/North-Holland, Amsterdam.

Harrison, B. D., and Robinson, D. J. (1986). *Plant Viruses* **2**, 339–369.

Herman, R. C. (1987). *Biochemistry* **26**, 8346–8350.

Hiebert, E., and Dougherty, N. G. (1988). *Plant Viruses* **4**, 159–178.

Hirth, L., and Richards, K. (1981). *Adv. Virus Res.* **26**, 145–199.

Hohn, T., Hohn, B., and Pfeiffer, P. (1985). *Trends Biochem. Sci. (Pers. Ed.)* **10**, 205–209.

Howarth, A. J., Caton, J., Bossert, M., and Goodman, R. M. (1985). *Proc. Natl. Acad. Sci. U.S.A.* **82**, 3572–3576.

Hughes, G., Davies, J. W., and Wood, K. K. (1986). *J. Gen. Virol.* **67**, 2125–2133.

Huisman, M. J., Sarachu, A. N., Ablas, F., Broxterman, H. J. G., Van Vloten-Doting, L., and Bol, J. F. (1986). *Virology* **154**, 401–404.

Huisman, M. J., Linthorst, H. J. M., Bol, J. F., and Cornelissen, B. J. C. (1988). *J. Gen. Virol.* **69**, 1789–1798.

Hull, R. (1989). *Annu. Rev. Phytopathol.*, in press.

Hull, R., and Covey, S. N. (1985). *BioEssays* **3**, 160–163.

Hull, R., and Maule, A. J. (1985). *Plant Viruses* **1**, 83–116.

Hull, R., Sadler, J., and Longstaff, M. (1986). *EMBO J.* **5**, 3083–3090.

Jang, S. K., Krausslich, H. G., Nicklin, M. G. H., Duke, G. M., Palmenberg, A. G., and Wimmer, E. (1988). *J. Virol.* **62**, 2636–2643.

Jockusch, H. (1968). *Virology* **35**, 94–101.

Joshi, S., Pleij, C. W., Haenni, A. L., Chapeville, F., and Bosch, L. (1983). *Virology* **127**, 100–111.

Kaplan, I. B., Kozlov, Y. V., Pshennikova, E. S., Taliansky, M. E., and Atabekov, J. G. (1982). *Virology* **118**, 317–323.

Kiberstis, P. A., Loesch-Fries, L. S., and Hall, T. C. (1981). *Virology* **112**, 804–808.

Kitajima, E. W., and Lauritis, J. A. (1969). *Virology* **37**, 681–685.

Kondakova, O. A., Malyshenko, S. I., Mozhaeva, K. A., Vasilieva, T. Y., Taliansky, M. E., and Atabekov, J. G. (1989). *J. Gen. Virol.*, in press.

Kozak, M. (1986a). *Cell (Cambridge, Mass.)* **47**, 481–483.

Kozak, M. (1986b). *Adv. Virus Res.* **31**, 229–292.

Kraev, A. S., Morozov, S. Y., Lukasheva, L. I., Rozanov, M. N., Chernov, B. K., Simonova, M. L., Golova, Y. B., Belzhelarskaya, S. N., Pozmogova, G. E., Skryabin, K. G., and Atabekov, J. G. (1988). *Dokl. Akad. Nauk SSSR* **300**, 711–716 (in Russian).

Kubo, S., and Takanami, Y. (1979). *J. Gen. Virol.* **42**, 387–398.

Kuszala, M., Ziegler, V., Bouzoubaa, S., Richards, K., Putz, C., Guilley, H., and Jonard, G. (1986). *Ann. Appl. Biol.* **109**, 155.

Lawson, R. H., and Hearon, S. S. (1971). *Virology* **44**, 454–456.

Lawson, R. H., and Hearon, S. S. (1974). *J. Ultrastruct. Res.* **48**, 201–215.

Lawson, R. H., Hearon, S. S., and Smith, F. F. (1979). *Virology* **46**, 453–463.

Leonard, D. A., and Zaitlin, M. (1982). *Virology* **117**, 416–424.

Linstead, P. J., Hills, G. J., Plaskitt, K. A., Wilson, I. G., Hurker, C. L., and Maule, A. J. (1988). *J. Gen. Virol.* **69**, 1809–1818.

Lommel, S. A., Weston-Fina, M., Xiong, Z., and Lomonossoff, G. P. (1988). *Nucleic Acids Res.* **16**, 8587–8602.

Lomonossoff, G. P., and Shanks, M. (1983). *EMBO J.* **2**, 2253–2258.

Lot, H., Marchoux, G., Marrou, J., Kaper, J. M., West, C. K., Van Vloten-Doting, L., and Hull, R. (1974). *J. Gen. Virol.* **22**, 81–93.

MacKenzie, D. J., and Tremane, J. H. (1988). *J. Gen Virol.* **69**, 2387–2395.

Malyshenko, S. I., Taliansky, M. E., Kondakova, O. A., Ulanova, E. F., and Atabekov, J. G. (1987). *Izv. Akad. Nauk SSSR, Ser. Biol.* **5**, 680–685 (in Russian).

Malyshenko, S. I., Lapchic, L. G., Kondakova, O. A., Kuznetzova, L. L., Taliansky, M. E., and Atabekov, J. G. (1988). *J. Gen. Virol.* **69**, 407–412.

Malyshenko, S. I., Kondakova, O. A., Taliansky, M. E., and Atabekov, J. G. (1990). Submitted for publication.

Martinez-Izquierdo, J. A., Futterer, J., and Hohn, T. (1987). *Virology* **160**, 527–530.

Mason, W. S., Taylor, J. M., and Hull, R. (1987). *Adv. Virus. Res.* **32**, 35–96.
Matthews, R. E. F. (1981). "Plant Virology," 2nd ed. Academic Press, New York.
Mayo, M. A., Fritsch, C., and Hirth, L. (1976). *Virology* **69**, 408–415.
Mayo, M. A., Robinson, D. J., Jolly, C. A., and Hyman, L. (1989). *J. Gen. Virol.*, in press.
Meshi, T., and Okada, Y. (1987). *Plant–Microbiol. Interact., Mol. Gen. Perspect.* **2**, 285–304.
Meshi, T., Ohno, T., and Okada, Y. (1982a). *J. Biochem. (Tokyo)* **91**, 1441–1444.
Meshi, T., Ohno, T., and Okada, Y. (1982b). *Nucleic Acids Res.* **10**, 6111–6117.
Meshi, T., Kiyama, R., Ohno, T., and Okada, Y. (1983). *Virology* **127**, 54–64.
Meshi, T., Watanabe, J., Saito, T., Sugimoto, A., Maeda, T., and Okada, Y. (1987). *EMBO J.* **6**, 2557–2563.
Meshi, T., Motoyoshi, F., Maeda, T., Yoshiwoka, S., Watanabe, H., and Okada, Y. (1989). Submitted for publication.
Meyer, M., Hemmer, O., Mayo, M. A., and Fritch, C. (1986). *J. Gen. Virol.* **67**, 1257–1271.
Miller, W. A., Waterhouse, P. M., and Gerlach, W. L. (1988). *Nucleic Acids Res.* **6**, 6097–6111.
Morch, M.-D., Boyer, J.-C., and Haenni, A.-L. (1988). *Nucleic Acids Res.* **16**, 6157–6173.
Morozov, S. Y., Gorbulev, V. G., Novikov, V. K., Agranovsky, A. A., Kozlov, Y. V., Atabekov, J. G., and Baev, A. A. (1981). *Dokl. Akad. Nauk SSSR* **259**, 723–725 (in Russian).
Morris-Krsinich, B. A. M., Forster, L. S., and Mossop, D. W. (1983). *Virology* **124**, 349–356.
Moser, O., Gagey, M.-J., Godefroy-Colburn, T., Stussi-Garaud, G., Ellwart-Tschiirtz, H., Nitschko, H., and Mundry, K.-W. (1988). *J. Gen. Virol.* **69**, 1367–1373.
Motoyoshi, F., and Oshima, N. (1975). *J. Gen. Virol.* **29**, 81–91.
Motoyoshi, F., and Oshima, N. (1977). *J. Gen. Virol.* **34**, 499–506.
Murant, A. F. (1981). *In* "Handbook of Plant Virus Infections and Comparative Diagnosis" (E. Kurstak, ed.), pp. 198–238. Elsevier/North-Holland, Amsterdam.
Murant, A. F., and Roberts, I. M. (1971). *J. Gen. Virol.* **10**, 65–70.
Murthy, M. R. N. (1983). *J. Mol. Biol.* **168**, 469–475.
Mushegyan, A. R., Malyshenko, S. I., Taliansky, M. E., and Atabekov, J. G. (1986). *Mol. Biol. (Moscow)* **20**, 1371–1376 (in Russian).
Mushegyan, A. R., Malyshenko, S. I., Taliansky, M. E., and Atabekov, J. G. (1990). Submitted for publication.
Nassuth, A., and Bol, J. F. (1983). *Virology* **124**, 75–85.
Nassuth, A., Alblas, F., and Bol, J. F. (1981). *J. Gen. Virol.* **53**, 207–214.
Nishiguchi, M., Motoyoshi, F., and Oshima, N. (1978). *J. Gen. Virol.* **39**, 53–61.
Nishiguchi, M., Motoyoshi, F., and Oshima, N. (1980). *J. Gen. Virol.* **46**, 497–500.
Nutter, R. C., Scheets, K., Panganiban, L. C., and Lommel, S. A. (1989). *Nucleic Acids Res.*, submitted for publication.
Ohno, T., Takamatsu, N., Meshi, T., Okada, Y., Nishiguchi, M., and Kiho, Y. (1983). *Virology* **131**, 255–258.
Ohno, T., Aoyagi, M., Yamanashi, Y., Saito, H., Ikawa, S., Meshi, T., and Okada, Y. (1984). *J. Biochem. (Tokyo)* **96**, 1915–1923.
Oliver, M. J., Deom, K. M., De, B. K., and Beachy, R. N. (1986). *Virology* **155**, 277–283.
Ooshika, I., Watanabe, Y., Meshi, T., Okada, Y., Igano, K., Inouye, K., and Yoshima, N. (1984). *Virology* **132**, 71–78.
Osman, T. A. M., and Buck, K. W. (1987). *J. Gen. Virol.* **68**, 289–296.
Paje-Manalo, L. L., and Lommel, S. A. (1989). *Mol. Plant Microbe Interact.*, in press.
Palukaitis, P. (1987). *Virology* **158**, 239–241.

Palukaitis, P., and Zaitlin, M. (1986). *Plant Viruses* **2**, 105–131.

Pappu, H. R., and Hiruki, C. (1988). *Can. J. Plant Pathol.* **10**, 110–115.

Peden, K. W. C., and Symons, R. H. (1973). *Virology* **53**, 487–492.

Pfeiffer, P., and Hohn, T. (1983). *Cell (Cambridge, Mass.)* **33**, 781–789.

Purcifull, D. E., and Edwardson, J. R. (1981). *In* "Handbook of Plant Virus Infections and Comparative Diagnosis" (E. Kurstak, ed.), pp. 627–693. Elsevier/North Holland, Amsterdam.

Randles, J. W., Harrison, B. D., Murant, A. F., and Mayo, M. A. (1977). *J. Gen. Virol.* **36**, 187–193.

Rezaian, M. A., Williams, R. H. V., and Symons, R. H. (1985). *Eur. J. Biochem.* **150**, 331–339.

Rezelman, G., Franssen, H. J., Goldbach, R., Je, T. S., and van Kammen, A. (1982). *J. Gen. Virol.* **60**, 335–342.

Richards, K., Jonard, D., Gulley, H., Ziegler, V., and Putz, C. (1985). *J. Gen. Virol.* **35**, 397–401.

Richins, R. D., Schlothof, H. B., and Shepherd, R. J. (1987). *Nucleic Acids Res.* **15**, 8451–8466.

Riesner, D., and Gross, H. J. (1985). *Annu. Rev. Biochem.* **54**, 531–564.

Robinson, D. J., Barker, H., Harrison, B. D., and Mayo, M. A. (1980). *J. Gen. Virol.* **51**, 317–326.

Robinson, D. J., Mayo, M. A., Fritsch, C., Jones, A. T., and Raschke, J. H. (1983). *J. Gen. Virol.* **64**, 1591–1599.

Rogers, S. G., Bisaro, D. M., Horsch, R. B., Fraley, R. T., Hoffmann, N. L., Brand, L., Elmer, J., and Lloyd, A. M. (1986). *Cell (Cambridge, Mass.)* **45**, 593–600.

Rosenberg, N., Pines, M., and Sela, I. (1982). *FEBS Lett.* **137**, 105–107.

Saito, T., Imai, Y., Meshi, T., and Okada, Y. (1988). *Virology* **167**, 653–656.

Savithri, H. S., and Murthy, M. R. N. (1983). *J. Biosci.* **5**, 183–183.

Schoelz, J. E., and Shepherd, R. J. (1988). *Virology* **162**, 30–37.

Schroder, H. C., Bachman, M., Diehl-Seifert, B., and Muller, W. E. G. (1987). *Prog. Nucleic Acid Res. Mol. Biol.* **34**, 89–142.

Severin, E. S., and Kochetkova, M. N. (1985). *In* "Role of Phosphorylation in the Regulation of Cell Activity." Nauka, Moscow (in Russian).

Shin, D. S., Park, I. W., Evane, C. L., Jaynes, J. M., and Palmenberg, N. (1987). *J. Virol.* **61**, 2033–2037.

Skryabin, K. G., Morozov, S. Y., Kraev, A. S., Rozanov, M. N., Chernov, B. K., Lukasheva, L. I., and Atabekov, J. G. (1988). *FERS Lett.* **240**, 33–40.

Stanley, J., and Gay, M. R. (1983). *Nature (London)* **301**, 260–262.

Stanley, J., and Townsend, R. (1986). *Nucleic Acids Res.* **14**, 5981–5998.

Stanley, J., Markham, P. G., Cails, R. J., and Pinner, M. S. (1986). *EMBO J.* **5**, 1761–1767.

Steven, A. C., Trus, B. L., Putz, C., and Wurtz, M. (1981). *Virology* **113**, 428–438.

Stobbs, L. W., and MacNeill, B. H. (1980). *Can. J. Plant Pathol.* **2**, 5–11.

Stussi-Garaud, C., Garaud, J.-C., Berna, A., and Godefroy-Colburn, T. (1987). *J. Gen. Virol.* **68**, 1779–1784.

Sulzinski, M. A., and Zaitlin, M. (1982). *Virology* **121**, 12–19.

Sunter, G., Gardiner, W. E., Rushing, A. E., Rogers, S. G., and Bisaro, D. M. (1987). *Plant Mol. Biol.* **8**, 477–484.

Symons, R. H. (1985). *Plant Viruses* **1**, 57–81.

Takamatsu, N., Ohno, T., Meshi, T., and Okada, Y. (1983). *Nucleic Acids Res.* **11**, 3767–3778.

Takamatsu, N., Ishikawa, M., Meshi, T., and Okada, Y. (1987). *EMBO J.* **6**, 307–311.

Takatsuji, H., Hirochika, H., Fukushi, T., and Ikeda, J. E. (1986). *Nature (London)* **319**, 240–243.

Taliansky, M. E., Atabekova, T. I., Kaplan, I. B., Morozov, S. Y., Malyshenko, S. I., and Atabekov, J. G. (1982a). *Virology* **118**, 301–308.

Taliansky, M. E., Kaplan, I. B., Jarvekulg, L. V., Atabekova, T. I., Agranovsky, A. A., and Atabekov, J. G. (1982b). *Virology* **118**, 309–316.

Taliansky, M. E., Malyshenko, S. I., Pshennikova, E. S., Kaplan, I. B., Ulanova, E. F., and Atabekov, J. G. (1982c). *Virology* **122**, 318–326.

Taliansky, M. E., Malyshenko, S. I., Pshennikova, E. S., and Atabekov, J. G. (1982d). *Virology* **122**, 327–332.

Thomas, P. E., and Mink, G. I. (1979). *CMI/AAB Descriptions of Plant Viruses* **210**.

Thornbury, D. W., Hellman, G. M., Rhoads, R. E., and Pirone, T. P. (1985). *Virology* **144**, 260–267.

Toh, H., Hayashida, H., and Miyata, T. (1983). *Nature (London)* **305**, 827–829.

Tomenius, K., Claphat, D., and Meshi, T. (1987). *Virology* **160**, 363–371.

Townsend, R., Stanley, J., Ourson, S. J., and Short, M. N. (1985). *EMBO J.* **4**, 33–37.

Tu, J. C. (1977). *Physiol. Plant Pathol.* **10**, 117–123.

van der Wilk, F., Huisman, M. J., Cornelissen, B. J. C., Huttinga, H., and Goldbach, R. (1989). *FEBS Lett.,* in press.

Van Telgen, H. J., Van der Zaal, E. J., and Van Loon, L. C. (1985a). *Physiol. Plant Pathol.* **26**, 83–98.

Van Telgen, H. J., Van der Zaal, E. J., and Van Loon, L. C. (1985b). *Physiol. Plant Pathol.* **26**, 99–109.

Van Telgen, H. J., Goldbach, R. W., and Van Loon, L. C. (1985c). *Virology* **143**, 612–616.

Van Vloten-Doting, L. (1975). *Virology* **65**, 212–225.

Van Vloten-Doting, L. (1985). *Plant Viruses* **1**, 117–161.

Van Vloten-Doting, L. (1987). *In* "Molecular Basis of Virus Disease," pp. 303–318. Cambridge Univ. Press, London.

Van Vloten-Doting, L., and Jaspars, E. M. J. (1977). *Compr. Virol.* **11**, 1.

Van Vloten-Doting, L., Hasrat, J. A., Oosterwijk, E., Van't Sant, P., Schoen, M. A., and Roosien, J. (1980). *J. Gen. Virol.* **46**, 415–426.

Van Wezenbeek, P., Verver, J., Harmsen, J., Vos, P., and van Kammen, A. (1983). *EMBO J.* **2**, 941–946.

Veidt, I., Lot, H., Leiser, M., Scheidecker, D., Guilley, H., Richards, K., and Jonard, G. (1988). *Nucleic Acids Res.* **16**, 9917–9932.

Watanabe, Y., Emori, Y., Ooshika, I., Meshi, T., Ohno, T., and Okada, Y. (1984). *Virology* **133**, 18–24.

Watanabe, Y., Ooshika, I., Meshi, T., and Okada, Y. (1986). *Virology* **152**, 414–420.

Weber, I., Stanarius, A., Kegler, H., and Kleinhempel, H. (1985). *Arch. Phytopathol. Pflanzenschutz* **21**, 347.

Wellink, J., and van Kammen, A. (1988). *Arch. Virol.* **98**, 1–26.

Wellink, J., and van Kammen, A. (1989). Submitted for publication.

Wellink, J., Rezelman, G., Goldbach, R., and Beyrenther, K. (1986). *J. Virol.* **59**, 50–58.

Wellink, J., Jaegle, M., Prinz, H., van Kammen, A., and Goldbach, R. (1987). *J. Gen. Virol.* **68**, 2577–2585.

Woolston, C. J., Covey, S. N., Penswick, J. R., and Davies, J. W. (1983). *Gene* **23**, 15–23.

Wu, S., Rinchart, C. A., and Kaesberg, P. (1987). *Virology* **161**, 73–80.

Zimmern, D. (1983). *J. Mol. Biol.* **171**, 345–352.

Zimmern, D., and Hunter, T. (1983). *EMBO J.* **2**, 1893–1900.

Zuidema, D., Linthorst, H. J. M., Huisman, M. J., Asjer, C. J., and Bol, J. F. (1989). *J. Gen. Virol.* **70**, 267–276.

ADVANCES IN VIRUS RESEARCH, VOL. 38

STRUCTURAL AND FUNCTIONAL PROPERTIES OF PLANT REOVIRUS GENOMES

Donald L. Nuss and David J. Dall

Department of Molecular Oncology and Virology
Roche Institute of Molecular Biology
Roche Research Center
Nutley, New Jersey 07110

I. Introduction

Plant-infecting members of the family Reoviridae provide appealing experimental systems with which to study a range of biological and molecular interactions and processes. The fact that they replicate both in plant hosts and in their insect vectors, coupled with the availability of cultured cell lines derived from the insect vector, makes these viruses particularly well suited for examining the intimate nature of virus–insect vector–plant host interactions. Infection of the insect vector is noncytopathic and persistent, while infection of the plant host is tissue specific and results in numerous symptoms, including neoplasia. Consequently, these viral systems also provide opportunities for examining the molecular basis of viral persistence, cytopathology, and disease symptom expression in both the plant and animal kingdoms. Interkingdom differences in the expression, post-translational modification, transport, and secretion of specific viral

249

polypeptides are also now readily approachable with the advent of molecular reagents in the form of functional cDNA clones of viral genomic RNAs and antisera to individual viral gene products. Recent analyses of plant reovirus genomes have exposed a number of structural similarities to the genomes of other segmented RNA viruses, including other members of the Reoviridae, influenza viruses, arenaviruses, and bunyaviruses. Furthermore, characterization of defective interfering RNAs associated with the genome of a plant reovirus has revealed a number of basic principles involved in the recognition, sorting, and packaging of segmented RNA genomes. This information, coupled with functional and conformational studies of synthetic viral transcripts, has led to the identification of sequence elements potentially involved in the sorting and packaging events. Recent advances in the characterization of plant reovirus genomes as they relate to these biological and molecular processes are the focus of this chapter.

II. Plant-Infecting Reoviridae: An Overview

The association of reoviruses with plant disease has been recognized for many years, and a considerable literature has been assembled. The studies forming the basis of these reports have, for the most part, directly implicated the viruses as causal agents of disease. On occasion, however, it has been shown that the presence of reoviruses in unhealthy plants is coincidental. Thus, for example, leafhopper A virus, long thought to be the cause of maize wallaby ear disease (see Grylls, 1979), has lately been recognized as an insect virus which incidentally enters plants during feeding of its cicadellid host (Ofori and Francki, 1985), the salivary toxins of which actually cause the observed syndrome (Ofori and Francki, 1983). This chapter does not include data from studies of such viruses and is restricted in scope to those known to multiply in plant hosts. Furthermore, it is not intended that the following section should provide a minutely detailed description of all plant reoviruses; such coverage is available in various recent reviews (e.g., Shikata, 1981; Francki and Boccardo, 1983; Francki et al., 1985). Rather, the general characteristics of the plant reoviruses are outlined here, with the aim of providing a backdrop against which to judge the interest and importance of studies of their genomes.

A. Virus Classification and Characters of Taxonomic Importance

Many physical and biological characteristics are shared by all plant-infecting reoviruses; these are described in Sections II,B and C. A number of other characteristics, described in this section, show patterns of variation which allow their use as taxonomic criteria for all known plant reoviruses. Specifically, consideration of (1) the number of genomic segments and their electrophoretic profile, (2) viral capsid morphology, (3) serological relationships, and (4) the identity of the insect vectors of the viruses allows their division into the two genera, *Phytoreovirus* and *Fijivirus,* and a third unclassified group (Table I).

1. Genus Phytoreovirus

The genus *Phytoreovirus* contains three members—wound tumor virus (WTV), rice dwarf virus (RDV), and rice gall dwarf virus (RGDV)—each of which has a genome consisting of 12 segments of double-stranded (ds) RNA with a total molecular weight of approximately 16×10^6 (Matthews, 1982). All three viruses are transmitted by leafhopper vectors (Francki and Boccardo, 1983). Antisera prepared against intact purified particles of each virus fail to recognize the other two members of the genus (Omura et al., 1982; Francki and Boccardo, 1983), but antiserum prepared against the proteins of sodium dodecyl sulfate-dissociated RDV particles has been found to react with all RDV proteins and with the 45-kDa protein of RGDV (Matsuoka et al., 1985). High-resolution microscopy of purified particles shows that capsids of

TABLE I

PLANT REOVIRUSES[a]

Genus	Virus
Phytoreovirus	Wound tumor virus
	Rice dwarf virus
	Rice gall dwarf virus
Fijivirus	
Subgroup I	Fiji disease virus
Subgroup II	Maize rough dwarf virus
	Cereal tillering disease virus[b]
	Rice black-streaked dwarf virus
	Pangola stunt virus
Subgroup III	Oat sterile dwarf virus
	Arrhenatherum blue dwarf virus[c]
	Lolium enation virus[c]
Unclassified	Echinochloa ragged stunt virus
	Rice ragged stunt virus

[a] Summarized from Francki et al. (1985), with additional information from Chen et al. (1986, 1989).

[b] An isolate of maize rough dwarf virus.

[c] Isolates of oat sterile dwarf virus.

all three viruses are morphologically similar and that each lacks the capsid spikes seen on particles of other plant reoviruses (Omura *et al.*, 1982; Francki *et al.*, 1985). As described in Section V,B, the validity of this generic grouping is strongly supported by recent molecular biology studies which suggest a common evolutionary origin for WTV and RDV.

2. Genus Fijivirus

Members of the genus *Fijivirus* have genomes consisting of 10, rather than 12, segments of dsRNA, with a total molecular weight of $18–20 \times 10^6$ (Matthews, 1982) and are transmitted by planthopper, rather than leafhopper, vectors (Francki and Boccardo, 1983). The genus contains a total of five viruses, which are further categorized into three subgroups. Members of each subgroup are serologically related to each other, but not to members of other subgroups. Furthermore, each subgroup has a characteristic electrophoretic genome profile (Francki *et al.*, 1985). Subgroups I and III are both monotypic, containing Fiji disease virus (FDV) and oat sterile dwarf virus (OSDV), respectively; Subgroup II contains maize rough dwarf virus (MRDV), rice black-streaked dwarf virus (RBSDV), and pangola stunt virus (PSV). Although as many as eight viruses have previously been assigned to this genus, it is now considered that several of these represent isolates of the same virus. Thus, cereal tillering disease virus is now regarded as a strain of MRDV (Francki *et al.*, 1985), and arrhenatherum blue dwarf virus and lolium enation virus are both considered isolates of OSDV (Boccardo and Milne, 1980). Capsids of *Fijivirus* members are unique among the nonoccluded reoviruses in having both A spikes on their outer shells and B spikes on the subviral particles which are frequently observed in purified preparations (Milne *et al.*, 1973; Hatta and Francki, 1977).

3. Unclassified Plant Reoviruses

Two plant reoviruses, echinochloa ragged stunt virus (ERSV) and rice ragged stunt virus (RRSV), remain unclassified. In many respects both viruses appear to be similar to members of the genus *Fijivirus*. Each has a genome of 10 segments of dsRNA and a similar electrophoretic profile; total genomic molecular weights of ERSV and RRSV have been estimated, respectively, as 17.9×10^6 and 16.5×10^6 or 18.2×10^6 (Kawano *et al.*, 1984; Chen *et al.*, 1989b). Each is transmitted by planthopper vectors to graminaceous plant hosts (Chen *et al.*, 1986; Boccardo and Milne, 1984). Furthermore, like members of other established subgroups within the genus, the two viruses are serologically related to each other, but not to other fijiviruses.

While ERSV has particles with typical *Fijivirus* morphology, including the presence of A and B capsid spikes (Chen *et al.*, 1989a), RRSV has previously been separated from the other plant reoviruses on the basis of its apparent morphological differences. Although B-spiked subviral RRSV particles have frequently been observed (Milne, 1980; Kawano *et al.*, 1984; Chen *et al.*, 1989a), these examinations of purified particle preparations have failed to demonstrate the presence of an outer capsid shell, as found in other plant reoviruses. However, other studies make negative evidence of this type difficult to interpret. Observations of crystalline arrays of ERSV and RRSV in insect vector cells suggest that particles of both viruses have similar size and morphology *in vivo* (Chen *et al.*, 1989a).

Furthermore, production of antiserum against purified preparations of RRSV suggests that the virus is considerably more labile than other plant reoviruses. While antiserum against the other viruses generally reacts only to a subset of viral proteins (Omura *et al.*, 1985) that raised against RRSV recognizes all component proteins of the particle (Hagiwara *et al.*, 1986), insinuating that RRSV particles break down rapidly following injection. Thus, it could be that previous failure to observe an outer viral shell *in vitro* reflects an extreme lability, rather than an absence. Recent reports (C. C. Chen, personal communication) of visualization of both an outer shell and A-spike structures on RRSV particles support the suggestion that ERSV and RRSV should be included in one taxon (Chen *et al.*, 1989b) and argue for the inclusion of this group in the genus *Fijivirus*.

B. Physical Properties

1. Particle Morphology

Like all members of the Reoviridae (Tyler and Fields, 1985a), plant reoviruses consist of an icosohedral core particle comprising the dsRNA genome tightly associated with several polypeptides, surrounded by an outer capsid shell. While estimates of particle dimensions vary, it is generally accepted that the inner core structure is 50–60 nm in diameter and that, with the outer capsid shell, the complete particle has a diameter of about 70 nm. An additional amorphous protein layer surrounds the capsid of WTV (Reddy and MacLeod, 1976) and probably also that of RDV (Francki *et al.*, 1985). Exposure of plant reovirus particles to a variety of enzymes, organic solvents, and ionic or salt milieux has been found to result in their partial or complete disintegration. Specific methodologies based on this general approach have allowed detailed examination of the morphology and ultrastructure of both intact and subviral particles. Results of such studies have been reviewed elsewhere (Francki *et al.*, 1985).

2. Viral Proteins

a. Structural Polypeptides. Plant reovirus particles purified under nondegradative conditions consist of six or seven virally encoded polypeptides. The relative stability of *Phytoreovirus* members in organic solvents and their propensity to progressively degrade following digestion with trypsin or chymotrypsin, or centrifugation in CsCl gradients at the appropriate pH, have allowed detailed analysis of their component polypeptides and the assignment of many of these to specific sites within the particle (Reddy and MacLeod, 1976; Nakata *et al.*, 1978; Omura *et al.*, 1985; Kimura *et al.*, 1987). As shown in Table II, the three phytoreoviruses have two or three major protein components with molecular weights of 108,000–120,000 and 42,000–45,000. For each virus the smaller protein(s) has been shown to be a component of the viral capsid, while for WTV and RGDV, at least, the larger apparently forms part of the viral core structure. Expression of individual WTV structural polypeptides has been demonstrated both in infected vector cells and in cell-free translation systems (Nuss and Peterson, 1980), and a nomenclature has been proposed (Nuss and Peterson, 1980; Nuss, 1984; Table II).

Table III contains data for three other plant reoviruses. As shown, particles of RRSV and ERSV also contain seven polypeptide components. These are apparently present in molar ratios rather different from those of the phytoreoviruses, each virus having four or five major constituent proteins, the others being present in trace amounts. Some

TABLE II

STRUCTURAL PROTEINS OF *Phytoreovirus* MEMBERS[a]

	WTV		RDV			RGDV	
Protein	M_r ($\times 10^3$)	Location	M_r ($\times 10^3$)	M_r ($\times 10^3$)	Location	M_r ($\times 10^3$)	Location
I	155[1] (P1)	Core[2]	210[3]	193[4]	Core[4]	183[5]	Core[5]
II	130 (P2)	Outer coat	185	152	Outer coat	165	Core
III	**108** (P3)	Core	165	131	Core	150	Capsid
IV	76 (P5)	Outer coat	**120**	110	Outer coat	143	—
V	57 (P6)	Core	60	62	Core	**120**	Core
VI	**42** (P8)	Capsid	**43**	46	Capsid	56	Core
VII	**41.5** (P9)	Capsid	—	45	Capsid	**45**	Capsid

[a] Proteins in bold type have been identified as major viral components by the cited authors: [1]Nuss and Peterson, 1980; [2]Reddy and MacLeod, 1976; [3]Kimura *et al.*, 1987; [4]Nakata *et al.*, 1978; [5]Omura *et al.*, 1985. The apparent molecular weight (M_r) values are based on relative electrophoretic mobility measurements.

TABLE III

STRUCTURAL PROTEINS OF MRDV, RRSV, AND ERSV[a]

Protein	MRDV M_r ($\times 10^3$)	Location	RRSV M_r ($\times 10^3$)	M_r ($\times 10^3$)	Location	ERSV M_r ($\times 10^3$)
I	**139**[1]	Core[1]	**129**[2]	**145**[3]	Core[3]	**127**[2]
II	**126**	Core	**123**	**137**	Core	**123**
III	**123**	B spike	113	118	—	103
IV	111	Capsid	88	**72**	Core	**63**
V	97	Capsid	**63**	50	—	50
VI	**64**	Capsid	**50**	47	B spike	49
VII	—	—	**35**	37	B spike	34

[a] Proteins in bold type have been identified as major viral components by the cited authors: [1]Boccardo and Milne, 1975; [2]Chen *et al.*, 1989b; [3]Hagiwara *et al.*, 1986. The apparent molecular weight (M_r) values are based on relative electrophoretic mobility measurements.

RRSV proteins have been assigned to subviral locations (Hagiwara *et al.*, 1986), but no such study has been made of ERSV. Only one *Fijivirus* member, MRDV, has been the subject of thorough examination. Boccardo and Milne (1975) reported that viral preparations containing a large proportion of intact MRDV particles showed only six structural proteins; degradation experiments allowed them to assign these proteins to various particle components. However, the validity of comparisons between MRDV, RRSV, and ERSV is questionable, since the taxonomic positions of the latter two are uncertain. Meaningful discussion of the MRDV data awaits further experimental evidence about other fijiviruses.

A number of enzymatic functions are associated with reovirus subviral particles and thus, presumably, with viral structural proteins. All are involved with either the synthesis or posttranscriptional modification of the viral genome. RNA-dependent RNA polymerase (i.e., transcriptase) activity, which is responsible for mRNA synthesis, has been found in preparations of all *Phytoreovirus* members and in RRSV subviral particles; this enzyme has also been detected in extracts from FDV-induced gall tissue on sugar cane (Black and Knight, 1970; Uyeda and Shikata, 1984; Yokoyama *et al.*, 1984; Ikegami and Francki, 1976; Uyeda *et al.*, 1987a). Optima for temperature, pH, and Mg^{2+} conditions have been defined for the enzymes from these viruses and in general lie in the range 25–40°C, pH 8.0–9.0, and 2.0–8.0 mM $MgCl_2$ (see Lee *et al.*, 1987). The presence of two other enzymes, each

associated with RNA capping (mRNA-guanyltransferase and gua-
nine-7-methyltransferase), and a second methyltransferase (mRNA-2'-
O-methyltransferase) has been demonstrated in WTV preparations (see
Nuss, 1984), but not in other plant reoviruses. There is no evidence to
allow assignment of these functions to individual virally encoded proteins (see Section V).

   b. *Nonstructural Polypeptides.* In addition to the structural proteins
of the virus particle, reovirus genomes also encode a number of nonstructural polypeptides which are expressed in infected cells. As
shown by Nuss and Peterson (1980), insect vector cell cultures can
greatly facilitate the identification of virally encoded nonstructural
polypeptides. Polyacrylamide gel analysis of [^{35}S]methionine-labeled
lysates prepared from WTV-infected *Agallia constricta* (AC20) cells
resulted in the identification of five presumptive nonstructural polypeptides, subsequently designated Pns4, Pns7, Pns10, Pns11, and
Pns12 (see Table IV). Cell-free translation of *in vitro* synthesized WTV
transcripts confirmed that these polypeptides were viral gene products, rather than virally induced host proteins. The establishment and
successful infection of cell cultures derived from the vectors of RDV
and RGDV (Kimura, 1986; Omura *et al.*, 1988) should also allow rapid
progress in the identification of nonstructural polypeptides encoded by
these viruses. Functions associated with the *Phytoreovirus*-encoded
nonstructural polypeptides are currently unknown.

C. Nucleic Acid

   As mentioned in Section II,A, all plant reoviruses have genomes
consisting of either 10 or 12 segments of dsRNA; their known characteristics are described in detail in the following sections. The approximate sizes and corresponding electrophoretic profiles of genomic segments of the recognized plant reoviruses have been presented
elsewhere (Francki *et al.*, 1985), with the exception of the newly characterized ERSV (Chen *et al.*, 1990), general characteristics of which
are described in Section II,A. The basic mechanism by which the individual genomic segments are selected for encapsidation is now under
active investigation (see Sections III,A,4 and 5), but little is known
about the "higher-order" form in which such segments are held. A
single report (Mizuno *et al.*, 1986) suggests that genomic dsRNA of
RDV is packaged within virus particles in a supercoiled manner, in a
fashion similar to that observed in cytoplasmic polyhedrosis virus
dsRNA (Yazaki and Miura, 1980). While it is possible that a similar
strategy is used by other plant reoviruses, further experimental evidence is required before conclusions can be made.

TABLE IV

PROPERTIES ASSOCIATED WITH NUCLEOTIDE AND DEDUCED AMINO ACID SEQUENCES OF WTV GENOMIC SEGMENTS[a]

Segment	Size (bp)	G + C:A + T ratio	Noncoding 5'	Noncoding 3'	% Coding	Polypeptide	Size (amino acids)	Molecular weight	pI	Conformation H	E	T	C
S4	2565	42.7:57.3	63 (2.4)	306 (11.9)	85.7	Pns4	732	81,137 (72,000)	7.55	43	39	8	10
S5	2613	36.1:63.9	25 (1.0)	176 (6.7)	92.3	P5	804	91,074 (76,000)	9.00	40	44	8	8
S6	1700	36.5:63.5	44 (2.6)	96 (5.6)	91.8	Pns7	520	58,755 (52,000)	6.73	48	39	7	7
S7	1726	37.7:62.3	20 (1.2)	149 (8.6)	90.2	P6	519	57,637 (57,000)	7.79	46	39	6	8
S8	1472	40.1:59.9	18 (1.2)	173 (11.8)	87.0	P8	427	48,068 (42,000)	6.02	4	71	15	11
S9	1182	40.0:60.0	25 (2.2)	122 (10.3)	87.5	Pns10	345	38,614 (39,000)	4.75	33	39	18	10
S10	1172	37.8:62.2	24 (2.1)	107 (9.1)	88.8	Pns11	347	38,971 (35,000)	8.54	41	38	12	10
S11	1128	41.9:58.1	21 (1.8)	168 (15.0)	83.2	P9	313	35,611 (41,500)	4.88	44	35	15	7
S12	851	38.7:61.3	34 (4.0)	283 (33.3)	62.7	Pns12	178	19,240 (19,000)	9.75	54	4	19	22

[a] The number of nucleotides (nt) contained within the 5'- and 3'-noncoding regions of each segment and the percentages of the total sequence which they represent (in parentheses) are presented in the fourth and fifth columns, respectively. The sixth column presents the percentage of the total sequence contained within the coding region of each segment. The column headed "Molecular weight" lists the calcuated molecular weight of each nonstructural polypeptide based on the deduced amino acid squence and, in parentheses, the apparent molecular weight of these polypeptides based on relative migration in polyacrylamide gels (Nuss and Peterson, 1980). The approximate isoelectric point for each polypeptide, based on the deduced amino acid sequence, is presented in the column designated "pI." The last four columns list the percentages of an amino acid sequence predicted to be in a helical (H), extended (E), turn (T), or coil (C) conformation. These values were generated by the algorithm of Garnier et al. (1978). Nucleotide sequences of eight WTV genomic segments have been deposited in the European Molecular Biology Laboratory/GenBank data base under the following accession numbers: S4, M24117; S5, J03020; S6, M24116; S7, X14218; S8, J04344; S9, M24115; S10, M24114; and S11, X14219.

D. Biological Properties

The biological properties of plant infecting reoviruses show an over-all uniformity, as might be expected for members of closely allied groups within a single virus family. However, as in all groupings of biological entities, there are variations and exceptions to these similarities. This section briefly describes the observed generalities and pays perhaps disproportionate attention to the variations, since it is hoped that further investigations will allow their correlation with functional differences related to changes in genomic sequence or structure.

With the exception of WTV, the recorded host range of plant reoviruses is limited to the Graminae, where several cause significant damage to agricultural crops, such as rice and sugar cane (Francki and Boccardo, 1983). In contrast, WTV is unable to infect monocot plants, has an experimental dicot host range which extends to 20 families, and has never been associated with a naturally occurring plant disease (Black, 1965).

The most common symptom of reovirus infection in graminaceous plants is severe stunting or dwarfing. This is frequently accompanied by marginal serration and other distortions of leaves, which can also show flecking or streaking patterns and the development of abnormally dark green coloration. As described below, the development of some form of vein swelling, enation, or hyperplastic growth on abaxial leaf surfaces is a consequence of infection with all but RDV, the only plant reovirus not associated with neoplastic growth of its host (Shikata, 1981).

While WTV infection can also cause stunting of host plants and distortion of leaves, its most notable symptom is the neoplastic growth of phloem tissue, which occurs both as a natural consequence of infection and in response to the wounding of infected plants. Other recorded symptoms of WTV infection include rosetting of leaves, suppression of flowering and increased production of axillary shoots (Black, 1972; Selsky, 1961). The latter symptom might be analogous to the excessive tillering frequently observed in plants infected with RDV and MRDV (Francki and Boccardo, 1983).

In all reovirus infections the degree and duration of hyperplastic phloem development are dependent on the interaction between a virus and its specific host. Thus, while WTV causes vein enlargement in *Trifolium* spp., infection results in the development of vein and leaf enations on *Nicotiana* spp. and the formation of massive stem and root tumors on susceptible clones of *Melilotus* spp. (Black, 1972). Such WTV-induced tumors can continue to enlarge for long periods on infected plants; in contrast, it has been reported (Hatta and Francki,

1981) that galls on FDV-infected plants develop for only a short time before the cessation of growth. However, as noted by Francki *et al.* (1985), it is unclear whether this variation reflects differences in viral pathogenicity or host plant physiology.

The examination of infected plant material by electron microscopy has shown that plant reoviruses can always be detected in phloem tissue and that, with the exception of RDV and ERSV, the viruses are limited to vascular bundle-derived cells which compose virally induced plant galls (Boccardo and Milne, 1984). RDV and ERSV have also been detected in mesophyll cells (Iida *et al.*, 1972; Chen *et al.*, 1989), the latter apparently in a distinctive distribution pattern in the outer chloroplast membrane. The finding that ERSV is capable of inducing neoplastic growth, despite not being limited to vascular tissue, indicates that these two traits are not necessarily correlated, as had previously been suggested by observation of RDV-infected plants (Omura *et al.*, 1985).

As mentioned in Section II,A,1 and 2, all known plant reoviruses have leafhopper or planthopper vectors. Insects become infected after feeding on infected hosts, and after a sufficient incubation time (generally 10–30 days; see, e.g., Maramorosch, 1950; Chen *et al.*, 1986) they transmit the virus in a persistent manner. With the exception of ERSV and RRSV (Milne *et al.*, 1982; Chen *et al.*, 1986), all of the viruses can be transmitted transovarially, with a frequency apparently determined by both the virus identity and specific culture of vector insects (Francki and Boccardo, 1983). In contrast to the strict tissue tropisms observed in infected plants, the viruses multiply in a wide range of cell types in the insect alimentary and nervous systems and also in fat body and musculature (e.g., Hirumi *et al.*, 1967; Omura *et al.*, 1985; Chen *et al.*, 1989).

Despite the cytological changes that accompany infection (see below), viruliferous vectors of WTV apparently show no "behavioral disorders" (Hirumi *et al.*, 1967) and display no alterations in fecundity or longevity (Maramorosch, 1975). Similarly, persistent productive WTV infections of AC20 cell cultures cause no apparent changes in cell growth rate or protein synthesis capabilities (Black, 1979; Peterson and Nuss, 1985). Reductions in fecundity of MRDV- and RDV-infected vectors have been reported (Harpaz, 1972; Fukushi, 1969), but no parallel studies are available on the effect of RDV infection on vector cell monolayers. The mechanisms responsible for inapparent persistent virus infections of insects are enigmatic (Podgwaite and Mazzone, 1986), but in cultured leafhopper cells, at least, they might be associated with alterations in the translational capabilities of viral mRNA (Peterson and Nuss, 1986).

Infection of plants and insects with these groups of viruses results in the formation of morphologically similar cytoplasmic inclusions in the cells of both hosts, which have been described and illustrated (Francki *et al.*, 1985). Most prominent of these inclusions is the viroplasm, a loosely structured collection of fibrils and amorphous material known to be the site of viral synthesis (Favali *et al.*, 1974). Elsewhere in the cytoplasm virus particles can be distributed in a variety of configurations. While those of FDV and RBSDV are usually seen as individual particles, cells infected with the other viruses frequently contain large particle aggregations or crystals and commonly show linear arrangements of particles within tubular membrane structures. The biological function(s) and importance of these tubules remain to be determined.

III. Genome Organization for Members
of the Genus *Phytoreovirus*

A. *Wound Tumor Virus*

The availability of cDNA clones of individual genomic segments has provided new experimental approaches for studying the molecular biology of plant reoviruses, as illustrated by recent progress with WTV. The original cDNA library generated and characterized by Asamizu *et al.* (1985) provided full-length cDNA clones of nine of the 12 WTV genomic segments and partial clones of the three largest segments. Over the last several years these cDNA clones have served as reagents for accurately measuring WTV gene expression in acutely and persistently infected insect vector cells (Peterson and Nuss, 1986), for assigning individual virally encoded polypeptides to cognate genomic segments (Xu *et al.*, 1989a), for investigating the mechanisms involved in sorting and packaging of a segmented RNA genome (Anzola *et al.*, 1987; Xu *et al.*, 1989b), and for deriving the complete nucleotide sequences of nine genomic segments (Asamizu *et al.*, 1985; Anzola *et al.*, 1987, 1989a,b; Xu *et al.*, 1989b; Dall *et al.*, 1989). The contributions that these studies have made to our understanding of the structure and function of WTV genomic RNA are the focus of this section. Potential uses of these cDNA clones for examination of the functional properties of WTV-encoded polypeptides are discussed in Section V.

1. Cloning

The protocol used to generate the WTV cDNA library (Fig. 1) was similar to that described by Cashdollar *et al.* (1982) for generating cDNA clones of human reovirus genomic RNAs, with several notewor-

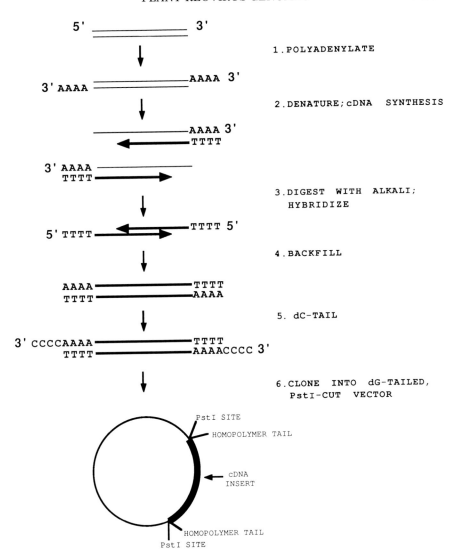

5' ─────────── 3'

                              ↓   1.POLYADENYLATE

3' AAAA ═══════════════AAAA 3'
                              ↓   2.DENATURE;cDNA   SYNTHESIS

        ───────────────AAAA 3'
                   ◄────────────TTTT

3' AAAA ───────────
   TTTT   ────────►
                              ↓   3.DIGEST WITH ALKALI;
                                                     HYBRIDIZE
5' TTTT ───────◄══════TTTT 5'

                              ↓   4.BACKFILL

AAAA═══════════════════TTTT
TTTT═══════════════════AAAA
                              ↓   5. dC-TAIL

3' CCCCAAAA════════════════TTTT
        TTTT════════════════AAAACCCC 3'

                              ↓   6.CLONE INTO dG-TAILED,
                                                     PstI-CUT VECTOR

                                    PstI SITE
                                      HOMOPOLYMER TAIL
                                      cDNA
                                      INSERT
                              HOMOPOLYMER TAIL
                           PstI SITE

Fig. 1.   Cloning strategy used for the generation of WTV cDNA library from total
genomic dsRNAs. Details of individual reaction steps have been described by Asamizu *et
al.* (1985) and in Section III,A. As indicated, the resulting cDNA clones are flanked by
homopolymer tails and regenerated *Pst*I restriction sites.

thy modifications. Whereas denaturation of human reovirus dsRNAs
was required for the efficient addition of 3'-polyadenosine priming
sites, nondenatured WTV dsRNAs were efficiently polyadenylated by
Escherichia coli poly(A) polymerase. Furthermore, cDNA synthesized

from WTV RNA templates that were denatured before polyadenylation was generally more heterogenous in size than that synthesized
from RNA which was polyadenylated without prior denaturation. This
difference could result from the generation of additional poly-
adenosine priming sites at single-stranded break points exposed, or
introduced, by denaturation. cDNA synthesis was initiated by addition
of the complete prewarmed (42°C) reverse transcriptase reaction mixture to a dried pellet containing the denatured polyadenylated RNA
template and the oligo(dT)12–18 oligonucleotide primer. This modification was introduced to allow the initiation of cDNA synthesis before
extensive renaturation of the complementary RNA template strands
and was suggested by observations that a lower proportion of full-
length cDNA was synthesized as the template concentration was increased, a condition that would promote renaturation of the complementary RNA strands.

An additional modification that significantly improved the yield of
full-length cDNA was the inclusion of 4 mM sodium pyrophosphate
(Murray et al., 1983). Finally, the presence of incomplete cDNA products ($\sim$200–700 nucleotides) was consistently observed to significantly
lower the final yield of full-length cDNA clones, perhaps as a result of
the formation of a molecular network during the step in which cDNA
strands are allowed to anneal (Fig. 1). Consequently, it is now recom-
mended that a sizing step be included before the annealing step to
remove small cDNA products.

2. Characterization of cDNA Clones

A prerequisite for characterization of individual WTV cDNA clones
was the determination of terminal nucleotide sequences and strand
polarity of the genomic RNAs and viral transcripts. Wandering-spot
analysis of individual [3'-^{32}P]pCp-labeled WTV genomic segments revealed that all 12 segments contained the sequence 3'-CCAUAA-5' at
one terminus and 3'-UAGU-5' at the other (Fig. 2). Using strand-
specific cDNA clones as hybridization probes, it was possible to show
that strands containing the latter sequence were of the same polarity
as the single-stranded viral transcripts. Sequence analysis of full-
length WTV cDNA clones subsequently confirmed that the coding
strand contained the 3'-terminal sequence 3'-UAGU-5'. However, as
indicated below, these results raised an inconsistency that has yet to
be resolved.

For several members of the Reoviridae (e.g., human reovirus,
cytoplasmic polyhedrosis virus, and rotavirus), the viral transcripts
have been shown to be identical copies of the positive-sense genomic
RNA strand (Furuichi et al., 1975a,b; Furuichi and Miura, 1975; Li et
al., 1980; Spencer and Garcia, 1984). For example, both the genomic

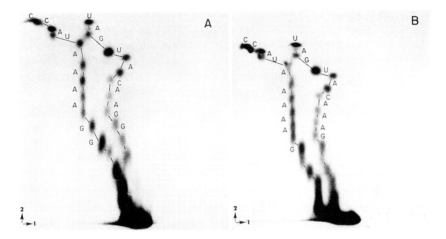

FIG. 2. Two-dimensional wandering-spot analysis of 3'-terminal [^{32}P]pCp-labeled WTV genomic segments S3 (A) and S9 (B). In each panel the left track represents the 3'-terminal sequence of the noncoding ($-$) strand, which contains a 3'-terminal cytidine, while the right track represents the 3'-terminal sequence of the coding ($+$) strand, which has a 3'-terminal uridine. Each of the 12 WTV genomic segments exhibited the same two-dimensional pattern for the four 3'-terminal nucleotides of the coding strand and the six 3'-terminal nucleotides of the noncoding strand. Reprinted with permission from Asamizu et al. (1985).

positive-sense RNAs and viral transcripts of human reovirus contain the terminal sequences 5'-m^{7}GpppG^{m}CUA------UCAUC-3'. Analyses of WTV transcripts synthesized in vitro in the presence of [^{3}H]methylS-adenosylmethionine have previously shown that the labeled transcripts contained the 5'-terminal structure m^{7}GpppA^m-- (Rhodes et al., 1977; Nuss and Peterson, 1981a), while nearest-neighbor analysis of post-transcriptionally labeled transcripts revealed a 3'-terminal uridine. However, wandering-spot analysis, cDNA sequence analysis, strand-specific hybridization analysis, and 3'-terminal nearest-neighbor analysis have all indicated that the terminal sequences of WTV positive-sense genomic RNAs are 5'-GGUAUU----UGAU-3'. Unfortunately, attempts to determine whether WTV genomic RNAs contain a 5'-terminal m^{7}G cap have given ambiguous results (D. L. Nuss, unpublished observations).

Irrespective of this information, it would appear that, unlike the situation with other reoviruses, the 5'-termini of WTV transcripts are not identical to those of the genomic positive-sense RNA strands. A similar inconsistency was recently reported for RDV (Omura et al., 1988). However, it should be emphasized that analyses have only been performed on in vitro synthesized [^{3}H]methyl-labeled WTV transcripts.

In another study (D. L. Nuss, unpublished observations) transcripts were synthesized *in vitro* in the presence of a saturating amount of S-adenosylmethionine, and their 5' termini were subsequently labeled with ^{32}P. Analysis of these transcripts revealed that 60% were uncapped, containing either a 5' guanosine or 5' adenosine residue. Whether this apparent heterogeneity extends to viral transcripts present in WTV-infected cells is currently unknown. Determination of the precise structural properties of the 5' termini of genomic RNAs and viral transcripts is required for a clear understanding of plant reovirus replication strategies and their relationships to strategies used by other members of the Reoviridae.

Once the terminal nucleotide sequences and strand polarities of the WTV genomic RNAs were determined, the terminal portions of candidate cDNA clones were subjected to sequence analysis to verify that they represented full-length copies of the genomic RNAs. This analysis was important for two reasons. First, it provided the sequence information required for the tailoring of cDNA clones for efficient *in vitro* expression; second, it exposed a common terminal structural motif present in each segment. This structural motif consisted of a six- to 14-nucleotide segment-specific inverted repeat located immediately adjacent to the conserved terminal hexanucleotide and tetranucleotide sequences initially identified by wandering-spot analysis. As indicated in Fig. 3, a similar terminal structural motif is found in the genomic segments of influenza virus, a segmented negative-stranded RNA virus. Potential functional properties of the terminal inverted repeats are discussed in detail in Section III,A,5.

Plant reoviruses do not cause local lesions on their plant hosts or plaques on vector cell monolayers. Consequently, they are not amenable to the genetic reassortment analysis so successfully applied to the study of vertebrate reoviruses (reviewed by Joklik, 1983). Fortunately, *in vitro* and *in vivo* expression of cDNA copies of genomic segments provide alternative means for defining functional, as well as structural, properties of plant reovirus genomic segments. In this context a series of *in vitro* expression studies was performed to test the functional integrity of candidate full-length WTV cDNA clones and to assign virally encoded polypeptides to cognate genomic segments. However, as described above and illustrated in Fig. 1, the cloning protocol used to generate WTV cDNA clones resulted in the addition of the flanking sequences $5'\text{-}(G)_n(T)_n\text{-----}(A)_n(C)_n\text{-}3'$, where n varied from approximately 15 to approximately 40 nucleotides. As expected, these homopolymer sequences severely restricted the *in vitro* translation of synthetic transcripts generated from cDNAs directly subcloned into transcription vectors (Xu *et al.*, 1987).

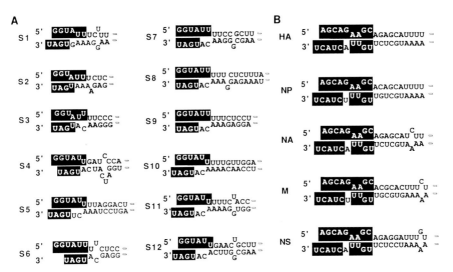

FIG. 3. (A) Terminal sequence domains of WTV genomic segments. The terminal nucleotide sequences (coding strand only) of all 12 WTV genomic segments are presented in the left and center columns. The conserved terminal 5'-hexanucleotide and 3'-tetranucleotide sequences shared by all 12 WTV genomic segments are shown in white on a black background. Segment-specific inverted repeats are oriented to indicate potential base-pairing interactions. (B) For purposes of comparison, available sequence information for terminal domains of influenza B genomic segments are presented in the right column. The influenza sequence information was taken from Stoeckle *et al.* (1987) and reformatted to illustrate similarities to WTV terminal structures. HA, hemagglutinin; NP, nucleoprotein; NA, neuraminidase; M, membrane protein; NS, nonstructural protein. Reprinted with permission from Anzola *et al.* (1987).

To circumvent this problem, Xu *et al.* (1987) devised a method which allowed the deletion of homopolymer flanking sequences and tailoring of the resulting termini prior to subcloning. Using this method, candidate full-length cDNA clones of segments S12–S4 lacking homopolymer flanking sequences were subcloned into transcription vectors (Fig. 4) (Xu *et al.*, 1989a). The relationships between the resulting synthetic transcripts and cognate genomic segments were confirmed by hybridization to the WTV genome, as indicated in Fig. 5. Electrophoretic analysis of *in vitro* translation reactions programmed with the individual synthetic transcripts showed that the major polypeptide synthesized in each reaction comigrated with a viral-specific polypeptide in the infected cell lysate (Fig. 6).

The results of the *in vitro* expression studies and terminal sequence analyses demonstrated that the candidate cDNAs were full-length functional copies of the WTV genomic segments and allowed the assignment of virally encoded polypeptides to their cognate genomic seg-

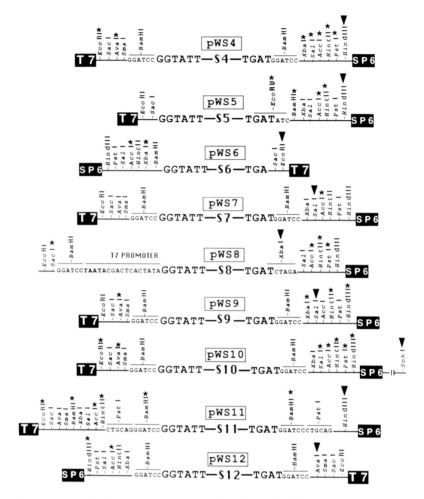

FIG. 4.   Organization of constructs containing tailored cDNA clones of WTV genomic segments. The plasmids are designated pWS4–pWS12 (open boxes), where the number preceded by the letter S indicates the corresponding WTV genomic segment; for example, pWS4 is the construct containing the tailored cDNA clone of WTV genomic segment S4. The conserved sequences 5'-GGTATT-3' and 5'-TGAT-3' found at the 5' and 3' termini, respectively, of the coding strand of each WTV genomic segment are indicated by large letters. Note that pWS6 lacks the 3'-terminal nucleotide. Sequences engineered to replace the homopolymer sequences flanking the original cDNA clones are indicated by small letters (e.g., the BamHI site, T7 promoter sequence, and the XbaI site in pWS8). An asterisk over a restriction site indicates that such a site also occurs in the viral cDNA sequence. The solid triangle indicates the cleavage site used to linearize the plasmid in preparation for the run-off transcription reaction. The SP6 and T7 promoters are indicated by the solid boxes. The tailored cDNA copy of genomic segment S8 was cloned into transcription vector pSP64, while the tailored cDNA copies of the remaining genomic segments were all cloned into pGEM-1. Reprinted with permission from Xu et al. (1989a).

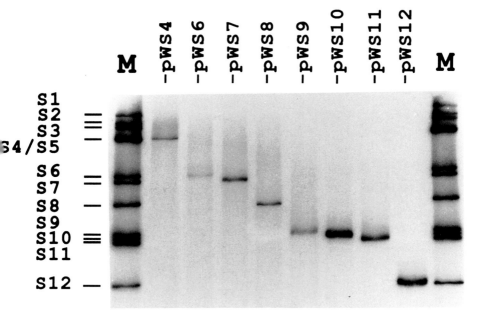

FIG. 5.   Hybridization analysis of synthetic transcripts to confirm their relationship
to WTV double-stranded genomic segments. ^{32}P-labeled synthetic transcripts generated
in run-off transcription reactions were hybridized individually to unlabeled WTV genomic RNA, as described by Nuss and Peterson (1981b), and the resulting hybrid molecules were subjected to electrophoresis on a 7.5% polyacrylamide gel. The plasmid used
to generate the transcript is indicated at the top of each lane. Lanes marked "M" contain
[^{32}P]pCp-labeled WTV genomic RNAs as markers (Nuss and Peterson, 1981b). The
migration position of individual WTV genomic segments are indicated at the left. Reprinted with permission from Xu et al. (1989a).

ments, as summarized in Fig. 7. The five virally encoded nonstructural
polypeptides Pns4, Pns7, Pns10, Pns11, and Pns12 were assigned to
cognate genomic segments S4, S6, S9, S10, and S12, respectively.
Three structural polypeptides—core protein P6 and capsomere proteins P8 and P9—were revealed as the products of genomic segments
S7, S8, and S11, respectively. In a separate set of experiments, the
outer coat proteins P5 and P2 were shown to be encoded by genomic
segments S5 and S2, respectively (Anzola et al., 1987; Reddy and
Black, 1977).

As indicated in Fig. 7, the relative gel migration positions of WTV
genomic segments correlate well with the relative electrophoretic mobilities of the WTV-specified polypeptides. Only segments S6 and S11
and their encoded polypeptides exhibit significant deviations. Although the core proteins P1 and P3 have yet to be formally assigned to
cognate genomic segments, the relationship between the gel migration
positions of segments S1 and S3 and the electrophoretic mobilities of

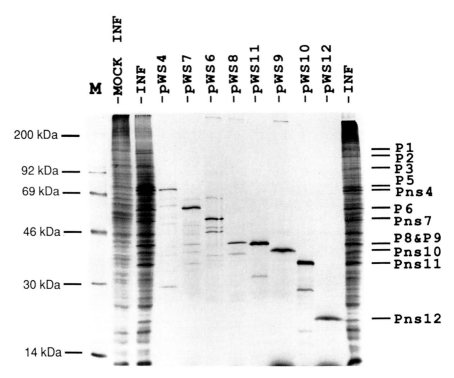

FIG. 6. Cell-free translation of synthetic transcripts generated from tailored cDNA clones of WTV genomic segments. Individual synthetic transcripts were used to program a nuclease-treated rabbit reticulocyte lysate, as described by Xu et al. (1987), and resulting [35S]methionine-labeled translation products were analyzed by gel electrophoresis in parallel with [35S]methionine-labeled lysates of mock-infected and WTV-infected AC20 cells (Nuss and Peterson, 1980). The lane designated "M" contains 14C-labeled protein molecular weight markers (Amersham). The lanes designated "MOCK INF" and "INF" contain [35S]methionine-labeled lysates from mock-infected and WTV-infected cells, respectively. The plasmids encoding the synthetic transcripts used to program the individual translation reactions are also indicated at the top of the gel. The migration positions of WTV structural (P) and nonstructural (Pns) polypeptides are indicated at the right, while the apparent molecular weights of the marker proteins are indicated at the left. Reprinted with permission from Xu et al. (1989a).

P1 and P3 suggests the assignment proposed in Fig. 7. Knowledge of the coding assignment allows a rational approach toward defining functions of the encoded polypeptides. For example, in an effort to identify the genetic information responsible for tumor formation, the genomic segments encoding nonstructural polypeptides have been targeted for *in vivo* expression studies in transgenic plants, while the segments encoding structural polypeptides have been targeted for *in*

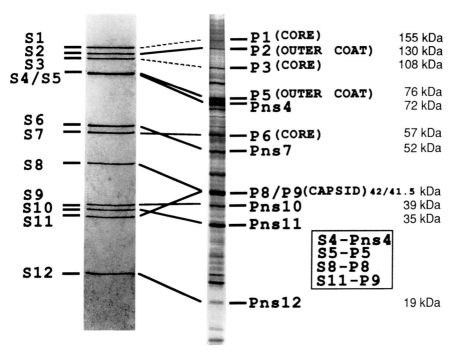

FIG. 7. Relationship of WTV-encoded polypeptides to cognate genomic segments. Silver-stained WTV genomic segments separated on a 7.5% polyacrylamide gel are presented in the left lane, while an autoradiograph of a 12.5% polyacrylamide gel containing a [35S]methionine-labeled lysate from WTV-infected AC20 cells is presented in the right lane. The assignments of WTV polypeptides Pns4, P6, Pns7, P8, P9, Pns10, Pns11, and Pns12, indicated by the connecting lines, are based on the results shown in Fig. 6. The assignments of polypeptides P2 and P5 to their cognate genomic segments are based on results presented by Reddy and Black (1977) and by Anzola et al. (1987), respectively. The tentative assignments (dashed line) of polypeptides P1 and P3 to their cognate genomic segments are based on relative electrophoretic mobilities and predicted coding capacities. The locations within the virus particle of the seven structural polypeptides are indicated in parentheses (Reddy and MacLeod, 1976). The apparent molecular weights of the 12 WTV primary gene products are indicated at the extreme right (Nuss and Peterson, 1980). The box at the lower right is included to clarify the assignments for Pns4, P5, P8, and P9. Reprinted with permission from Xu et al. (1989a).

vitro studies aimed at understanding protein–protein and protein–RNA interactions (discussed in Section V,C).

3. Sequence Organization of the WTV Genome

Complete nucleotide sequences are now available for nine of the 12 WTV genomic segments (Asamizu et al., 1985; Anzola et al., 1987, 1989a,b; Xu et al., 1989b; Dall et al., 1989). The basic organization of each segment (Fig. 8) includes the conserved terminal hexanucleotide and tetranucleotide sequences, a six- to 14-nucleotide segment-specific inverted repeat located within each terminal domain, and one long open reading frame (ORF) on the positive-sense strand, flanked by a relatively short (18- to 63-nucleotide) 5′-noncoding region and a slightly longer (93- to 303-nucleotide) 3′-noncoding region (Table IV). The exception is segment S12, which has a second short ORF consist-

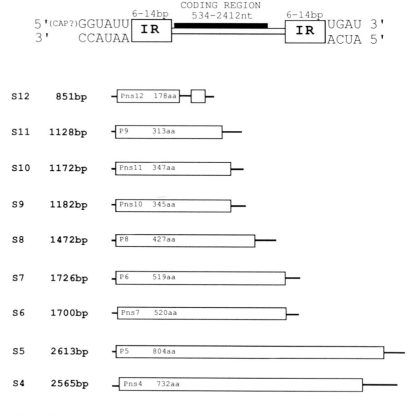

FIG. 8.   The general organization of WTV genomic segments is illustrated at the top. The terminal conserved hexanucleotide and tetranucleotide sequences are indicated in large letters, while the positions of the 6- to 14-bp terminal inverted repeats (IR) are indicated by the large open boxes. The solid rectangle indicates the general position of the coding region (534–2412 nucleotides). The remainder of the figure illustrates specific information regarding the size of individual genomic segments (bp) and their encoded polypeptides (aa) as well as the coding assignments. There is no evidence for expression of the second small coding region present in genomic segment S12.

ing of 40 codons located 102 nucleotides downstream from the large ORF which encodes Pns12 (Asamizu *et al.*, 1985). However, there is currently no evidence for expression of this second ORF.

The relative electrophoretic mobilities of the nine segments generally correlate with the relative base pair (bp) lengths of the segments, the notable exception being segments S6 and S7. Although segment S7 is 26 bp longer than segment S6, it migrates considerably faster (Fig. 7) than S6 on polyacrylamide gels. Segment S5, which is 48 bp longer than segment S4, was also shown to migrate slightly faster than segment S4 (Nuss and Peterson, 1981b). All nine segments have a similar nucleotide composition, being richer in A + T than G + C residues by a factor of 3 : 2. Furthermore, a comparison of the nucleotide distribution for segments S6 and S7 revealed no extensive A + T- or G + C-rich regions that could account for the difference in relative mobilities. These results emphasize the problems associated with the assignment of polypeptide products to cognate genomic segments based solely on relative electrophoretic mobilities.

For each of the nine sequenced WTV genomic segments, the coding region extends from the 5′-proximal AUG triplet. Each presumptive initiation codon lies within a favorable sequence context (Kozak, 1986a) with a purine at position -3 and, with three exceptions, a purine at position $+4$. There is generally good agreement among the calculated molecular weights of the nine viral polypeptides based on the deduced amino acid sequences and their apparent molecular weights based on relative migration in polyacrylamide gels (Nuss and Peterson, 1980). Of the nine deduced WTV amino acid sequences the calculated molecular weights exceeded the apparent molecular weights for five: by 19.7% for P5, 14.2% for P8, 12.5% for P4, 11.5% for Pns7, and 11.4% for Pns11. The apparent molecular weight exceeded the calculated molecular weight in only one case, by 16.5% for P9. The two values differ by approximately 1% for P6, Pns10, and Pns12.

The deduced amino acid sequences of the nine polypeptides were analyzed by the algorithm of Garnier *et al.* (1978) to predict general conformational properties of the proteins. Using similar methodology, Joklik and co-workers identified a conformational trend among human reovirus polypeptides, indicating that structural polypeptides have a considerably lower predicted α-helix content (13–27%) than nonstructural polypeptides (approximately 50%) (Wiener and Joklik, 1987, 1988; Bartlett and Joklik, 1988). This trend is not apparent for the deduced WTV polypeptides (Table IV). The α-helix content exceeds the β-sheet content for four nonstructural and two structural polypeptides: Pns4, Pns7, Pns11, Pns12, P6, and P9. The β-sheet content slightly exceeds the α-helix content for two of the WTV polypeptides:

Pns10 and P5. Only in the case of structural polypeptide P8, one of the two capsomere proteins, is the α-helix–β-sheet ratio significantly low: 4% : 71%.

4. Defective Interfering RNAs and Genome Function

A central unresolved question regarding the replication cycle of members of the Reoviridae concerns the mechanisms involved in the recognition and sorting of individual viral RNAs which lead to the packaging of only one copy of each genomic segment per virus particle. The characterization of remnants of genomic RNAs, generated by deletion events, has provided insights into the mechanisms of genome replication and encapsidation for members of other virus families (Nayak *et al.*, 1985; Re and Kingsbury, 1986; Levis *et al.*, 1986; DePolo *et al.*, 1987). In this regard characterization of deletion-generated remnants of WTV genomic RNAs has revealed a number of basic principles involved in the replication and packaging of segmented dsRNA genomes and has led to the identification of sequence elements potentially involved in the sorting and packaging events (Reddy and Black, 1974, 1977; Nuss and Summers, 1984; Anzola *et al.*, 1987). In general terms the WTV genomic remnant RNAs are analogous to influenza virus defective interfering (DI) RNAs (Nayak *et al.*, 1985). They are defective with respect to encoding authentic viral polypeptides, but remain functional with respect to transcription, replication, and packaging (Reddy and Black, 1974, 1977; Nuss, 1983a,b). In addition, they specifically interfere with the replication or packaging of the genomic segment from which they are derived (Reddy and Black, 1974, 1977).

WTV DI RNAs were discovered by Black and co-workers during the characterization of viral populations that had lost their ability to be transmitted by the leafhopper vector (Reddy and Black, 1974). Electrophoretic analysis of genomic RNAs isolated from transmission-defective populations revealed banding patterns that differed in several respects from wild-type genomic RNA banding patterns (Fig. 9). For most transmission-defective populations there was a reduction in the concentration of at least one wild-type genomic segment and a corresponding appearance of smaller dsRNA species. Moreover, the banding patterns were observed to change with time as the transmission-defective populations were maintained in the vegetatively propagated plant host (Reddy and Black, 1974, 1977). It soon became apparent that an increase in the molar concentration of a particular DI RNA always correlated with a concomitant decrease in the molar concentration of a particular wild-type genomic segment. Occasionally, a DI RNA appeared to completely replace a wild-type segment (Reddy and Black, 1977). These observations indicated that there was specific

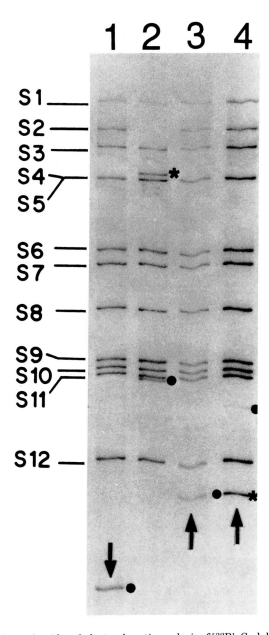

FIG. 9. Polyacrylamide gel electrophoretic analysis of [³²P]pCp-labeled genomic segments associated with several transmission-defective isolates of WTV. The migration positions of segments associated with wild-type transmission-competent virus are indicated at the left. Defective interfering (DI) RNAs are indicated by circles. Asterisks indicate DI RNAs, characterized by Nuss and Summers (1984), while arrows indicate three segment S5-related DI RNAs fully characterized by Anzola *et al.* (1987). The four transmission-defective isolates are -S5(60) (lane 1), -S2(70) (lane 2), 10%S1(60) (lane 3), and 10%S1(49) (lane 4). Nomenclature is described in detail by Reddy and Black (1977) and by Nuss and Summers (1984). Note that in lane 2 segment S2 has been completely replaced by a related DI RNA (Nuss and Summers, 1984). RNAs were isolated from purified virus particles. Reprinted with permission from Anzola *et al.* (1987).

competition between a DI RNA and the segment from which it was derived at some step in the sorting, packaging, or replication process and that the DI RNA was favored. In addition, by observing associated changes in the concentration of DI RNAs and genomic segments, it was possible to tentatively identify the segment from which a particular DI RNA was derived. Several of these assignments were later confirmed by molecular hybridization analysis (Nuss and Summers, 1984).

Since the DI RNAs are transcribed, replicated, and packaged into virus particles in systemically infected plants, they must retain the nucleotide sequence domains required for the efficient execution of these events. The DI RNAs can, therefore, be viewed as naturally occurring reagents useful for defining, to a first approximation, the location of these sequence domains. One of the first questions to be addressed in the molecular characterization of the WTV DI RNAs concerned the nature of the deletion event; that is, were the DI RNAs generated by internal deletion or truncation events? Characterization of several 3' end-labeled WTV DI RNAs and their corresponding wild-type genomic segments by partial T1 nuclease digestion and subsequent gel-electrophoretic analysis provided the first evidence that the DI RNAs retained the terminal portions of the wild-type genomic segment and therefore were generated by an internal deletion event(s) (Nuss and Summers, 1984).

To determine the precise relationship between a DI RNA and its progenitor genomic RNA segment, it was necessary to generate cDNA clones of individual DI RNAs and compare their nucleotide sequences with the cDNA-derived nucleotide sequences of the corresponding wild-type genomic segments. Since mutational events associated with a loss of transmissibility occur with a high frequency in genomic segment S5 (Reddy and Black, 1974), several S5-related DI RNAs were cloned and sequenced (Anzola et al., 1987). Sequence analysis of cDNA clones of segment S5-related DI RNAs associated with three different transmission-defective WTV populations revealed that each contained a single deletion break point with no adjacent sequence rearrangements, suggesting that each DI RNA was generated by a single internal deletion event. Two of the DI RNAs were 776 bp in length, differed in nucleotide sequence at only two positions, and contained a break point located 319 bp from the 5' terminus of the positive strand of segment S5 and 457 bp from the 3' end of the same strand (Fig. 10).

A third DI RNA was only 587 bp in length and contained a break point located 382 bp from the 5' terminus of the positive strand of segment S5 and 205 bp from the 3' end of the same strand; that is,

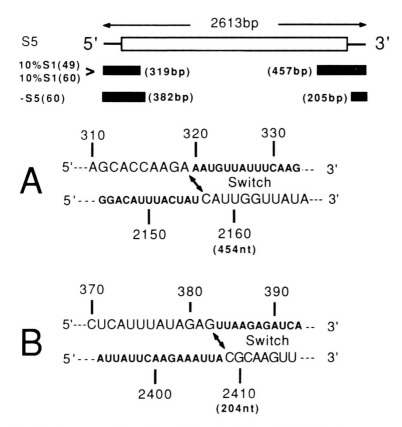

FIG. 10. Sequence information retained in S5-related DI RNAs. The sequences of cDNA clones of defective interfering (DI) RNAs were determined by analyzing several independent full-length clones of each DI RNA and by direct sequencing of denatured DI RNAs. The coding region of segment S5 is indicated by the open bar. The solid bars indicate the portions of segment S5 retained in the DI RNAs associated with isolates 10%S1(49), 10%S1(60), and -S5(60) in Fig. 9. The positions of the deletion boundaries relative to the 5' and 3' termini of the coding strand are indicated within parentheses inside the solid bars. (A and B) The sequence context around the deletion boundaries of the DI RNAs associated with isolates 10%S1(49) and -S5(60), respectively. Distances, in terms of nucleotides (nt) from the 3' end of the coding strand, are given in parentheses. Reprinted with permission from Anzola *et al.* (1987).

approximately 78% of the internal portion of segment S5 was deleted. Surprisingly, the nucleotide sequences of the 587-bp and one of the 776-bp DI RNAs were identical to that determined for the corresponding domains of the wild-type S5 segment. This was unexpected, since the virus populations containing the two DI RNAs had been in continuous passage in independent, systemically infected, vegetatively propagated plant hosts for 15 and 35 years, respectively. Furthermore, the genomic RNA used for cDNA synthesis and cloning of the S5 segment was obtained from an infrequently passaged transmission-competent virus stock. Thus, either replication of the WTV genome proceeds with a high degree of fidelity or mechanisms exist which ensure that a wild-type sequence predominates within the genome population.

The combined information gained from characterization of the WTV DI RNAs provides an emerging view of the principles that govern the replication and packaging of a segmented dsRNA genome, as recently presented by Anzola *et al.* (1987) and summarized below.

a. Sequence Information Required for Replication and Packaging of a Genomic Segment Is Located within the Terminal Domains. Sequence analysis of several DI RNAs clearly showed that sequence domains within the internal portion of a genomic segment are not required for transcription, replication, or packaging events. Furthermore, combined information indicated that the sequence elements required for the replication and packaging of segment S5 reside within 319 bp from the 5′ end of the positive strand and 205 bp from the 3′ end (Anzola *et al.,* 1987).

b. Packaging of a Genome Equivalent of Terminal Structures Is Favored over Packaging of a Genome Equivalent of Nucleotide Bases. This statement is supported by observations that purified transmission-defective WTV particles appear not to contain multiple copies of the DI RNAs, but contain an equimolar complement of the 12 pairs of terminal structures, even though some DI RNAs are only 20% of the length of the progenitor segment (Reddy and Black, 1974; Anzola *et al.,* 1987).

c. Packaging of One Pair of Terminal Structures Excludes the Packaging of a Second Copy of the Same Pair of Terminal Structures. Evidence for a selective exclusion mechanism is based on the observation that WTV DI RNAs displace only the genomic segment from which they were derived and do not interfere with the replication or packaging of unrelated genomic segments (Reddy and Black, 1974, 1977). Two corollaries follow from this statement. Firstly, each segment must contain at least two operational recognition domains (termed "sorting signals"), one that specifies that it is a viral—not a cellular—RNA, and a second that specifies that it is a particular segment. Anzola *et al.* (1987) postulated that the conserved terminal sequences might constitute the former signal, while the segment specific inverted repeat might constitute the latter. Second, packaging of the 12 individual WTV segments must involve 12 different specific protein–RNA or RNA–RNA interactions.

5. Evidence for Intramolecular Interactions between the 5'- and 3'-Terminal Domains of WTV Transcripts

In considering the temporal and physical nature of the macromolecular interactions leading to the production of WTV progeny virions, it is useful to review briefly what is known about the replication cycle of human reovirus, the best-characterized member of the Reoviridae (for reviews see Joklik, 1974; Zarbl and Millward, 1983; Tyler and Fields, 1985b). The initial event in the replication cycle is thought to involve an interaction between the σ1 protein located on the outer capsid and specific receptors on the cell surface. Once bound, the virus particle enters the cell by phagocytosis, where it is exposed to hydrolytic enzymes in the lysosome, resulting in partial digestion of the outer capsid and subsequent release into the cytoplasm. This process activates the virion-associated transcriptase and methyl- and guanylyltransferases, resulting in the transcription of the 10 dsRNA genomic segments to produce ssRNA copies of the coding strand of each segment. The viral transcripts must serve two functions: to act (1) as mRNA directing the production of viral polypeptides by host ribosomes and (2) as templates for negative-strand synthesis to yield progeny genomic dsRNAs. At some point a subset of viral polypeptides is able to recognize the viral transcripts in a process that leads to the formation of nascent subviral particles. It is during this step that the assortment of individual viral RNAs is thought to occur. Once sequestered in these subviral particles, the ssRNA transcripts serve as a template for one round of negative-strand synthesis to produce the 10 dsRNA genomic segments. The progeny subviral particles actively synthesize additional viral transcripts, leading ultimately to the accumulation of a pool of progeny subviral particles which mature and are released during cell lysis.

Assuming that WTV follows the same basic replication strategy as human reovirus, one would predict that the initial events leading to the replication and packaging of the WTV dsRNA genomic segments involve the specific recognition and sorting of viral transcripts. From analysis of WTV DI RNAs, it is clear that the structural elements required for efficient replication and packaging do not reside within the internal portion of viral transcripts. Consequently, the conserved 5'- and 3'-terminal oligonucleotides and the adjacent segment-specific inverted repeats represent obvious candidates for recognition signals. Also, the presence of terminal regions of inverted complementarity suggests the possibility that the single-stranded transcripts could exist in a conformation in which the 5'- and 3'-terminal regions are base paired. To test for intramolecular interactions involving the inverted repeats, Xu *et al.* (1989b) constructed a series of transcription vectors which allowed the synthesis of an exact copy of the coding strand of WTV genomic segment S8 and four analogs that differed from the authentic sequence only at the immediate 3' terminus. As described below, the transcripts were then used to determine whether such 3'-terminal modifications altered the conformational and functional properties associated with the 5' terminus.

The sequence of the 1472-bp WTV genomic segment S8 is presented in Fig. 11. The terminal inverted repeats are located between nucleotides 3 and 16 at the 5' terminus and between nucleotides 1456 and 1470 at the 3' terminus (indicated in the figure by solid circles). The presumptive initiation codon for the coding region of S8 (427 codons specifying the 42-kDa capsid protein, P8) is located just downstream from the 5'-terminal inverted repeat at nucleotides 19–21. To construct a transcription vector that would yield an exact copy of the S8 coding strand, a bacteriophage T7 promoter was fused to the 5' terminus of the S8 cDNA, so that transcription initiated with the first nucleotide of the authentic S8 sequence (Fig. 12). The sequence at the 3' terminus of the synthetic transcript was specified by introducing an *Xba*I site immediately adjacent to the 3' terminus of the S8 cDNA clone. Linearization of the construct with *Xba*I, followed by digestion of the resulting 5' overhang with mung bean nuclease, provided a run-off transcription site with the terminal sequence 5'-TGAT-3', corresponding to the conserved WTV 3'-terminal oligonucleotides.

Three of the analogs were designed to modify the sequence within the 3'-terminal inverted repeat at positions 1469–1464, 1469–1459, and 1469–1451 and were designated pWS8M-1, -2, and -3, respectively. One analog, pWS8M-4, was designed to extend the 3'-terminal inverted repeat by altering nucleotides 1470–1472 to complement the 5'-terminal nucleotides 1–3. Because of the sequence context adjacent to the engineered *Xba*I site, the pWS8M-4 transcript was one nucleotide longer than the other synthetic transcripts, terminating with a 5'-ACCU-3', rather than 5'-GAU-3' (Fig. 12). A series of S1 nuclease protection experiments using end-labeled oligonucleotide probes complementary to the 5' or 3' terminus of the transcripts showed that all transcripts initiated correctly and that the majority of the transcripts terminated at the desired run-off transcription site.

To determine whether alterations in the 3'-terminal nucleotide sequence of a viral transcript could influence the structural conformation of the 5' terminus, Xu *et al.* (1989b) exposed the 5' end-labeled synthetic transcripts to partial digestion with the guanosine-specific T1 nuclease to test the relative sensitivity of phosphodiester bonds

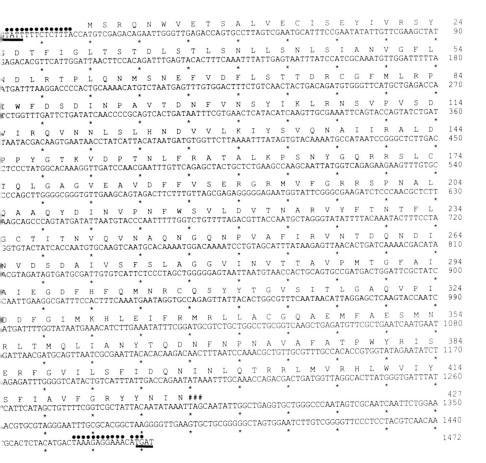

FIG. 11. Primary sequence of WTV genomic segment S8. The segment, which encodes the 42-kDa capsid protein P8 (Xu *et al.*, 1989a) is 1472 bp in length. The sequence contains one long open reading frame, which begins with the AUG codon at positions 19–21 of the positive strand, extends 427 codons, and terminates at positions 1300–1302 (hash marks). The 5′-terminal hexanucleotide and 3′-terminal tetranucleotide common to all WTV genome segments are underlined, while the inverted terminal repeats, including two additional residues predicted by computer analysis to be involved in G–U base pairing, are indicated by solid circles. Following convention, the nucleotide sequence presented is that of the coding strand of the cloned cDNA copy of the genomic dsRNA segment. Reprinted with permission from Xu *et al.* (1989b).

within the 5′-terminal domains. When the secondary structure of the synthetic transcripts was disrupted by heating at 65°C just before exposure to the nuclease, the partial digestion patterns of the different analogs and the authentic S8 sequence, pWS8, were indistinguishable

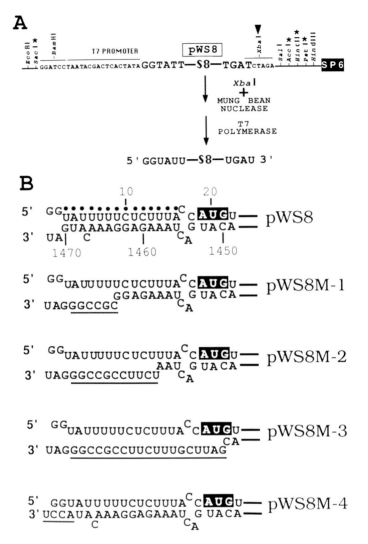

FIG. 12. Organization of transcription vectors constructed for synthesis of the WTV S8 transcript and analogs. (A) The basic elements of the transcription vectors, which include the complete cDNA copy of WTV genomic segment S8, or modified copies, fused at the coding strand 5′ terminus to the T7 polymerase promoter and at the coding strand 3′ terminus to an XbaI site. An asterisk over a restriction site indicates that such a site also occurs within the viral cDNA sequence. To generate the correct transcription run-off site, the plasmid is linearized with XbaI and digested with mung bean nuclease to remove the XbaI-generated 5′ overhang. Addition of the linearized mung bean nuclease-treated plasmid to a transcription reaction containing T7 polymerase yields transcripts with the desired 5′ and 3′ termini, as indicated in (B). Vector pWS8 specifies the authentic S8 transcript sequence. Transcripts specified by the other four vectors differ from the authentic S8 transcript sequence only at the 3′ terminus within the regions underlined. Note that the modifications introduced into vectors pWS8M-1, -2, and -3 were designed to alter the 3′-terminal inverted repeat, while the modification introduced into vector pWS8M-4 was designed to extend the 3′-terminal inverted repeat. Nucleotides are numbered as indicated in Fig. 11. Reprinted with permission from Xu et al. (1989b).

(data for pWS8 and pWS8M-4 are presented in lanes 3 and 4 of Fig. 13A).

However, when unheated transcripts were exposed to nuclease digestion, a number of significant differences were observed. For example, the three-nucleotide alteration in the 3'-terminal sequence designed to extend the 3'-terminal inverted repeat (pWS8M-4) resulted in the reduced sensitivities of G1, G2, and G21 to nuclease activity, but increased the sensitivities of G30 and G40 (lanes 5 and 6, Fig. 13A). In contrast, the alterations in the nucleotide sequence within the 3'-terminal inverted repeat (pWS8M-1, -2, and -3) resulted only in an increased sensitivity of G21, with no change in sensitivities at positions G30 and G40 (Fig. 13B; lanes 1–3 show pWS8, pWS8M-4, and pWS8M-3, respectively).

Moreover, G21 became more sensitive with increasing alterations in the 3'-terminal inverted repeat in the order pWS8 $\simeq$ pWS8M-1 < pWS8M-2 < pWS8M-3. Interestingly, the magnitude of the differences in nuclease sensitivities was reduced as the temperature at which nuclease digestion was performed was decreased. For example, the difference in the sensitivities of G30 and G40 in pWS8M-4 and pWS8 was much more apparent at 30°C than at 17°C (cf. Fig. 13A, lanes 5 and 6, and Fig. 13B, lanes 1 and 2). This result might reflect an increased stability of putative 5'-terminal–3'-terminal interactions for all of the synthetic transcripts as the temperature is lowered.

Two major conclusions were derived from the nuclease digestion studies. First, even minor alterations in the 3'-terminal sequence (e.g., a three-nucleotide extension of the 3'-terminal inverted repeat) can influence the conformation of the 5' terminus. Second, the change in 5'-terminal conformation is dependent on the type of alteration in the 3'-terminal sequence; that is, an alteration that extended the 3'-terminal inverted repeat resulted in a partial digestion pattern different from that of an alteration within the 3'-terminal inverted repeat.

In a separate series of experiments, Xu et al. (1989b) asked whether an extension or modification of the 3'-terminal inverted repeat would affect a functional property associated with the 5' terminus—specifically, the translational efficiency of the synthetic transcripts. As indicated in Fig. 14, the alteration of six, 11, and 19 nucleotides within the 3'-terminal inverted repeat (pWS8M-1, -2, and -3) increased in vitro translation by 20–40%, 50–100%, and 100–200%, respectively. Conversely, extending the 3'-terminal inverted repeat by three nucleotides (pWS8M-4) decreased translation by 15–30%. The observation that modifications in the nucleotide sequence within the terminal domain of the 3'-noncoding region in excess of 150 nucleotides downstream from the termination codon can affect the translational efficiency of a

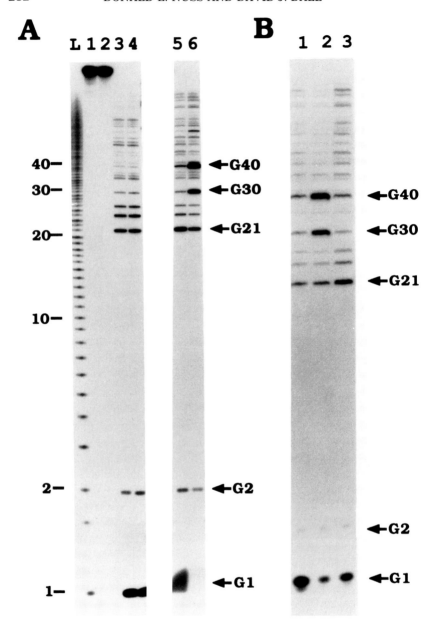

FIG. 13. Partial T1 nuclease digestion of synthetic transcripts. 5'-Terminal ^{32}P-labeled synthetic transcripts were partially digested with T1 nuclease at 30°C (A) or 17°C (B) and analyzed on a 20% polyacrylamide–7 M urea gel. (A) Lanes 1 and 2 contain undigested transcripts specified by pWS8 and pWS8M-4, respectively. Lanes 3 and 4 contain T1 digestion products of heat-treated pWS8 and pWS8M-4 transcripts, respec-

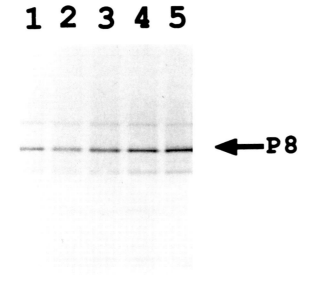

FIG. 14. *In vitro* translation of synthetic transcripts. [³⁵S]Methionine-labeled translation products specified by the WTVS8 synthetic transcripts in a reticulocyte translation system were analyzed on 12.5% polyacrylamide–sodium dodecyl sulfate gels. Lanes 1–5 were loaded with 10 μl of translation reaction mixture, programmed by transcripts specified by pWS8, pWS8M-4, pWS8M-1, pWS8M-2, and pWS8M-3, respectively. The migration position of the S8 gene product, P8, is indicated at the right. Similar results were consistently observed in seven different experiments with multiple sets of transcript preparations. Reprinted with permission from Xu *et al.* (1989b).

synthetic transcript provides additional evidence for 5'-terminal–3'-terminal interactions.

A model of the potential intramolecular interaction between the 5'- and 3'-terminal domains of the WTV transcripts was generated by computer-assisted secondary structure analysis (Xu *et al.*, 1989b). The minimal energy structures of transcripts corresponding to six of the WTV genomic segments were calculated using the program by Jacob-

---

tively. Lanes 5 and 6 contain T1 digestion products of unheated pWS8 and pWS8M-4 transcripts, respectively. The lane marked "L" contains an alkali-generated oligonucleotide ladder. (B) Lanes 1–3 contain the T1 digestion products of unheated pWS8, pWS8M-4, and pWS8M-3 transcripts, respectively. The relationship between oligonucleotide size and migration position is indicated by numbers at the left, while the numbered arrows indicate bands corresponding to positions of guanosine residues relative to the 5' end of the transcripts. Transcripts analyzed in (A) were labeled with [γ-³²P]GTP during transcription, while transcripts analyzed in (B) were labeled post-transcriptionally with [α-³²P]GTP and guanylyltransferase. Reprinted with permission from Xu *et al.* (1989b).

son *et al.* (1984), with results as illustrated in Fig. 15 for transcripts of segments S10, S8, and S6 (1172, 1472, and 1700 nucleotides, respectively). For all six of the transcripts examined in this study, the modeling program predicted a conformation in which the 5'- and 3'-terminal domains are base paired. In contrast, the program failed to predict 5'-terminal–3'-terminal interactions for transcripts consisting of random sequences generated from the WTV genomic segments.

Examination of the computer-predicted base pair interactions between the 5'- and 3'-terminal domains of the six WTV transcripts revealed several common features (Fig. 16). In each case the terminal inverted repeats and portions of the conserved terminal oligonucleotides were arranged in a base-paired panhandle structure. While the region immediately downstream from the 5'-terminal repeat was generally devoid of predicted stem–loop structures, the region immediately upstream from the 3'-terminal repeat was characterized by the presence of one or two loop or stem–loop structures consisting of as few as 8 nucleotides (S6) to as many as 134 (S5). With the exception of the S10 transcript, the presumptive AUG initiation codons were located within predicted base-paired regions. Note that, for the S8 transcript, the AUG triplet was located immediately adjacent to the projected panhandle structure.

Although the structures presented in Figs. 15 and 16 are first approximations of the secondary structural organization of the WTV transcripts, based solely on an analysis of the primary sequence through the application of thermodynamic and conformational rules specified by the modeling program, the predicted structures are surprisingly consistent with the results obtained from the *in vitro* translation and nuclease T1 accessibility analyses described above. There is clear evidence that the secondary structure within the 5'-noncoding region can affect the translational efficiency of an mRNA (Pelietier and Sonenberg, 1985; Kozak, 1986b; Lawson *et al.*, 1986; Baim and Sherman, 1988), presumably by impeding the migration of the 40 S ribosomal subunit (Kozak, 1986b). In addition, Spena *et al.* (1985) reported that a 5'-terminal–3'-terminal hybrid with a calculated stability of -20 kcal reduced the translational efficiency of A1 zien mRNA by one-half. The theoretical stability of the terminal panhandle structure in the S8 transcript from the 5' terminus through the initiation codon (position 21) is approximately -21.3 kcal. Therefore, by analogy with the A1 zien results, removal of the 3'- or 5'-terminal complementary sequences within the region preceding the initiation codon would be predicted to result in a maximal increase in translational efficiency of two- to threefold. Consistent with this prediction, sequence changes within the 3'-terminal inverted repeat which de-

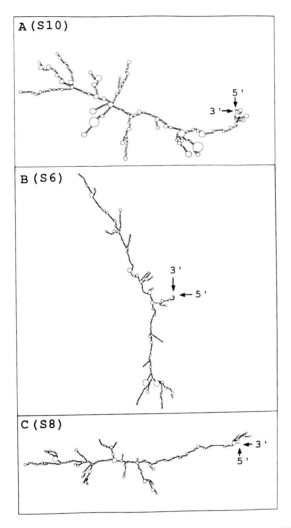

FIG. 15. Computer-generated folding patterns of WTV transcripts. Untangled folds representing the minimal energy structures of transcripts corresponding to WTV genomic segments S10, S6, and S8 are presented in panels A–C, respectively. The estimated minimal energies for the complete primary sequence of six WTV transcripts were predicted by the program by Jacobson *et al.* (1984) to be −284.1 kcal for the transcript of S12 (851 nucleotides), −383 kcal for the transcript of S10 (1172 nucleotides), −398.5 kcal for the transcript of S9 (1182 nucleotides), −514.5 kcal for the transcript of S8 (1472 nucleotides), −531.4 kcal for the transcript of S6 (1703 nucleotides) and −844.5 kcal for the transcript of S5 (2613 nucleotides). The locations of the 5′ and 3′ termini are indicated by arrows. Reprinted with permission from Xu *et al.* (1989b).

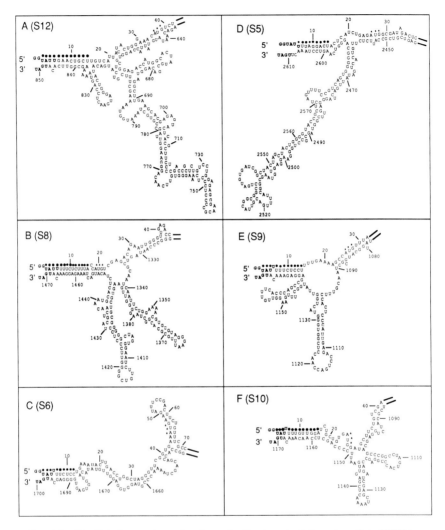

FIG. 16. Predicted secondary structures for the terminal domains of six WTV transcripts. (A–F) The predicted base pairing for the 5′- and 3′-terminal domains of six WTV transcripts corresponding to segments S12, S8, S6, S5, S9, and S10, respectively. The base-pairing assignments are based on analysis of the complete primary sequence of each transcript (see Fig. 15). Nucleotides are numbered from the 5′ terminus. The common 5′-terminal hexanucleotide and 3′-terminal tetranucleotide sequences are indicated in bold face. The solid circles indicate the position of the putative base pairing, including computer-predicted G–U base pairs, formed by the segment-specific terminal inverted repeats. Asterisks indicate the positions of the first AUG codon within the coding region of each transcript. Reprinted with permission from Xu et al. (1989b).

creased the theoretical stability of the terminal panhandle structure by 6.6 kcal (pWS8M-1), 11 kcal (pWS8M-2), and 19 kcal (pWS8M-3) increased the translational efficiency of the synthetic S8 transcript by 20–40%, 50–100%, and 100–200%, respectively. Conversely, a change in the 3'-terminal nucleotide sequence, which increased the theoretical stability of the terminal panhandle structure by 8.4 kcal (pWS8M-4), decreased translation by 15–30%. While the changes in translational efficiency that resulted from these modifications were modest, they were consistent, both qualitatively and quantitatively, with the predicted changes in the stability of the putative panhandle structure.

Due to difficulties in discriminating between primary and secondary nuclease T1 cuts (Favorova *et al.*, 1981), it is not possible to derive a detailed picture of the secondary structure involving the terminal domains based on the T1 nuclease accessibility studies. However, it is interesting to note (Fig. 16B) that 5'-terminal guanosine residues G30 and G40 of the S8 transcript are predicted to reside within loop structures which would be stabilized to present these residues in a more accessible form when the 5'-terminal–3'-terminal interactions are stabilized. This could explain the increased T1 nuclease sensitivity of these residues in the transcript with the extended 3'-terminal inverted repeat (pWS8M-4, Fig. 13). The increased accessibility of residue G21, with increasing modification of the 3'-terminal inverted repeat, is also consistent with projected destabilization of the putative panhandle structure.

As indicated previously, there is a striking similarity in the structural organization of the terminal domains of both WTV and influenza virus (Fig. 3) in the form of conserved terminal oligonucleotides and segment-specific inverted repeats. Although less apparent, a similar organization exists for other members of the Reoviridae. Within the family the conserved terminal sequences are similar for all members of a genus, but they differ among genera. For example, all members of the genus *Orthoreovirus* possess the terminal sequence (+) 5'-GCUA-----UCAUC-3' (Antczak *et al.*, 1982), while both WTV and RDV, members of the genus *Phytoreovirus,* contain the terminal sequences (+) 5'-GGU/CA--------U/CGAU-3' (Asamizu *et al.*, 1985; Uyeda *et al.*, 1987a, 1989; Omura *et al.*, 1988, 1989). For WTV and, as recently shown, for RDV (Uyeda *et al.*, 1987b; Omura *et al.*, 1988) the segment-specific inverted repeats are located immediately adjacent to the conserved terminal oligonucleotides, as also occurs in the influenza virus RNAs.

In the case of human reovirus, however, these elements are located at variable distances (three to 30 nucleotides) from the conserved terminal oligonucleotides (Joklik, 1981; the inspection of sequences is discussed by Antczak *et al.*, 1982, along with more recent cDNA-derived sequence information). It should also be noted that terminal inverted repeats have also been reported in the segmented genomic RNAs of arenaviruses (Auperin *et al.*, 1984) and bunyaviruses (Obijeski *et al.*, 1980; Cabradilla *et al.*, 1983). This similarity in terminal domain organization provides additional circumstantial evidence for the involvement of 5′-terminal–3′-terminal intramolecular interactions in the expression, replication, or assembly of segmented RNA genomes.

Hsu *et al.* (1987) recently provided experimental evidence that genomic RNAs of influenza virus A exist within virus particles in a configuration in which the 5′- and 3′-terminal inverted repeats are base paired. However, since they were able to demonstrate this interaction only with ribonucleoprotein complexes, but not purified RNAs, the authors suggested that protein–RNA interactions were required to stabilize the panhandle conformation. Xu *et al.* (1989b) provided evidence that purified WTV transcripts alone in solution can assume a conformation in which the 5′- and 3′-terminal domains interact, and they suggested that it is this conformation that is recognized and subsequently stabilized by protein–RNA interactions, leading to assembly and replication.

As an initial hypothesis to explain the apparent specificity operating during this interaction, Xu *et al.* suggested the possibility that the presence of the conserved terminal oligonucleotide sequences and the absence of a 3′-poly(A) tail could serve to specify that the RNA transcript is viral, not cellular, in origin. Further, the base-paired structure involving the terminal inverted repeats and the adjacent loop and stem–loop structures within the 3′-terminal domain were suggested to function as a structural signature, distinguishing one segment from another. In this regard D. J. Dall, J. V. Anzola, Z. Xu, and D. L. Nuss (unpublished observations) recently demonstrated the interaction of synthetic WTV transcripts with a complex present in vector cell lysates. The development of a convenient and specific binding assay, coupled with the availability of synthetic viral transcripts and their analogs, will provide the first opportunity to experimentally test these predictions. These studies are in progress.

B. Rice Dwarf Virus

The complete nucleotide sequences of three RDV genomic segments have recently been reported: segments S10 (1319 bp), S9 (1305 bp), and S8 (1424 bp). In each case sequence information was derived by analysis of cDNA clones selected from libraries generated essentially as described for WTV (Asamizu *et al.*, 1985). The sequence of RDV segment S10 was the first to be completed and was reported by two independent research groups (Uyeda *et al.*, 1987b; Omura *et al.*, 1988). Although the two nucleotide sequences differed at five positions, the deduced amino acid sequences were identical. Not unexpectedly, the overall sequence organization was similar to that described for WTV. The strand having the same polarity as the viral transcript contained one long ORF comprising 80% of the nucleotide sequence (Table V), while the 5'- and 3'-noncoding regions consisted of 26 and 234 nucleotides, respectively. In addition, the terminal nucleotides were identical to those of WTV genomic segments [i.e., (+) 5'-GGUA----------UGAU-3']. This represents the first biochemical evidence that RDV and WTV are closely related viruses.

Sequence analysis of RDV segment S9 (Uyeda *et al.*, 1989) and segment S8 (Omura *et al.*, 1989) confirmed the general organizational properties of the RDV genomic segments (Table V). However, the reported sequence composition of the terminal oligonucleotides was unexpected: (+) 5'-GGUAA-------CGAU-3' for segment S9 and (+) 5'-GGCAA-------UGAU-3' for segment S8. That is, the terminal oligonucleotide sequences for segments S9 and S8 differed from those of segment S10 and the WTV sequences at the fourth nucleotide from the (+) 3' terminus and at the third nucleotide from the (+) 5' terminus, respectively. This differs from the pattern exhibited by other reoviruses, in which at least four nucleotides at each terminus are conserved in each of the segments comprising the genome. It will be interesting to determine the extent of heterogeneity within the RDV terminal oligonucleotide sequences.

In describing the sequence of RDV segment S9, Uyeda *et al.* (1989) noted the presence of segment-specific terminal inverted repeats adjacent to the terminal oligonucleotides. Inspection of the sequences of RDV segments S10 and S8 also revealed the presence of segment-specific terminal inverted repeats similar in length and location to those described for WTV (Anzola *et al.*, 1987). The implications of these findings are discussed in Section V,B.

The product of RDV segment S8 was identified by comparing the predicted amino acid sequence with the experimentally derived oligopeptide sequences of the 43-kDa virion capsomere protein. Since the RDV nonstructural polypeptides have not been identified and formal assignment of the RDV structural polypeptides to cognate genomic segments has not been made, there was insufficient information to allow assignment of the RDV S10 and S9 sequences to specific viral polypeptides. However, as described in Section V,B, comparison of the RDV S9 and S10 sequences with the WTV sequence data base revealed

TABLE V

PROPERTIES ASSOCIATED WITH NUCLEOTIDE AND DEDUCED AMINO ACID SEQUENCES OF RDV GENOMIC SEGMENTS

Segment	Size (bp)	G + C:A + T ratio[a]	Noncoding 5'	Noncoding 3'	% Coding	Polypeptide	Size (amino acids)[a]	Molecular weight[a]
S8[b]	1424	45.5 : 54.5	23 (1.6)	141 (9.9)	88.5	Capsid	420	46,429
S9[c]	1305	46.3 : 53.7	24 (1.8)	228 (17.5)	80.7	Unassigned (structural)[e]	351	38,769
S10[d]	1319	41.3 : 58.7	26 (2.0)	234 (17.7)	80.3	Unassigned (non-structural)[f]	353	39,230

[a] Values were determined with the aid of the DNASTAR sequence analysis program.
[b] From Omura et al., 1989.
[c] From Uyeda et al., 1989.
[d] From Omura et al., 1988. This sequence differs slightly from that reported by Uyeda et al. (1987b) and by Matsumura et al. (1988).
[e] Probable assignment based on sequence similarity to WTV structural polypeptide P9.
[f] Probable assignment based on sequence similarity to WTV nonstructural polypeptide Pns11.

significant sequence similarity with WTV segments S11 and S10, respectively. Based on the coding assignments of these two WTV segments, it was predicted that RDV S9 encodes a virion structural component and that RDV S10 encodes a nonstructural polypeptide (Xu *et al.*, 1989b; Dall *et al.*, 1989).

IV. GENOME ORGANIZATION FOR MEMBERS OF THE GENUS *Fijivirus*

No nucleotide sequence information presently exists for members of the genus *Fijivirus*. However, a limited cDNA library has been generated from MRDV genomic RNAs, and several cDNA clones are currently being characterized (C. Marzachi, G. Boccardo, and D. L. Nuss, unpublished observations). Nearest-neighbor analysis has identified the MRDV 3'-terminal nucleotides as cytidine and uridine, the same as found for WTV and RDV. Comparative analysis of MRDV sequence data with that of WTV and RDV should prove to be informative with respect to the evolutionary, biological, and molecular relationships of the plant reoviruses.

V. DEFINING FUNCTIONS OF VIRALLY ENCODED POLYPEPTIDES

As mentioned in Section II,A, a genetic approach for determining functional properties of virally encoded polypeptides is not readily available for the plant reoviruses. Alternative approaches rely on the availability of cDNA clones of genomic segments and derived sequence information and include general data base searches, direct sequence comparisons with other members of the Reoviridae, and *in vitro* and *in vivo* expressions of the cloned cDNAs. Progress in the application of these methods to the identification of possible functions of WTV-encoded polypeptides is discussed in the following three sections.

A. Data Base Searches

The literature documents numerous cases in which comparison of a nucleotide or amino acid sequence with an appropriate data base has provided the first indication of a possible functional property associated with that sequence. Unfortunately, to date, comparisons of WTV nucleotide and derived amino acid sequences with sequences in GenBank have failed to provide convincing evidence of functional or evolutionary relationships. In an effort to identify the WTV genome product(s) responsible for tumor formation in infected plants, derived amino acid sequences corresponding to nine WTV genomic segments were compared with the derived amino acid sequences of other microbial genes known to be involved in plant tumor formation. These included the cytokinin and auxin biosynthetic genes from *Agrobacterium tumefaciens* and *Pseudomonas savastanoi* (Hiedekamp *et al.*, 1983; Klee *et al.*, 1984; Yamada *et al.*, 1985; Powell and Morris, 1986) and the *rol* genes of *Agrobacterium rhizogenes* (Slightom *et al.*, 1986). No convincing evidence for relationships to these sequences was found. Moreover, no sequence similarity has been detected between any WTV-encoded polypeptide and the polypeptides specified by members of other genera of Reoviridae. A similar result was reported for human reovirus (Wiener *et al.*, 1989). Although these results have been disappointing, comparison of the WTV and RDV sequence information has, as discussed in the next section, revealed a significant level of homology at both the nucleotide and amino acid levels.

B. Comparison of Plant Reovirus Sequence Information

Given the similarities exhibited by the plant-infecting reoviruses at the molecular level and the diversities exhibited at the biological level (Section II,C), it is anticipated that a comparative analysis of nucleotide and derived amino acid sequence information will eventually provide insight into a variety of biological and molecular processes, ranging from disease symptom expression in infected plants to the recognition, sorting, and replication of segmented RNA genomes. In this regard comparison of the three available RDV genomic RNA sequences with the combined WTV sequence information has revealed significant levels of sequence similarity. At the nucleotide level RDV segment S10 was found to be 54.9% similar to WTV genomic segment S10 (Fig. 17) (Xu *et al.*, 1989b), while RDV segments S9 and S8 were 53.3% and 54.1% similar to WTV segments S11 and S8, respectively (Dall *et al.*, 1989; Figs. 18 and 19).

This contrasts with the 20–25% sequence similarity that each RDV shares with unrelated WTV sequences, a level presumed to be equivalent to the 25% homology expected for random sequences. As expected, the similarities at the amino acid level were more impressive: 30.6% for RDV S10 and WTV S10, 31.9% for RDV S9 and WTV S11, and 48.3% for RDV S8 and WTV S8.

As mentioned in Section III,B, RDV S10 was found to contain the same terminal oligonucleotide sequences shared by all 12 WTV genomic segments: 5'-GGUA-------UGAU-3' (Fig. 17) (Anzola *et al.*, 1987). In addition, the coding regions of both RDV S10 and WTV S10 began and terminated at similar positions relative to the 5' terminus: resi-

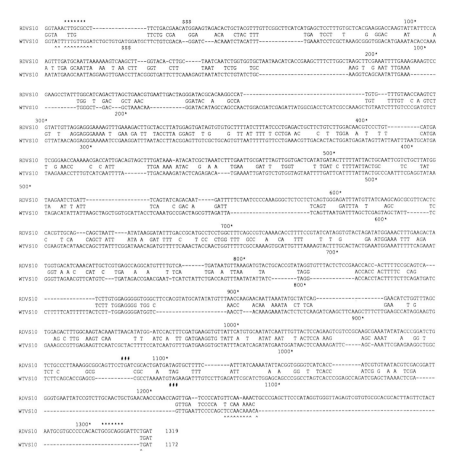

FIG. 17. The complete nucleotide sequence of RDV segment S10 (top line), aligned with the complete sequence of WTV segment S10 (bottom line). Predicted translation initiation and termination codons are indicated by $$$ and ###, respectively. Asterisks indicate the terminal inverted repeats present in the RDV sequence, while carets indicate these elements in the WTV sequence. The length (bp) of each segment is indicated at the end of sequence information. Alignment analysis was performed with the DNASTAR sequence analysis program. The WTV S10 sequence was taken from Anzola *et al.* (1989a), while the RDV S10 sequence appeared in work by Omura *et al.* (1988). Reprinted with permission from Anzola *et al.* (1989a).

dues 27–29 and 1086–1088 for RDV S10 versus residues 25–27 and 1065–1067 for WTV S10. While the overall nucleotide sequence similarity between the two segments was 54.9% with 44 gaps, several significant continuous stretches with similarities greater than 70% were located throughout the coding region. Sequence divergence was

```
         ••• ••••••••••      $$$                                                           100•
RDVS9  GGTAAAAATCGTGTGTCCTCCGTGATGGGTAAGCTCCAAGATGGAATCGCCATCAAGCGGATCAACGACGCGATTACCACTTTCAAGAATTACAAG--------CTTGGTGAACTGGAACA-
       GGTA    TC     CC C TGA  GG AA T CAAGATGGA T GC ATC  G GGAT    GACGCGAT    TTTCA AATTACA G        CTT TGA
WTVS11 GGTATTTTTCTACCTACCGCGATGAGTGGCAAAATACAAGATGGAGTTGCTATCCGGAGGATGTCGGACGCGATCCTATTTTTCACTAATTACACGAGTAGAAATCTTATTGA---------
       ^^^^^^^^^   $$$                                                               200•

                                                                                          200•
RDVS9  GGGCGGCTCAATGGCCATCAACACATTGAGTAACGTCCGCGCCCATG-------------TTGGGCTG--GCTTGGCCGGCCATCTTGCGAAATTGTTTGATACACACTTCATCCCATCTTG
         CCA C  ACAT           CGC  CAT          TTGGG  GCT G  C   TCT   GAA TTGTT GA  A ACT   TC CA  TG
WTVS11 --------------CCAGCGTGACATCACCCTTTCCACGCTA-CATACAATAAGAAGAAATTTGGGTACCTGCTGGAGCATAGCTCTATTGAA-TTGTTGGAATGAAACTAGCTCACACGCTG

                                                300•
RDVS9  GGTTCATGAAGTTTATG----------ATTGATATTGCTACTACCTGGAAAGTTGGTGCTTTCACCCTTCTGGGCAGCGTCGGTGACGAAGATCCTTTCACTGACGTTGACTTGATTTACAC
       G T ATGA TTTAT          ATT  ATTGC A        TTGG G TTT ACC T CT GG    CGG  A G GATCC TT  TGA G  G    AT T
WTVS11 GTGTGATGAGATTTATATTGGACATCGCATTTTCATTGCGAT----------TTGGAGATTTTACCATGCTAGGTGCGTGCGGCAATGTGGATCCGTTTGATGATGCCGGTCAAATATTTCT
                                                300•

                                                400•
RDVS9  TAAGACCTGCTTGCATTTGGGTCTTAAAGACAATGATTTTCTGCAATTTCCAGAAGAGTTTGCCTATGAGGCGAATTCTTTTCTAG--AAGCGCAGTCGATGAATGCTAGGGTGGACATGCTC
       AA  C TGC  G T  GG C AA GAC    TT CT     C GA  A TTTG TAT A      TTC TTT TA  AAG GCAGT   TGAA  T  GT GACATG  C
WTVS11 AAAATCATGCAAGGCTACAGGACGCAATGACTCATGCTTCCTTACCCCATCGGATAACTTTGGTTATTACCTTGTTTCATTTTTAAACAAAGAGCAGT--TGAAATGTGTGTTGACATGAAC
                                                400•

                                                500•
RDVS9  ACTGGTGTCCACAATATTGAAGATAAATATGTCTTTAGAATAGAGTCTATATCTAAGTTTTTGAAAGCTTACTATACTGCTTCAGAA--------GACGTTGCTTA----CTTGA-------
       GG  TCCACAA ATTGAAGA A TATGT  AGAAT GA TC ATA   A TT  T A    AGAA      GACGTTG    CTTGA
WTVS11 GTCGGAATCCACAACATTGAAGACATTTATGTTACAAGAATGGAATCCATAATGGAATTCATATATTATTATTATAC------AGAATCCGGCCGCGACGTGTTGTCAATTGGCTTGAAAAATTG
                                                500•

                                                600•
RDVS9  ------------CTGGATTTATAAAGCCTGACGGCTCTAAG-------------------GAGTCAATCTTGAGTGCCGAACT----CTTGAAAGCGCAGGTCACATCCGAGGTGCTACGC
                CTGGATT     AGC    GCT AAG                  GAG    TTGA         CTTGAAAG C  GTC  TC            TA
WTVS11 GAATCGGCTGATGCGGATTGGC--AGCTCAT--GCTAAAAGCCAAAAGATTAATGCGTGCAGAGATCGATTTGATCAGGCGGGAGATACTTGAAAGGACCCGTCTGTTCATCAACAATAACA
                                                600•

                                                700•
RDVS9  GTGCGTAATTTAATTACCACCAAGATTCAGCAGTACATTAATTTGTACAAGAATTCGCAGTTACCGCATTTTCGGCGAGCGGCTTTGTCCTCACCAGGATTGGCATTGTTGATGG------
       G   T TT A ACCACCA  T          AATT GT CGAAGAT                              ACA   G  TGGG    TGATG
WTVS11 GAAATTCGTTCCACGACCACCATCGTG-----------AATTAGTCCGAAGAT----------------------------ACAGAACGATCTGGGC---TGATGTAATATC
       700•

                                                800•
RDVS9  ---CGGTGTGCCCGCTGCACTCCCCACAGCCTGATACAACTGACGATGAAAGTCCCGTCCATAAGCCTGGCGCTAGTGCACCAACAG------------TGAGTAAAGGTGCTGATCAGCCA
          CGGTG   G G C   CTGA CAACT C    CCC    CAT C G    AGTGCA  AAC           TGAGTA  AG     GT
WTVS11 CGACGGTGATGTGGTAGAAGAGACGTCAACTGACCTGAGGCAACTACCTCAG------CCCAA--CATTCAACAGCACTAAGTGCAG--AACTTGACGAAGTTGATGAGTA-------TGATCATCCA
                                                800•

                                                900•
RDVS9  GA--------------------AGACGAGGAGATAATACATAAAAAGGTGATAGCTTCGAAAGATGCTCCACCTAAGGCAGTTTCTTCTTGGAAATGTAAGCGCCAGAGGTATTCCTGCCT
       A                    AGA AG AGAT T CAT AAA     TG T TT       GAT       CTA          TCT TG A ATG     CCA GGTA    T
WTVS11 AATGATGGTCTGCTCACATTTAGGGAGAGAAGAAGAGCTGCTGCATCCAAACCTTGATTCATTACTTGGATCA-----CTA-------TCTGGTGAAGATGCGTT--CCAAGGGTAAATATCATG
                                                900•      ###

                                                1000•
RDVS9  TTTTAGAAGATGATATG-------------------AGTGAGATGGACGCACCTGATGGCTTCCATGATTACTTAACGAGGGAACATGAGAACAACTTCGACTTGGCGCAGTTGGGACTCGC
       T T GAA ATG A                      AGTG   G  G C T ATG T CA          TAA               AACA CTT    GGC        T       C
WTVS11 TAATTGAAAATGGAAGAGTAGACACACTCGGTCATTAGTTGTCCAGTGGGGTCGTCATGTTTCCCAAAGGGTTGTAAT------------AACATCTTGTGGAGGGCCACTGTACCCACGTC
                                                1000•

                     ###    1100•
RDVS9  ACCCTCAGTCTGACTTTACGCTGGAGTAGATGAATGCCTCACCAACATTTGTTACTCCATAATGATGTTATCATGTCTGCATGGATTCTTCAGAATGTTGATTCGTGTATTAGGTTGCTTGT
       TCAG          AGT        CCT   CCA    ACTC
WTVS11 GTGTTCAG--------------AGT---------CCTGGTCAA-------ACTCTG-------------------------
                    1100•

       1200•                                                                    ••••••••••• ••• 1305
RDVS9  AAACAACGGAGTGGTGATCGGTAGATCCGTCTGGCGTGGAAGTTTTGACAGCGAACCTGTGTCCTAACGCGGATGCGGGAGGGTCATTAACTTCCATGTCACACGATTTATACGAT
                          GTT T  GAA                                   ACA                        GAT
WTVS11 --------------------------GTTGTCTGGTGGAAA-------------------ACA-------T--GAT
                             ^^^^^^^^^                       ^              1128
```

FIG. 18. The complete nucleotide sequence of RDV segment S9 (top line), aligned with the complete sequence of WTV segment S11. Predicted translation initiation and termination codons are indicated by $$$ and ###, respectively. The WTV S11 sequence is taken from Dall et al. (1989), while the RDV S9 sequence appeared in work by Uyeda et al. (1989).

greatest within the 3'-noncoding region; this differed in length by 127 nucleotides (234 nucleotides for RDV S10 versus 107 for WTV S10) and contained only two regions of significant homology—a 22-nucleotide stretch between RDV S10 positions 1208 and 1229 and the 3'-terminal tetraoligonucleotide. There was also little sequence similarity within the 5'-noncoding regions, with the exception of the 5'-terminal tetranucleotides.

Alignment of RDV S9 and WTV S11 nucleotide sequences revealed significant sequence similarity at the extreme 5' terminus and also within most of the coding region (Fig. 18), even though the WTV S11 coding region terminated 118 nucleotides upstream from the RDV S9 termination codon, and the RDV S9 3'-noncoding region was 60 nu-

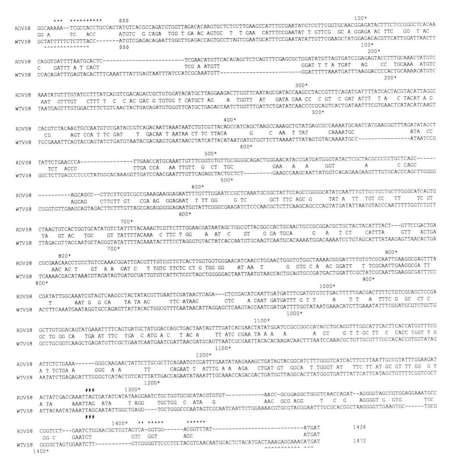

FIG. 19. The complete nucleotide sequence of RDV segment S8 (top line), aligned with the complete sequence of WTV segment S8. Predicted translation initiation and termination codons are indicated by \$\$\$ and ###, respectively. The WTV S8 sequence is taken from Xu *et al.* (1989b), while the RDV S8 sequence appeared in work by Omura *et al.* (1989).

cleotides longer than that of WTV S11. As in the comparison of RDV S10 and WTV S10, the 3′-noncoding regions of these two segments showed little sequence homology. As indicated in Section III,B, the 3′-terminal sequence of RDV S9 was 5′-CGAU-3′, rather than 5′-UGAU-3′, as found in all 12 WTV segments and in RDV S10 and RDV S8.

The sequence context downstream from the presumptive initiation codon and upstream from the presumptive termination codon of the RDV S8 and WTV S8 coding regions exhibited a significant level of similarity (Fig. 19), and, in addition, each segment contained an amber termination codon. In this case the 3'-noncoding region of the WTV segment was 32 nucleotides longer than the RDV 3'-noncoding region, and the 3'-terminal pentanucleotides were identical. However, the 5'-terminal tetraoligonucleotide sequence of RDV S8 differed from WTV and other RDV terminal sequences at the third position, having a cytosine residue in place of the uracil residue.

An additional point of potential significance concerns the terminal inverted repeats. Each of the three RDV segments contain terminal inverted repeats located at the same position, relative to the conserved terminal oligonucleotides, as those found in the related WTV segments (Fig. 20). However, they exhibit no sequence similarity. For example, WTV S10 contains a 12-bp imperfect inverted repeat rich in A + U, while RDV S10 contains a G + C-rich 7-bp imperfect inverted repeat. Segments WTV S11 and RDV S9 contain a 9-bp imperfect and a 14-bp imperfect inverted repeat, respectively, consisting of completely unrelated sequences. A similar relationship exists for WTV S8 and RDV S8. The fact that the terminal structural motif has been conserved despite divergence of the terminal nucleotide sequences provides additional evidence that the inverted repeats represent important functional elements.

The relatedness of the three pairs of RDV and WTV segments is clearly apparent on alignment of the amino acid sequences (Figs. 21–23). The derived amino acid sequences of RDV S8 and WTV S8, both capsomere proteins (Omura et al., 1989; Xu et al., 1989a), exhibited the greatest degree (i.e., 48.3%) of sequence similarity (Fig. 21). In particular, the amino- and carboxy-terminal domains are highly conserved. For example, 18 of the first 20 amino-terminal amino acids and 28 of the last 32 carboxy-terminal amino acids are identical or related. As expected from the sequence similarity, both proteins have similar predicted physical properties (Table VI). As indicated in Section III,B, the polypeptide products of RDV S10 and RDV S9 have not been assigned as structural or nonstructural viral components. However, based on the similarities between RDV S9 and WTV S11, the latter of which encodes the capsomere protein P9, and between RDV S10 and WTV S10, which encodes the nonstructural polypeptide Pns11, it is likely that RDV S9 and RDV S10 encode a structural and a nonstructural polypeptide, respectively. In this context it is interesting that the sequence diversity pattern for these two segments is much further advanced than that of RDV S8 and WTV S8. This is most apparent for the polypeptide products of RDV S9 and WTV S11, which exhibit a significant level of sequence similarity within the amino-terminal and middle portions, but little similarity within the carboxy-terminal re-

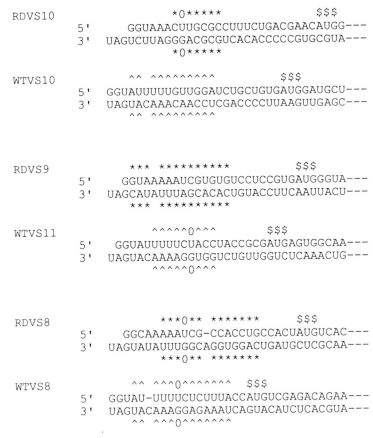

```
RDVS10                    *0*****                  $$$
              5'  GGUAAACUUGCGCCUUUCUGACGAACAUGG---
              3'  UAGUCUUAGGGACGCGUCACACCCCCGUGCGUA---
                          *0*****


WTVS10              ^^  ^^^^^^^^^          $$$
              5'  GGUAUUUUUGUUGGAUCUGCUGUGAUGGAUGCU---
              3'  UAGUACAAACAACCUCGACCCCUUAAGUUGAGC---
                   ^^  ^^^^^^^^^



RDVS9             ***  **********          $$$
              5'  GGUAAAAAUCGUGUGUCCUCCGUGAUGGGUA---
              3'  UAGCAUAUUUAGCACACUGUACCUUCAAUUACU---
                  ***  **********


WTVS11             ^^^^^0^^^             $$$
              5'  GGUAUUUUUCUACCUACCGCGAUGAGUGGCAA---
              3'  UAGUACAAAAGGUGGUCUGUUGGUCUCAAACUG---
                   ^^^^^0^^^



RDVS8            ***0** *******          $$$
              5'  GGCAAAAAUCG-CCACCUGCCACUAUGUCAC---
              3'  UAGUAUAUUUGGCAGGUGGACUGAUGCUCGCAA---
                  ***0** *******


WTVS8           ^^  ^^^0^^^^^^^^  $$$
              5'  GGUAU-UUUUCUCUUUACCAUGUCGAGACAGAA---
              3'  UAGUACAAAGGAGAAAUCAGUACAUCUCACGUA---
                   ^^  ^^^0^^^^^^^^
```

Fig. 20. Terminal nucleotide organization for related RDV and WTV genomic
RNAs. The terminal nucleotide sequences (coding strand only) are oriented to indicate
potential base-pairing interactions. The terminal inverted repeats present in the RDV
sequences are indicated by asterisks, while the locations of these elements within the
WTV sequences are indicated by carets. The symbol 0 indicates computer predicted U–G
base pairing. The locations of predicted translation initiation codons are indicated by
$$$. Note that, although there is extensive sequence diversity within the terminal
domains of the related RDV and WTV genomic segments, the terminal structural orga-
nization consisting of conserved terminal oligonucleotides and inverted terminal repeats
has been conserved.

gion. In addition, the derived amino acid sequence of RDV S9 contains
a highly negatively charged 39-residue carboxy-terminal region not
present in the WTV sequence. In spite of the rather low level of se-
quence similarity for these polypeptides, they retain similar physical
properties (Table VI).

```
                                 50*
RDVS8  MSRQMWLDTSALLEAISEYVVRCNGDTFSGLTTGDFNALSNMFTQLSVSSAGYVSDPRVPLQTMSNMFVSFITSTDRCGYMLRKTWFNSD
       MSRQ W::TSAL:E ISEY:VR. GDTF GLT:.D:::LSN:::LS::.G:::D R.PLQ:MSN FV:F:::TDRCG:MLR .WF:SD
WTVS8  MSRQNWVETSALVECISEYIVRSYGDTFIGLTSTDLSTLSNLLSNLSIANVGFLNDLRTPLQNMSNEFVDFLSTTDRCGFMLRPIWFDSD
                                 50*
       100*                           150*
RDVS8  TKPTVSDDFITTYIRPRLQVPMSDTVRQLNNLSLQPSAKPKLYERQNAIMKGLDIPYSEPIEPCKLFRSVAGQTGNIPMMGILATP---P
       .:P:V:D:F:::YI: R .VP:SD.:RQ:NNLSL::. K:Y: QNAI:::LD PY:. ::P.:LFR:.A ...N :  . L T:    :
WTVS8  INPAVTDNFVNSYIKLRNSVPVSDVIRQVNNLSLHNDVVLKIYSVQNAIIRALDPPYGTKVDPTNLFRATALKPSNYGQRRSLCTQLGAG
       100*                           150*
       200*                           250*
RDVS8  AQQQPFFVAERRRILFGIRSNAAIPAGAYQFVVPAWASVLSVTGAYVYFTNSFFGTIIAGVTATATAADAATTFT-VPTDANNLPVQTDS
       .:. .FFV:ER R::FG RS.:A::A:.Y:. VP:: SVL:VT.A VYFTN:F:G..I:.V..:A.::::::.F. V.TD.N::.V::D:
WTVS8  VEAVDFFVSERGRMVFGRRSPNALQAAQYDINVPNFWSVLDVTNARVYFTNTFLGCTITNVQVNAQNGQNPVAFIRVNTDQNDINVDSDA
       200*                           250*
       300*                           350*
RDVS8  RLSFSLGGGNINLELGVAKTGFCVAIEGEFTILANRSQAYYTLNSIT-QTPTSIDDFDVSDFLTTFLSQLRACGQYEIFSDAMDQLTNSL
       :SFSL:GG IN:. :V: TGF :AIEG:F . NR.Q:YYT  SIT .:::..IDDF.:  . L..F  :L ACGQ E:F:::M::LT .L
WTVS8  IVSFSLAGGVINVTTAVPMTGFAIAIEGDFHFQMNRCQSYYTGVSITLGAQVPIDDFGIMKHLEIFRMRLLACGQAEMFAESMNRLTMQL
       300*                           350*
                     400*
RDVS8  ITNYMDPPAIPAGLAFTSPWFRFSERARTILA--LQNVDLNIRKLIVRHLWVITSLIAVFGRYYRPN        420
       I:NY.:.. P:::AF::PW:R.SER .IL:    QN:::L:.R:L:VRHLWVI S:IAVFGRYY. N
WTVS8  IANYTQDNFNPNAVAFATPWYRISERFGVILSFIDQNINLQTRRLMVRHLWVIYSFIAVFGRYYNIN        427
                     400*
```

Fig. 21. The deduced amino acid sequence of the RDV S8-encoded protein (top line), aligned with the deduced amino acid sequence of WTV P8 (bottom line), which is encoded by WTV genomic segment S8. Conserved amino acid substitutions are denoted with a colon, while neutral substitutions are denoted with a period. The sizes of the deduced polypeptides are indicated at the end of the sequence information.

The high degree of sequence similarity exhibited by the RDV S8- and WTV S8-derived amino acid sequences suggests the operation of stringent structural constraints consistent with the role of a capsomere component and provides convincing evidence that WTV and RDV share a common origin. However, the high degree of amino acid

```
                                 50*
RDVS10  MEVDTATFVRLHHELLCAHEGPSIISKFDAIKKVKLGTLANQSGGANNITEAFLAKLRNFERKSEAYLASDLAERELTRDTHKAIVFVTK
        M:.:.: ...LH E:L . :G :I .K::AI:K::L    :.::::.NI: :  :.: ::: ...:Y:AS:L::R: .  . KA::FV.
WTVS10  MDASVDRITNLHFEILAKAGGHEIHQKYEAIRKLNL----TGDSSKSNISVSARSAILKWADAKQGYIASQLDDRDYGDLIAKAVIFVPM
                                 50*
        100*                          150*
RDVS10  SVLLGGKSLKDLLPYGVIVCAFIFIPETASVLDNVPVMIGNQKRPLTVALIKYIAKSLNCDLVGDSYDTFYYCNSSAYGKNLISVSDNDF
        SV: GGK: KDL:PYGV:. .:IF:PET ::LD:: : : ::K:PL:. L:. I ::::.D:.G:::D:FYYC. S Y.:::I.::.:
WTVS10  SVITGGKNPKDLIPYGVVAAVLIFVEPTLTLLDEIVINLMHDKKPLSSILLTKILRDMKIDVCGSNFDSFYYCPISRYNRHIIKLAGALP
        100*                          150*
        200*                          250*
RDVS10  SNPQRALLSVGDLCYQAARSLHVAAANYIRIFDRMPPGFQPSKHLFRIIGVLDMETLKTMVTSNIAREPGMFCHDNVKDVLHRIGVYSPN
        . P. . LSV.DL  A ::H .  :  ::F ::P.GF.P. H ::::  .:ME.:.. V : :: P. F. ::  ::L.R.. :S.:
WTVS10  QMPTSVRLSVNDLARVAISEVHNQLISDKQMFFKLPTGFSPKVHCLKVLCTTEMEIFQKWVRTFMSDRPNEFIYSDQFNILSRTTYFSSD
        200*                          250*
        300*                          350*
RDVS10  HHFSAVILWRGWASTYAYMFNQEQLNMLSGTSGLAGDFGKYKLTYGSTFDEGVIHVQYQFVTPEVVRKRNIYPDLSALKGGSS        353
        . FS  .LWRGW STY. :::Q:Q : : :: G : .:    T::S.FDEG.I ::Y:::TP. . :  ::  ::  :: .:
WTVS10  DPFSFFTLWRGW-STYKEILSQDQASSFLEAIGSGKPLRSSIATFPSMFDEGAIYIRYEWITPKDSANSKKAGSSAPSAPKM          347
                                 300*
```

Fig. 22. The deduced amino acid sequence of the RDV S10-encoded protein (top line), aligned with the deduced amino acid sequence of WTV Pns11 (bottom line), which is encoded by WTV genomic segment S10. Conserved amino acid substitutions are denoted with a colon, while neutral substitutions are denoted with a period. Reprinted with permission from Anzola et al. (1989b).

```
                                     50*
RDVS9   MGKLQDGIAIKRINDAITTFKNYKLGELEQGGSMAINTLSNVRAHVGLAWPAILRNCLIHTSSHLGFMKFMIDIATTWKVGAFTLLGSVG
        GK:QDG:AI:R::DAI  F.NY.  :L :. :::::TL.::R ::G  W.  L NC   TSSH G M:F::DIA : : G:FT:LG:.G
WTVS11  MSGKIQDGVAIRRMSDAILFFTNYTSRNLIDQRDITLSTLHTIRRNLGTCWSIALLNCWNETSSHAGVMRFILDIAFSLRFGDFTMLGAQG
                                     50*
               100*                          150*
RDVS9   DEDPFTDVDLIYTKTCLHLGLKDNDFLQFPEEFAYEANSFLEAQSMNARVDMLTGVHNIEDKYVFRIESISKFLKAYYTASE--DVAYLTG
        : DPF.D.. I: K:C   G :D: FL. .::F:Y  SFL:.:.:: VDM .G:HNIED YV R:ESI .F: YYT:S:   V::L.
WTVS11  NVDPFDDAGQIFLKSCKATGRNDSQFLTPSDNFGYYLVSFLNKEQLKCVVDMNVGIHNIEDIYVTRMESIMEFIYYYYTESGRDVVNWLEK
               100*                          150*
                   200*                          250*
RDVS9   FIKPDGSKESILSAELLKAQVTSEVLRVRNLITTKIQQYINLYEDSQLPHFRRAALSYTQDWDVDGGVPAALPQPDTTDDESPVTKPGASA
        : .:D:: ::   .:. L  :.. :::R R:::. :.: :IN   :S  .H R . Y . W:   : ::.::::.:T.:..S:  ..: SA
WTVS11  LESADAGLAAHAKSKRLM-RAEIDLIR-REILE-RTRLFINNNRNSFHDHHRELVRRYRTIWADVISDGDVVEETSTEATTSAQHSTALSA
                   200*                          250*
                       300*                          350*
RDVS9   PTVSKGA-DQPEDEEIIHKKVDASKDAPPKAVSSGNVSARGIPAFLEDDMSEMDAPDGFHDYLTREHENNFDLAQLGLAPSV        351
        : : .: D:P:D: :. :: :::.:: ...::S : ::    :
WTVS11  ELDEVDEYDHPNDGLLTFRREEDAASNLDSLLGSLSGEDAFQG                                         313
                       300*
```

FIG. 23. The deduced amino acid sequence of the RDV S9-encoded protein (top line), aligned with the deduced amino acid sequence of WTV P9 (bottom line), which is encoded by WTV genomic segment S11. Conserved amino acid substitutions are denoted with a colon, while neutral substitutions are denoted with a period. Reprinted with permission from Dall *et al.* (1989).

sequence divergence exhibited by the two other related segments suggests that divergence was an evolutionarily distant event, a proposition consistent with the different biological properties of the two viruses and with the apparently high degree of WTV genomic stability noted in Section III,A,4.

It is interesting to speculate briefly on possible evolutionary paths that could have led to such divergence. It is clear that the relationship between reoviruses and insects is both close and well developed; four of the six virus genera are closely associated with invertebrates as either definitive or vector hosts, and the relationship between each virus and its insect vector is one that would usually be regarded as highly evolved. Furthermore, WTV transcripts possess the 2'-O-methylated penultimate 5' nucleotides usually associated with mRNAs of insects and insect viruses (Banerjee, 1980), and at least one known insect reovirus (i.e., leafhopper A virus) is similar to plant reoviruses in many respects (Boccardo *et al.*, 1980), but incapable of replication in vegetative hosts.

On these grounds it is tempting to suggest that WTV and RDV evolved from a common insect virus ancestor which adapted to replicate in plant species on which the insect host fed. This development would have set the stage for accelerated divergence of the virus population as insect–plant interactions evolved, leading to the different biological properties currently exhibited by these two related viruses.

TABLE VI

COMPARISON OF THE PHYSICAL PROPERTIES OF RELATED WTV- AND RDV-ENCODED POLYPEPTIDES

Related segments	Size (amino acids)	Molecular weight[a]	Amino acids[b]				Charge at		Similarity (%)[b]	
			Basic	Acidic	Hydrophobic	Polar	pI[b]	pH 7.0[b]	Nucleotides	Amino acids
RDV S8	420	46,429	32	32	159	137	6.94	−0.06		
WTV S8	427	48,068	32	35	167	140	6.02	−2.58	54.1	48.3
RDV S9	351	38,769	32	52	125	85	4.73	−18.44		
WTV S11	313	35,611	32	48	109	86	4.88	−14.72	53.3	31.9
RDV S10	353	39,230	37	33	132	95	8.38	5.57		
WTV S10	347	38,971	40	36	128	93	8.54	4.96	54.9	30.6

[a] Molecular weights are based on deduced amino acid sequences, rather than relative gel migrations.
[b] Values were determined with the aid of the DNASTAR sequence analysis program.

C. Expression of Cloned Genomic Segments

Cultured insect vector cells have provided a valuable experimental system for examining the details of plant reovirus gene expression (Nuss and Peterson, 1980; Peterson and Nuss, 1985, 1986). However, the availability of cDNA clones which direct independent expression of individual gene products, either via synthetic transcripts in cell-free translation systems or *in vivo* in microorganisms or transgenic plants, provides a new dimension for study of the functional properties of viral gene products. While such studies are in the early stages of development, they have already proved useful in the assignment of viral gene products to cognate genomic segments (Xu *et al.*, 1989a). Additional *in vitro* studies with WTV cDNA clones, now in progress, include an examination of protein–protein and protein–RNA interactions. By expressing combinations of cDNA clones of genomic segments, we aim to determine which viral polypeptides interact, and in what specific order. Moreover, it might be possible to reconstitute specific RNA binding intermediates *in vitro*. This development would provide an opportunity to examine detailed aspects of protein–RNA interactions by site-directed mutagenesis.

Progress has also been made with *in vivo* expression of WTV cDNA clones. Nine WTV gene products have been fused to β-galactosidase and expressed in *E. coli* (J. V. Anzola, Z. Xu, and D. L. Nuss, unpublished observations). Monospecific antisera to several of the fusion products, including two nonstructural polypeptides, have recently been obtained, and it is anticipated that these will have wide application in future studies of WTV molecular biology.

Studies using transgenic plants have also been initiated, with the aim of identifying the WTV gene products responsible for disease symptom expression. We have chosen to begin this work using tobacco (*Nicotiana tabacum*), a host which is amenable to regeneration (Horsch *et al.*, 1988), shows clear symptom development when infected with WTV (Selsky, 1961), and in which expression of foreign genes from the bacterial plant pathogens *A. tumefaciens* and *A. rhizogenes* cause marked phenotypic changes (Schmulling *et al.*, 1988; Medford *et al.*, 1989). To date, cDNAs derived from each WTV segment encoding a nonstructural polypeptide have been individually subcloned into a plant expression vector (i.e., pBI121) (Jefferson *et al.*, 1987), and a series of transconjugant *A. tumefaciens* strains has been developed. Fertile kanamycin-resistant transgenic plants containing the gene encoding WTV Pns7 (see Table IV) have been grown, and their progeny are now being evaluated for gene expression and symptom development (D. J. Dall, Z. Xu, and D. L. Nuss, unpublished observations). It is anticipated that the construction of plants containing other viral genes will allow crossing between transgenic lines and enable us to determine the individual and interactive gene functions of WTV.

VI. Concluding Remarks

Our laboratory was originally attracted to the plant reoviruses, and WTV in particular, because their properties offered unique opportunities for studying a number of the fundamental biological and molecular processes outlined in this chapter. However, it soon became apparent that the full potential of this virus group as an experimental system could not be attained without the aid of a bank of molecular reagents and a basic understanding of the genome organization. Consequently, over the past few years we have concentrated on examination of the structural and functional properties of the WTV genome. In the process we have gained insight into the mechanisms involved in the recognition, sorting, and packaging of a segmented RNA genome—information which we anticipate will have general application for understanding protein–RNA interactions in other viral and cellular systems. In addition, we now have access to molecular reagents which should be useful in probing biological processes ranging from disease symptom expression in infected plants to viral persistence. Unfortunately, the number of research groups working on plant reoviruses is relatively small. It is hoped that the recent advances in the molecular biology of this virus group, as described in this chapter, will interest other molecular biologists and virologists in the study of plant reoviruses.

Acknowledgments

This review could not have been written without the talented efforts and hard work of John Anzola, Tetsuya Asamizu, Andrew Peterson, and Zhengkai Xu. The authors are also grateful to I. Uyeda, E. Shikata, T. Omura, and C. C. Chen for providing preprints of their work prior to publication.

References

Antczak, J. B., Chimelo, R., Pickup, D. J., and Joklik, W. K. (1982). *Virology* 121, 307–319.
Anzola, J. V., Xu, Z., Asamizu, T., and Nuss, D. L. (1987). *Proc. Natl. Acad. Sci. U.S.A.* 84, 8301–8305.
Anzola, J. V., Dall, D. J., Xu, Z., and Nuss, D. L. (1989a). *Virology* 171, 222–228.
Anzola, J. V., Xu, Z., and Nuss, D. L. (1989b). *Nucleic Acids Res.* 17, 3300.
Asamizu, T., Summers, D., Motika, M. B., Anzola, J. V., and Nuss, D. L. (1985). *Virology* 144, 398–409.

Auperin, D. D., Romanowski, V., Galinski, M., and Bishop, D. H. L. (1984). *J. Virol.* **52**, 897–904.

Baim, S. B., and Sherman, F. (1988). *Mol. Cell. Biol.* **8**, 1591–1601.

Banerjee, A. K. (1980). *Microbiol. Rev.* **44**, 175–205.

Bartlett, J. A., and Joklik, W. K. (1988). *Virology* **167**, 31–37.

Black, L. M. (1965). *In* "Encyclopedia of Plant Physiology" (W. Ruhland, ed.), pp. 236–266. Springer-Verlag, Berlin.

Black, L. M. (1972). *Prog. Exp. Tumor Res.* **15**, 110–137.

Black, L. M. (1979). *Adv. Virus Res.* **25**, 191–271.

Black, D. R., and Knight, C. A. (1970). *J. Virol.* **6**, 194–198.

Boccardo, G., and Milne, R. G. (1975). *Virology* **68**, 79–85.

Boccardo, G., and Milne, R. G. (1980). *CMI/AAB Descriptions of Plant Viruses* **217**.

Boccardo, G., and Milne, R. G. (1984). *CMI/AAB Descriptions of Plant Viruses* **294**.

Boccardo, G., Hatta, T., Francki, R. I. B., and Grivell, C. J. (1980). *Virology* **100**, 300–313.

Cabradilla, C. D., Holloway, B. P., and Obijeski, J. F. (1983). *Virology* **128**, 463–468.

Cashdollar, L. W., Esparza, J., Hudson, G. R., Chimelo, R., Lee, P. W. K., and Joklik, W. K. (1982). *Proc. Natl. Acad. Sci. U.S.A.* **79**, 7644–7648.

Chen, C. C., Chen, M. J., and Chiu, R. J. (1986). *Chih Wu Pao Hu Hsueh Hui Hui K'an* **28**, 371–381 (in Chinese).

Chen, C. C., Chen, M. J., Chiu, R. J., and Hsu, H. T. (1989a). *Phytopathology* **79**, 235–241.

Chen, C. C., Hsu, Y. H., Chen, M. J., and Chiu, R. J. (1989b). *Intervirology* **30**, 278–284.

Dall, D. J., Anzola, J. V., Xu, Z., and Nuss, D. L. (1989). *Nucleic Acids Res.* **17**, 3599.

DePolo, N. J., Giachetti, C., and Holland, J. J. (1987). *J. Virol.* **61**, 454–464.

Favali, M. A., Bassi, M., and Appiano, A. (1974). *J. Gen. Virol.* **24**, 563–565.

Favorova, O. O., Fasiolo, F., Keith, G., Vassilenko, S. K., and Ebel, J. P. (1981). *Biochemistry* **20**, 1006–1011.

Francki, R. I. B., and Boccardo, G. (1983). *In* "The Reoviridae" (W. K. Joklik, ed.), pp. 505–563. Plenum, New York.

Francki, R. I. B., Milne, R. G., and Hatta, T. (1985). *In* "An Atlas of Plant Viruses," pp. 47–72. CRC Press, Boca Raton, Florida.

Fukushi, T. (1969). *In* "Viruses, Vectors and Vegetation" (K. Maramorosch, ed.), pp. 279–301. Wiley (Interscience), New York.

Furuichi, Y., and Miura, K. (1975). *Nature (London)* **253**, 374–375.

Furuichi, Y., Morgan, M. A., Muthukrishnan, S., and Shatkin, A. J. (1975a). *Proc. Natl. Acad. Sci. U.S.A.* **72**, 362–366.

Furuichi, Y., Muthukrishnan, S., and Shatkin, A. J. (1975b). *Proc. Natl. Acad. Sci. U.S.A.* **72**, 742–745.

Garnier, J., Osguthorpe, D. J., and Robson, B. (1978). *J. Mol. Biol.* **120**, 97–120.

Grylls, N. E. (1979). *In* "Leafhopper Vectors and Plant Disease Agents" (K. Maramorosch and K. F. Harris, eds.), pp. 179–214. Academic Press, New York.

Hagiwara, K., Minobe, Y., Nozu, Y., Hibino, H., Kimura, I., and Omura, T. (1986). *J. Gen. Virol.* **67**, 1711–1715.

Harpaz, I. (1972). "Maize Rough Dwarf: A Planthopper Virus Disease Affecting Maize, Rice, Small Grain and Grasses." Israel Universities Press, Jerusalem.

Hatta, T., and Francki, R. I. B. (1977). *Virology* **76**, 797–807.

Hatta, T., and Francki, R. I. B. (1981). *Physiol. Plant Pathol.* **19**, 337–346.

Heidekamp, F., Dirkse, W. G., Hitle, J., and van Ormondt, H. (1983). *Nucleic Acids Res.* **11**, 6211–6223.

Hirumi, H., Granados, R. R., and Maramorosch, K. (1967). *J. Virol.* **1**, 430–444.

Horsch, R. B., Fry, J., Hoffmann, N., Neidermeyer, J., Rogers, S. G., and Fraley, R. T.

(1988). *In* "Plant Molecular Biology Manual" (S. B. Gelvin and R. A. Schilperoort, eds.), pp. A5:1–9. Kluwer, Dordrecht, The Netherlands.

Hsu, M.-T., Parvin, J. D., Gupta, S., Krystal, M., and Palese, P. (1987). *Proc. Natl. Acad. Sci. U.S.A.* **84,** 8140–8144.

Iida, T. T., Shinkai, A., and Kimura, I. (1972). *CMI/AAB Descriptions of Plant Viruses* **102.**

Ikegami, M., and Francki, R. I. B. (1976). *Virology* **70,** 292–300.

Jacobson, A. B., Good, L., Simonetti, J., and Zuker, M. (1984). *Nucleic Acids Res.* **12,** 45–52.

Jefferson, R. A., Kavanagh, T. A., and Bevan, M. W. (1987). *EMBO J.* **6,** 3901–3907.

Joklik, W. K. (1974). *Compr. Virol.* **2,** 231–334.

Joklik, W. K. (1981). *Microbiol. Rev.* **45,** 483–501.

Joklik, W. K. (ed.) (1983). "The Reoviridae," pp. 1–563. Plenum, New York.

Kawano, S., Uyeda, I., and Shikata, E. (1984). *J. Fac. Agric., Hokkaido Univ.* **61,** 408–418.

Kimura, I. (1986). *J. Gen. Virol.* **67,** 2119–2124.

Kimura, I., Minobe, Y., and Omura, T. (1987). *J. Gen. Virol.* **68,** 3211–3215.

Klee, H., Montoya, A., Horodyski, F., Lichtenstein, C., Garfinkel, D., Fuller, S., Flores, C., Peschon, J., Nester, E., and Gordon, M. (1984). *Proc. Natl. Acad. Sci. U.S.A.* **81,** 1728–1732.

Kozak, M. (1986a). *Cell (Cambridge, Mass.)* **44,** 283–292.

Kozak, M. (1986b). *Proc. Natl. Acad. Sci. U.S.A.* **83,** 2850–2854.

Lawson, T. G., Ray, B. K., Dodds, J. T., Grifo, J. A., Abramson, R. D., Merrick, W. C., Betsch, D. F., Weith, H. L., and Thach, R. E. (1986). *J. Biol. Chem.* **261,** 13979–13989.

Lee, S. Y., Uyeda, I., and Shikata, E. (1987). *Intervirology* **27,** 189–195.

Levis, R., Weiss, B. G., Tsiang, M., Huang, H., and Schlessinger, S. (1986). *Cell (Cambridge, Mass.)* **44,** 137–145.

Li, J., Scheibel, P. P., Keene, J. D., and Joklik, W. K. (1980). *Virology* **105,** 282–286.

Maramorosch, K. (1950). *Phytopathology* **40,** 1071–1093.

Maramorosch, K. (1975). *In* "Invertebrate Immunity" (K. Maramorosch and R. E. Shope, eds.), pp. 49–53. Academic Press, New York.

Matsumura, T., Uyeda, I., Sano, T., and Shikata, E. (1988). *Hokkaido Daigaku Nogakubu Hobun Kiyo* **16,** 205–211 (in Japanese).

Matsuoka, M., Minobe, Y., and Omura, T. (1985). *Phytopathology* **75,** 1125–1127.

Matthews, R. E. F. (1982). *Intervirology* **17,** 1–160.

Medford, J. I., Horgan, R., El-Sawi, Z., and Klee, H. J. (1989). *Plant Cell* **1,** 403–413.

Milne, R. G. (1980). *Intervirology* **14,** 331–336.

Milne, R. G., Conti, M., and Lisa, V. (1973). *Virology* **53,** 130–141.

Milne, R. G., Boccardo, G., and Ling, K. C. (1982). *CMI/AAB Descriptions of Plant Viruses* **248.**

Mizuno, A., Sano, T., Fujii, H., Miura, K., and Yazaki, K. (1986). *J. Gen. Virol.* **67,** 2749–2755.

Murray, M. G., Hoffman, L. M., and Jarvis, N. P. (1983). *Plant Mol. Biol.* **2,** 75–84.

Nakata, M., Fukunaga, K., and Suzuki, N. (1978). *Nippon Shokubutsu Byori Gakkaiho* **44,** 288–296 (in Japanese).

Nayak, D. P., Chambers, T. M., and Akkina, R. K. (1985). *Curr. Top. Microbiol. Immunol.* **114,** 103–151.

Nuss, D. L. (1983a). *In* "Double Stranded RNA Viruses" (D. H. L. Bishop and R. W. Compans, eds.), pp. 415–423. Elsevier, Amsterdam.

Nuss, D. L. (1983b). *In* "Plant Infectious Agents: Viruses, Viroids, Virusoids, and Satellites" (H. D. Robertson, S. H. Howell, M. Zaitlin, and R. L. Malumberg, eds.) pp. 111–116. Cold Spring Harbor Lab., Cold Spring Harbor, New York.

Nuss, D. L. (1984). *Adv. Virus Res.* **29,** 57–93.
Nuss, D. L., and Peterson, A. J. (1980). *J. Virol.* **34,** 532–541.
Nuss, D. L., and Peterson, A. J. (1981a). *J. Virol.* **39,** 954–957.
Nuss, D. L., and Peterson, A. J. (1981b). *Virology* **114,** 399–404.
Nuss, D. L., and Summers, D. (1984). *Virology* **133,** 276–288.
Obijeski, J. F., McCauley, J., and Skekel, J. J. (1980). *Nucleic Acids Res.* **8,** 2431–2438.
Ofori, F. A., and Francki, R. I. B. (1983). *Ann. Appl. Biol.* **103,** 185–189.
Ofori, F. A., and Francki, R. I. B. (1985). *Virology* **144,** 152–157.
Omura, T., Morinaka, T., Inoue, H., and Saito, Y. (1982). *Phytopathology* **72,** 1246–1249.
Omura, T., Minobe, Y., Matsuoka, M., Nozu, Y., Tsuchizaki, T., and Saito, Y. (1985). *J. Gen. Virol.* **66,** 811–815.
Omura, T., Minobe, Y., and Tsuchizaki, T. (1988). *J. Gen. Virol.* **69,** 227–231.
Omura, T., Ishikawa, K., Hirano, H., Ugaki, M., Minobe, Y., Tsuchizaki, T., and Kato, H. (1989). *J. Gen. Virol.* **70,** 2759–2764.
Pelietier, J., and Sonenberg, N. (1985). *Cell (Cambridge, Mass.)* **40,** 515–526.
Peterson, A. J., and Nuss, D. L. (1985). *J. Virol.* **56,** 620–624.
Peterson, A. J., and Nuss, D. L. (1986). *J. Virol.* **59,** 195–202.
Podgwaite, J. D., and Mazzone, H. M. (1986). *Adv. Virus Res.* **31,** 293–320.
Powell, G. K., and Morris, R. D. (1986). *Nucleic Acids Res.* **14,** 2555–2565.
Re, G. G., and Kingsbury, D. W. (1986). *J. Virol.* **58,** 578–582.
Reddy, D. V. R., and Black, L. M. (1974). *Virology* **61,** 458–473.
Reddy, D. V. R., and Black, L. M. (1977). *Virology* **80,** 336–346.
Reddy, D. V. R., and MacLeod, R. (1976). *Virology* **70,** 274–282.
Rhodes, D. P., Reddy, D. V. R., MacLeod, R., Black, L. M., and Banerjee, A. K. (1977). *Virology* **76,** 554–559.
Schmulling, T., Schell, J., and Spena, A. (1988). *EMBO J.* **7,** 2621–2629.
Selsky, M. I. (1961). *Phytopathology* **51,** 581–582.
Shikata, E. (1981). *In* "Handbook of Plant Virus Infections: Comparative Diagnosis" (E. Kurstak, ed.), pp. 423–451. Elsevier, Amsterdam.
Slightom, J. L., Durand-Tardif, M., Jouanin, L., and Tepfer, D. (1986). *J. Biol. Chem.* **261,** 108–121.
Spena, A., Krause, E., and Dobberstein, B. (1985). *EMBO J.* **4,** 2153–2158.
Spencer, E., and Garcia, B. I. (1984). *J. Virol.* **52,** 188–197.
Stoeckle, M. Y., Shaw, M. W., and Chapin, P. W. (1987). *Proc. Natl. Acad. Sci. U.S.A.* **84,** 2703–2707.
Tyler, K. L., and Fields, B. N. (1985a). *In* "Virology" (B. N. Fields, D. M. Knipe, R. M. Chanock, J. L. Melnick, B. Roizman, and R. E. Shope, eds.), pp. 817–821. Raven, New York.
Tyler, K. L., and Fields, B. N. (1985b). *In* "Virology" (B. N. Fields, D. M. Knipe, R. M. Chanock, J. L. Melnick, B. Roizman, and R. E. Shope, eds.), pp. 823–862. Raven, New York.
Uyeda, I., and Shikata, E. (1984). *Virus Res.* **1,** 527–532.
Uyeda, I., Lee, S. Y., Yoshimoto, H., and Shikata, E. (1987a). *Nippon Shokobutsu Byori Gakkaiho* **53,** 60–62 (in Japanese).
Uyeda, I., Matsumura, T., Sano, T., Oshima, K., and Shikata, E. (1987b). *Proc. Jpn. Acad. Ser. B* **63,** 227–230.
Uyeda, I., Kudo, H., Takahashi, T., Sano, T., Ohshima, K., Matsumura, T., and Shikata, E. (1989). *J. Gen. Virol.* **70,** 1297–1300.
Wiener, J. R., and Joklik, W. K. (1987). *Virology* **161,** 332–339.
Wiener, J. R., and Joklik, W. K. (1988). *Virology* **163,** 603–613.
Wiener, J. R., Bartlett, J. A., and Joklik, W. (1989). *Virology* **169,** 293–304.

306 DONALD L. NUSS AND DAVID J. DALL

Xu, Z., Anzola, J. V., and Nuss, D. L. (1987). *DNA* **6**, 505–513.
Xu, Z., Anzola, J. V., and Nuss, D. L. (1989a). *Virology* **168**, 73–78.
Xu, Z., Anzola, J. V., Nalin, C. M., and Nuss, D. L. (1989b). *Virology* **170**, 511–522.
Yamada, T., Palm, C., Brooks, B., and Kosuge, T. (1985). *Proc. Natl. Acad. Sci. U.S.A.* **82**, 6522–6526.
Yazaki, K., and Miura, K. (1980). *Virology* **105**, 467–479.
Yokoyama, M., Nozu, Y., Hashimoto, J., and Omura, T. (1984). *J. Gen. Virol.* **65**, 533–538.
Zarbl, H., and Millward, S. (1983). *In* "The Reoviridae" (W. K. Joklik, ed.), pp. 107–196. Plenum, New York.

REGULATION OF
TOBAMOVIRUS GENE EXPRESSION

William O. Dawson and Kirsi M. Lehto

Department of Plant Pathology
University of California, Riverside
Riverside, California 92521

I. Introduction

The tobamoviruses make up a group of plant viruses whose type member is tobacco mosaic virus (TMV) (see Gibbs, 1977). The group is characterized by virions, which are straight tubes of approximately 300 × 18 nm with a 4-nm-diameter hollow canal, made up of about 2000 units of a single structural protein surrounding one molecule of single-stranded RNA of approximately 2×10^6 Da. Tobamoviruses are widespread throughout the world, generally have wide host ranges, and cause substantial crop losses. No natural vectors have been identified, but these viruses are easily transmitted mechanically and move quickly through crops that are handled; sometimes they are also spread in the soil.

Tobamoviruses infect almost all cells within the plant, reaching high titers. They replicate in the cytoplasm, but virions are often found in chloroplasts and other organelles. The viruses usually cause disease by preventing proper chloroplast development, resulting in leaves with a mosaic pattern of light and dark green on stunted plants. For detailed reviews of the history, members, virion structure, replication, epidemiology, and cytopathological effects of this virus group, see Volume 2 of

307

The Plant Viruses, edited by Van Regenmortel and Frankel-Conrat (1986).
There are numerous viruses within the tobamovirus group (Gibbs, 1986). Although this is one of the best-studied virus groups, the taxonomic relationships among viral strains or distinct viruses are not always clear. The best-examined tobamoviruses often are referred to as strains of TMV, but they are sufficiently different that in other virus groups they would be considered distinct viruses. This situation is compounded by the historical tendency of plant virologists to consider all 300-nm viruses that are unusually stable and have a wide host range to be strains of TMV.

In this chapter we discuss TMV strains U1, OM, L, CGMMV, O, and Cc. The U1 and OM strains are the American and Japanese isolates of what is referred to as the common, *vulgare,* or type strain of TMV. Their sequences differ by only a few nucleotides (Meshi *et al.,* 1982b). The L strain, tomato strain, or tomato mosaic virus (Brunt, 1986) is a closely related virus that is approximately 80% similar to TMV-U1 at the nucleotide level (Ohno *et al.,* 1984). This virus is characterized by having a distinct host range and symptomatology, although most of its host range overlaps that of the common strain. Thus, TMV-L should be considered a strain of the type virus.

Cucumber green mottle mosaic virus (CGMMV) (Okada, 1986) has a host range consisting primarily of cucurbits and differs enough to be considered a separate tobamovirus. Strain O is a field isolate that we (M. E. Hilf and W. O. Dawson, unpublished observations) obtained from orchids. Its characteristics overlap those described for *Odontoglossum* ringspot virus and TMV-O (Edwardson and Zettler, 1986). It has a host range that includes monocotyledonous species and is sufficiently different from the type strain to be considered a separate tobamovirus. The cowpea (Cc) strain, or sunn-hemp mosaic virus (Varma, 1986), is the most distantly related virus in this group (Gibbs, 1986). However, because of custom we refer here to these viruses as strains of TMV.

A. Genome Organization

The genomes of tobamoviruses consist of one molecule of plus-sense RNA of approximately 6400 nucleotides. The entire sequences of the genomes of two tobamoviruses (i.e., TMV-U1, Goelet *et al.,* 1982; TMV-L, Ohno *et al.,* 1984) and parts of several others have been determined: Cc (Meshi *et al.,* 1981, 1982a), OM (Meshi *et al.,* 1982b), CGMMV (Meshi *et al.,* 1983; Saito *et al.,* 1988), and TMV-O (M. E. Hilf and W. O. Dawson, unpublished observations). The tobamovirus genomes examined have four large open reading frames (ORFs) (Fig. 1). The first, which begins near nucleotide 70 and is terminated near nucleotide 3420, with an amber stop codon (UAG), encodes an approximately 126K protein. Read-through of this amber stop codon and termination with a stop codon (UAA) near residue 4920 results in a 183K protein.

Additionally, within this read-through region in the same ORF, there is a start codon that could encode a 54K protein. The 183K ORF is followed by the 30K (nucleotides 4900–5700) and 17.5K (nucleotides 5700–6200) ORFs. The 183K/54K ORF overlaps the 30K ORF by eight to 23 nucleotides in the different tobamoviruses. The tobamoviruses differ by whether the 30K and 17.5K (i.e., coat protein) ORFs overlap. These ORFs do not overlap in the U1, OM, L, and O strains, having two or three nucleotides between the ORFs, while those of strains Cc and CGMMV overlap by 26 nucleotides. In all tobamoviruses the ORFs are followed by approximately 200 nucleotides that are not translated.

The 126K/183K proteins are thought to be translated from genomic RNAs. These proteins are produced by *in vitro* translation of virion RNA, whereas the other proteins are not (Knowland, 1974). The 30K and coat proteins are translated from subgenomic mRNAs that are

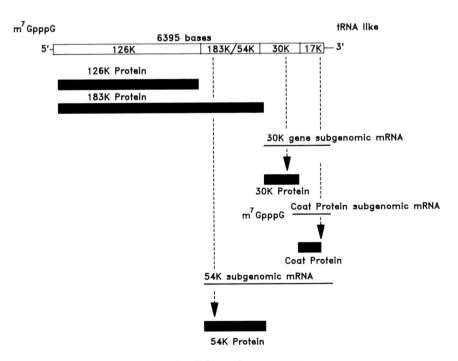

FIG. 1. Tobamovirus genomes.

produced during replication (Hunter *et al.*, 1976; Siegel *et al.*, 1976; Bruening *et al.*, 1976; Higgins *et al.*, 1976; Beachy and Zaitlin, 1977). These mRNAs initiate upstream from the ORFs and continue to the 3′ terminus of the genome, but, as with the genomic RNA, only the 5′-most ORF is translated. An mRNA for the 54K protein has been found in infected cells associated with polyribosomes (Sulzinski *et al.*, 1985), but this protein has not been found *in vivo*.

The coat protein mRNA (corresponding to nucleotides 5703–6395 of TMV-U1) has an eight- or nine-nucleotide leader that is AU rich and has a 7-methylguanosine (m^7G) cap. The 30K mRNA (nucleotides 4828–6395 of TMV-U1) also contains the coat protein ORF. The 30K mRNA leader is reported to be 65 nucleotides for the OM strain (Watanabe *et al.*, 1984b) and 75 nucleotides for the U1 strain (Lehto *et al.*, 1990b). Available evidence suggests that this mRNA is not capped (Hunter *et al.*, 1983; Joshi *et al.*, 1983; Lehto *et al.*, 1990b). The putative 54K mRNA initiates at nucleotide 3405 of TMV-U1 and continues to the 3′ terminus (Sulzinski *et al.*, 1985). This mRNA has 90 nucleotides upstream from the probable ORF. There is no information concerning whether it is capped.

The noncoding areas of the virus are prime candidates for involvement in the regulation of replication and gene expression. The 5′-nontranslated region consists of about 70 nucleotides that compose the leader of the 126K/183K mRNA. It contains a m^7G cap (Keith and Fraenkel-Conrat, 1975) and is AU rich. These sequences are thought to be involved in regulation of the expression of the 5′ genes. The 3′-nontranslated region contains a tRNA-like sequence that specifically accepts histidine (or valine for the Cc strain) *in vitro* (Oberg and Philipson, 1972; Beachy *et al.*, 1976).

Considerable sequence similarity exists among the different strains, including a 30-nucleotide sequence that is almost identical in all strains examined and is also contained in the 3′ region of the satellite of TMV (Mirkov *et al.*, 1989). Both terminal regions are presumed to be required for recognition by the replicase to initiate minus- and plus-sense RNA syntheses, as has been demonstrated for bromoviruses (Bujarski *et al.*, 1985; Miller, *et al.*, 1986; French and Ahlquist, 1987). The TMV replicase is assumed to recognize specific viral sequences and to replicate only TMV RNAs. However, tobamovirus hybrids with 3′ termini and replicase genes from different viruses replicate efficiently (Ishikawa *et al.*, 1988; M. E. Hilf and W. O. Dawson, unpublished observations), demonstrating that this specificity must extend to other tobamoviruses.

In contrast to bromoviruses, no internal sequences appear to be required for TMV RNA replication. Viral mutants with the 30K and/or coat protein genes deleted are able to replicate. Mutants with the 30K and/or coat protein ORFs deleted can replicate in protoplasts, but not in intact plants (Meshi *et al.*, 1987). Mutants with the coat protein ORF removed replicate in plants and protoplasts as free-RNA viruses (Takamatsu *et al.*, 1987; Dawson *et al.*, 1988), while those with most of the 126K/183K ORFs deleted replicate if the wild-type virus is present as a helper (A. J. Raffo and W. O. Dawson, unpublished observations). The latter mutants consist only of the 5' 250 nucleotides and the 1500 3' nucleotides of the U1 genome.

One type of internal regulatory sequence is the subgenomic RNA promoters. The RNA species produced by *in vitro* replicase preparations (Watanabe and Okada, 1986) and the apparent similarities with bromoviruses (Marsh *et al.*, 1987, 1988) indicate that the virus replication complex recognizes a specific sequence of the genomic-length minus-sense RNA to initiate synthesis of plus-sense subgenomic RNAs, which continues to the genomic 3' terminus. With tobamoviruses three different subgenomic RNA promoters should exist, one each for the 54K, 30K, and coat protein mRNAs (Fig. 1). The bromovirus RNA 4 promoter consists of the 60–70 nucleotides upstream from the coat protein ORF (Marsh *et al.*, 1988; French and Ahlquist, 1988). The tobamovirus subgenomic RNA promoters have not been defined, but based on sequence similarities to a proposed RNA virus core promoter sequence and the results of experiments using deletion mutants (Meshi *et al.*, 1987) and hybrid–virus constructs (Dawson *et al.*, 1989), the coat protein promoter of TMV appears to be within the 100 nucleotides upstream from the ORF. Since each promoter probably is within the preceding ORF, sequences comprising the promoter must be bifunctional.

There is no obvious similarity in primary sequence in these three putative subgenomic RNA promoter regions, which might be due to protein coding constraints on the sequence; alternatively, the replicase complex might recognize three-dimensional features that are dictated by more than one primary sequence. An additional possibility is that the three promoters are different to allow differential regulation.

Another potential internal regulatory region is the origin of assembly. This sequence initiates virion assembly, which begins at an internal area of the RNA, proceeds to the 5' terminus, and then continues to the 3' terminus (Butler *et al.*, 1977; Lebeurier *et al.*, 1977; see Bloomer and Butler, 1986). The origin of assembly is within the coding sequences of the 30K ORF of some strains (i.e., U1, OM, L, and O), but within the coat protein ORF of others (i.e., Cc and CGMMV). Subgenomic RNAs that contain the origin of assembly are also encapsidated and make up a minor component of the virion population. It is possible that assembly might function in the regulation of gene expression by making mRNAs unavailable for translation. Not only would this remove genomic RNA, which is the message for the 126K and 183K proteins, but also the 54K and 30K mRNAs in all tobamoviruses and the coat protein mRNAs of the Cc and CGMMV strains.

Several groups of plant RNA viruses, including the tobamoviruses, have a series of "pseudoknots" in the 3'-nontranslated region, generally between the tRNA-like region and the 3' ORF (Pleij et al., 1985, 1987). Pseudoknots are involved in translational regulation (Tang and Draper, 1989) and in ribosomal frameshifting in coronaviruses (Brierley et al., 1989). TMV has three pseudoknots immediately 3' from the coat protein ORF. There is no information concerning whether they are involved in the regulation of TMV gene expression or replication. Coat protein deletion mutant cp35 has the first pseudoknot deleted, but retains the ability to replicate, demonstrating that all three are not absolutely necessary for infectivity (Dawson et al., 1988).

B. Ability to Manipulate Tobamovirus Genomes

The life cycles of two tobamoviruses have been artificially extended through a DNA phase that allows manipulation by recombinant DNA techniques. The complete genomes of TMV-U1 (Dawson et al., 1986) and TMV-L (Meshi et al., 1986) were cloned as cDNAs behind a λ phage promoter (Ahlquist and Janda, 1984) so that precise replicas of virion RNA could be produced in vitro. The m^7G cap is added to the 5' end by initiating transcription with the capped dinucleotide, m^7Gpp-pG. The cap is required for infectivity when RNA is used as the inoculum (Dawson et al., 1986), but a low level of infectivity is achieved when the uncapped RNA is assembled into virions prior to inoculation (Meshi et al., 1986). RNA with a nonmethylated cap is essentially as infectious as that with the methylated cap. A nearly precise 5' sequence is required for infectivity. In contrast, the addition of up to 10 nucleotides to the 3' end has little effect on infectivity. On replication initiated by RNAs with additional nucleotides on either end, the progeny viral RNA does not contain the extra nucleotides.

The development of infectious cDNA clones of tobamoviruses has greatly increased the scope of possible experiments with which to examine the molecular genetics of these viruses. Entire genomes, even those of variants carrying lethal mutations, can be maintained in bacterial plasmids and later transcribed into full-genomic RNAs, providing a uniformity of inoculum not previously available. This has allowed the examination of primary and secondary functions of specific genes (Ishikawa et al., 1986; Meshi et al., 1987; Dawson, et al., 1988), mapping of specific mutant phenotypes to specific nucleotides (Meshi *et al.*, 1988; Saito *et al.*, 1987; Knorr and Dawson, 1988; Watanabe *et al.*, 1987; Culver and Dawson, 1989a,b), or the manipulation of genome organization (Dawson *et al.*, 1989; Lehto *et al.*, 1990b; Lehto and Dawson, 1990b; Beck and Dawson, 1989).

C. Replication

Tobamoviruses replicate through double-stranded intermediates in a manner similar to that of other plus-sense RNA viruses. RF molecules, which consist of intact plus and minus strands, and larger RI molecules, which are partially double stranded and partially single stranded, are found in infected tissues (Nilsson-Tillgren, 1970; Jackson *et al.*, 1971). Labeling kinetics under numerous different conditions show that RF and RI appear to interconvert rapidly, each being different transient states of the same replicating unit. From short to long labeling periods, the amount of label in each of these double-stranded RNAs is always approximately equal (Kielland-Brandt, 1974; W. O. Dawson, unpublished observations). Incorporation during longer labeling periods (i.e., 2–4 hours) results in about 90% incorporation into genomic-length single-stranded RNA and about 4–6% each into RF and RI. Free minus-sense strands have not been detected. Genomic RNAs are quickly assembled into virions, and the concentration of free plus-sense strands is never more than a small percentage of the final amount of viral RNA produced (Dawson and Schlegel, 1976b).

The first step of replication is assumed to be the infection process, which requires entry of the viral RNA into the cell and its translation into proteins required for the initial events of replication. Whether virions or viral RNA enters cells and how the virions disassemble have been controversial (de Zoeten, 1981). Recently, it has been shown *in vitro* that virions can disassemble in association with ribosomes as the genomic RNA is translated (Wilson, 1984a,b). This cotranslation/disassembly process might occur as the first step in replication. Structures that resemble those created *in vitro* by virions associated with ribosomes have been isolated from newly infected cells, supporting this concept (Shaw *et al.*, 1986).

One of the earliest events observed after infection is a proliferation of membranes in the cytoplasm (Nilsson-Tillgren *et al.*, 1969). Most evidence suggests that replication occurs in this region, in membranous vesicles within the cytoplasm (Hills *et al.*, 1987; Okamoto *et al.*, 1988).

Synthesis of viral proteins and viral RNA are first detected at approximately the same time: 3–6 hours after infection (Sakai and Takebe, 1974; Paterson and Knight, 1975; Aoki and Takebe, 1975; Dawson and Schlegel, 1976b; Siegel *et al.*, 1978). Each initially increases exponentially, but then assumes a linear rate of synthesis at approximately 12–16 hours. Synthesis continues at or near this maximal rate for a number of hours, which varies among studies. The duration of synthesis in protoplasts often depends on the longevity of the protoplasts, in which growth curves have continued longer when better methods were developed to maintain protoplasts. Overall, it appears that replication continues for about 3–4 days, after which synthesis decreases to 1–2% of the maximal rate (W. O. Dawson, unpublished observations).

Other undefined steps of replication have been identified indirectly by measuring the times that an infection is sensitive to inhibition by specific chemicals. One of the earliest steps is an actinomycin D-sensitive step that occurs prior to viral protein or RNA synthesis (Lockhart and Semancik, 1969; Dawson, 1978). 2-Thiouracil inhibits a step that occurs later than the actinomycin D-sensitive step, but earlier than viral protein or RNA synthesis (Dawson and Schlegel, 1976a; Dawson and Grantham, 1983). Ribovirin and low concentrations of guanidine inhibit with the same kinetics and might inhibit the same process (Dawson, 1975; Dawson and Lozoya-Saldana, 1984). Cycloheximide and arabinofuranosyladenine inhibit a later step that coincides with protein and RNA syntheses (Dawson and Schiegel, 1976a; Dawson and Lozoya-Saldana, 1986). When added to an infection in which virus replication is at the maximal rate, arabinofaranosyladenine inhibits the syntheses of single-stranded, but not double-stranded, RNA and viral proteins. These experiments emphasize that there must be viral functions for which we have not been able to associate biochemical events and that at least two inhibitor-sensitive functions occur before viral protein and RNA syntheses.

The above discussions describe the progression of events in individual cells. In young rapidly growing plants the virus infection spreads from individual cells to almost all other cells. Within an inoculated leaf the infection spreads from initially infected cells to other cells of the leaf both by cell-to-cell movement through plasmodesmata and by long-distance movement in vascular tissues. Free-RNA mutants can move from cell to cell as well as wild-type virus (Dawson *et al.*, 1988). However, the coat protein greatly facilitates long-distance movement. This can be seen clearly by examining free-RNA mutants in inoculated leaves. Infection by the free-RNA mutants proceeds radially at the same rate as by wild-type virus, but wild-type virus spreads quickly to distant parts of the inoculated leaf and fully infects the leaf within 10–12 days, whereas the free-RNA mutant will have moved only 2–3 cm. After about 2 days infectious virus begins moving via phloem cells into the rest of the plant, utilizing both cell-to-cell and long-distance movement.

D. Gene Functions

1. Coat Protein Gene

The primary function of the coat protein is as a structural unit of virions. Additionally, several secondary characteristics have been associated with this protein. In plants of the genus *Nicotiana* with a specific gene (*N'*) for resistance to most strains of TMV, the host specifically recognizes the coat protein, and a hypersensitive resistance response is actuated (Saito *et al.*, 1987; Knorr and Dawson, 1988; Culver and Dawson, 1989a,b). In addition, the coat protein has been implicated in altering chloroplast structure and preventing normal photosynthesis in susceptible plants (Dawson *et al.*, 1988; Reinero and Beachy, 1989; Hodgson *et al.*, 1989). There is evidence that the coat protein is involved in long-distance movement within the plant in some manner other than encapsidation (Dawson *et al.*, 1988; W. O. Dawson, unpublished observations). Additionally, the coat protein is involved in cross-protection (Sherwood and Fulton, 1982; Register and Beachy, 1988).

2. 30K Gene

The 30K protein is required for cell-to-cell movement of the virus (Nishiguchi *et al.*, 1978; Meshi *et al.*, 1987). The protein becomes tightly bound to cell wall membrane fractions, but its precise mode of action is not understood. The movement function can be provided by other unrelated viruses (Dodds and Hamilton, 1972; Malyshenko *et al.*, 1988), and the tobamovirus 30K protein can mediate movement of other viruses. Transgenic plants expressing the 30K protein allow the movement of both TMV mutants with a defective 30K gene (Deom *et al.*, 1987) and some other unrelated viruses (C. Holt and R. N. Beachy, personal communication). The ability of the movement protein to function in association with the host component appears to be a determinant of the viral host range. TMV has been shown to replicate in inoculated cells of several hosts in which it cannot move from cell to cell (Sulzinski and Zaitlin, 1982). This suggests that the 30K protein must specifically interact with a host component(s) to facilitate cell-to-cell movement.

3. 126K/183K Genes

Most evidence indicates that the 126K and 183K proteins are required for replication. A protein approximately the size of the 126K protein is found in partially purified replicase preparations (Zaitlin *et al.*, 1973). Virus deletion mutants with only the 126K/183K ORFs are capable of replicating in protoplasts (Meshi *et al.*, 1987). Viral deletion mutants without the 126K or 183K ORF replicate only in association with a helper virus that contains these genes (A. J. Raffo and W. O. Dawson, unpublished observations). When mutants are engineered to prevent production of the 183K protein, no replication occurs (Ishikawa *et al.*, 1986). When the stop codon is removed, so that only the 183K protein is produced, replication occurs at a low level until a stop codon is regenerated. This result implies that the 183K protein alone, which contains all of the functional domains of this region, can replicate viral RNA, but that the production of both proteins together results in much more efficient replication.

Another line of evidence suggesting that the 126K and 183K proteins are replicase proteins comes from comparison of the amino acid sequences of different viruses. Members of a supergroup, including tobamoviruses, alphaviruses, bromoviruses, cucumoviruses, ilarviruses, and tobraviruses (Haseloff *et al.*, 1984; Ahlquist *et al.*, 1985; Cornelissen and Bol, 1984; Hamilton *et al.*, 1987), have similarities in three domains, all within similar regions of the genomes (Fig. 2).

The D3 domain is thought to function as an RNA polymerase, based on the GDD motif (Kamer and Agros, 1984). Recent work with alphaviruses suggests that the D1 domain is required for the initiation of minus-sense RNA synthesis (Hahn *et al.*, 1989) and has a methyltransferase activity (Mi *et al.*, 1989). D2 appears to be involved in subgenomic RNA synthesis, and D3 appears to function as the viral polymerase (Hahn *et al.*, 1989).

Some evidence suggests that the tobamovirus replicase complex contains host proteins, as shown for other plant virus replicases (Mouches *et al.*, 1984; Dorssers *et al.*, 1984). Tomato plants with the *Tm-1* gene for resistance to TMV allow only minimal levels of replication of wild-type TMV-L, even in protoplasts. Reduction of the ability to replicate is greater in *Tm-1/Tm-1* plants than in *Tm-1/+* (heterozygous) plants.

A mutant virus was found that was able to overcome resistance, multiply to a high titer, and cause a mosaic-type disease. This mutant has two nucleotide substitutions, which result in amino acid changes Gln-979 to Glu and His-984 to Tyr within the D2 domain of the 126K and 183K proteins (Meshi *et al.*, 1988). One possibility is that this viral protein must interact with a host protein to form the replicase complex and the *Tm-1* allele encodes that host protein. The host protein in plants without the *Tm-1* allele might interact with the wild-type replicase, but the replicase must be altered to function with the Tm-1 protein. In heterozygous (*Tm-1/+*) plants there might be reduced

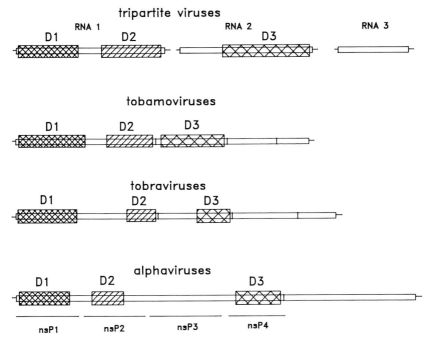

FIG. 2. Viral genomes showing regions having amino acid sequence similarities (Haseloff *et al.*, 1984; Cornelissen and Bol, 1984; Ahlquist *et al.*, 1985; Hamilton *et al.*, 1987). The three similar domains are labeled D1, D2, and D3 and are diagrammed as hatched boxes.

amounts of host protein with which the wild-type virus protein can interact.

Other features of TMV replication have been identified by examining replication-deficient temperature-sensitive mutants (Dawson and Jones, 1976). One group of mutants was deficient in the synthesis of all viral RNAs on shift to the restrictive temperature (Dawson and White, 1978); in the other group of mutants double-stranded RNA synthesis continued, but single-stranded RNA synthesis ceased after the temperature shift (Dawson and White, 1979). These results suggest that the first set of mutants, at the restrictive temperature, lost a function required for the synthesis of all RNAs. It is possible that these mutants had a defect in the D3 domain for polymerization. The second set of mutants was capable of synthesizing double-stranded, but not single-stranded, RNA; perhaps these mutants were defective, at the nonpermissive temperature, in a function that regulates plus-sense–minus-sense RNA ratios or is required to initiate plus-sense strands at the 3′ terminus of minus-sense strands. Arabinofuranosyladenine, when added to an ongoing infection, inhibits similarly (Dawson and Lozoya-Saldana, 1986). Other functions are known to be involved in viral replication. Genomic and coat protein subgenomic RNAs are capped. This process, which apparently occurs in the cytoplasm, might require a virus-encoded protein. Some function must control the ratio of double-stranded RNA produced relative to single-stranded RNA and ratios of plus-sense to minus-sense molecules. Finally, some mechanism must exist for shutting off replication after a specific amount of virus accumulates within the infected cell.

II. REGULATION OF GENE EXPRESSION

A. Synthesis of Tobamovirus Proteins Is Individually Regulated

The production of each TMV protein is regulated differently, both in amounts and times of production. Theoretically, after infection the 126K and 183K proteins should be produced first to provide the replicase complexes required to produce subgenomic RNAs for the syntheses of 30K and coat proteins. However, because RNA viruses replicate by self-saturation kinetics, in which initial progeny RNAs become templates for new replication centers, even within a single cell, replication does not occur synchronously and transcription of the infecting RNA producing only 126K and 183K proteins has not been detected. In practice, all of the viral proteins are detected initially at about the same time, usually between 3 (Watanabe *et al.*, 1984a) and 7 (Siegel *et al.*, 1978) hours after infection. The syntheses of 126K, 183K, and coat proteins reach maximal rates at 16–24 hours. After a sharp peak of maximal synthesis that lasts for only a few hours, the rates of synthesis of 126K and 183K proteins begin declining and reach a low level by 72–96 hours. In contrast, synthesis of the coat protein continues at maximal rates for more than 40 hours. However, during the course of infection, the relative proportions of these proteins vary. Initially, the coat protein is produced at approximately one-half the rate of the 126K protein, but at 70 hours coat protein synthesis exceeds that of the 126K protein by 20-fold (Siegel *et al.*, 1978; Ogawa and Sakai, 1984).

The 183K protein is produced in parallel with the 126K protein, but at approximately 10% as much, suggesting that about 10% of the time the stop codon is read through (Siegel *et al.*, 1978). However, syntheses of the 126K and 183K proteins vary somewhat at different periods of the infection. Early in the infection, the 183K–126K ratio is higher than at later times (Siegel *et al.*, 1978; Watanabe *et al.*, 1984a).

In earlier studies of TMV protein synthesis, the 30K protein was not detected, because this protein is tightly bound to a particulate fraction and is not extracted with the soluble proteins during sodium dodecyl sulfate extraction. It was first detected by Ooshika *et al.* (1984) using an antiserum produced against a synthetic polypeptide corresponding to the 16 carboxy-terminal amino acids of the 30K protein.

Although the 30K protein accumulates in the nuclei of infected protoplasts (Watanabe *et al.*, 1986), *in situ* localization demonstrated that in intact leaves the 30K protein accumulates in plasmodesmatal areas of the cell wall (Tomenius *et al.*, 1987). Similar movement proteins of other plant viruses have also been localized in the cell wall by *in situ* labeling (Stussi-Garaud *et al.*, 1987; Linstead *et al.*, 1988). Procedures that have effectively extracted the movement proteins were designed to extract proteins that were tightly associated with the nonsoluble cell wall fraction (Godefroy-Colburn *et al.*, 1986; Lehto *et al.*, 1990a). In a recent subcellular fractionation study Moser *et al.* (1988) confirmed the association of the 30K protein with the nonsoluble cell wall material and, additionally, found that the protein occurs transiently in the cytoplasmic membrane fraction. The localization of the protein in cell walls suggests that in intact tissue it is actively transported out of the cell.

The 30K protein levels detected in protoplasts (Ooshika *et al.*, 1984; Watanabe *et al.*, 1984a; Blum *et al.*, 1989) are substantially lower than in intact leaves (Moser *et al.*, 1988; Lehto *et al.*, 1990a). These data suggest that accumulation and possibly regulation of the 30K protein are different in the cells of intact leaves than in protoplasts. The difference in levels of this protein in intact cells compared to protoplasts could be due to the protein's being transported out of protoplasts (and degraded) during the experiments. Alternatively, the lack of deposition of the protein into cell walls might lead to the shutting down of synthesis earlier in protoplasts than in leaves.

The timing of synthesis of the 30K protein is quite different from that of the other viral proteins. The 30K protein is produced transiently during the early stage of infection (2–10 hours after inoculation) in synchronously infected protoplasts (Watanabe *et al.*, 1984a; Blum *et al.*, 1989). In near-synchronously infected leaves, the production of 30K protein also occurs during the early period of infection, but synthesis continues longer than in protoplasts, until approximately 24 hours (Lehto *et al.*, 1990a). In leaves the 30K protein is stably associated with a particulate fraction and appears to undergo little turnover (Moser *et al.*, 1988; Lehto *et al.*, 1990a). The maximal concentration of the 30K protein produced in leaves appears to be approximately the same as that of the 126K protein, but higher than that of the 183K protein.

The coat protein of TMV is one of the most highly produced proteins in plants. During maximal synthesis coat protein synthesis can constitute up to 70% of the total cellular protein synthesis, even though there is little reduction in the host protein synthesis (Siegel *et al.*, 1978); during the 2- to 3-day period of rapid synthesis, the coat protein can accumulate as much as 10% of the total cellular protein (Fraser, 1987).

B. *Regulation of 5' Proximal Genes*

Several lines of evidence suggest that in established infections virions or previrion RNA does not function as mRNA for TMV protein synthesis. Polysomes producing the 126K or coat protein contain associated double-stranded RNAs (Beachy and Zaitlin, 1975; Ogawa *et al.*, 1983). Additionally, TMV protein synthesis is correlated with double-stranded, but not single-stranded, RNA synthesis (Dawson, 1983). TMV protein synthesis was examined in leaves infected with a *ts* (temperature-sensitive) mutant that, on shift to the restrictive temperature (35°C), stopped genomic single-stranded RNA synthesis, but continued double-stranded RNA synthesis. Syntheses of 126K, 183K, and coat proteins continued uninhibited for at least 16 hours in the absence of detectable single-stranded RNA synthesis, even though other experiments demonstrated that protein synthesis quickly declined after the inhibition of all TMV RNA synthesis.

Also, when the replication machinery was partially disrupted by a heat treatment (40°C), on return of the samples to the permissive temperature, replication resumed only at a minimal rate, and protein synthesis recovered after several hours at the permissive temperature, in parallel with double-stranded RNA synthesis (Dawson, 1983). This recovery occurred several hours before the recovery of single-stranded RNA synthesis. These data suggest that a specific function for mRNA synthesis exists that is different from the function that produces progeny virion RNA. This result was not unexpected for subgenomic mRNA synthesis, but it also appears to be the case for the mRNA for the 126K and 183K proteins.

1. *126K Leader Enhances Translation*

The 5' leader of the 126K mRNA is probably involved in regulation of the level of expression of this gene. TMV virion RNA is an exceptionally efficient translation template *in vitro,* and this high efficiency has been related to the leader sequence (Gallie *et al.,* 1987a,b, 1988; Sleat *et al.,* 1987). Addition of the 126K leader to other RNAs greatly stimulates the translation of both eukaryotic and prokaryotic mRNAs *in vitro* and *in vivo* (Gallie *et al.*, 1987a,b). The efficient translation of TMV mRNAs in prokaryotic *in vitro* translation systems is exceptional, because most eukaryotic mRNAs are not correctly initiated in bacterial cells (Kozak, 1983).

The 126K leader of TMV, as well as some other viral leaders, contains binding sites (AUU) for a second 80 S ribosome upstream from the start codon (Gallie *et al.*, 1987a; Ahlquist *et al.*, 1979; Filipowicz and Haenni, 1979), thus providing a putative second in-frame translation initiation site (Tyc *et al.*, 1984). The simultaneous binding of two ribosomes (i.e., disome formation) has been suggested to contribute to the stimulation of translation by the leader of the 126K mRNA. However, disome formation by different viral mRNA leaders does not correlate with their efficiency of translation (Gallie *et al.*, 1987a), and deletion of the upstream ribosome binding site does not abolish the enhancement effect of the leader (Gallie *et al.*, 1988).

Although the 126K mRNA has an extraordinarily effective leader for translation, the 126K protein is not produced at extraordinarily high levels. The high efficiency might be necessary to enhance translation of the few molecules of RNA introduced into the cell on inoculation to insure that infection is established. Other factors (e.g., the availability of the mRNA) might reduce the expression of this gene later in the infection.

2. Translation of Tobamovirus Proteins after Heat Shock

TMV protein synthesis is translationally regulated differently than host protein synthesis (Dawson and Boyd, 1987). After a heat shock most host protein synthesis is suppressed, and heat-shock proteins begin to be produced (Key *et al.*, 1981). However, after a heat shock TMV 126K, 183K, and coat protein syntheses continue at their normal rates. This result suggests that TMV mRNAs are recognized differently or that they might function on a different set of ribosomes than the majority of plant mRNAs. The discrimination between host normal mRNAs and heat-shock protein mRNA is due to differences in their leader sequences (McGarry and Lindquist, 1985). However, TMV mRNA leaders do not appear to be similar to the heat-shock protein mRNA leaders.

3. Read-through of the Stop Codon

The 183K protein is produced by read-through of the 126K amber stop codon (Pelham, 1978). The transcription of mRNA for this protein and the efficiency of initiation of translation are thought to be the same as for the 126K protein. Regulation is determined by the frequency of read-through, which occurs about 10% of the time *in vivo*

(Siegel *et al.*, 1978) and provides an effective mechanism to produce the 183K protein in lower amounts.

The read-through in cell-free translation systems can be mediated by the wild-type tyrosine tRNA, amounting to 2–5% read-through. A tyrosine tRNA was isolated from tobacco and other plant species that can enhance suppression of the stop codon up to 30% (Bier *et al.*, 1984). Also, the ratio of production of 126K and 183K proteins *in vitro* is affected by the concentration of message. At lower concentrations of mRNA, read-through occurs more frequently than with higher template concentrations (Joshi *et al.*, 1983). This observation could explain the higher ratio of 183K protein produced early in the infection, when mRNA concentrations are low (Siegel *et al.*, 1978).

Regulation of protein synthesis by read-through is a common phenomenon among numerous RNA virus groups, including tobra-, carmo-, tymo-, and furoviruses from plants and alphaviruses and retroviruses from animals. Regulation appears to depend not only on suppressor tRNAs, but also on the sequence context of the stop codon (Valle and Morch, 1988). However, when the amber stop codon of TMV-L was replaced with an ochre stop codon, the mutant was viable and produced normal amounts of 183K protein, suggesting that the ochre stop codon was suppressed as efficiently as the amber one (Ishikawa *et al.*, 1986). Tobamoviruses provide an ideal system to examine whether the sequence context of the stop codon affects read-through and could at the same time allow the examination of mutants that produce different ratios of 126K to 183K proteins.

Read-through of the stop codon and consequent syntheses of both the 126K and 183K proteins are needed for the efficient replication of TMV (Ishikawa *et al.*, 1986). Mutations of TMV-L that prevent production of the 183K protein are lethal. A mutation that changed the amber codon to a tyrosine codon to allow production of the 183K, but not the 126K, protein resulted in a virus population that initially replicated poorly and later began to replicate rapidly. The progeny virus that replicated rapidly contained revertants with an ochre stop codon. This suggests that the mutant was able to replicate minimally with only the 183K protein. However, a mutant with the stop codon deleted entirely and lacking one amino acid in the 183K protein was not infectious. Neither protein alone is sufficient for efficient replication. It will be interesting to determine the effects of altered ratios of these proteins on replication and regulation.

C. Regulation of Internal Genes

Particularly interesting is the regulation of the genes expressed through subgenomic mRNAs, because they are expressed so differently. The 30K protein is an early gene product produced in minimal amounts, while the coat protein is a late gene product produced at extraordinarily high levels. The 54K protein has not been found in infected tissues, perhaps because of controlled expression during a limited period or because of low amounts of the protein. Although the genes are expressed by similar mechanisms (i.e., subgenomic mRNAs), regulation of the mechanisms appears to occur independently.

Expression of these genes could be controlled at the transcriptional, posttranscriptional, and/or translational levels. Syntheses of the coat protein and 30K mRNAs temporally correlate with production of the corresponding proteins (Ogawa and Sakai, 1984; Watanabe *et al.*, 1984a), suggesting transcriptional regulation. Some of the 30K mRNA is encapsidated by the coat protein, perhaps resulting in posttranslational regulation. Translational regulation is also suggested by the distinctly different leaders of the mRNAs. The coat protein mRNA has a m^7G cap at its 5' terminus and a short (nine-nucleotide) AU-rich leader (Guilley *et al.*, 1979). The leader of the 30K mRNA is not capped (Hunter *et al.*, 1983; Joshi *et al.*, 1983) and is substantially longer (i.e., 75 nucleotides, as found by Lehto *et al.*, 1990c). The leader of the 54K mRNA is even longer, containing approximately 90 nucleotides (Sulzinski *et al.*, 1985).

1. Translational Regulation

Most eukaryotic mRNAs are capped. The cap is a strong determinant of mRNA stability and also strongly enhances the binding of 49 S ribosomal subunits to the 5' end of mRNAs (Shatkin, 1976; Kozak, 1983). The cap-binding protein complex melts the secondary structure of the mRNA leader and facilitates ribosome binding and/or migration to the initiation codon (see Sonenberg, 1987). Removal of the 5' cap impairs ribosomal binding and translation. The addition of cap-binding protein complex to a translation system can relieve translational competition between mRNAs, suggesting that it is a limiting factor in translation initiation (Sarkar *et al.*, 1984). This difference between the coat protein and 30K mRNAs might greatly affect the expression of these genes. The coat protein continues to be produced at relatively high rates several hours after RNA synthesis stops, demonstrating the stability of this mRNA (Dawson, 1983). No information is available concerning the stability of the 30K mRNA.

Start Codon Sequence Context. Another possible type of translational regulation could result from differences in the efficiency of translation initiation, due to differences in the start codon sequence contexts. Kozak's modified scanning model suggests that eukaryotic ribosomes bind to the 5' end of mRNAs and scan in the 3' direction until a start codon within the proper sequence context is found, at which time translation is initiated (Kozak, 1981, 1983, 1984a,b, 1986a). Usually translation initiates at the first start codon. However, the surrounding sequences are thought to determine whether translation initiates at the first start codon and, if so, with what efficiency. The most optimal consensus sequence, as defined by the Kozak model, is ACCAUGG, purines at positions −3 and +4 being the most important regulatory signals. This model has been supported by experimental and sequence data from animal, plant, and viral mRNAs. Recently, however, Lutcke *et al.* (1987) suggested that plant ribosomes might not strictly recognize the above consensus sequence as optimal. They suggested that the −3 position was less important in the wheat germ *in vitro* translation system, while the +4 position was more important than that proposed by Kozak. However, some of the most strongly expressed mRNAs in plant systems fit the original consensus.

Comparison of the translation initiation sites of the different tobamoviruses for which sequences are available shows remarkable similarities (Table I).

All of the known 126K start codon contexts are identical and would be considered strong by both the Kozak and Lutcke models. The leader for the 126K/183K mRNA has already been shown to strongly enhance translation and must be near-optimal *in vivo*, supporting both models. All of the coat protein start codons have an A at the −3 position and, in contrast with both models, U1, L, OM, and O have a U at the +4 position. The coat protein is produced at such extraordinarily high levels that this start codon context also must be near-optimal. At the 30K start codon the U1, L, OM, and CGMMV strains have similar contexts with −3 U and +4 G, which is defined by the Kozak model as a "weak" start codon context. The TMV-O 30K start codon context is defined as "strong" and that of the Cc strain is intermediate. The

TABLE I

COMPARISON OF THE TRANSLATION INITIATION SITES OF DIFFERENT TOBAMOVIRUS GENES[a]

Strain	126K Protein	54K Protein	30K Protein	Coat protein
U1	ACAAUGG	GAUAUGC	UAGAUGG	AAUAUGU
L	ACAAUGG	GACAUGU	UUGAUGG	AAUAUGU
OM	ACAAUGG		UAGAUGG	AAUAUGU
CGMMV			UAGAUGG	ACGAUGG
Cc		GUGAUGU	UUGAUGAUGG
O			ACAAUGG	AAUAUGU

[a]Underlined residues indicate the start codons.

theoretical nonoptimal start codon context might be one factor responsible for down-regulation of the expression of this gene and the production of 30K protein at minimal levels. We examined the start codon context of the 30K gene by changing the weak start codon context (UAGAUGG) to stronger contexts by site-directed mutagenesis and examination of the expression of the 30K gene (Lehto and Dawson, 1990a). We chose to examine the mutants *in planta,* because similar experiments have resulted in different translation efficiencies *in vitro* than those observed *in vivo* (Roner *et al.,* 1989). Two mutants were produced:

wild type	GUUUAUAGAUGGCUCUAG
KK1	GUUUAUAGACUCGAGAUGGCUCUAG
KK2	GUUUAUAGACGAUGGCUCUAG

Mutant *KK1* had a −3 G, which, according to the Kozak model, is a strong start codon context; mutant *KK2* had an even stronger context. Complicating this experiment was that the sequences in front of the 30K start codon could not be modified without also altering the 183K gene, which overlaps the 30K ORF by 14 nucleotides, as well as the subgenomic promoter/leader sequences that control the 30K gene. Mutant *KK1,* which contained an insertion of seven nucleotides, possessed an altered 183K ORF, leading to changes in four amino acids at the carboxy terminus. Mutant *KK2* was designed to contain the strongest possible start codon context and minimal modifications to the 183K protein and promoter/leader sequence. The mutation in *KK2* modified the 183K protein by the insertion of only one amino acid at position 5 from the carboxy terminus; this amino acid was identical to the adjacent amino acid. Three nucleotides were added to the leader of *KK2.* Neither of the 183K protein alterations appeared to have an effect on replication of these mutants in terms of the amounts of progeny RNA and viral proteins produced.

Although start codon contexts often affect translation efficiency in eukaryotic systems severalfold, alteration of the start codon contexts of the TMV-U1 30K gene to more "optimal" contexts, as defined by Kozak's ribosome scanning model, did not enhance the expression of the 30K gene. Mutant *KK2,* with what should have been an optimal start codon context, produced amounts of the 30K protein approximately equal to those of the wild-type virus. Mutant *KK1,* with G at the −3 position, produced only about 30% as much 30K protein as the wild-type virus. This result suggests that the consensus sequence recognized by most eukaryotic ribosomes is not a major factor in the regulation of TMV 30K protein synthesis.

The virus strains with a -3 U in the 30K start codon sequence have additional in-frame (potential) start codons within stronger contexts. The common strains, U1 and OM, have additional start codons at positions 20, 43, and 97, each of which has a -3 A and a $+4$ G, and TMV-L has similar start codons at positions 34, 42, and 96. CGMMV has one start codon at position 100. A possibility is that the 5' start codons are weaker, so that ribosomes occasionally progress to the internal start codons before translation initiates, producing shorter proteins. In fact, *in vitro* translation of the 30K protein mRNA results in multiple proteins with common carboxy termini (Hunter *et al.*, 1983; Joshi *et al.*, 1983). It is not known whether these truncated proteins are active *in vivo;* in fact, they have not been found *in vivo*. The predominant protein produced *in vivo* appears to be the protein initiated at the first start codon. However, it cannot be excluded that a few of the amino-terminal truncated proteins are produced to provide a specific function.

A mutant was produced to examine whether the putative internally initiated carboxy-coterminal 30K proteins could mediate cell-to-cell movement. The 5' AUG of this ORF was changed to ACG to prevent initiation at the first start codon. However, this mutant was not able to infect intact plants, suggesting that the intact 30K protein is needed for viability in the plant (Lehto and Dawson, 1990a).

2. Effect of Subgenomic RNA Promoter/Leader Sequences on Gene Expression

The kinetics of the syntheses of 30K and coat protein mRNAs parallel those of their respective proteins (Ogawa and Sakai, 1984; Watanabe *et al.*, 1984a), which indicates that the regulation of these two genes occurs at least partially at the level of transcription. Bromoviruses replicase preparations are able to initiate the synthesis of subgenomic RNA 4 from a specific internal promoter sequence on the full-length minus-sense RNA (Miller *et al.*, 1985). The replicase recognizes a specific sequence on this RNA to initiate synthesis of the subgenomic RNA. The bromovirus promoter is contained within the 60–70 nucleotides upstream from and including the translation initiation site (Marsh *et al.*, 1987; Marsh *et al.*, 1988; French and Ahlquist, 1988). In contrast to bromoviruses, tobamoviruses should have three subgenomic RNA promoters, one each for the 54K, 30K, and coat protein mRNAs. The lack of similarities in these regions suggests that these promoters might be regulated differently. One possibility is that the genes are expressed at a level determined by an efficiency of transcription initiation characteristic of the specific promoter (i.e., that the coat protein promoter might be stronger than the 30K and 54K promoters).

The precise sequences of the subgenomic RNA promoters of TMV have not been characterized, but the coat protein subgenomic RNA promoter appears to be within approximately 100 nucleotides upstream from the coat protein ORF. The insertion of 250 nucleotides from upstream from the coat protein ORF in front of the chloramphenicol acetyltransferase (CAT) ORF resulted in the production of a new subgenomic RNA (Dawson et al., 1989). In other experiments, when the 30K gene was deleted, leaving only 96 nucleotides of the putative coat protein subgenomic RNA promoter region, the coat protein was still produced, although in reduced amounts (Meshi et al., 1987). No work examining the 30K and 54K promoter regions has been described.

We examined the regulation of gene expression by the coat protein subgenomic RNA promoter/leader by inserting this sequence in front of the 30K ORF and determining its effect on 30K protein synthesis (Lehto et al., 1990c). If the coat protein subgenomic RNA promoter, which controls the highest expressed gene, causes a much higher level of transcription, much higher production of 30K protein should have occurred. Studies of the bromovirus subgenomic RNA promoter show that sequences that contain the subgenomic RNA promoter also contain the leader for the mRNA. A complication is that it probably is not possible to use the coat protein subgenomic RNA promoter to produce an mRNA with the wild-type 30K leader, because insertion of a subgenomic RNA promoter in front of an ORF also changes the leader of the mRNA. Since the minimal unit of the coat protein promoter was not known, we made two mutants, each containing different amounts of the promoter region, anticipating that the insertion of the minimal active unit would perturb the virus less. Mutant KK7 contained 49 nucleotides of the sequence 5' from the coat protein ORF plus 16 non-TMV nucleotides inserted upstream from its 30K ORF. Mutant KK6 contained 253 nucleotides from the coat protein promoter region plus the same 16 non-TMV nucleotides (Fig. 3).

Mutant KK6, with the large insertion (Fig. 3), produced a new 30K protein mRNA with a shorter leader, with similarities to the coat protein mRNA leader. The KK6 30K protein mRNA leader was 24 nucleotides, compared to a 75-nucleotide leader of the wild-type 30K mRNA and nine nucleotides of the wild-type coat protein mRNA leader. The KK6 30K protein mRNA leader contained the sequences of the coat protein mRNA, as would be expected if promoted by the coat protein promoter. Additionally, it contained nine extra nucleotides

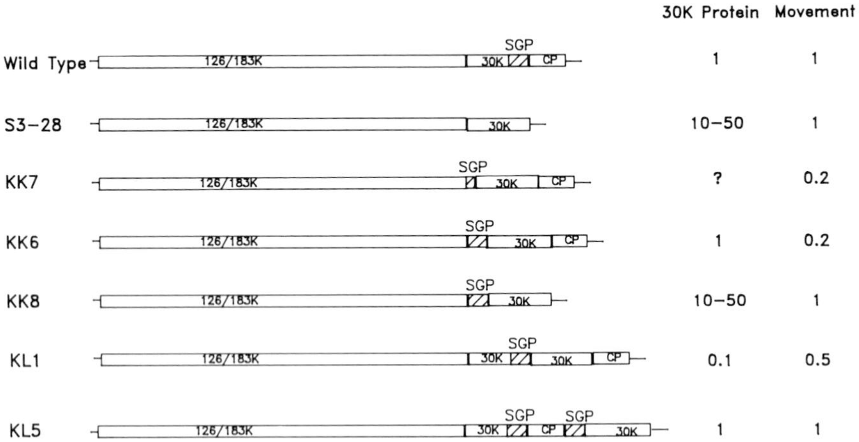

FIG. 3. TMV mutants and relative amounts of movement and amounts of 30K protein produced. Hatched areas show the coat protein (CP) subgenomic RNA promoter region (SGP). The amount of 30K protein was estimated by densitometric measurements of Western immunoblots and compared to that produced by wild-type TMV, arbitrarily set a 1.0. The relative amounts of movement were estimated by the diameter of local lesions produced in Xanthi nc, with wild-type TMV equal to 1.0.

from the *Xho*I site created to make the construct and six extra nucleotides added because the subgenomic RNA was initiated six nucleotides upstream from the normal coat protein mRNA initiation site.

Another difference between the wild-type 30K and coat protein mRNAs is that the latter is capped, while the former is not. What determines whether a virus RNA becomes capped is not known. In fact, we do not know how cytoplasmic viruses such as TMV are capped, since host enzymes involved in this process are thought to be limited to the nucleus. Possible controls of capping of the TMV RNAs are the subgenomic RNA promoter and leader. However, the new 30K protein mRNA of *KK6* that was initiated by the coat protein subgenomic RNA promoter and contained a hybrid coat protein mRNA leader was not capped (Lehto *et al.*, 1990c). Yet, since both the promoter and the leader were modified by being repositioned in front of the 30K ORF, this does not exclude that they control capping in the native position within the genome.

Insertion of the coat protein subgenomic RNA promoter/leader sequence into mutant *KK6* allowed efficient replication of the virus, but expression of the 30K protein was not greatly increased to a level similar to that of the coat protein. Instead, the time course of production of the 30K protein was altered. Instead of being produced earlier than the 126K protein, the 30K protein of *KK6* was produced later than the 126K protein. In mechanically inoculated tobacco leaves the wild-type TMV 30K protein accumulated to a maximal level (2–3 days after infection) earlier than the 126K protein (3–5 days). In contrast, the maximal accumulation of *KK6* 30K protein occurred 4–7 days later (6–10 days) than that of the 126K protein (4–5 days). Thus, insertion of the coat protein subgenomic promoter/leader sequence greatly delayed production of the 30K protein.

Mutant *KK7*, with the 49-nucleotide coat protein subgenomic RNA promoter/leader region insert (Fig. 3), initially replicated slowly, but later began replicating like wild-type TMV. Progeny virus that replicated well had the insert precisely deleted, resulting in wild-type virus. A comparison of mutant *KK6*, which was stably maintained as progeny virus, to *KK7*, which quickly lost the inserted sequences, suggests that the 49 inserted nucleotides did not contain a functional subgenomic RNA promoter. If the insert failed to promote a subgenomic mRNA, it would be expected to lengthen the leader of the mRNA induced by the native 30K protein subgenomic mRNA promoter. Apparently, this mutant replicated in this manner until the inserted sequences were deleted.

Insertion of the promoter/leader sequences into mutant *KK6* allowed efficient replication of the virus, but greatly altered the time of accumulation of the 30K protein. We do not know whether the delay of 30K protein synthesis by mutant *KK6* is due primarily to the inserted promoter with delayed transcription of the mRNA or the modified leader, resulting in translational regulation, or both. Determining whether insertion of the 30K promoter/leader sequence in front of the coat protein ORF will cause the coat protein to become an early product is important. It is possible that the sequences upstream from each of the internal (i.e., 54K, 30K, and coat protein) ORFs determine their times of expression during infection.

3. Control of 30K Protein Subgenomic RNA Synthesis

The D1 and/or D2 domain of the 126K or 183K protein (Fig. 2) might be involved in the production of the 30K protein mRNA. An attenuated strain of TMV-L was isolated that is produced in reduced amounts in tomato plants and induces only mild symptoms. However, this virus replicates like wild-type virus in protoplasts (Nishiguchi *et al.*, 1982). Sequencing demonstrated that the attenuated strain differed from TMV-L by 10 nucleotides within the 126K gene, three of which result in amino acid substitutions, one in the D1 domain and two in the D2 domain (Nishiguchi *et al.*, 1985). During replication in protoplasts, the synthesis of the 30K protein and its mRNA are specifically reduced in isolate $L_{11}A$, which has all three amino acid alterations, but the reduction is less in isolate L_{11}, which has only the D1 alteration (Watanabe *et al.*, 1987). Syntheses of genomic RNA, coat protein mRNA, and 126K, 183K, and coat proteins were not reduced. This finding suggests that alterations in these domains of the 126K protein can be involved in the production of 30K protein, but not coat protein, mRNA. If so, the specific subgenomic RNA promoters are recognized independently.

4. Effect of Actinomycin D on 30K Protein Synthesis

A recent observation is that actinomycin D appears to alter the regulation of the 30K gene. It selectively enhances the synthesis of this protein up to 100-fold, while stimulating the syntheses of other viral proteins no more than twofold (Blum *et al.*, 1989). Actinomycin D treatment of TMV-infected protoplasts not only greatly increases production of the 30K protein, but also causes the protein to be produced for longer periods. Instead of synthesis peaking at 8–10 hours and then declining, in actinomycin D-treated protoplasts, the maximal enhanced rate of synthesis continued 16–24 hours after infection, the latest time at which samples were taken.

5. Effect of Position Relative to the 3' Terminus on Gene Expression

Positioning of the 30K gene nearer to the 3' terminus by the deletion of portions of the coat protein gene proportionally increases the amount of the 30K protein produced (Lehto *et al.*, 1990b). Mutants with the coat protein gene completely deleted produce approximately 10–50 times as much 30K protein as the wild-type virus. This occurred with mutants with the native 30K protein subgenomic RNA promoter in front of the 30K ORF (mutant *S3-28,* wild-type TMV with the entire coat protein gene deleted) and a mutant with the coat protein subgenomic RNA promoter/leader sequences in front of the 30K ORF (*KK8*, mutant *KK6* with the coat protein gene removed) (Fig. 3). This increase in 30K protein synthesis is similar to that observed in actinomycin D-treated protoplasts (Blum *et al.*, 1989).

Initially, we thought that the coat protein might be a negative regulator that repressed 30K protein synthesis. However, mutants with the start codons altered so that no coat protein was produced, but with the rest of the coat protein ORF left intact, produced only wild-type amounts of the 30K protein. When mutants with different sizes of deletions in their coat protein genes were examined, the increased production of the 30K protein was always proportional to the number of nucleotides removed, demonstrating that the position of the 30K gene determines its level of expression. The level of expression was affected little by which promoter/leader sequence controlled the gene. This suggests that one reason the coat protein is produced in greater amounts than the 30K protein is because of their relative positions to the 3' terminus. The same logic suggests that the 54K protein, if it exists, would be produced in amounts proportionally less than the 30K protein. This argument is consistent with the observation by French and Ahlquist (1988), who showed that the level of production of subgenomic RNA from bromovirus RNA 3 was progressively greater when the promoter was inserted into different positions nearer the 3' terminus.

Although 30K protein synthesis was markedly increased by removing the coat protein gene and positioning the 30K gene nearer the 3' terminus, the resulting level of synthesis of the 30K protein was still substantially less than wild-type levels of coat protein synthesis. Two contributing factors might be that the 30K gene is larger than the coat protein gene and its 5' end is still positioned farther from the 3' terminus in the mutant than that of the coat protein gene in the wild-type genome. However, even another 10-fold increase in the 30K protein would still be substantially less than the amount of coat protein produced. This suggests that other differences between these genes affect their expression. For example, the 5' cap of the coat protein mRNA probably gives it more longevity than the mRNA of the 30K gene (Dawson, 1983).

6. Effects of Coat Protein Gene Deletions on Coat Protein Synthesis

Factors other than distance from the 3' terminus, subgenomic promoter, and subgenomic mRNA leader are involved in the regulation of the internal genes of TMV. We examined a series of mutants with deletions in the coat protein gene and determined the amount of altered coat protein they produced (Dawson *et al.*, 1988). The mutants varied widely in their production of altered coat proteins, from amounts equivalent to the wild-type level of coat proteins to levels too low to detect. There was no correlation between the amount of coat protein produced and the number of nucleotides deleted. Often, mutants differing by less than 10 nucleotides in length differed in coat protein production by several orders of magnitude. In general, mutants that maintained the coat protein ORF through the deletion and produced the normal carboxy terminus produced more coat protein. Pulse-labeling of proteins in mutant infected cells demonstrated that the observed differences in the accumulation of proteins were due to reduced synthesis of the truncated proteins, rather than degradation of proteins without normal carboxy termini (unpublished observations).

7. Effects of Coat Protein on Gene Expression

Initially, we suspected that coat protein might be a regulatory molecule, perhaps binding to specific areas of the RNA to positively or negatively affect replication. We constructed the series of coat protein deletion mutants (Dawson et al., 1988) to examine this phenomenon, but we have found no evidence of altered regulation of these mutants. Mutants with the coat protein start codon changed to ACG produce no coat protein, but produce normal amounts of 126K, 183K, and 30K proteins (J. N. Culver and W. O. Dawson, unpublished observations). This also suggests that the encapsidation of mRNAs does not reduce gene expression by removing either 126K/183K or 30K mRNAs, which could support the argument that there are separate functions for the production of mRNAs and virion RNAs (Dawson, 1983).

D. Effects of Genome Organization on Gene Expression

We examined the effects of different genomic organizations on virus replication, genome stability, and the level of gene expression by constructing different chimeric mutants of TMV. A hybrid, *CAT-CP*, with a gene cartridge consisting of the CAT ORF fused behind the coat protein subgenomic RNA promoter inserted into the TMV genome between the 30K and coat protein ORFs replicated efficiently and produced additional subgenomic RNA and CAT activity (Dawson et al., 1989). However, this hybrid was not stably maintained. Large amounts of the hybrid virus were produced in inoculated leaves, but progeny virus with the inserted sequences deleted predominated in systemically infected tissues. Virus hybrids with two coat protein genes were even less stable (Beck and Dawson, 1990). In fact, they were too transient to detect in their original form: Progeny virus contained only one coat protein gene. These results indicated that there is strong selection against propagating viruses with unnecessary sequences in their genomes. In contrast, however, mutant *KK6* (Fig. 3), which contained the 269-nucleotide insertion, was propagated stably for months (Lehto et al., 1990c). Thus, the selection pressures appear to operate differentially on differently altered genomes.

At present we have little information concerning why the stability of genome organizations differed so greatly during propagation. Viral protein synthesis is precisely regulated, both temporally and quantitatively. Apparently, the gene products are needed in different amounts and at different times for optimal virus replication. Genomic organization, along with specific regulatory sequences, must provide the regulation of individual genes. However, another factor could be the efficiency of replication of the genomic RNAs. There appears to be selection based on how well the genomic RNA molecule is replicated, indicating that the genomic organization must provide for both efficient replication and effective gene expression. The optimal genome organization might balance gene expression against efficient replication of the RNA.

We created a series of TMV hybrids with two 30K ORFs to examine how well they would replicate and what effect genome position has on gene expression (Lehto and Dawson, 1990b). We have seen that inserted or altered sequences can be deleted or rearranged quickly and that selection is a strong force determining the constitution of the progeny population. A major question, then, is what determines whether a virus can compete with its altered progeny. Can modified viruses be propagated and maintained as the major component of a progeny population?

A mutant with a second 30K ORF fused to the coat protein ORF, *CP30K,* produced more fusion protein than wild-type 30K protein but substantially less than wild-type coat protein (Fig. 3). Simple fusion of an ORF to the amino-terminal two-thirds of the coat protein ORF did not provide the same level of expression as that of the native coat protein. Expression of the fusion protein gene might have been partially decreased by its position relative to 3′ end, since the insertion of 30K ORF moved the coat protein subgenomic RNA promoter approximately 700 nucleotides farther from the 3′ terminus than that of the wild-type virus. This insertion also decreased the amount of native 30K protein produced by this mutant. This reduction in 30K protein apparently was due to positioning the native gene farther from the 3′ terminus.

Another hybrid, *KL1,* has two 30K ORFs in tandem, the 5′ ORF driven by the native 30K protein subgenomic RNA promoter/leader sequences and the 3′ ORF driven by the coat protein subgenomic RNA promoter/leader sequences (Fig. 3). At the time this mutant was built, we expected that with two 30K genes, one driven by its native promoter and the other controlled by the coat protein subgenomic RNA promoter, more 30K protein would be produced. However, this mutant produced greatly reduced amounts of the 30K protein, at least one-tenth of that produced by wild-type TMV. This result might be due in part to decreased expression of the native 30K gene, because of its being positioned farther from the 3′ terminus by the size of the insertion (i.e., 800 nucleotides). However, the position of the inserted 30K ORF relative to the 3′ terminus was identical to that of *KK6,* which produced normal amounts of the 30K protein, but later in the infection (Lehto *et al.,* 1990c). *KL1* produced only barely detectable amounts of the 30K protein at any time. This suggests that factors other than increased distance from the 3' terminus caused the decreased production of the 30K protein by this mutant.

Mutant *KL5* has the 30K ORF fused behind the coat protein subgenomic RNA promoter/leader region and inserted between the coat protein ORF and the 3'-nontranslated region (Fig. 3). The second 30K gene was in the position of the coat protein gene relative to the 3' terminus. *KL5* produces increased amounts of the 30K protein compared to the other mutant with two 30K ORFs, but this occurred at the expense of reduced abilities for replication and stable propagation. This mutant replicated poorly and did not move out of inoculated leaves, and most progeny virus had altered sequences. The hybrids that we have examined with insertions between the coat protein ORF and the 3'-nontranslated region—*KL5* and *CP-CAT* (Dawson et al., 1989)—replicated poorly, and virus with the inserted sequences removed predominated the progeny population. However, mutant *KL5* reaffirmed that genes positioned nearer the 3' terminus could be expressed at higher rates.

III. Effects of Alteration of Production of 30K Protein on Phenotype

The alteration of gene expression also allows examination of the effects of protein levels on gene function. For example, a wide range of levels of 30K protein production appears to be sufficient for normal cell-to-cell movement within the plant. Mutant *KK1* produced only about 30% as much 30K protein as wild-type TMV, but there was no detectable decrease in the mutant's ability to move (Lehto and Dawson, 1990a). Mutants with greater decreases in the 30K protein resulted in a decreased ability to move in inoculated, but not into systemically infected, leaves. Mutant *KL1* (Fig. 3) produced barely detectable amounts of the 30K protein, which apparently resulted in a reduction in the size of local lesions to approximately one-half the diameter of those produced by wild-type TMV (Lehto and Dawson, 1990b). However, *KL1* systemically infected upper leaves as well as wild-type TMV. Thus, even this greatly reduced amount of 30K protein was sufficient for long-distance systemic movement. A surprising result was that, with this greatly reduced amount of 30K protein and its reduced ability to spread in inoculated leaves, *KL1* was able to compete effectively during the course of infection of tobacco plants with the wild-type virus that arose by deletion of the inserted sequences. The wild-type virus only gradually overtook the *KL1* population.

Increases in the production of the 30K protein did not increase the ability of mutants to move from cell to cell. The coat protein deletion mutants (Fig. 3), which had up to 50 times more 30K protein than did wild-type TMV, moved in inoculated leaves identically to wild-type virus (Lehto et al., 1990b). Apparently, an amount of 30K protein above a threshold is sufficient for normal movement, and increased amounts have no effect.

Delayed production of the 30K protein by mutant *KK6* (Fig. 3) greatly affected cell-to-cell as well as long-distance movement (Lehto et al., 1990c). The final accumulation of the 30K protein of *KK6* was equal to that of wild-type TMV, but mutant *KK6* local lesions in inoculated leaves were much smaller than wild-type lesions and *KK6* moved more slowly and to only limited areas in upper leaves. The movement defect of *KK6* was due to to delayed production of the 30K protein, rather than to reduced levels, which suggests that the 30K protein is needed during the early hours of the replication cycle to properly mediate cell-to-cell movement.

IV. CONCLUSIONS AND SPECULATION

Tobamoviruses appear to use several strategies to control gene expression:

1. Different subgenomic RNA promoter/leader sequences control timing of expression of genes.
2. Genes expressed via subgenomic mRNAs are expressed in decreasing amounts with increasing distances from the 3' terminus.
3. TMV mRNAs appear to be translationally regulated differently from host mRNAs.
4. Capped mRNAs probably are translated at higher levels than noncapped mRNAs.

However, there certainly is much regulation that we do not understand.

Genome organization also affects gene expression, but it appears to be equally important for the efficiency of replication and the ability of the genomic structure to be stably propagated. What advantages do different genome organizations have? Different virus groups have evolved different gene arrangements. Examples of similar viruses that express genes via subgenomic mRNAs that have different genome organizations are shown in Fig. 4. We have little understanding of why the particular genomic structure of tobamoviruses arose. Within the tobamoviruses we have established that different types of reg-

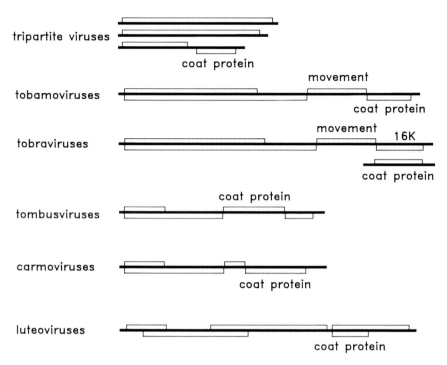

FIG. 4. Generalized genomes of plant virus groups. Open reading frames are shown as open boxes (Ahlquist *et al.*, 1984; Goelet *et al.*, 1982; Hamilton *et al.*, 1987; Rochon and Tremaine, 1989; Carrington *et al.*, 1989; van der Wilk *et al.*, 1989).

ulation exist, but our understanding of regulatory mechanisms is meager. However, we have observed enough to make it tempting to speculate on the relationships of RNA virus genomic organizations to the regulation of gene expression.

Genes that need to be expressed first to produce enzymes required for later production of subgenomic mRNAs and late gene products are expected to be at the 5′ terminus. We assume that genes expressed through subgenomic RNAs would require the prior production of replicase and could not be expressed initially. However, within the Sindbis supergroup that includes tobamoviruses, tripartite viruses, and alphaviruses, even the replicase domains are expressed differently. With alphaviruses the D1, D2, and D3 domains (Fig. 2) are expressed as a polyprotein that cleaves into proteins containing these domains singularly. The D1 and D2 domains of tripartite viruses and tobamoviruses are contained within a single protein. Tripartite viruses express the D3 domain as a single protein, but tobamoviruses express this domain in a protein that contains all three domains, or, if the 54K protein is produced, expresses it both ways. With tobamoviruses the initial expression of the 126K and 183K genes allows production of all three domains without the requirement of RNA synthesis. Does the 183K protein function, which differs from the 126K protein function, require the D1 and D2 domains, or is this simply a method to produce the D3 domain without requiring RNA synthesis (Fig. 2)?

Also, tripartite viruses express their putative movement protein from a genomic RNA instead of from a subgenomic mRNA, as do tobamoviruses. Does this mean that this protein must be produced initially? It will be interesting to determine whether a TMV hybrid with the D3 and 30K domains on separate genomic RNAs in the 5' position will replicate.

Viruses that express genes via subgenomic mRNAs generally produce structural proteins in greater amounts than nonstructural proteins. In most of these viruses, the structural protein gene(s) is positioned at the 3' terminus of the genome (Fig. 4). Tobamovirus genes expressed via subgenomic mRNAs appear to be expressed in increasing amounts when positioned nearer the 3' terminus. This has similarities to rhabdovirus transcriptional regulation, in which amounts of transcript are progressively decreased as genes are positioned farther from the genomic 3' end (see Banerjee, 1987). If this observation can be extended to other virus groups (e.g., tobraviruses), the 16K gene which is positioned at the 3' terminus (Fig. 4) should be expressed in greater amounts than the movement protein, as recently demonstrated (Angenent *et al.*, 1989). Tombusviruses (Fig. 4) do not have their coat protein genes positioned at the 3' terminus, and whether their 3' genes are expressed at higher levels than their coat protein genes remains to be seen. Tombusviruses and carmoviruses (Fig. 4) appear to be similar, except for their gene order (Carrington *et al.*, 1989; Rochon and Tremaine, 1989). Do they differ in levels of expression of the coat protein? What advantages do each of these genome organizations provide? We have only begun to understand the regulation of RNA viruses and how it relates to the evolution of genome organizations.

REFERENCES

Ahlquist, P., and Janda, M. (1984). *Mol. Cell. Biol.* **4**, 2876–2882.
Ahlquist, P., Dasgupta, R., Shih, D. S., Zimmern, D., and Kaesberg, P. (1979). *Nature (London)* **281**, 277–282.
Ahlquist, P., Bujarski, J. J., Kaesberg, P., and Hall, T. C. (1984). *Plant Mol. Biol.* **3**, 37–44.

Ahlquist, P., Strauss, E. G., Rice, C. M., Strauss, J. H., Haseloff, J., and Zimmern, D. (1985). *J. Virol.* **53**, 536–542.
Angenent, G. C., Verbeek, H. B. M., and Bol, J. F. (1989). *Virology* **169**, 305–311.
Aoki, S., and Takebe, I. (1975). *Virology* **65**, 343–354.
Banerjee, A. K. (1987). *Microbiol. Rev.* **51**, 66–87.
Beachy, R. N., and Zaitlin, M. (1975). *Virology* **63**, 84–97.
Beachy, R. N., and Zaitlin, M. (1977). *Virology* **81**, 160–169.
Beachy, R. N., Zaitlin, M., Bruening, G., and Israel, H. (1976). *Virology* **73**, 498–507.
Beck, D. A., and Dawson, W. O. (1990). Manuscript in preparation.
Bier, H., Barciszewska, M., Krupp, G., Mitnacht, R., and Gross, H. J. (1984). *EMBO J.* **3**, 351–356.
Bloomer, A. C., and Butler, P. J. G. (1986). *In* "The Plant Viruses: 2. The Rod-Shaped Plant Viruses" (M. H. V. Van Regenmortel and H. Fraenkel-Conrat, eds.), pp. 19–57. Plenum, New York.
Blum, H., Gross, H. J., and Beir, H. (1989). *Virology* **169**, 51–61.
Brierley, I., Digard, P., and Inglis, S. C. (1989). *Cell (Cambridge, Mass.)* **57**, 537–547.
Bruening, G., Beachy, R. N., Scalla, R., and Zaitlin, M. (1976). *Virology* **71**, 498–517.
Brunt, A. A. (1986). *In* "The Plant Viruses: 2. The Rod-Shaped Plant Viruses" (M. H. V. Van Regenmortel and H. Fraenkel-Conrat, eds.), pp. 181–204. Plenum, New York.
Bujarski, J., Dreher, T., and Hall, T. C. (1985). *Proc. Natl. Acad. Sci. U.S.A.* **82**, 5636–5640.
Butler, P. J. G., Finch, J. T., and Zimmern, D. (1977). *Nature (London)* **265**, 217–219.
Carrington, J. C., Heaton, L. A., Zuidema, D., Hillman, B. I., and Morris, T. J. (1989). *Virology* **170**, 219–226.
Cornelissen, B. T. C., and Bol, J. F. (1984). *Plant Mol. Biol.* **3**, 379–384.
Culver, J. N., and Dawson, W. O. (1989a). *Mol. Plant–Microbe Interact.* **2**, 209–213.
Culver, J. N., and Dawson, W. O. (1989b). *Virology* **173**, 755–758.
Dawson, W. O. (1975). *Intervirology* **6**, 83–89.
Dawson, W. O. (1978). *Intervirology* **9**, 304–309.
Dawson, W. O. (1983). *Virology* **125**, 314–323.
Dawson, W. O., and Boyd, C. (1987). *Plant Mol. Biol.* **8**, 145–149.
Dawson, W. O., and Grantham, G. L. (1983). *Intervirology* **19**, 155–161.
Dawson, W. O., and Jones, G. E. (1976). *Mol. Gen. Genet.* **145**, 308–309.
Dawson, W. O., and Lozoya-Saldana, H. (1984). *Intervirology* **22**, 77–84.
Dawson, W. O., and Lozoya-Saldana, H. (1986). *Intervirology* **26**, 149–155.
Dawson, W. O., and Schlegel, D. E. (1976a). *Phytopathology* **66**, 177–181.
Dawson, W. O., and Schlegel, D. E. (1976b). *Phytopathology* **66**, 437–442.
Dawson, W. O., and White, J. L. (1978). *Virology* **90**, 209–213.
Dawson, W. O., and White, J. L. (1979). *Virology* **93**, 104–110.
Dawson, W. O., Beck, D. L., Knorr, D. A., and Grantham, G. L. (1986). *Proc. Natl. Acad. Sci. U.S.A.* **83**, 1832–1836.
Dawson, W. O., Bubrick, P., and Grantham, G. L. (1988). *Phytopathology* **78**, 783–789.
Dawson, W. O., Lewandowski, D. J., Hilf, M. E., Bubrick, P., Raffo, A. J., Shaw, J. J., Grantham, G. L., and Desjardins, P. R. (1989). *Virology* **173**, 285–292.
Deom, C. M., Oliver, M. J., and Beachy, R. N. (1987). *Science* **237**, 389–393.
de Zoeten, G. A. (1981). *In* "Plant Diseases and Vectors: Ecology and Epidemiology" (K. Maramorosch and K. F. Harris, eds.), pp. 221–239. Academic Press, New York.
Dodds, J. A., and Hamilton, R. I. (1972). *Virology* **50**, 404–411.
Dorssers, L., van der Krol, S., van der Meer, J., van Kammen, A., and Zabel, P. (1984). *Proc. Natl. Acad. Sci. U.S.A.* **81**, 1951–1955.
Edwardson, J. R., and Zettler, F. W. (1986). *In* "The Plant Viruses: 2. The Rod-Shaped

Plant Viruses" (M. H. V. Van Regenmortel and H. Fraenkel-Conrat, eds.), pp. 233–247. Plenum, New York.

Filipowicz, W., and Haenni, A.-L. (1979). *Proc. Natl. Acad. Sci. U.S.A.* **76**, 3111–3115.

Fraser, R. S. S. (1987). "Biochemistry of Virus-Infected Plants," pp. 1–7. Res. Stud. Press, Letchworth, England.

French, R., and Ahlquist, P. (1987). *J. Virol.* **61**, 1457–1465.

French, R., and Ahlquist, P. (1988). *J. Virol.* **62**, 2411–2420.

Gallie, D. R., Sleat, D. E., Watts, J. W., Turner, P. C., and Wilson, T. M. A. (1987a). *Nucleic Acids Res.* **15**, 8693–8711.

Gallie, D. R., Sleat, D. E., Watts, J. W., Turner, P. C., and Wilson, T. M. A. (1987b). *Nucleic Acids Res.* **15**, 3257–3273.

Gallie, D. R., Sleat, D. E., Watts, J. W., Turner, P. C., and Wilson, T. M. A. (1988). *Nucleic Acids Res.* **16**, 883–894.

Gibbs, A. J. (1977). *CMI/AAB Descriptions of Plant Viruses* **184**.

Gibbs, A. J. (1986). *In* "The Plant Viruses: 2. The Rod-Shaped Plant Viruses" (M. H. V. Van Regenmortel and H. Fraenkel-Conrat, eds.), pp. 167–180. Plenum, New York.

Godefroy-Colburn, T., Gagey, M., Berna, A., and Stussi-Garaud, C. (1986). *J. Gen. Virol.* **67**, 2233–2239.

Goelet, P., Lomonossoff, G. P., Butler, P. J. G., Akam, M. E., Gait, M. J., and Karn, J. (1982). *Proc. Natl. Acad. Sci. U.S.A.* **79**, 5818–5822.

Guilley, H., Jonard, G., Kukla, B., and Richards, K. E. (1979). *Nucleic Acids Res.* **6**, 1287–1307.

Hahn, Y. S., Strauss, E. G., and Strauss, J. H. (1989). *J. Virol.* **63**, 3142–3150.

Hamilton, W. D. O., Boccara, M., Robinson, D. J., and Baucombe, D. C. (1987). *J. Gen. Virol.* **68**, 2563–2575.

Haseloff, J., Goelet, P., Zimmern, D., Ahlquist, P., Dasgupta, R., and Kaesberg, P. (1984). *Proc. Natl. Acad. Sci. U.S.A.* **81**, 4358–4362.

Higgins, T. J. V., Goodwin, P. B., and Whitfield, P. R. (1976). *Virology* **71**, 486–497.

Hills, G. J., Plaskitt, K. A., Young, N. D., Dunigan, D. D., Watts, J. W., Wilson, T. M. A., and Zaitlin, M. (1987). *Virology* **161**, 488–496.

Hodgson, R. A. J., Beachy, R. N., and Pakrasi, H. B. (1989). *FEBS Lett.* **245**, 267–270.

Hunter, T. R., Hunt, T., Knowland, J., and Zimmern, D. (1976). *Nature (London)* **260**, 759–764.

Hunter, T., Jackson, R., and Zimmern, D. (1983). *Nucleic Acids Res.* **11**, 801–821.

Ishikawa, M., Meshi, T., Motoyoshi, F., Takamatsu, N., and Okada, Y. (1986). *Nucleic Acids Res.* **14**, 8291–8305.

Ishikawa, M., Meshi, T., Watanabe, Y., and Okada, Y. (1988). *Virology* **164**, 290–293.

Jackson, A. O., Mitchell, D. M., and Siegel, A. (1971). *Virology* **45**, 182–191.

Joshi, S., Pleij, C. W. A., Haenni, A. L., Chapeville, F., and Bosch, L. (1983). *Virology* **127**, 100–111.

Kamer, G., and Agros, P. (1984). *Nucleic Acids Res.* **12**, 7269–7282.

Keith, J., and Fraenkel-Conrat, H. (1975). *FEBS Lett.* **57**, 31–33.

Key, J. L., Kin, C. Y., and Chen, Y. M. (1981). *Proc. Natl. Acad. Sci. U.S.A.* **78**, 3526–3530.

Kielland-Brandt, M. C. (1974). *J. Mol. Biol.* **87**, 489–503.

Knorr, D. A., and Dawson, W. O. (1988). *Proc. Natl. Acad. Sci. U.S.A.* **85**, 170–174.

Knowland, J. (1974). *Genetics* **78**, 383–394.

Kozak, M. (1981). *Nucleic Acids Res.* **9**, 5233–5252.

Kozak, M. (1983). *Microbiol. Rev.* **47**, 1–45.

Kozak, M. (1984a). *Nucleic Acids Res.* **12**, 857–872.

Kozak, M. (1984b). *Nucleic Acids Res.* **12**, 3873–3893.

Kozak, M. (1986a). *Cell (Cambridge, Mass.)* **44**, 283–292.

Kozak, M. (1986b). *Proc. Natl. Acad. Sci. U.S.A.* **83**, 2850–2854.

Kozak, M. (1986c). *Adv. Virus Res.* **31**, 229–292.

Lebeurier, G., Nicolaieff, A., and Richards, K. E. (1977). *Proc. Natl. Acad. Sci. U.S.A.* **74**, 149–153.

Lehto, K., Bubrick, P., and Dawson, W. O. (1990a). *Virology* **174**, 290–293.

Lehto, K., and Dawson, W. O. (1990a). *Virology* **174**, 169–176.

Lehto, K., and Dawson, W. O. (1990b). *Virology,* in press.

Lehto, K., Culver, J. N., and Dawson, W. O. (1990b). Manuscript in preparation.

Lehto, K., Grantham, G. L., and Dawson, W. O. (1990c). *Virology* **174**, 145–157.

Linstead, P. J., Hills, G. J., Plaskitt, K. A., Wilson, I. G., Harker, C. L., and Maule, A. J. (1988). *J. Gen. Virol.* **69**, 1809–1818.

Lockhart, B. E. L., and Semancik, J. S. (1969). *Virology* **39**, 362–365.

Lutcke, H. A., Chow, K. C., Mickel, F. S., Moss, K. A., Kern, H. F., and Scheele, G. A. (1987). *EMBO J.* **6**, 43–48.

Malyshenko, S. I., Lapchic, L. G., Kondakova, O. A., Kuznetzova, L. L., Taliansky, M. E., and Atabekov, J. G. (1988). *J. Gen. Virol.* **69**, 407–412.

Marsh, L. E., Dreher, T. W., and Hall, T. C. (1987). *In* "Positive Strand RNA Viruses" (M. A. Brinton and R. R. Reukert, eds.), pp. 327–336. Liss, New York.

Marsh, L. E., Dreher, T. W., and Hall, T. C. (1988). *Nucleic Acids Res.* **16**, 981–995.

McGarry, T. J., and Lindquist, S. (1985). *Cell (Cambridge, Mass.)* **42**, 903–911.

Meshi, T., Ohno, T., Iba, H., and Okada, Y. (1981). *Mol. Gen. Genet.* **184**, 20–25.

Meshi, T., Ohno, T., and Okada, Y. (1982a). *Nucleic Acids Res.* **10**, 6111–6117.

Meshi, T., Ohno, T., and Okada, Y. (1982b). *J. Biochem. (Tokyo)* **91**, 1441–1444.

Meshi, T., Kiyama, R., Ohno, T., and Okada, Y. (1983). *Virology* **127**, 54–64.

Meshi, T., Ishikawa, M., Motoyoshi, K., Semba, K., and Okada, Y. (1986). *Proc. Natl. Acad. Sci. U.S.A.* **83**, 5043–5047.

Meshi, T., Watanabe, Y., Saito, T., Sugimoto, A., Maeda, T., and Okada, Y. (1987). *EMBO J.* **6**, 2557–2563.

Meshi, T., Motoyoshi, F., Adachi, A., Watanabe, Y., Takamatsu, N., and Okada, Y. (1988). *EMBO J.* **7**, 1575–1581.

Mi, S., Durbin, R., Huang, H. V., Rice, C. M., and Stollar, V. (1989). *Virology* **170**, 385–391.

Miller, W., Dreher, T., and Hall, T. (1985). *Nature (London)* **313**, 68–70.

Miller, W. A., Bujarski, J. J., Dreher, T. W., and Hall, T. C. (1986). *J. Mol. Biol.* **187**, 537–546.

Mirkov, T. E., Matthews, D. M., Du Plessis, D. H., and Dodds, J. A. (1989). *Virology* **170**, 139–146.

Moser, O., Gagey, M.-J., Godefroy-Colburn, T., Ellwart-Tschurtz, M., Nitschko, H., and Mundry, K.-W. (1988). *J. Gen. Virol.* **69**, 1367–1373.

Mouches, C., Candresse, T., and Bove, J. M. (1984). *Virology* **134**, 78–90.

Nilsson-Tillgren, T. (1970). *Mol. Gen. Genet.* **109**, 246–256.

Nilsson-Tillgren, T., Kohenmainen-Seveus, L., and van Wettstein, D. (1969). *Mol. Gen. Genet.* **104**, 124–141.

Nishiguchi, M., Motoyoshi, F., and Oshima, N. (1978). *J. Gen. Virol.* **39**, 53–61.

Nishiguchi, M., Motoyoshi, F., Oshima, N., and Kiho, Y. (1982). *In* "Plant Tissue Culture" (A. Fujiwa, ed.), pp. 665–666. Marunzen, Tokyo.

Nishiguchi, M., Kikuchi, S., Kiho, Y., Ohno, T., Meshi, T., and Okada, Y. (1985). *Nucleic Acids Res.* **13**, 5585–5590.

Oberg, B., and Philipson, L. (1972). *Biochem. Biophys. Res. Commun.* **48**, 927–932.

Ogawa, M., and Sakai, F. (1984). *Phytopathol. Z.* **109**, 193–203.

Ogawa, M., Sakai, F., and Takebe, I. (1983). *Phytopathol. Z.* **107**, 146–158.

Ohno, T., Aoyagi, M., Yamanashi, Y., Saito, H., Ikawa, S., Meshi, T., and Okada, Y. (1984). *J. Biochem. (Tokyo)* **96,** 1915–1923.

Okada, Y. (1986). *In* "The Plant Viruses: 2. The Rod-Shaped Plant Viruses" (M. H. V. Van Regenmortel and H. Fraenkel-Conrat, eds.), pp. 267–281. Plenum, New York.

Okamoto, S., Machida, Y., and Takebe, I. (1988). *Virology* **167,** 194–200.

Ooshika, I., Watanabe, Y., Meshi, T., Okada, Y., Igano, K., Inouye, K., and Yoshida, N. (1984). *Virology* **132,** 71–78.

Paterson, R., and Knight, C. A. (1975). *Virology* **64,** 10–22.

Pelham, H. R. B. (1978). *Nature (London)* **272,** 469–471.

Pleij, C. W. A., Rietveld, K., and Bosch, L. (1985). *Nucleic Acids Res.* **13,** 1717–1731.

Pleij, C. W. A., Abrahams, J. P., van Belkum, A., Rietveld, K., and Bosch, L. (1987). *In* "Positive Strand RNA Viruses" (M. A. Brinton and R. R. Ruekert, eds.), pp. 299–316. Liss, New York.

Register, J. C., III, and Beachy, R. N. (1988). *Virology* **166,** 524–532.

Reinero, A., and Beachy, R. N. (1989). *Plant Physiol.* **89,** 111–116.

Rochon, D., and Tremaine, J. H. (1989). *Virology* **169,** 251–259.

Roner, M. R., Gaillard, R. K., Jr., and Joklik, W. K. (1989). *Virology* **168,** 292–301.

Saito, T., Meshi, T., Takamatsu, N., and Okada, Y. (1987). *Proc. Natl. Acad. Sci. U.S.A.* **84,** 6074–6077.

Saito, T., Imai, Y., Meshi, T., and Okada, Y. (1988). *Virology* **167,** 653–656.

Sakai, F., and Takebe, I. (1974). *Virology* **62,** 426–433.

Sarkar, G., Edery, I., Gallo, R., and Sonenberg, H. (1984). *Biochim. Biophys. Acta* **783,** 122–129.

Shatkin, A. J. (1976). *Cell (Cambridge, Mass.)* **9,** 645–653.

Shaw, J. G., Pladkitt, K. A., and Wilson, T. M. A. (1986). *Virology* **148,** 326–336.

Sherwood, J. L., and Fulton, R. W. (1982). *Virology* **119,** 150–158.

Siegel, A., Montgomery, V. H. I., and Kolacz, K. (1976). *Virology* **73,** 363–371.

Siegel, A., Hari, V., and Kolacz, K. (1978). *Virology* **85,** 494–503.

Sleat, D. E., Gallie, D. R., Jefferson, R. A., Beven, M. W., Turner, P. C., and Wilson, T. M. A. (1987). *Gene* **60,** 217–225.

Sonenberg, N. (1987). *Adv. Virus Res.* **31,** 175–204.

Stussi-Garaud, C., Garaud, J., Berna, A., and Godefroy-Colburn, T. (1987). *J. Gen. Virol.* **68,** 1779–1784.

Sulzinski, M. A., and Zaitlin, M. (1982). *Virology* **121,** 12–19.

Sulzinski, M. A., Gabaro, K. A., Palukaitis, P., and Zaitlin, M. (1985). *Virology* **145,** 132–140.

Takamatsu, N., Ishikawa, M., Meshi, T., and Okada, Y. (1987). *EMBO J.* **6,** 307–311.

Tang, C. K., and Draper, D. E. (1989). *Cell (Cambridge, Mass.)* **57,** 531–536.

Tomenius, K., Clapham, D., and Meshi, T. (1987). *Virology* **160,** 363–371.

Tyc, K., Konarska, M., Gross, H. J., and Filipowicz, W. (1984). *Eur. J. Biochem.* **140,** 503–511.

Valle, R. P. C., and Morch, M.-D. (1988). *FEBS Lett.* **235,** 1–15.

van der Wilk, F., Huisman, M. J., Cornelissen, B. J. C., Huttinga, H., and Goldbach, R. (1989). *FEBS Lett.* **245,** 51–56.

Van Regenmortel, M. H. V., and Fraenkel-Conrat, H. (eds.) (1986). "The Plant Viruses: 2. The Rod-Shaped Plant Viruses." Plenum, New York.

Varma, A. (1986). *In* "The Plant Viruses: 2. The Rod-Shaped Plant Viruses" (M. H. V. Van Regenmortel and H. Fraenkel-Conrat, eds.), pp. 249–266. Plenum, New York.

Watanabe, Y., and Okada, Y. (1986). *Virology* **149,** 64–73.

Watanabe, Y., Emori, Y., Ooshika, I., Meshi, T., Ohno, T., and Okada, Y. (1984a). *Virology* **133,** 18–24.

Watanabe, Y., Meshi, T., and Okada, Y. (1984b). *FEBS Lett.* **173,** 247–250.

Watanabe, Y., Ooshika, I., Meshi, T., and Okada, Y. (1986). *Virology* **152,** 414–420.

Watanabe, Y., Morita, N., Nishiguchi, M., and Okada, Y. (1987). *J. Mol. Biol.* **194,** 699–704.

Wilson, T. M. A. (1984a). *Virology* **137,** 255–265.

Wilson, T. M. A. (1984b). *Virology* **138,** 353–356.

Zaitlin, M., Duda, C. T., and Petti, M. A. (1973). *Virology* **53,** 300–311.

ADVANCES IN VIRUS RESEARCH, VOL. 38

THE "MERRY-GO-ROUND": ALPHAVIRUSES BETWEEN VERTEBRATE AND INVERTEBRATE CELLS

Hans Koblet

Institute for Medical Microbiology
University of Berne
CH-3010 Berne, Switzerland

I. Introduction[1]

In this chapter, I treat recent findings concerning the biosynthetic machinery in mosquito and vertebrate cells infected with alphatogaviruses and attempt to integrate fusion phenomena into the overall

[1] Abbreviations: DIDS, 4,4′-Diisothiocyanostilbene-2,2′-disulfonate; dMM, 1-deoxymannojirimycin, 1,5-dideoxy-1,5-imino-D-mannitol; dNM, 1-deoxynojirimycin, 1,5-dideoxy-1,5-imino-D-glucitol; DTNB, 5,5′-dithiobis(nitrobenzoic acid), Ellman reagent;

343

picture. Alphaviruses are fusogenic at a mildly acidic pH. However, fusion from without (FFWO) and fusion by liposomes are not discussed here at any length.

The infectious chain in nature is a kind of merry-go-round involving mosquito and vertebrate cells. However, the outcome of the infection is not the same in these cell types, being far apart from one another in evolution. Therefore, I compare the intracellular events finally leading to cytopathic effects (CPEs) and fusion from within (FFWI) and show that FFWI is not a phenomenon which can be considered a natural CPE. However, it is an important reaction, and we must examine how it is related to endosomal fusion (EF).

Since there is evidence that proton channels are opened in several virus families if fusogenic conformational changes take place at the plasma membrane (PM), we have to discuss which reactions of the biochemical fusion cascade need energy. Finally, because channels might have to do with quaternary protein structures in the PM, we must comment on the virion surface. Important reviews concerning fusion have been written by Poste and Nicolson (1978), Gallaher *et al.* (1980), White *et al.* (1983), Evered and Whelan (1984), Blumenthal (1987, 1988a), Sowers (1987), Spear (1987), Hoekstra *et al.* (1988, 1989), Hoekstra (1988), Hoekstra and Kok (1989), and Stegmann *et al.* (1989).

II. ALPHAVIRUSES

A. Taxonomy

Alphaviridae is a family of spherical, enveloped, positive-stranded RNA viruses. The genera are (1) *Alphavirus,* with 26 species; (2) *Rubivirus,* with one species; (3) *Pestivirus,* with three species; and (4) *Arterivirus,* with one species (Westaway *et al.,* 1985). It is probable that the classification of pestiviruses and arteriviruses will have to be reevaluated (Collett *et al.,* 1989). Only alphaviruses are reviewed here.

Endo-D, endo-β-*N*-acetylglucosaminidase D (EC 3.2.1.96); Endo-H, endo-β-*N*-acetyl-glucosaminidase H (EC 3.2.1.96); HA, hemagglutinin; HEPES, hydroxyeth-ylpiperazinesulfonic acid; m⁷G, 7-methylguanosine; ns P, nonstructural protein; PMSF, phenylmethylsulfonyl fluoride; SITS, 4-acetamido-4′-isothiocyanostilbene-2,2′-disulfo-nate; TAME, *p*-tosyl-L-arginine methylester; TLCK, *N*-α-tosyl-L-lysyl chloromethyl-ketone, L-1-chloro-3-(4-tosylamido)-7-amino-2-heptanone hydrochloride; TNBS, tri-nitrobenzenesulfonic acid; TPCK, *N*-α-tosyl-L-phenylalanyl chloromethylketone, L-1-chloro-3-(4-tosylamido)-4-phenyl-2-butanone.

The laboratory prototypes are Semliki Forest virus (SFV) and Sindbis virus strain AR 339.

Alphaviruses have a large host range, infecting in nature and in the laboratory many vertebrates and invertebrates (e.g., mosquitoes) (Steele, 1981a,b). The distribution of the viruses is worldwide, and they cause clinical syndromes. Encephalitis is the most common and most dangerous complication (Grimstad, 1983). Alphaviruses comprise six serological complexes based on protein (see below) E_1 (hemagglutination inhibition) and E_2 (neutralization) serology (Dalrymple *et al.*, 1976; Calisher *et al.*, 1980; for a review see Strauss and Strauss, 1985).

B. Virion

Alphaviruses have a diameter of 60–65 nm and carry surface projections (i.e., spikes) of 6–10 nm length. In SFV the outer radius is 32 nm. The density of alphaviruses in sucrose is 1.18–1.19 g/cm^3 and the sedimentation coefficient is 260–300 S, corresponding to an overall weight of about 5×10^7. The percentage of weight of the genomic RNA is 6%; for proteins it is 62%; for lipids, 26%; and for glycans, 6%.

The envelope consists of a bilayer membrane associated with the transmembrane proteins E_1 and E_2. The radius including the outer leaflet of the membrane is 23 nm in SFV. In some viruses, as in SFV, a third protein, E_3, is present on the outside of the spike. The spike proteins are heavily modified, carrying glycans and fatty acids. The molecular weights of E_1 and E_2 are 50–60 $\times 10^3$; of E_3 it is 10 $\times 10^3$. The lipid and glycan compositions are host dependent. E_1 is the hemagglutinin and the fusogenic protein (Omar and Koblet, 1988, 1989a). E_2 fixes complement and induces the production of neutralizing antibodies (Dalrymple *et al.*, 1976; Roehrig *et al.*, 1988).

The envelope encloses the nucleocapsid (the "core") with a diameter of 35–39 nm and a sedimentation coefficient of 140 S. This ribonucleoprotein contains a single protein type, the core (C) protein (molecular weight 30–34 $\times 10^3$) and one single strand of linear plus-sense (i.e., infectious) RNA. The RNA has a sedimentation coefficient of 49 S and a molecular weight of 4×10^6, and it comprises about 12,000 bases, with a base composition of 28% A, 26% C, 25% G, 21% U (Sindbis virus), excluding the 3' poly(A) tract. This poly(A) tract is rather short (Wittek *et al.*, 1977). At the 5' end there is a m^7G cap. The gene sequence is 5'–ns P1–ns P2–ns P3–ns P4–C–E_3–E_2–6K–E_1–3'. The 6K peptide has the function of a signal and is described later.

Complete or partial nucleotide sequences have been determined for the following alphaviruses: Sindbis virus (Rice and Strauss, 1981;

Strauss *et al.*, 1984), SFV (Garoff *et al.*, 1980a,b; Takkinen, 1986), Ross River virus (Dalgarno *et al.*, 1983), Middelburg virus (Strauss *et al.*, 1983), Eastern equine encephalitis (Chang and Trent, 1987), Venezuelan equine encephalitis (Kinney *et al.*, 1986), and Western equine encephalitis (Hahn *et al.*, 1988).

Differences in the composition of the virions grown in vertebrate or invertebrate cells can be detected. Alphaviruses grown in mosquito cells are devoid of sialic acid (Stollar *et al.*, 1976). Differences in phospholipid composition have also been described (Luukkonen *et al.*, 1976, 1977). However, all of these differences seem to be irrelevant with regard to antigenicity, infectivity (Stollar *et al.*, 1976), and other biological activities. More information has been provided by Burge and Strauss (1970), Laine *et al.* (1973), Garoff *et al.* (1982), and Strauss and Strauss (1983).

The alphavirus envelope and nucleocapsid are believed to be icosahedral (Rossmann and Johnson, 1989). Icosahedrons have a surface of 20 equilateral triangles arranged in a symmetrical structure, with 30 edges and 12 corners (i.e., vertices), where five triangles meet. Results of conventional electron microscopy supported a $T = 4$ arrangement of the spikes (the structure of the envelope proteins is visible in the electron microscope) (Horzinek and Mussgay, 1969; von Bonsdorff, 1973; von Bonsdorff and Harrison, 1975). Vogel *et al.* (1986) confirmed the $T = 4$ symmetry of the SFV envelope and showed for the first time the trimeric structure of the spikes. Thus, SFV contains $20 \times 4 = 80$ spikes and $80 \times 3 = 240$ copies of each membrane protein. The spike is a heterotrimeric trimer structure consisting of nine polypeptides with the formula $(E_1E_2E_3)_3$. Therefore, formally, six $E_1E_2E_3$s are clustered around an edge and five around a corner, resulting in the formula $(30 \times 6) + (12 \times 5) = 240$ $E_1E_2E_3$s.

The symmetry of the nucleocapsid is not necessarily the same. Classical electron microscopy has produced equivocal results of $T = 3$ (Horzinek and Mussgay, 1969; Enzmann and Weiland, 1979), $T = 4$ (von Bonsdorff, 1973; Söderlund *et al.*, 1975; Coombs and Brown, 1987a,b), and $T = 9$ (Brown *et al.*, 1972; Brown and Gliedman, 1973). With cryoelectron microscopy of Sindbis virus in vitrified solution, it has now been shown that the nucleocapsid exhibits a $T = 3$ symmetry. Thus, a Sindbis virion contains 240 copies of each of the spike proteins and 180 copies of the capsid protein. Two types of spike–capsid interactions result. Spikes near the fivefold axes interact tightly with triplets of capsid proteins. Spikes near the threefold axes interact more loosely (Fuller, 1987; Fuller and Argos, 1987).

Peripentonal spikes have a concave surface, and the spikes on the threefold axes are convex (Vogel *et al.*, 1986). Each spike has the shape of an upside-down mushroom, the hats nearly covering the bilayers. The caps of the mushrooms penetrate deeper into the membrane at their circumferences than at their centers. Spike–spike interactions might influence the envelope geometry (Vogel *et al.*, 1986). Thus, there are reasons to believe in spike cooperativity during adsorption and fusion.

The isolated capsid is smooth and fenestrated (Fuller and Argos, 1987). One-quarter of the spikes of SFV can be extracted by low levels of n-octyl-β-D-glucopyranoside (Helenius and Kartenbeck, 1980). In isolated nucleocapsids RNA is easily degraded by ribonuclease (Söderlund *et al.*, 1972, 1979). The exposure of nucleocapsids to a low pH leads to a contraction of 60 Å (Söderlund *et al.*, 1972, 1975). In the case of SFV, 31 amino acids of E_2 and two amino acids of E_1, both carboxy terminal, are on the inside of the membrane (Garoff *et al.*, 1982). Six transmembrane helices lie in a bundle beneath the stalk of a spike. This is probably essential for the trimer formation, since the proteolytically released ectodomains of the spike polypeptides are monomeric (Kielian and Helenius, 1985).

Specific interactions between the nucleocapsid and the transmembrane roots of the spikes have long been assumed (Garoff and Simons, 1974; Helenius and Kartenbeck, 1980). The 80 regularly spaced depressions in the surface of the nucleocapsids (Fuller, 1987) might provide the binding sites. In fact, the nucleocapsid of SFV contains binding sites for the cytoplasmic tails of the E_2 protein (Vaux *et al.*, 1988).

Coombs and Brown (1987a,b) suggest a hollow core in which all of the C proteins are arranged in such a way that at least one protein domain toward the carboxy-terminal side is exposed on the surface. The amino-terminal 88 amino acids contain 20 basic amino acids (Boege *et al.*, 1981; Rice and Strauss, 1981). Thus, the RNA binding region might be in the amino-terminal region and the RNA might be distributed primarily on the surface. A similar conclusion is reached by digesting cross-linked nucleocapsids with ribonuclease, whereby RNA-free protein shells with intact morphology are generated (Coombs and Brown, 1987a,b).

Trimeric spike arrangements which might form channels seem to be quite common among different viral families. The hemagglutinin of influenza virus is a trimeric transmembrane glycoprotein in which each subunit consists of the disulfide-linked glycoproteins HA1 and HA2 (Wiley *et al.*, 1981; Wilson *et al.*, 1981). A mildly acidic pH causes an irreversible conformational change which triggers fusion activity and exposes previously buried hydrophobic groups (Rott *et al.*, 1984; Doms *et al.*, 1985; Gething *et al.*, 1986a). Various data suggest that the domains at the top of the spike dissociate from each other (Doms *et al.*, 1985). Isolated trimeric bromelain-solubilized hemagglutinin fragments, lacking the carboxy-terminal membrane anchor, dissociate to monomers when exposed to an acidic pH (Nestorowicz *et al.*, 1985). This dissociation does not take place in virions; rather, hemagglutinin molecules might form larger oligomeric structures in the plane of the virion membrane (Doms and Helenius, 1986).

From the surface of protease-treated West Nile virus (i.e., flavivirus), an E-protein trimer can be isolated (Wengler *et al.*, 1987). Flaviviruses are fusogenic at a mildly acidic pH (H. Koblet and A. Igarashi, unpublished observations).

The G protein of vesicular stomatitis virus (VSV) (i.e., rhabdovirus) is also trimeric (Kreis and Lodish, 1986; Dubovi and Wagner, 1977). It is the only transmembrane protein type in this virus which is fusogenic at a mildly acidic pH (Florkiewicz and Rose, 1984). The 14K envelope protein of vaccinia virus (i.e., poxvirus) is involved in cell fusion and forms disulfide-bonded trimers on the surface of the virion (Rodriguez *et al.*, 1987). All of these viruses, which are different with respect to family, information content, and structure, exhibit a covalent or noncovalent trimerization of the fusogenic surface protein. In addition, we know that SFV, influenza virus, and VSV provoke FFWO and FFWI at a mildly acidic pH and form proton channels (see Section VI,H) in the PM (Kempf *et al.*, 1987c). Sendai virus (i.e., paramyxovirus) (Asano and Asano, 1985), fusogenic at pH 7.0, might be different. Of the three envelope proteins HN, F, and M, the HN glycoprotein with the hemagglutination and sialidase activities forms the HN spike, and the F glycoprotein composed of the two disulfide-bonded F_1 and F_2 subunits with the hemolytic and fusion (F_1) activities forms the F spike. The F spike is probably composed of four identical F_1–F_2 dimers. The dimers are linked by noncovalent associations (Sechoy *et al.*, 1987).

Ion channels might be a common feature even of viral capsids. Theoretical considerations in the case of icosahedral plant viruses point in this direction (Silva *et al.*, 1987). Ion channels leading through the bilayer and eventually into the nucleocapsid could play an important role in fusion and uncoating.

III. Mosquito Cells

Many recent books and reviews cover the topics of mosquito cell growth, invertebrate media, and the relationships between arboviruses and invertebrate hosts (Weiss, 1971; Kurstak and Maramorosch, 1976; Kurstak *et al.*, 1980; Schlesinger, 1980; Maramorosch and Mitsuhashi, 1982; Schlesinger and Schlesinger, 1986; Fallan and Stollar,

1987; Yunker, 1987a,b; Kuroda *et al.*, 1988). Specifically, Gould and Clegg (1985) treat the practical aspects. Mitsuhashi (1982) indicates the formulas of many insect media, among them that of the famous Mitsuhashi-Maramorosch (1964) medium. Kurtti and Munderloh (1984) describe cell lines and clones. Brown (1984) compares alphavirus growth in cultured vertebrate and invertebrate cells. Igarashi (1985) comments on persistently infected mosquito cells, defective interfering particles, CPEs, and applications of mosquito cells in epidemiology and diagnostics. Stollar (1987a) discusses in depth relationships between host cells and viruses, especially CPE of Sindbis virus in *Aedes albopictus* cells and host range mutants. Thus, a few remarks concerning mosquito cells in culture might suffice.

The main reasons for studying arboviruses in arthropod cells are the following (Stollar, 1987b): (1) The biochemical composition of the viruses depends on whether they are grown in vertebrate or arthropod hosts (Stollar *et al.*, 1976; Luukkonen *et al.*, 1976, 1977). Therefore, it is important to know possible biological differences of the viruses. (2) Alphaviruses, usually cytopathogenic for vertebrate cells, can grow to high levels in mosquito cells without a CPE. Comparative studies might shed light on the mechanism of cell killing. (3) The study of viral infection of mosquito cells in culture might answer questions concerning the transmission of disease in nature. (4) Host range viral mutants might be selected. (5) Mosquito cells are useful in investigating cell–cell fusion. This is due mainly to their insensitivity to pH and concentration shifts in the medium.

The pH values of dipteran hemolymph, which fluctuate during the life cycle, range from 6.3 to 7.7. The pH of most cell culture media is fixed to 6.8–7.0, but there is no need for this. It is a remarkable advantage that, for example, *Culex tritaeniorhyncus* cells grow equally well from pH 6.3 to 7.6 (Hsu *et al.*, 1972). Likewise, the osmotic pressure of hemolymph varies during the life cycle, and the osmolality of the media is not critical (Sarver and Stollar, 1977). We can confirm this for Igarashi's (1978) *A. albopictus* clone C6/36, which grows from pH 6.0 to 8.0 in vertebrate (290 mOsm/liter) medium and in invertebrate (400 mOsm/liter) Mitsuhashi–Maramorosch medium. This clone is routinely used in our laboratory.

Clones were selected to yield high levels of certain arboviruses (Igarashi, 1978; Gillies and Stollar, 1980; Tooker and Kennedy, 1981). Some high virus-producing clones show a CPE in the form of an aggregation, but no FFWI at pH 7.0. Such a clone is, for example, Igarashi's clone C6/36 from *A. albopictus* (Singh). This clone is particularly useful (Koblet *et al.*, 1987) for several reasons: (1) It can be stored at room temperature (25–28°C) for up to 2 weeks without a change in the medium. (2) With a doubling time of about 27 hours, there is an excellent split ratio of 1 : 8 to 1 : 10; when seeded at 5 × 10⁵ cells per milliliter, a complete monolayer will form in 3–4 days. (3) It can be grown in vertebrate or invertebrate media, in monolayers and in spinner culture. (4) Systematic surveys show a remarkable number of prototype viruses which grow in this cell line (White, 1987). (5) Arboviruses such as SFV reach titers of 10^9–10^{10} plaque-forming units (pfu)/ml in 24–36 hours postinfection with a multiplicity of infection (MOI) of 1–10 per cell when the titer is assayed with chick embryo fibroblasts or Vero cells. (6) During the time of most massive virus (i.e., SFV) production a CPE (i.e., aggregation of cells or starlike contacts between cells) appears (14–36 hours postinfection), but many cells survive in a state of persistent infection (Stalder et al., 1983). (7) After infection with SFV, the flavivirus Japanese encephalitis, or VSV, the cells readily fuse at a mildly acidic pH, and the syncytia survive for prolonged periods. (8) There is no evidence that the fundamental organellar organization of such clones is different from that of the conventional vertebrate cell lines; however, there are clear-cut differences in posttranslational modifications (Hsieh and Robbins, 1984) and cholesterol biosynthesis (Mitsuhashi et al., 1983). Thus, it is assumed in the next section that, in principle, traffic and sorting problems as well as viral infection are the same as in the well-studied vertebrate cell lines.

IV. REPLICATION MACHINERY

A. Adsorption

Entry starts with the binding of the virus through its spikes to cell surface structures. The exact nature of this event is poorly understood. The E_1 protein seems to be sufficient for binding (Omar and Koblet, 1988). Adsorption is dependent on ionic strength (Pierce et al., 1974). Sindbis virus binds to liposomes; cholesterol and phosphatidylethanolamine enhance the binding (Mooney et al., 1975). The number of particles bound per cell increases with a growing free concentration of virus, but saturation is barely achieved (Fries and Helenius, 1979) with baby hamster kidney (BHK), HeLa, Daudi, and P815 cells. Signs of uptake saturation were seen only when more than 50,000 SFV particles per cell were bound (Marsh et al., 1982). Thus, on the PM of BHK cells, there are at least 10^5 binding sites for Sindbis virus (Birdwell and Strauss, 1974). C6/36 cells behave similarly. By cross-linking Sindbis virus to the cell surface of lymphoblastoid cell lines, a 90K protein was identified as a major protein located near the binding site of the virus (Maassen and Terhorst, 1981).

Together, consecutive binding steps of the virus could be possible, first, to charged phospholipid head groups of the outer leaflet of the PM, and second, to proteins in the neighborhood of the binding site. The structural differences of the glycans of the viral membrane proteins seem to play a minor role (Naim and Koblet, 1988). Indeed, lipids as alphavirus receptors have been suggested (Kääriäinen and Söderlund, 1978). This is attractive in view of the broad host range of alphaviruses. There is evidence that binding of the virus at pH 7.0 to a hydrophobic support forces the spike into a configuration which might allow the fusogenic conformation to be adopted at pH 6.0 (Omar and Koblet, 1988).

B. Penetration and Uncoating

1. General Remarks on Membrane Traffic

Transport of macromolecules and lipids into and out of the cells has been investigated extensively. A discussion of the relevant literature by far exceeds the scope of this chapter. However, since the topic is of paramount importance in virology, a few remarks are mandatory. The network interlinked in intracellular movements—"an intracellular merry-go-round"—has been summarized by Hoekstra et al. (1989). We distinguish an endocytotic pathway leading into the cell, several recycling pathways, and an exocytotic pathway transporting newly formed proteins from the endoplasmic reticulum (ER) into lysosomes, into secretory vesicles (i.e., a signal-mediated regulated pathway), or directly into the outside world (i.e., constitutive, or signal-independent default, pathway), respectively. Thereby, all parts of the assembly line are constantly being repaired (e.g., housekeeping proteins).

It is generally assumed that all intracellular transports are performed by vesicles with a bilayer membrane. In Fig. 1 the different phases (a–i, k–t) are shown and numbered (1–14). After binding to cell surface receptors, many extracellular ligand molecules (i.e., virions which fuse at an acidic pH) are internalized through a process termed "receptor-mediated endocytosis." This process starts with migration of the ligand–receptor complexes to pits (a) coated on the cytoplasmic side with a clathrin shell (Pearse, 1982, 1987). The invagination of the coated pit yields a coated vesicle (b) (Linden and Roth, 1983), losing the coat rapidly (Anderson et al., 1977) to form an uncoated vesicle (c). Fusion of such vesicles with one another (c) results in the creation of the early endosome (d) (Tycko and Maxfield, 1982; Helenius et al., 1983; Geuze et al., 1983).

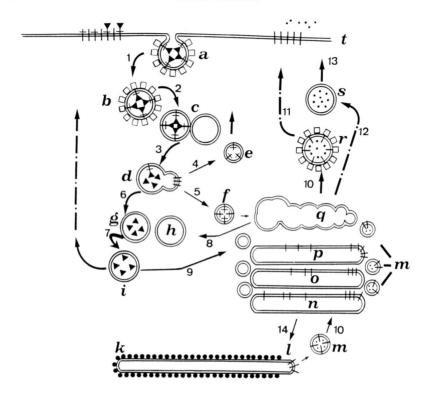

FIG. 1. The intracellular merry-go-round. a, Coated pit with receptors and ligands; b, coated vesicle; c, uncoated vesicle; d, early, or sorting, endosome; e, receptor recycling vesicle; f, merging traffic vesicle; g, late endosome; h, primary lysosome (i.e., pre-lysosome); i, secondary lysosome; k, rough endoplasmic reticulum; l, smooth endoplasmic reticulum; m, transport vesicles with housekeeping and secretory proteins; n, cis-Golgi cisterna; o, medial-Golgi cisterna; p, trans-Golgi cisterna; q, trans-Golgi network; r, clathrin-coated secretory vesicle of the constitutive pathway; s, uncoated storage vesicle of the regulated pathway; t, plasma membrane; u, nucleus. 1–3, Pathway to sorting endosomes; 4, direct recycling pathway to the plasma membrane; 5, indirect recycling pathway (via the trans-Golgi network) to the plasma membrane; 6–7, pathway to secondary lysosomes; 8, transport pathway for lysosomal enzymes; 9, recycling of mannose 6-phosphate receptors; 10, transport pathway leading to exocytotic vesicles; 11, constitutive secretory pathway and plasma membrane turnover; 12–13, regulated pathway; 14, recycling of endoplasmic reticulum housekeeping proteins. †, Cell surface receptor; ▼, ligand; □, clathrin; |, housekeeping protein; ·, secretory protein; ↑, leads outside; ↓, leads inside; ●, ribosome; =, endo- and exoplasmic leaflets of a membrane.

Through the action of proton pumps (Yamashiro and Maxfield, 1988), the intravesicular pH gradually decreases to about 6.0, and ligands and receptors start to dissociate. A mathematical model has been constructed (Linderman and Lauffenburger, 1988). Receptors are sorted into distinct areas where receptor-enriched vesicles are pinched off and return to the PM (e) (receptor recycling) (Geuze et al., 1983, 1987). Thus, the early endosomes develop into late endosomes (g) (White et al., 1983; Davoust et al., 1987; Schmid et al., 1988). Finally, they fuse with primary lysosomes (i.e., prelysosomes) (h). These are products of the trans-Golgi network (q) (Griffiths and Simons, 1986; Kornfeld, 1987; Griffiths et al., 1988) and carry the lysosomal enzymes, which, in turn, are bound to the mannose 6-phosphate receptors (von Figura and Hasilik, 1986) on their luminal side. The unification of late endosomes and primary lysosomes gives rise to a final product, the secondary lysosome (i), in which ligands are degraded. Products are then released from the cell (von Figura and Hasilik, 1986; Lloyd and Forster, 1986).

There are probably several pathways where endocytotic and exocytotic pathways meet ("merging traffic"): (1) Some receptors (f) might recycle from the early endosomes (d) to the trans-Golgi network (q) or the Golgi apparatus (van Deurs et al., 1987) and from there back to the PM (Roth et al., 1985; Hedman et al., 1987; Stoorvogel et al., 1988). In most cases the recycling and the exocytotic vesicles remain distinct (Hedman et al., 1987). (2) After delivery of the lysosomal enzymes, the mannose 6-phosphate receptors return to the trans-Golgi compartment to sort the next wave of lysosomal enzymes (9) (Brown et al., 1986; Duncan and Kornfeld, 1988).

Acidification of the contents of the immigrating vesicles proceeds in phases. In early endosomes the pH is 6.2 to 6.3; in sorting endosomes, lower than 6.2; in late endosomes, 5.2–5.8; in lysosomes, 5.2–5.3. These steps determine where and when certain processes take place and are important with respect to the uptake of viruses. An ATP-dependent electrogenic (Mellman et al., 1986) proton pump has been identified in early endosomes (Yamashiro and Maxfield, 1988), together with a Na^+,K^+-ATPase (Fuchs et al., 1986). The latter cannot be found in late endosomes; possibly, it recycles away from the sorting endosome. If the Na^+,K^+-ATPase creates a positive potential inside the endosome, it inhibits the acidification; if it is lost from the endosome, additional acidification by the proton pump would be allowed. Thus, early acidification might be determined by the inside positive membrane potential, whereas later acidification might be dependent on the absolute pH (Cain and Murphy, 1988).

Logically, one is tempted to assume that at least the early endosome is nothing other than a piece of internalized PM. Therefore, it is most intriguing that the PM H⁺-ATPase belongs into the P (phosphorylated enzyme intermediate) class, whereas the H⁺-ATPase of coated vesicles and endosomes represents a V (nonphosphorylated) type (Pedersen and Carafoli, 1987; Nelson, 1987). Thus, it is not clear at what stage the endosomal membrane becomes a membrane of its own. There is increasing evidence that endosomal subpopulations each contain unique sets of proteins not found on the PM. Therefore, endosomes cannot derive entirely from the cell surface (Schmid *et al.*, 1988). The advent of cell-free systems (Davey, 1987; Braell, 1987; Marsh *et al.*, 1988; Beckers and Balch, 1989; Beckers *et al.*, 1989; Diaz *et al.*, 1989; Wilson *et al.*, 1989) for studying the endosomal and exocytotic pathways will soon clarify this situation.

In Chinese hamster ovary cells reduced temperature (17°C) strongly inhibits transport from early endosomes. pH elevators such as NH₄Cl, monensin, and HEPES buffer (i.e., lysosomotropic drugs), as well as the respiratory inhibitors NaF or KCN, inhibit transport most strongly at the early endosomal stage (Sullivan *et al.*, 1987).

Clathrin-coated membranes are also involved in the exocytotic pathway from the Golgi apparatus (Fig. 1, r and s). However, differences have been described between the clathrin assembly protein complexes of the endocytotic and the exocytotic pathways. The tentative conclusion drawn is that the assembly proteins play a role in targeting (Ahle *et al.*, 1988).

2. Entry of Alphaviruses

Alphaviruses enter cells by the endosomal route as intact virions (i.e., pinocytosis) (White *et al.*, 1983; Marsh, 1984; Asano and Asano, 1984; Kielian, 1987; Marsh and Helenius, 1989). In endosomes, at about pH 6.0, the viral membrane fuses with the endosomal membrane and the nucleocapsid is ejected into the cytoplasm. This process represents part of the uncoating and starts infection. The viral spike proteins are now components of the endosomal membrane and are later degraded in lysosomes. Three main findings led to this view: (1) Lysosomotropic agents inhibit SFV infectivity by dissipating the proton gradient (Helenius *et al.*, 1980, 1982; Talbot and Vance, 1982). (2) Efficient FFWO occurs at pH 6.0 or lower between phospholipid–cholesterol liposomes and isolated SFV, whereby the nucleocapsid is translocated into the liposomes (Helenius *et al.*, 1980). (3) Infection can be induced by a brief low-pH treatment of cells with bound SFV under conditions in which the normal infection route is blocked (FFWO at pH 6) (White *et al.*, 1980; Matlin *et al.*, 1981). Using a high MOI of 1000 pfu per cell and more, such treatment will lead to syncytium formation of the C6/36 cells (Omar *et al.*, 1986). However, this is not the natural pathway of infection.

Radiolabeled Sindbis virus does not penetrate into the lysosomes of BHK cells treated with NH_4Cl for at least 40 minutes post infection (Talbot and Vance, 1982). This shows that endocytic vesicles become acidic before they reach the lysosomes. The first virus particles reach a low-pH (i.e., lower than 6.1) compartment within 3–4 minutes (Helenius and Marsh, 1982). Accordingly, the SFV genome release into the cytosol of BHK cells at 37°C can be detected within 5–7 minutes after the start of endocytosis, whereas delivery of the residual virus particles to secondary lysosomes occurs within only 15–20 minutes. The SFV fusion mutant *fus-1* reaches its fusion pH of 5.3 in late endosomes after 8–10 minutes (Schmid *et al.*, 1989).

Degradation of the viral envelope proteins takes place in the lysosomes and degradation products appear in the medium (Marsh and Helenius, 1980). At temperatures of 15–20°C, virus particles are internalized by endocytosis, but they are not transported to secondary lysosomes (Marsh *et al.*, 1983). However, uncoating and infection are possible. This, again, indicates that lysosomes are not required for penetration of the nucleocapsid (Marsh *et al.*, 1983). Yoshimura and Ohnishi (1984) came to a similar conclusion with respect to influenza WSN virus in Madin–Darby canine kidney cells. The process can be visualized in the electron microscope (Helenius, 1984). There is good evidence that the principles of adsorptive endocytosis are valid in all permissive cells investigated (i.e., mammalian, avian, piscine, and arthropod) (Superti *et al.*, 1987; Hase *et al.*, 1989). It must be mentioned here that alternative entry pathways might exist. Using a temperature-sensitive mutant of Chinese hamster ovary cells, defective in its ability to acidify endosomes, J. Edwards and D. T. Brown (in preparation) demonstrated that Sindbis virus RNA synthesis was initiated in wild-type parent cells and mutant cells at the same time after the addition of virus at either a permissive or nonpermissive temperature. Thus, exposure to the acidic environment of an endosome seems not always a prerequisite for alphavirus infection.

Uptake is rapid (half-time on the BHK cell surface is about 10 minutes) and is not dependent on the MOI (Marsh and Helenius, 1980). At a high MOI (37°C) BHK cells ingest up to 3000 SFV particles per minute. About 2400 coated vesicles per minute are internalized (Marsh and Helenius, 1980). This corresponds, in terms of coated vesicles, to 1% of the cell surface area and 0.04% of the cell volume. Uptake is not inhibited by methylamine, chloroquine, NH_4Cl, or the ionophores monensin and nigericin. However, these agents inhibit endosomal penetration by increasing the pH of the vesicle content above the threshold value to activate the fusion reaction (Marsh *et al.*, 1982; Helenius and Marsh, 1982).

The candidate endosome to release the nucleocapsid is the sorting endosome. If this were true, the fusing membrane of the endosome would still be quite similar to the PM, and FFWI at the PM (as described in Section VI) could be considered a valid model for EF. Agents disrupting the cytoskeleton (e.g., colchicine and cytochalasin B) have only a marginal effect on uptake, whereas inhibitors of oxidative phosphorylation (e.g., NaN_3, dinitrophenol, and 2-deoxy-D-glucose) are quite effective inhibitors.

In summary, endosomal fusion requires a critical pH of 6.1 or lower, intact viral spike glycoproteins, cholesterol in the target membrane, and a temperature above 10°C. It is not clear whether sterols play a role in cell tropism in mosquitoes, which are dependent on dietary sterols.

At low pH the spike proteins undergo cholesterol-dependent (Kielian and Helenius, 1984, 1985) conformational changes (Edwards et al., 1983) and the E_1 protein loses its trypsin sensitivity (Omar and Koblet, 1988). In media with a low sodium ion concentration (10 mM) (e.g., BHK cells), virus endocytosis and degradation are nearly normal and trypsin resistance of E_1 occurs. However, there is no fusion and no release of the nucleocapsid into the cytosol. There is a close correlation between conditions which inhibit infection and which cause depolarization of the cells. Thus, fusion of SFV with the endosomal membrane in intact cells also depends on the maintenance of a potential (Helenius et al., 1985), which may be established by the action of the electrogenic H^+-ATPase.

Not much is known about the structure of the freshly released nucleocapsid and the transport to ribosomes, where the genomic RNA must be liberated to function as a message. An elegant hypothesis has been formulated by Wengler (1987), according to which dissociation and formation of nucleocapsids are governed by mass action. The replicase is located in ribonucleoprotein structures associated with the cytoplasmic surface of modified endosomes. These structures often form bridges between the vacuoles and the rough ER (RER). Apparently, such "cytopathic vacuoles" are sites for viral RNA synthesis, translation of structural proteins, and assembly of nucleocapsids (Froshauer et al., 1988).

C. Synthesis of the SFV Components

1. RNA

In invertebrate and vertebrate cells a 49 S (4.5×10^6) and 26 S RNA (1.6×10^6 D) are formed (Simmons and Strauss, 1972). The 49 S RNA (plus strand; SFV, 11,442 nucleotides; Sindbis virus, 11,703 nucleotides) corresponds to the genomic RNA. The 5'-terminal codes for the 4 nonstructural proteins composing the RNA-dependent RNA polymerase. This enzyme transcribes and then replicates the viral RNAs (Michel and Gomatos, 1973; Gomatos *et al.*, 1980; Sawicki and Sawicki, 1980).

Several activities of the Sindbis virus ns proteins have recently been deduced (Hahn *et al.*, 1989a,b). ns P1 (complementation group B) seems to be involved in the initiation of minus-strand RNA synthesis and could function as an RNA methyltransferase (Mi *et al.*, 1989); ns P2 (groups A and G) appears to be required for the initiation of 26 S RNA synthesis and could, in addition, be the protease that processes the nonstructural polyproteins (Ding and Schlesinger, 1989). Furthermore, ns P2 might regulate minus-strand RNA synthesis. ns P4 (group F) seems to function as the polymerase or as an RNA elongation factor (Sawicki *et al.*, 1990). Mutations in any of the four nonstructural proteins can result in an RNA-negative phenotype.

The 3' one-third of the genomic RNA codes for the structural proteins; however, it is not used for protein biosynthesis. There is no overlap between the genes for nonstructural proteins and structural proteins (Riedel *et al.*, 1982). The 26 S RNA serves as a subgenomic messenger for the biosynthesis of all structural proteins (Wengler and Wengler, 1974; Clegg and Kennedy, 1975; Wirth *et al.*, 1977; Bonatti *et al.*, 1979). At the 3' end of this RNA, containing 4074 and 4106 nucleotides, there are 264 and 322 untranslated nucleotides in SFV and Sindbis virus, respectively (Garoff *et al.*, 1980a; Rice and Strauss, 1981; Strauss *et al.*, 1984). The 3' ends of 49 S and 26 S RNA are colinear (Kennedy, 1976; Wengler and Wengler, 1976a; Ou *et al.*, 1981). There is also a short stretch of polyadenylic acid (Wittek *et al.*, 1977). The 5' end is capped (Dubin *et al.*, 1979); 50 (49, Sindbis) untranslated bases then follow before the initiation codon, which is the first AUG.

The replication of alphaviruses has been extensively treated by Strauss and Strauss (1977, 1983), Kääriäinen and Söderlund (1978), and Kääriäinen *et al.* (1987). It may suffice to say that the infecting genomic plus strand is first transcribed into a minus-strand 49 S RNA; the latter, in turn, is the template for the replication of either progeny plus-strand 49 S RNA or 26 S mRNA. The regulation involved in the synthesis of the two plus-stranded RNAs is not fully understood. The formation of 26 S RNA starts internally on the minus-strand 49 S RNA (Wengler *et al.*, 1979; Petterson *et al.*, 1980). The result is the synthesis of about 75,000–150,000 molecules per cell (e.g., BHK and HeLa) of both 49 S and 26 S RNA at 8 hours postinfection (Tuomi *et al.*, 1975).

In mosquito cells (Enzmann, 1987a) the replication proceeds according to the same rules. In some cell lines the amount of 26 S RNA might be much smaller than that of 49 S RNA (Eaton, 1978), but in clones of

A. *albopictus* cells, Tooker and Kennedy (1981) found no difference in molar ratios. Later, the inhibition of viral RNA synthesis represents the first demonstrable step on the road to persistent infection (Davey and Dalgarno, 1974; Eaton, 1978). The formation of defective RNA could play a role in this inhibition (Stalder *et al.*, 1983).

2. Protein Biosynthesis

A most interesting situation arises because the messenger for the viral structural proteins is polycistronic and codes for proteins destined for different cellular compartments. The C protein forming the nucleocapsid, together with the progeny genomic RNA, travels in the cytoplasm, whereas the spike proteins E_1, E_2, and E_3 are directed to the PM. Several approaches have shown the gene order to be 5′–C protein–p62 protein–E_1–protein (Lachmi *et al.*, 1975; Clegg, 1975). In the precursor p62 (Simons *et al.*, 1973; Schlesinger and Schlesinger, 1973) E_3 is amino terminal and E_2 is carboxy terminal. Between E_2 and E_1 there is a small (i.e., 6K) peptide not expressed in the mature virion (Welch and Sefton, 1980; Garoff *et al.*, 1982). There is only one initiation site (Clegg and Kennedy, 1975; Cancedda *et al.*, 1975; Glanville *et al.*, 1976).

Therefore, the translation product is a polyprotein, and the final products must be generated by endoproteolytic cleavages between C and E_3, E_3 and E_2, E_2 and 6K, and 6K and E_1. The cleavages between C and p62, p62 and 6K, and 6K and E_1 are cotranslational. The capsid protein is the first to be translated on "cytoplasmic" polysomes and is efficiently cleaved between the carboxy-terminal tryptophan (amino acid 267) and the amino-terminal serine of E_3 (amino acid 268) soon after synthesis, when translation of the polyprotein is still occurring.

The newly synthesized and cleaved capsid protein is bound to the large ribosomal subunit; after that it complexes with progeny 49 S RNA to give nucleocapsids in the cytoplasm (Söderlund and Ulmanen, 1977; Ulmanen, 1978). Nucleocapsid assembly must be fast and efficient, because incomplete nucleocapsids and free components are not found in infected cells. However, in view of the equimolar biosynthesis and the 180 C : 240 E protein relationship in the mature virion, an intracellular pool of nonused C protein must exist. This free C protein might migrate into the cell nucleus (Jalanko, 1985) and then into the nucleolus (M. R. Michel, M. Elgizoli, Y. Daig, R. Jacob, P.-A. Arrigo, and H. Koblet, in preparation).

Quite a stringent regulation of a 180 C protein to one 49 S RNA stochiometry must be postulated. It is unknown how the packaging and the transport of the nucleocapsid to the site of virus maturation take place. A region close to the 5′ end of the genomic RNA seems to contain the sequence essential for specific binding to the C protein (Weiss *et al.*, 1989). In mosquito cells the cytoskeleton is not involved in this transport (C. Kempf, A. Omar, U. Kohler, and H. Koblet, unpublished observations).

The biosynthesis of the membrane proteins proceeds on membrane-bound polysomes, where these proteins are segregated in part to the luminal side of the RER (Wirth *et al.*, 1977; Garoff and Söderlund, 1978). The luminal side is topologically equivalent to the outside of the cell and the virion. Protease digestion in cell-free translation assays containing "microsomal membranes" proves that the carboxy-terminal ends (i.e., endodomains) of p62 and E_1 sit on the cytoplasmic side, which corresponds to the nucleocapsid side of the mature virion. The newly emerged p62 and the E_1 proteins appear to form a spikelike structure in the RER membrane (Ziemiecki and Garoff, 1978; Ziemiecki *et al.*, 1980), also in mosquito cells (Naim and Koblet, 1990). This seems to be the prerequisite for trimerization, correct conformation, and transport. Whereas E_2 expressed alone can be transported to the cell surface, E_1 must be in noncovalent association with p62 to move out of the RER (Kondor-Koch *et al.*, 1983).

The proteases involved in the processing of viral polyproteins have recently evoked much interest (for reviews see Kräusslich and Wimmer, 1988; Wellink and van Kammen, 1988). Processing of alphavirus polyproteins has been reviewed by Strauss *et al.* (1987). Using an *in vitro* transcription–translation assay, Melançon and Garoff (1987) showed that the release of the C protein is catalyzed by the core itself. A change of the sequence Gly–Asp–Ser-219–Gly, a conserved tetrapeptide of several serine proteases, to Gly–Asp–Arg–Ser–Thr abolishes this *in vitro* cleavage completely.

This highly specific chymotrypsinlike activity can act in trans (Aliperti and Schlesinger, 1978) or in cis (Melançon and Garoff, 1987). However, the C protease is not involved in the cleavages of p62 or the 6K peptide, which are probably performed by membrane-associated host proteases. The processing of the nonstructural polymerase (Ishihama and Nagata, 1988) polyproteins ns P1–ns P4 will not be treated here. A cleavage model in the case of Sindbis virus was recently published by Hardy and Strauss (1988). A common amino acid sequence motif shared by many nonviral, plant, and animal viral polymerases is noteworthy (Argos, 1988).

D. Protein Import into the RER

The segregation of the membrane proteins into the RER membranes after cleavage of the C protein starts with the newly exposed signal containing the amino terminus (E_3) of the p62 protein. Translocation continues until it is arrested at the transmembrane stop–transfer segment near the carboxy terminus of p62. Translocation is resumed when the next signal (Hashimoto et al., 1981) appears in the last 26 amino acids of the 6K peptide (Melançon and Garoff, 1986). The movement of E_1 again comes to a halt when its transmembrane segment is formed. It is not clear whether the two cleavages releasing the 6K occur before, during, or after translocation. The cleavage between 6K and E_1 is probably catalyzed by the host signal peptidase. Whereas the signal of E_1 (i.e., 6K) is lost in the cell, the signal of p62 is conserved in the mature SFV as the E_3 protein without known additional function. In Sindbis virus the E_3 protein is lost into the extracellular fluid (Welch and Sefton, 1979). It is obvious that the fusogenic sequence of E_1 cannot reside in a signal sequence.

Biosynthesis of SFV transmembrane proteins follows the accepted rules of RER protein import (for reviews see Garoff, 1985; Wickner and Lodish, 1985; Lesser et al., 1987; Robinson and Austen, 1987; Singer et al., 1987a,b; Dreyfuss et al., 1988; von Heijne, 1988a; von Heijne and Gavel, 1988; Tsou, 1988; Verner and Schatz, 1988). There is no evidence that these rules should not apply to mosquito cells. Whether translocation occurs through lipid or through a proteinaceous pore is not known. No transmembrane electrochemical potential and no proton motive force are needed. However, ATP (or GTP) is mandatory for the cotranslational and posttranslational transport of certain precursors across the ER membrane and might participate in the unfolding of these proteins (Eilers and Schatz, 1988).

There are common structural features in signal sequences. They range from 15 to 30 residues in length, with an average of about 20. There is a central region of 7–15 uncharged consecutive residues (von Heijne, 1988b). Membrane-spanning (stop–transfer) regions might contain longer stretches of uncharged residues. As a general rule, on the amino-terminal side, positively charged residues flank the apolar regions on the cytoplasmic nontranslocated side of cleaved and uncleaved signal peptides, whereas stop–transfer signals have basic amino acids on the cytoplasmic carboxy-terminal side (i.e., the "positive-inside rule"). According to the positive-inside rule, signal sequences insert in such a way that the growing protein forms a loop through the ER membrane (Shaw et al., 1988). The alpha p62 is, therefore, an unusual class I membrane protein with a translocated, glycosylated, and noncleaved signal peptide (Mayne et al., 1984) and an internal stop–transfer (e.g., transmembrane or anchor) signal.

Class I proteins are defined by an amino-terminal cleavable signal peptide and an internal stop–transfer sequence. They are left in the mature state, with the carboxy terminus facing the cytoplasm. In contrast, E_1 is a class I membrane protein with a cleaved signal peptide and an internal stop–transfer domain. In addition, there is an internal fusogenic sequence. The resulting topology is, by definition, N_{out}–C_{in}. Signal sequences and transmembrane domains might be interchangeable in some situations. For example, the membrane anchor of SFV p62 can function as both a translocation signal and a membrane anchor in place of the normal transmembrane domain of the transferrin receptor (Zerial *et al.*, 1987).

Fusogenic viral proteins such as SFV E_1 (see Section VI,F) have rather long apolar segments (i.e., fusion peptides) that are translocated across the RER membrane, with no stop–transfer function. These sequences must be close to the stop–transfer threshold. They do not halt translocation in an internal position; however, when placed at the carboxy terminus, they act as a stable anchor (Paterson and Lamb, 1987). This might explain why most fusogenic sequences are translocated in the depth of a precursor protein and, by a cleavage phenomenon, are exposed late only during maturation on the amino-terminal side of the mature fusogenic protein. 70K heat shock-related cytoplasmic proteins stimulate the protein translocation into microsomes (Chirico *et al.*, 1988). The significance of this finding for the heat shock-prone mosquito cells is unknown.

E. Oligomerization

Many secreted and PM proteins in eukaryotic cells are homo- or heterooligomeric. For several viral membrane proteins trimerization seems to be a constructional principle (see Section II,B). VSV G protein forms G_3 (Doms *et al.*, 1987); influenza hemagglutinin forms $(HA1–HA2)_3$ (Wiley *et al.*, 1981; Wilson *et al.*, 1981); Sindbis virus oligomerizes into the heterotrimer $(E_1E_2)_3$ (Fuller, 1987), and SFV, $(E_1E_2E_3)_3$ (Vogel *et al.*, 1986); and the flavivirus West Nile aggregates into E_3 (Wengler *et al.*, 1987). In these cases the associations are noncovalent.

Trimerization is an early event after translocation and after folding of the monomers (Gething *et al.*, 1986b; Kreis and Lodish, 1986; Copeland *et al.*, 1986, 1988; Doms *et al.*, 1987). For example, trimerization of the uncleaved influenza precursor $[(HA0)_3]$ starts within a few minutes after translocation of HA0 at random and from a common pool (Boulay *et al.*, 1988). Monomers are not transported to the cis-Golgi cisterna. Thus, correct protein folding and subunit assembly are linked to the transport (Copeland *et al.*, 1988; Doms *et al.*, 1988).

N-Glycosylation plays an indirect role in this transport process out of the RER. G Protein lacking glycans or having no glycans at the normal positions is subject to aberrant intermolecular disulfide bonds and accumulates large complexes in the RER (Machamer and Rose, 1988a,b). The most convincing role of the cotranslational glycosylation is, therefore, an influence on the tertiary and quaternary structures of a protein.

The topographical site of the trimerization of SFV is not known. The early heterodimer formation between p62 and E_1 (Bracha and Schlesinger, 1976; Ziemiecki and Garoff, 1978; Ziemiecki et al., 1980; Naim and Koblet, 1990) might be sufficient for transport to take place. If trimerization were coupled to p62 cleavage, it would occur near the PM in higher eukaryotes. In view of the intracellular budding in mosquito cells (see Section IV,H), it could be assumed that trimerization occurs early in the transport pathway of these cells.

All of these proteins integrated into heterotrimeric or homotrimeric spikes are fusogenic at a mildly acidic pH. However, there is no dominant icosahedral building principle. Thus, oligomerization might also be an important element for viral fusion, due to a cooperative effect and because trimers could form channels (see Sections VI,H and I).

F. Transport Pathway and the Sorting Problem

In eukaryotic cells the number of destinations for newly synthesized proteins is large (e.g., the cytoplasm, nucleus, peroxisomes, chloroplasts, mitochondria, RER, smooth ER, Golgi apparatus, lysosomes, PM, and the extracellular space). A large number of reviews have appeared covering this topic (Rodriguez-Boulan et al., 1985; Caplan et al., 1986, 1987; Griffiths and Simons, 1986; Matlin, 1986; Njus et al., 1986; Burgess and Kelly, 1987; Pfeffer and Rothman, 1987; Rothman, 1987; Warren, 1987; Anderson and Orci, 1988; Bartles and Hubbard, 1988; Bourne, 1988; Brodsky, 1988; Cutler, 1988; Fisher and Scheller, 1988; Lodish, 1988; Rose and Doms, 1988; Hoekstra et al., 1989; Hubbard et al., 1989; Klausner, 1989; Roth, 1989).

As protein biosynthesis proceeds on the cytoplasmic side of the RER, most destinations must be reached with a translocation or insertion of a given protein through or into an apolar lipid bilayer. It has not been elucidated whether housekeeping proteins are kept back by retention signals or whether sorting and transport signals lead to movement. However, such signals reside in the protein itself or are added by posttranslational modifications. A well-known case is the mannose 6-phosphate modification for targeting proteins to the lysosomes (von Figura and Hasilik, 1986). Several soluble proteins residing in the lumen of the ER possess the carboxy-terminal sequence Lys–Asp–Glu–Leu (Pelham, 1988). The trans-Golgi network is probably the major area for sorting proteins passing through the Golgi stacks (Fig. 1, q).

There is strong evidence that mutations in the cytoplasmic tails of transmembrane proteins affect transport by influencing interactions at or near the cytoplasmic side of the membrane (Guan et al., 1988). Correspondingly, it has been suggested that short linear sequences at the extreme COOH-terminal position of transmembrane ER proteins serve as retention signals (Nilsson et al., 1989). Many viral proteins use this common pathway of proteins destined for the PM or for the extracellular space; they have to travel by vesicular transport from the RER to the Golgi apparatus and from the Golgi apparatus to the PM. Green et al. (1981) have shown that SFV membrane proteins move this way. In BHK cells they need about 15 minutes per compartment.

The motor for vesiculation is unknown. The process could be driven by the biosynthesis of lipid molecules on the cytoplasmic side of the ER, the movement of the lipid molecules across the bilayer, and the lateral movement of lipids in the endomembranes of the cells (for reviews see Bishop and Bell, 1988; Simons and van Meer, 1988; van Meer and Simons, 1988). However, maturation steps have to be included in these viral transports, since the viral membrane proteins with their transmembrane pieces behave as housekeeping, not secretory, proteins (see Section IV,H). It is tempting to assume that the differences in lipid composition found in vertebrate- and invertebrate-born SFV (Luukkonen et al., 1976) are due to dissimilarities of lipid biosynthesis in the ER.

The rate of synthesis in BHK cells is about 1.2×10^5 copies of each SFV glycoprotein per cell per minute, which corresponds to less than 1% of the endogenous ER protein, and 3×10^2 spike protein complexes per square micron. A sevenfold concentration is found in the Golgi complex. An additional nearly 15-fold concentration is observed in the virion (Quinn et al., 1984; Griffiths et al., 1984, 1989; Griffiths and Hoppeler, 1986).

G. Other Posttranslational Modifications

Besides oligomerization, the viral membrane proteins are modified in several ways. Oligosaccharides which are added cotranslationally mature into the forms found in the virion. Sulfate, phosphate, and fatty acids might be integrated. In alphaviruses p62 is cleaved into E_2 and E_3. The general role of glycosylation in sorting and targeting is not clear. Viral proteins containing altered glycans, if any, could reach the PM with a reduced or normal speed. Not much is known about the posttranslational modifications in mosquito cells, except with regard to glycans and fatty acylation. The SFV membrane proteins are not sulfated (H. Naim and H. Koblet, unpublished observations).

1. Glycan Processing

Excellent reviews have appeared concerning glycan processing (Hubbard and Ivatt, 1981; Kobata, 1984; Kornfeld and Kornfeld, 1985; Kornfeld, 1986; West, 1986; Rademacher *et al.*, 1988). In all cases all types of asparagine-linked oligosaccharides are believed to be the product of a common precursor glycan transferred cotranslationally from dolichyl–phosphate–phosphate–glycan. Complex oligosaccharides are generated after trimming in the transport pathway. The trimming of the original glycan-type $Glc_3 Man_9 (GlcNAc)_2$ to a glucose-free type starts in the RER by glucosidases I and II. Mannose residues are then removed by ER mannosidase. Subsequently, Golgi α-mannosidases I and II cleave several additional mannose residues. The result is a low-mannose-type glycan. After that Golgi transferases of vertebrates form a complex-type glycan containing mannose, *N*-acetylglucosamine, galactose, sialic acid, and fucose.

There is no evidence that viral genes code for the corresponding enzymes. Thus, the structures of viral oligosaccharides reflect those of the host, and a viral glycoprotein can be localized in the transport pathway, provided its glycan type is known. The glycosylation of alphatogavirus envelope proteins has been studied in vertebrates and invertebrates (Stollar, 1987a). The processing follows the outline (Pesonen and Renkonen, 1976; Rasilo and Renkonen, 1979; Pesonen *et al.*, 1981). However, differences between the late parts of the processing pathways of insect and vertebrate cells have been described (Stollar *et al.*, 1976; Butters and Hughes, 1981; Butters *et al.*, 1981; Hsieh and Robbins, 1984).

According to these authors, neither galactose nor sialic acid is added, no complex types are formed, and, in essence, the products of trimming are mainly high-mannose- and low-mannose-type glycans linked to asparagine (Hsieh and Robbins, 1984). However, Naim and Koblet (1988, 1990) have evidence that in C6/36 cells the SFV E_1 (carrying one glycan chain) and SFV E_2 (carrying two glycan chains) proteins each contain one primitive complex or hybrid type in addition to the high- and low-mannose structures. The simplest explanation for the formation of such glycans would be to assume an immature type of Golgi apparatus. However, electron micrographs donated to our laboratory by M. L. Ng (Singapore) show a well-developed Golgi apparatus in C6/36 cells containing at least four cisternae.

Despite the differences of the glycans in virions grown in higher eukaryotes and invertebrates mainly in the E_1 protein, which is sufficient for infection to occur (Omar and Koblet, 1988), cross-infection proceeds rapidly and without adaptive delay. This is in line with the fact that the trimming inhibitors dNM, dMM, and swainsonine— which are biochemically active in C6/36 cells—allow the production of fusogenic infectious virions in normal amounts without any shift in the ratio of particles to infectious units (Naim and Koblet, 1988). Only tunicamycin inhibits virus formation, but it has little effect on FFWI; even cleavage of p62 is visible in the presence of tunicamycin in C6/36 cells. These results indicate that correct glycosylation is not a prerequisite for the biological activities of SFV, whereas glycosylation per se is needed for virus production.

In higher eukaryotes (e.g., BHK cells) maturation of Sindbis virus is affected by such inhibitors (McDowell et al., 1987). Thus, we assume that glycosylation influences discrete conformational states of the SFV proteins within the cell in an unpredictable manner. This is in line with the finding that a variant of Sindbis virus in E_2 with two extra glycan chains can replicate only in mosquito cells, not in vertebrate cells (Durbin and Stollar, 1986). It is difficult to assess the influence of temperature, which is so different for the growth of higher eukaryotes and invertebrates. No general rules can be deduced from these findings, as they are valid only for alphaviruses.

2. Fatty Acid Acylation

Several reviews cover the problem of fatty acylation of proteins (Sefton and Buss, 1987; Olson, 1988; Schultz et al., 1988; Towler et al., 1988; Schmidt, 1989). The alphatoga envelope proteins are part of a class of proteins containing palmitic or stearic acid linked (to cysteine) via a (thio) ester bond (Schmidt et al., 1979; Schmidt, 1982). E_1 of Sindbis virus contains 1–2 mol of fatty acid per mole of glycoprotein; E_2, 5–6 mol per mole. E_2 might become acylated nonenzymatically. SFV E_1 protein carries 1 palmitic acid in position 433 (i.e., at the only cysteine of the membrane-spanning region) (Schmidt et al., 1988). Acylation is one of the factors, besides glycosylation, leading to the heterogeneity of E_1 visible in electrophoresis (Bonatti et al., 1989). The 6K protein of Sindbis virus, containing the sequence –Cys–Cys–Ser– Cys–Cys–, is heavily acylated.

In chick embryo fibroblasts attachment occurs about 20 minutes after synthesis of the proteins (Schmidt and Schlesinger, 1980), before cleavage of p62. In VSV and SFV evidence suggests that the modification takes place between the ER and the Golgi apparatus or in the cis-Golgi cisterna (Dunphy et al., 1981; Quinn et al., 1983). In cell-free systems using a microsomal preparation for the acylation of exogeneous SFV E_1, little selectivity was observed among the long-chain fatty acids (Berger and Schmidt, 1984, 1985; Schmidt, 1984). Sindbis virus proteins, nonacylated after application of cerulenin, are glycosylated and transported to the PM, but the formation of virions is drastically inhibited (Schlesinger and Malfer, 1982).

Covalently bound fatty acids have been implicated in fusogenic properties; however, the evidence for a general role in fusion is not convincing, since Sendai virus possesses a nonacylated fusion protein (Lambrecht and Schmidt, 1986). It is not clear whether ester-linked acyl groups are involved in protein transport and subcellular targeting. Protein acylation–deacylation cycles have recently been implicated in transport through the Golgi apparatus (Glick and Rothman, 1987). All insects examined are able to synthesize 16:0, 18:0, and 18:1 fatty acids *de novo* (Stanley-Samuelson *et al.*, 1988). In C6/36 cells p62, E_1, and E_2 are rapidly labeled after the addition of [^{3}H]palmitate to the cell cultures (C. Schärer and H. Koblet, unpublished observations).

3. Phosphorylation

In SFV and Sindbis virus (Waite *et al.*, 1974) low amounts of phosphate can be detected in all proteins. However, nothing is known about the significance of the phosphorylation–dephosphorylation reactions and the topographical sites of these reactions. The C protein has been implicated as a protein phosphokinase (Tan and Sokol, 1974). A recent review covers the topic of viral protein phosphorylation (Leader and Katan, 1988).

4. Cleavage of the p62 to E_3 and E_2

The last known processing event is the extracytoplasmic p62 cleavage in SFV between His–Arg–Arg-66 (E_3, carboxy terminal) and Ser-1 (at position 67) (E_2, amino terminal). E_2 and E_3 remain associated after the cleavage (Simons *et al.*, 1973; Ziemiecki *et al.*, 1980). The carboxy-terminal arginine is then removed (Garoff *et al.*, 1980a). In BHK cells the Golgi fraction contains p62 and E_1; the PM contains E_1, E_2, and E_3 (Green *et al.*, 1981). Efficient cleavage can occur only when p62 is coexpressed with E_1 (Garoff *et al.*, 1983; Kondor-Koch *et al.*, 1983; Cutler and Garoff, 1986; Cutler *et al.*, 1986). The cleavage takes place about 30–35 minutes after synthesis of the protein, about 10 minutes after Endo-H resistance is acquired and before insertion into the PM.

Recent analysis has shown that in mechanically permeabilized BHK cells cleavage of p62 occurs before arrival at the PM in a Ca^{2+}- and glucose-dependent fashion (De Curtis and Simons, 1988, 1989). On the other hand, Knipfer and Brown (1989) detected significant amounts of

E_2 protein at the end of a 5-minute pulse in Sindbis virus-infected BHK cells. The E_2 produced early was still in an Endo-H-sensitive state. Several prohormones, proproteins, and glycoproteins of enveloped viruses (e.g., influenza, retroviruses, human immunodeficiency virus, and paramyxoviruses) undergo similar cleavage reactions at a pair of basic amino acids at late stages of intracellular transport. These cleavages are essential for fusogenicity and infectivity. However, in alphaviruses this cleavage does not concern the fusogenic protein, which is E_1, and does not create a new hydrophobic amino terminus. Secretory and viral membrane proteins might be cleaved by the same host endoprotease (Hynes, 1987; Melançon and Garoff, 1987) in a slightly acidic compartment of clathrin-coated secretory vesicles (proinsulin, Orci et al., 1987; Davidson et al., 1988; F_0 of New Castle disease virus, Yoshida et al., 1989). However, the pH is probably too high to trigger the fusogenic conformational change of a viral protein (Boulay et al., 1987a). In addition, the p62 E_1 precursor complex is more resistant to low-pH-induced dissociation than E_2E_1 (Wahlberg et al., 1989).

Today the evidence is conflicting that p62 cleavage is an event which is always related to maturation and occurs late only in the exocytotic pathway, and that it is a morphogenetic cleavage (Jones et al., 1974; Keränen and Kääriäinen, 1975; Enzmann, 1987a). Cleavage of p62 can be prevented by antiserum against E_1 (Bracha and Schlesinger, 1976; Smith and Brown, 1977), and antigenicity of p62 is changed during maturation (Kaluza et al., 1980; Kaluza and Pauli, 1980). Cleavage of p62 occurs in the absence of maturation in media of low ionic strength (Bell et al., 1978). However, Sindbis virus E_3 is still released rapidly into the medium. Moderate concentrations of monensin prevent cleavage of pE_2 (p62 of Sindbis virus); thereby, this precursor is incorporated into infectious virions (Presley and Brown, 1989). Also, a rapid penetration mutant of Sindbis virus with a change of serine to asparagine at position 1 of E_2, creating a new putative glycosylation site, allowed the formation of virions containing pE_2 (Russell et al., 1989). Taken together, there are relative arguments against tight coupling between p62 cleavage and budding (Mayne et al., 1984). In Sindbis virus-infected BHK cells N-methyl-dNM interferes with the cleavage of p62 and the release of viral particles. Allowing glucose removal but not mannose trimming by dMM leads to the cleavage of p62 and intracellular budding (McDowell et al., 1987).

In C6/36 cells cleavage of p62 is certainly not tightly coupled to budding and virus release. The precursor can be cleaved to nonglycosylated products in the presence of tunicamycin or deoxynojirimycin, although no virions are formed in the presence of tunicamycin. In these cells p62 cleavage is an early but ongoing phenomenon, probably starting before the Golgi complex. E_2, one of the products, is faintly visible after a [^{35}S]methionine pulse of 5 minutes and clearly visible after 20 minutes, whereas Endo-D sensitivity of E_1, which points to the arrival of the p62–E_1–E_2–E_3 complex in the Golgi apparatus, is established with a half-time of 30–40 minutes. Endo-H resistance of E_1 (Naim and Koblet, 1988), which is acquired in the late Golgi stacks, is detectable after 55–60 minutes, and virions appear in the medium only after 75 minutes (Naim and Koblet, 1990).

H. Maturation

Maturation is defined here as the interaction of the endodomains of the spikes (after cleavage of p62) with the nucleocapsid on the cytoplasmic side of a membrane, usually the PM. Thereby, the nucleocapsid is enveloped by a piece of host membrane. The spikes are transported by vesicles which fuse with the membrane where maturation occurs. The fusion is such that the large amino-terminal domains facing the vesicles' interior are looking outside of the cell and the virion surface. The small carboxy-terminal sequences are still on the cytoplasmic side, as before. The carboxy-terminal of E_2 forms the same cytoplasmic domain as p62 before cleavage (Ziemiecki et al., 1980). The assembly principles at cellular membranes have recently been reviewed (Stephens and Compans, 1988).

The nucleocapsid first binds to the inner leaflet of the PM. Thereby, a bud is formed with the membrane, which increasingly surrounds the nucleocapsid (Acheson and Tamm, 1967; Brown et al., 1972). The new virions are finally only connected to the cell with a fine membranous stem which breaks by another fusion phenomenon as soon as the nucleocapsid is completely enveloped. At this step the acylated 6K protein might play an important role; this peptide seems to be essential for assembly and budding. Budding is based on a direct multipoint interaction between the cytoplasmic portions of several spikes and the underlying nucleocapsid (Garoff and Simons, 1974).

In view of recent findings (see Section II,B), this is a noncovalent binding between the carboxy terminus of E_2 and the carboxy-terminal side of the C protein. This could explain the fact that in mature virions host PM proteins are excluded; the specificity of the binding would perform sorting without the use of energy. However, even mutants in the ectodomain of Sindbis virus E_2 could lead to defective assembly (Hahn et al., 1989c). On the other hand, infection could lead to the redistribution of PM host proteins (Pakkanen et al., 1988). It remains an open question as to whether cleavage of p62 and budding are closely linked in vertebrate cells (Mayne et al., 1984), but certainly budding is connected to conformational changes (Bell and Waite, 1977; Smith and Brown, 1977; Kaluza et al., 1980; Kaluza and Pauli, 1980; Rice and Strauss, 1982; Gates et al., 1982; Wolcott et al., 1984; Wust et al., 1989). In mosquito cells spontaneous intracellular maturation is found, a phenomenon which can be induced in vertebrate cells by blocking the transport into the PM with certain trimming inhibitors (McDowell et al., 1987) or monensin (Johnson and Schlesinger, 1980) or by using transport mutants of SFV which accumulate spike glycoproteins at nonpermissive temperatures (Saraste et al., 1980). In C6/36 cells about 50% of the total infectious units can be visualized intracellularly at 17–24 hours postinfection (Naim and Koblet, 1988). However, SFV proteins can also be detected by immunofluorescence on the cell surface (Omar et al., 1986). Similarly, Simizu and Maeda (1981) showed budding of Western equine encephalitis virus in intracellular vesicles as well as in the PM of C6/36 cells. This would explain the rapidity of the SFV-induced FFWI in C6/36 cells, as described in Section VI,C.

Dependent on the mosquito cell clone, budding of virions at the PM is rarely seen (Pudney et al., 1982). In these clones virions are visible in internal vesicles of complex structure, and nucleocapsids reside in vacuoles in various stages of envelopment (Brown and Gliedman, 1973; Raghow et al., 1973a,b; Gliedman et al., 1975; Brown et al., 1976). We assume that transport vesicles allow maturation and carry the virions by an exocytotic fusion event to the outside of the cell. The glycan analysis, showing partial Endo-D sensitivity and a loss of Endo-H sensitivity of the E_1 population of virions, suggests that the segregation of virions does not take place before the Golgi apparatus (Naim and Koblet, 1990).

There is, per se, in view of the topology of the spikes in the transport vesicles, no a priori reason against intracellular budding. However, why intracellular budding does not usually occur in vertebrate cells is not known. One can speculate that intracellular transport rates are slower in mosquito cells because of the lower incubation temperature (28°C) and that intracellular budding is dependent on spike concentration. One can also speculate that there is a more or less loose coupling between p62 cleavage and budding. We have presented evidence that in C6/36 cells p62 cleavage might occur all along the transport pathway (Naim and Koblet, 1988, 1990). Certain flaviviruses (Hase et al., 1987; Ng, 1987) seem to mature in the RER and in vesicles of mosquito cells. The interesting idea has therefore been put forward (Hase et al., 1987) that, in a trans-type maturation, virions can be assembled within the cisternae of the RER, after which they pass through the host secretory pathway, including the Golgi apparatus. We do not think that maturation in C6/36 cells is fundamentally different from that observed in vertebrates; it is only topographically more extended.

370 HANS KOBLET

I. Virus Production in Vertebrate and Mosquito Cells: A Comparison

In vertebrate cells alphavirus infection is short term and cytocidal. In cultured mosquito cells the infection is long term and inapparent. However, there is no evidence that the replication strategy differs in major respects. The virus yield in high-producer mosquito clones might be even higher than in vertebrate cells (Stollar, 1980). In C6/36 cells we find titers of 5×10^9 pfu/ml at 24 hours postinfection. With well-selected virus strains and cell clones all cells are infected in the acute stage (Igarashi et al., 1977; Omar et al., 1986). In mosquito cells the production of viral components is delayed as compared to vertebrate cells; this might be explained by the relatively low temperature (i.e., 28°C) used for cultivation.

In C6/36 cells the products can be detected as follows (Omar et al., 1986; Koblet et al., 1987): At 3 hours postinfection, 49 S and 26 S RNA (by hybridization with a nick-translated cDNA probe on Northern blots); at 4 hours, first signs of changes occurring at the cell surface (by aggregation of cells at pH 7.0 in Spinner culture); at 6 hours, discrete amounts of structural proteins (by [35S]methionine labeling and polyacrylamide gel electrophoresis); at 7 hours, viral structural proteins on the cell surface of some cells (by surface immunofluorescence) and virions in the medium (by plaque assay); at 9–11 hours, [3H]uridine-labeled viral RNA (by agarose gel electrophoresis and autoradiography); and at 16 hours, positive surface immunofluorescence of all cells and the highest rate of virus formation (by plaque assay). Similar findings have been published (Eaton and Regnery, 1975; Richardson et al., 1980).

The intracellular transport times in C6/36 cells (see Section IV,G,4) can roughly be estimated, again, using the data indicated above. Thus, structural proteins travel from the RER to the PM in about 1 hour. According to the results of a [35S]methionine pulse of 15 minutes and a chase of at least 2 hours, when viral proteins can no longer be seen in the cell extracts, one would calculate a residence time of some envelope proteins of about 1 hour in the PM. However, fusogenicity of the cells is only lost about 8 hours after the application of cycloheximide (which blocks protein biosynthesis in C6/36 cells within a few minutes) (Omar et al., 1986).

After peak titers have been released from mosquito cells, these cells enter an indefinite phase of chronic infection (Yunker, 1971; Stollar, 1980; Stalder et al., 1983; Enzmann, 1987b).

In vertebrate cells host mRNA is increasingly substituted by mRNA of viral origin, and, finally, virtually only viral proteins are made (Wengler, 1980). Concomitantly, overall protein biosynthesis is restricted more and more. In mosquito cells this shut-off phenomenon is not regular (Richardson et al., 1980; Ng and Westaway, 1979). However, in the high-producer clone C6/36 host protein synthesis is transiently, but strongly, shut off. Therefore, viral proteins can easily be visualized in polyacrylamide gel electrophoresis (Omar et al., 1986).

This corresponds to the findings by Tooker and Kennedy (1981), who analyzed the SFV proteins in A. albopictus cell clones. High-virus-producer clones exhibited a CPE, all virus-specified proteins were detectable, and host protein biosynthesis was depressed.

In cell-free systems of higher eukaryotes, it has been shown that the capsid protein is responsible for this shut-off and that translation of host and early 49 S RNA—but not of 26 S RNA—is inhibited (van Steeg et al., 1984).

Low virus producers had no signs of a CPE, had a low level of viral protein synthesis, and no shut-off occurred. In contrast to vertebrate cells, mosquito cells must provide host functions in order for alphavirus replication to take place. Production can be inhibited with actinomycin D; host nuclear functions can be demonstrated directly (Scheefers-Borchel et al., 1981; Erwin and Brown, 1983; Condreay et al., 1988).

V. Cytopathic Effects

Inhibition of cell functions by RNA virus infections has been reviewed by Kääriäinen and Ranki (1984). As a rule, arthropod cells infected with alphaviruses in culture do not exhibit a CPE (Davey et al., 1973), whereas infection of permissive vertebrate cells with the same viruses is highly cytopathogenic and cytolytic (Stollar, 1980, 1987b; Ulug et al., 1987). The reasons for this discrepancy are not clear. As virus yields are dependent on virus strains and cell clones (Sarver and Stollar, 1977), we think that infection of a high-producer mosquito clone in which all cells are infected in a "synchronous" fashion, as in C6/36 cells, invariably leads to a CPE. Additional factors seem to be the cell culture medium and the growth temperature (Gillies and Stollar, 1982).

Dominance of a "lethal" CPE-positive phenotype in hybrid A. albopictus cells infected with Sindbis virus was expressed mainly at 34°C (Tatem and Stollar, 1986). At this temperature viral RNA synthesis is increased. Regulation phenomena could be detected in clones reacting with a CPE, as in C6/36 cells. These phenomena concern protein secretion during the time of the virus burst (24–48 hours postinfection) (Reigel and Koblet, 1981).

We distinguish a nonlethal CPE from syncytium formation. This CPE is visible from 14–16 hours postinfection in C6/36 cells infected with SFV. A starlike appearance of the cells that have contacted one another at the edges or in an aggregation is typical. This occurs at pH 7.0, at 28°C, in Mitsuhashi–Maramorosch, Eagle's, Roswell Park Memorial Institute, and 199 media. It is considered to be the microscopic reflection of the binding of viral spikes expressed locally at the PM to the neighboring cell membrane. This aggregation is the prerequisite for fusion, occurring only at a medium pH value of 6.0. Fusion takes place spontaneously only in cell lines that lower the pH of the medium by metabolic activity (Späth and Koblet, 1980). This is not the case in C6/36 cells.

As reviewed by Garry (1988) for mammalian cells, permeability changes resulting from viral protein integration into the PM might provoke many reactions, from cell swelling to the inhibition of host protein synthesis (Carrasco, 1977). Garry et al. (1979a) suggested that alphaviruses inhibit the translation of host mRNA by increasing inside sodium and lowering inside potassium. Host protein synthesis is inhibited by high intracellular sodium ion concentrations, whereas viral protein synthesis is quite insensitive (Waite and Pfefferkorn, 1970; Clegg and Kennedy, 1975; Bell et al., 1978). This interference is on the level of initiation.

Alterations in ion fluxes have been linked with toxinlike effects of viruses reminiscent of the action of certain toxins (Pasternak, 1987a,b). Changes in intracellular concentrations of sodium and/or potassium ions are common in many RNA and DNA viruses (Kohn, 1979; Garry et al., 1979a; Ulug et al., 1984; Muñoz et al., 1985). This could be due to inhibition of the ion transport systems and interference with the Na^+,K^+-ATPase (Garry et al., 1979b; Garry and Bostick, 1987; Ulug et al., 1989). Carrasco (1978) suggested that increased permeabilities at the PM generally explain the shut-off phenomena. However, this view has been questioned (Cameron et al., 1986).

We have no evidence for such phenomena in C6/36 cells at pH 7.0 (although protein synthesis heavily decreases in favor of viral protein synthesis). This might be explained either by the sequestration of virus assembly (see Section IV,H) or by efficient pumps and high-energy stores. Support for this stems from the fact that altered ion fluxes in SFV-infected C6/36 cells can only be detected under fusion conditions [external (i.e., extracellular or medium) pH (pH_e) of 6.0 or lower] with the addition of oxidative phosphorylation inhibitors (see Section VI,H) (Kempf et al., 1987a–c, 1988a,b; Koblet et al., 1988).

So far as the transient inhibition of protein biosynthesis in C6/36 cells is concerned, we, therefore, rather accept the suggestion (Wengler and Wengler, 1976b; Wengler, 1980) that host mRNA is simply replaced by viral mRNA in polysomes. In addition, we have evidence that SFV capsid protein acts as a pleiotropic regulator of host protein synthesis, inducing host cellular protein synthesis at low concentrations and inhibiting it at high concentrations (Elgizoli *et al.*, 1989).

Many years ago "cytopathic vacuoles" were described in avian cells (Acheson and Tamm, 1967; Grimley *et al.*, 1968; Friedman *et al.*, 1972). Only now has this phenomenon begun to be understood in that these vacuoles could be sites of viral RNA and protein syntheses and nucleocapsid assembly (Froshauer *et al.*, 1988).

VI. FUSION

A. Definitions and Introduction

Fusion designates the dynamic phase of the recombination of two membrane bilayers into one. Fusion is a fundamental reaction occurring wherever membranes separate compartments. The compartments can be cells (i.e., cell–cell fusion); extracellular "vesicles" such as liposomes, virosomes, or membranous viruses (i.e., cell–virion fusion); or intracellular organelles such as the nucleus, transport vesicles, endosomes, lysosomes (i.e., intracellular fusion). The product of cell–cell fusion is a syncytium, which represents a multinucleated giant cell, or polykaryocyte.

Most fusion reactions are physiological. Examples on the cellular level are myotube formation (Konigsberg, 1971; Yaffe, 1971) or sperm–egg membrane fusion (Primakoff *et al.*, 1987). Examples at the subcellular level are endocytosis and fusion of lipid and protein transport vesicles with various organelles of the exocytotic transport pathways.

Little is known about the biochemistry of these fusion processes (Warren *et al.*, 1988). Considerable progress, however, has been made in studying pathological fusion events (White *et al.*, 1983; Spear, 1987) provoked by membranous viruses. The following types are distinguished.

1. Fusion from Without

FFWO (Bratt and Gallaher, 1969, 1970) is the fusion of the envelope of a virion with the PM at a physiological or mildly acidic pH, whereby infection may start. FFWO is an early phenomenon of the infectious cycle and cannot be inhibited by inhibitors of macromolecular biosynthesis. In FFWO cell–cell fusion might follow, because the viral membrane integrated into the PM renders the latter fusion competent. The classical example is the direct and rapid cell–cell fusion by Sendai virus added at high MOI, even after inactivation of the infectivity by ultraviolet irradiation (Okada, 1958, 1962a,b). Probably all enveloped viruses can fuse from without under appropriate conditions.

2. Endosomal Fusion

The viral membrane fuses with the endosomal membrane at low pH (see Section IV,B,2) so that the nucleocapsid carrying the genetic information can enter the cytoplasmic compartment (Marsh, 1984; Kielian et al., 1986; Kielian, 1987).

3. Fusion from Within

FFWI (Bratt and Gallaher, 1969, 1970) is the fusion of the PM of an infected cell containing, in transit, newly formed viral envelope proteins with the PM of a neighboring infected or uninfected cell. Thereby, syncytia are formed. Depending on the virus family, FFWI occurs at a physiological or mildly acidic pH of the cell culture medium. FFWI is a late phenomenon of the infectious cycle; it can be inhibited by inhibitors of macromolecular biosynthesis.

These definitions do not describe the mechanisms of fusion. They only indicate that the fusogenic material is either on the outside of a membrane, as in FFWO or in EF, or is delivered from the inside of the compartment, as in FFWI.

From a mechanistic point of view, fusion reactions will follow three consecutive stages: (1) The membranes of the compartment(s) come in close contact, overcoming several opposing forces, such as water and charge barriers (i.e., initial approach). (2) The membranes of the compartment(s) fuse by opening or closing a narrow channel (i.e., coalescence). (3) The compartments are combined by enlargement of the channel or are separated at the constriction (i.e., separation) (White et al., 1983; Spear, 1987).

Logically, then, there are two geometric forms of fusions: planar and annular (Silverstein, 1978; White et al., 1983). In planar fusion two compartments become one; in annular fusion one compartment yields two. FFWO, EF, and FFWI are examples of planar fusions of two separate compartments, whereas the endosome itself is the product of an annular fusion resulting in a compartment enclosed (i.e., the endosome) in another (i.e., the cell). As stated, FFWI induced by low pH, occurring in vertebrate and invertebrate cells, has to be distinguished from CPE.

FFWI is an important reaction. In essence, only two important factors exist in the case of alphaviruses: The endogenous factor is the infection resulting in the insertion of newly formed spike proteins into the PM; the exogenous factor is a medium pH of 6.1 or lower (Späth and Koblet, 1980; Koblet et al., 1987). Much can be learned about all three fusion types mentioned above. It should be borne in mind that in all of these cases the primary interaction site is the exoplasmic leaflet of each membrane, whereas in all intracellular vesicle–vesicle fusions the endoplasmic leaflet of the compartments is the primary site. In our opinion FFWI resembles quite closely the EF which is so important in the viral infection process. For such studies mosquito cells are an outstanding tool (see Section III).

B. Course of Events after Infection

In C6/36 cell monolayers the correlation between fusion and the appearance of viral products of SFV can easily be assayed. The infection is initiated at pH 7.0, and then the pH of the medium is adjusted to 6.0 at hourly intervals or the pH of the medium is lowered to 6.0 at 1 hour postinfection (an experiment which is not possible with vertebrate cells). In both cases discrete foci of fusion appear after 7 hours. From this time the number of fusion-competent cells increases until 16 hours, when more than 90% of the cells fuse. This course of events suggests an FFWI (Omar et al., 1986).

The interpretation is as follows: Expression of the fusogenic factor on the cell surface—probably SFV E_1 (see Section VI,F)—requires at least 7 hours. However, expression in different cells is asynchronous. At 16 hours all cells carry sufficient amounts of the fusogenic factor in the PM. The reason for the asynchrony is not known. C6/36 cells can obviously support biosynthesis and PM expression of viral proteins at an outside pH of 6.0, because the cytoplasmic pH can be kept constant (Kempf et al., 1987c). As soon as a fusogenic molecule appears on the cell surface, it becomes engaged in fusion, due to the low pH. The assembly of progeny virions is drastically hampered, and p62 remains partially uncleaved, whereas E_1 and C proteins are properly formed (F. Reigel and H. Koblet, unpublished observations).

C. Course of Events after Lowering the pH

FFWI seems to proceed slowly. If the pH is lowered after 16 hours postinfection, as visualized in the light microscope, it takes 30 minutes until the fusion of about 90% of the area of a monolayer is completed. In fact, fusion is a fast process, as perceived with the electron microscope (Knutton, 1980). Small intercellular bridges become visible within 5–30 seconds after lowering the pH (Koblet et al., 1987). With the novel method of microinjection of the strongly fluorescent dye lucifer yellow (Stewart, 1981), these data could be confirmed (Kempf *et al.*, 1987b). FFWI, then, proceeds in an asynchronous zipperlike fashion, which renders biochemical analysis quite difficult.

D. Fusion of SFV-Infected C6/36 Cells at Low pH Is a Fusion from Within

Apart from the time courses described above and the close correlation with the appearance of viral proteins at the PM, there are many additional arguments to designate the fusion of SFV-infected C6/36 cells as an FFWI (Omar *et al.*, 1986; Koblet *et al.*, 1987): (1) Extraneous virions bound to cells can be digested by proteinase K (Marsh and Helenius, 1980). This approach was used to remove bound virus at 16 hours postinfection from infected cells. After lowering the pH, fusion occurred immediately in the remaining cell sheets, clearly demonstrating that the extracellular viruses were not involved in syncytium formation.

(2) In contrast to FFWO, FFWI can be prevented by inhibitors of macromolecular biosynthesis (Gallaher *et al.*, 1980; Bratt and Gallaher, 1969). Cycloheximide was added to infected cultures from 2 to 16 hours postinfection at 2-hour intervals, and the pH was lowered to 6.0 at 16 hours. Syncytia were produced only in those cultures in which the drug had been added after 8 hours postinfection. This result shows that a defined incubation period is mandatory for the establishment of fusion conditions which correlate with an FFWI.

(3) On the basis of all of these experiments, it is reasonable to assume that the fusogenic factor is a viral protein expressed on the cell surface. Therefore, trypsin, α-chymotrypsin, thermolysin, proteinase K, and bromelain were applied to establish whether the fusogenic factor could be removed. Only bromelain prevented fusion of the cells.

(4) Lysosomotropic agents which block the initial steps of viral infection at the endosomal stage (Cassell *et al.*, 1984; Helenius *et al.*, 1982) proved to be quite inefficient in C6/36 cells. Even at the extremely high concentration of 30 mM ammonium chloride, the virus titer reached 2×10^7 pfu/ml 16 hours postinfection, and fusion was possible. With 10 mM chloroquine a low titer of 30 pfu/ml was obtained after 16 hours; however, fusion was not inhibited (Omar *et al.*, 1986).

(5) All fusogenic proteins, with rare exceptions (Wilcox and Compans, 1982), have been found to be glycoproteins (Gallaher *et al.*, 1980). Therefore, tunicamycin, which inhibits glycosylation (Schwarz and Datema, 1980; Klenk and Schwarz, 1982; Hsieh and Robbins, 1984; Naim and Koblet, 1988) also in C6/36 cells, was added. Although

virus production was drastically reduced, fusion was never completely blocked by tunicamycin (Omar et al., 1986; Naim and Koblet, 1988). Processing of the nonglycosylated p62 precursor was only partially inhibited. E_1, at 47 kDa, could be seen; its persistent presence, again, indicated that it might be the fusogenic protein.

(6) Several cell–cell fusions (e.g., fertilization and myotube formation) seem to be dependent on proteolytic processes. Therefore, the protease inhibitors TPCK, TLCK, PMSF, and TAME were applied at 16 hours postinfection at pH 7.0 before lowering the pH. Only TPCK (1 mM) was effective in inhibiting the fusion completely and irreversibly. However, p62 was properly cleaved after a treatment for 1 hour prior to pulse-labeling for 15 minutes and chasing for 1 hour.

E. Conformational Changes

Vertebrate cells infected with Sindbis virus fuse after a 1-minute exposure to low pH (Mann et al., 1983). The minimal time of exposure to pH 6.0 required for FFWI of C6/36 cells at 28°C is not more than 15 seconds. Again, the fusion is complete within 30 minutes (as seen by light microscopy). This brief exposure time to a critical pH of 6.0 (or below) indicates that a triggering mechanism is likely to exist for FFWI. From data showing conformational changes in intact virions induced by low pH (Edwards et al., 1983; Yewdell et al., 1983) and the suggestion that such changes could be involved in the fusion process (Mann et al., 1983), an attempt was made to dissect the process. The fusion is inhibited at temperatures below 17°C. If the cells are exposed for 15 seconds to pH 6.0 at 4°C and then replaced in a medium of pH 6.0 or 7.0 at 28°C, fusion occurs within 30 minutes.

The conclusion might be drawn that the fusion takes place in at least two steps: an initial step which is rapid, pH dependent, and temperature independent, and a second slower process which is pH independent but temperature dependent. In mosquito cells, in contrast to BHK cells, at a temperature above 15–17°C syncytium formation proceeds, irrespective of whether the cells are left at pH 6.0 or are brought back to pH 7.0 after a brief exposure to pH 6.0 (Koblet et al., 1985). A similar result is obtained by utilizing the ionophore (Pressman, 1976) monensin.

The simplest interpretation of these results is that a conformational change, probably of a viral PM protein, takes place at pH 6.0. Low temperature and monensin do not inhibit this conformational change, but they do inhibit the fusion proper of the PM as a consequence of the conformational change. It is unknown why fusion is impossible below 17°C. Import of the fusogenic protein into the neighboring membrane (see Section VI,I) might be hampered or the membrane lipids do not cooperate. Whatever the explanation, the blocking procedures provide a simple means for studying the conformational change (Koblet et al., 1985).

For example, covalent chemical modification of functional groups on the surface of infected cells with hydrophilic reagents can be performed. Table I shows the results. Thiol and disulfide groups are involved in the conformational change. All negatively charged reagents, either cleaving or modifying disulfide and thiol groups, are effective in inhibiting the fusion of the cells when applied after exposure to low pH. On the other hand, the positively charged reagents do not affect the fusion capacity whether applied before or after exposure to low pH. NH_2 and guanidino groups are also expressed at the cell surface following exposure to low pH. Newly exposed disulfide and thiol groups probably have an amino acid residue with a positive charge as a neighbor which repels positively charged reagents (Koblet et al., 1985, 1987; Omar and Koblet, 1989a,b).

Several reports have appeared presenting evidence for the role of SH— and —SS—/—SH interchange reactions at the PM. Sulfhydryl groups have been implicated in chemically induced membrane fusion (Ahkong et al., 1980). At this time we assume that the conformational change triggered by low pH is accompanied by a disulfide–sulfhydryl exchange reaction within the E_1 molecule (see Section VI,I) which stabilizes the new fusogenic conformation (Omar and Koblet, 1989b). Indeed, under the conditions tested, the new conformation is stable and irreversible, as shown by cell-mixing experiments (Koblet et al., 1985).

However, sulfide ions could lock the fusogenic protein onto glycoproteins of the neighboring PM, similar to what is observed for Sendai virus (Ozawa et al., 1979), insulin binding (Clarke and Harrison, 1983; Morgan et al., 1985), lectin binding (Roberts and Goldstein, 1984), or transport processes (Klip et al., 1979; May, 1985). Once this interlocking has occurred, it could be followed by penetration of a fusogenic hydrophobic peptide segment into the lipid bilayer of the neighboring cell, thereby inciting membrane–membrane fusion (Omar and Koblet, 1989b). Formation of heavy aggregates of viral proteins during fusion, as seen in polyacrylamide gel electrophoresis under nonreducing conditions, might support such an assumption.

The conformational possibilities seem to be manifold and complicated. Maturation during budding at pH 7.0 is connected with conformational changes (see Section IV,H). Virions behave differently when they are brought together with a hydrophobic support at pH 7.0 or 6.0 or at pH 7.0 after exposure to pH 6.0 (Omar and Koblet, 1988). Thus,

TABLE I

CHEMICAL MODIFYING AGENTS AND FUSION[a]

Reagent	Charge	Reaction with	Inhibition of FFWI with respect to low-pH exposure	
			Before	After
Sodium tetrathionate	−	Sulfide ions	No	Yes
Thiosulfate	−	Sulfide ions	No	Yes
DTNB (20 mM)	−	Sulfide ions	No	Yes
Sulfite	−	Cleaves —S—S—	No	Yes
Butanedione (1–10 mM)		Guanidino groups	No	Yes
Cyclohexanedione (1–10 mM)		Guanidino groups	No	Yes
Phenylglyoxal (1–10 mM)		Guanidino groups	No	Yes
TNBS (5 mM)	−	Amino groups	No	Yes
Cystamine	+	Sulfide ions	No	No
Iodoacetamide		Sulfide ions	Yes	Yes
N-Ethylmaleimide		Sulfide ions	Yes	Yes
Mercaptoethanol		Sulfhydryl-reducing agents	No	Yes, during presence
Dithiothreitol (1–10 mM)		Sulfhydryl-reducing agents	No	Yes, during presence
Cysteamine	+	Cleaves —S—S—	No	No
Phenylmethylsulfonyl fluoride			No	No
TAME			No	No
TLCK			No	No
TPCK			Yes	Yes

[a] The effects of various chemical modifying reagents (1 mM if not indicated otherwise) on fusion when added to SFV-infected C6/36 cells at 16–17 hours postinfection either before or after exposure to pH 6.0. The reagents were removed after a 20-minute incubation at 28°C and pH 7.0, and fusion was monitored by light-microscopic examination. FFWI, Fusion from within.

the conformation of the fusogenic E_1 protein (see Section VI,F) in a free virion (surrounded by water) in suspension at pH 6.0 would differ from that at pH 7.0 and, again, differ from that at pH 6.0, when the virion is bound to a hydrophobic support (or a PM).

From recent reports the unescapable conclusion must also be drawn that several conformational states exist in fusogenic proteins of various viruses (West Nile virus: Kimura and Ohyama, 1988; influenza virus: Stegmann et al., 1987; Wharton, 1987; White and Wilson, 1987;

Vesicular Stomatitis virus: Clague *et al.*, 1990). Similar conformational states could govern FFWI. We think that in alphaviruses there is a true precursor–product relationship between unique conformational states of E_1 going from a bound state at pH 7.0 to a hydrophobic interaction state at pH 6.0. This could explain the need for cholesterol (Kielian and Helenius, 1984) as a hydrophobic support.

All fusogenic viral proteins, during FFWO, EF, or FFWI, must undergo conformational changes. In paramyxoviruses the binding conformation is created by an endoproteolytic cleavage at pH 7.0 (Gething *et al.*, 1978; Hsu *et al.*, 1981), liberating a new hydrophobic amino terminus. Binding itself rapidly induces the fusogenic conformation at pH 7.0. In influenza viruses an endoproteolytic cleavage at pH 7.0, liberating a new hydrophobic amino terminus, is also needed for infectivity and fusogenicity, but the fusogenic conformational change takes place at low pH (Yewdell *et al.*, 1983; Skehel *et al.*, 1982). In alphaviruses the conformational change at low pH seems to be a sufficient condition for fusion. No additional cleavage creating a new hydrophobic amino terminus has been detected (Koblet *et al.*, 1987). This has hampered greatly identification of the fusogenic protein (E_1 or E_2).

F. Identification of the SFV Fusogenic Protein

The data provide evidence that SFV-induced FFWI is triggered by viral spike proteins behaving transiently as PM housekeeping proteins. The possibility seems remote that the viral infection—the absolute prerequisite for fusion—induces fusogenic host proteins. FFWI could also be obtained in cells microinjected with SFV spike glycoprotein genes which did not produce virions (Kondor-Koch *et al.*, 1983).

The question remains as to whether E_1 or E_2 is the fusogenic protein. E_3 cannot be responsible, since it is lost, in the case of Sindbis virus, without ever being integrated into the virion (Welch and Sefton, 1979). Only negative evidence can be obtained from genetic engineering experiments. E_2 alone can be expressed on the cell surface when no fusion occurs. E_1 alone cannot be expressed on the cell surface (Kondor-Koch *et al.*, 1983).

In several reports the E_1 protein had been taken to be the fusogenic protein (Garoff *et al.*, 1982; Kondor-Koch *et al.*, 1982, 1983). The evidence was circumstantial: (1) The precursor of E_2, p62, might remain uncleaved, yet fusion is possible (Mann *et al.*, 1983; Omar *et al.*, 1986; Koblet *et al.*, 1987; Naim and Koblet, 1988). (2) E_1 is the hemagglutinin (White *et al.*, 1983; Simizu *et al.*, 1984) and also contains the hemolytic activity. The former determinant is located near the apex of the spike and the latter near the virion membrane (Chanas *et al.*, 1982a,b; Schmaljohn *et al.*, 1983; Boere *et al.*, 1984). (3) The external domains of E_1 in Sindbis virus and SFV possess a conserved hydrophobic sequence (Garoff *et al.*, 1982), like other fusogenic proteins of various viruses. (4) Fusogenic proteoliposomes containing E_1 only of Western equine encephalitis virus have been constructed (Yamamoto *et al.*, 1981); they provoke low-pH-dependent hemolysis. (5) Mutants of Sindbis virus leading to a lowered fusion pH reside in the E_1 gene (Boggs *et al.*, 1989).

E_1 has now been identified as the fusogenic protein of SFV. The evidence stems from two lines: (1) Virions which contain the E_1 protein only, are infectious and can induce cell–cell fusion (Omar and Koblet, 1988). (2) E_1 changes the conformation at the cell surface when exposed to low pH. The covalently reacting fusion inhibitors [^{35}S]sulfite and [^{14}C]TNBS label mainly E_1 when applied at pH 7.0 after exposure of the cells to pH 6.0 (Omar and Koblet, 1989a,b). In all of these cases, the protein was bilayer bound, which is a prerequisite for fusion (Marsh, 1984).

Thus, there is clear evidence that E_1 has several biological functions. E_1 is one of several examples of viral proteins with more than one function (Herrler *et al.*, 1988). It is also one of several examples in which a single protein type is fusogenic (White *et al.*, 1982; Lapidot *et al.*, 1987; Gibson *et al.*, 1988; Herrler *et al.*, 1988).

G. Fusogenic Sequence

The fusogenic peptide within the E_1 sequence is not known. One strongly conserved hydrophobic peptide segment of E_1 is located between amino acid residues 79 and 96 (Garoff *et al.*, 1982). This peptide segment might be the mediator for fusion (White *et al.*, 1983). The repositioning of such a segment from an "intraspike" into a membranous environment at an acidic pH might be governed by a sulfhydryl–disulfide bond exchange, transforming the protein from an afusogenic into a fusogenic structure that is stable and irreversible. The amino acid sequence between positions 61 and 114 of E_1 contains seven cysteine residues, composing 41% of the total cysteine content.

Thus, unlike the fusogenic proteins of other viruses, in which the hydrophobic domain is located at the amino terminus, this putative domain of SFV E_1 is thought to be located about 80 amino acids away from the amino terminus (Garoff *et al.*, 1980a). There are, however, four additional hydrophobic domains in the E_1 molecule (Omar and Koblet, 1988) located between amino acids 153 and 173, 268 and 283,

352 and 372, and 387 and 400, as calculated from the hydropathic averages (Kyte and Doolittle, 1982). The significance of these sequences for the fusogenic ability of E_1 is unknown.

The positively charged groups of lysines and arginines mentioned above play a role in the fusion reaction. They are rather flanking and are not found in the hydrophobic peptides, indicating that the fusion process is not solely determined by hydrophobic interaction and that there are other segments involved in mediating membrane fusion. Indeed, analysis of Sindbis virus mutants exhibiting a lowered fusion pH shows that at least two sites in the region of positions 72 and 313 of E_1 participate in the fusion reaction (Boggs et al., 1989).

Many fusogenic viral sequences have been published and compared (Rott, 1982; White et al., 1983; Asano and Asano, 1984; Gallaher, 1987; Spear, 1987). The amino terminus of F_1 of paramyxoviruses contains the fusion peptide with the tripeptide Phe–X–Gly (Varsanyi et al., 1985). Similarly, in influenza A and B viruses the cleavage of the precursor HA0 yields HA2, with the inverse fusion peptide Gly–X–Phe (Skehel and Waterfield, 1975) in the hydrophobic amino terminus. In the fusogenic gp41 of human immunodeficiency virus the tandem repeat Phe–Leu–Gly–Phe–Leu–Gly can be found within the amino-terminal hydrophobic region (Gallaher, 1987). However, there is still a question as to what extent these sequence homologies can usefully be interpreted. Homologies might be traits for function in related families, but hydrophobicity per se could be sufficient for fusion.

H. Energy and Potential

Aedes cells can be grown and maintained at both pH 7.0 and 6.0. Correspondingly, FFWI is not halted when the cells are left at low pH, mimicking a situation prevailing in the endosomes (Helenius et al., 1985). Reverting back to pH 7.0 following exposure to low pH is not mandatory, as observed in Sindbis-infected BHK cells (Cassell et al., 1984; Edwards and Brown, 1984; Edwards et al., 1983; Mann et al., 1983). In the following section I provide a brief explanation for this interesting discrepancy. It is linked to an observation by Okada (1962b): Sendai virus-induced FFWO of Ehrlich's ascites tumor cells was inhibited by the addition of 2,4-dinitrophenol.

1. Fusion and ATP in C6/36 and BHK Cells

In C6/36 cells the intracellular ATP level decreases slowly (i.e., there is no leakage) only under fusion conditions, to reach 10–15% of the initial value after 1 hour at pH 6.0. If a pH of 7.0 is restored after 1 minute at pH 6.0 (to release the stress of the artificial proton gradient across the PM), the ATP is consumed in an identical fashion. This indicates that fusion per se requires ATP.

If fusion conditions are established in the presence of the inhibitors of oxidative phosphorylation, KCN, NaN_3, or 2,4-dinitrophenol, ATP is exhausted within a few minutes and FFWI is prevented. However, if the cells are exposed to pH 6.0 only for 1–5 minutes in the presence of the drugs and the pH then is brought back to 7.0 (in the presence of the inhibitors), the ATP concentration is restored within 5 minutes, decreasing slowly thereafter over 1 hour, when fusion takes place. Since oxidative phosphorylation is blocked during the restoration of ATP in these experiments, I assume that this regeneration is due to a rapid mobilization of energy stores.

Based on these and similar experiments, I conclude that (1) low pH of the medium (pH_e) is readily tolerated by C6/36 cells; (2) inhibition of oxidative phosphorylation can be overcome, to a certain degree, by transphosphorylations; (3) fusion per se needs energy; (4) only the combined burden of infection, low pH_e, and inhibition of oxidative phosphorylation leads to a breakdown of ATP so rapidly that fusion cannot take place; (5) without inhibitors fusion proceeds to completion at pH 6.0, because infected C6/36 cells are able to provide enough ATP for a sufficient period to drive the fusion and concomitantly sustain a pH_e of 6.0 (Kempf et al., 1987a; Koblet et al., 1987, 1988).

Conclusion (1) explains why soon after infection the pH of the medium can be lowered without harming the expression of the fusion competence in C6/36 cells (see Section VI,B).

KCN inhibits the fusion process only when added during the first 4 minutes after lowering the pH (Kempf et al., 1987b). Therefore, ATP seems to be used for an early event in the fusion cascade.

Infected BHK cells behave differently. As observed by others (Cassell et al., 1984; Edwards et al., 1983; Mann et al., 1983; Coombs and Brown, 1987c), no FFWI could be seen at pH 6.0. Under these conditions ATP immediately collapses and is at zero after 5 minutes. No inhibitors of oxidative phosphorylation are needed. Exposure of infected BHK cells to low pH for 2 minutes allows only a partial restoration of the ATP to 75%; this, in turn, allows fusion to proceed. We conclude that (1) low pH of the medium is tolerated in noninfected BHK cells; (2) low pH of the medium is not tolerated in infected BHK cells; (3) the combined burden of low pH_e and fusion demands leads to the breakdown of ATP; in contrast to C6/36 cells, BHK cells cannot provide enough energy to sustain the pH gradient across the PM and to drive fusion (Kempf et al., 1988a).

2. Fusion and Intracellular pH in C6/36 and BHK Cells

Whereas different energy turnover or stores satisfactorily explain the different behavior of BHK and mosquito cells with respect to fusion, the reason for the sudden breakdown of ATP in infected cells only is not given. Is it the large artificial proton gradient across the membrane which elicits ATP-driven exchange processes? However, the rate of diffusion of protons through lipid bilayers is not abruptly increased when the pH_e is lowered (Boron, 1983; Deamer and Bramhall, 1986; Deamer, 1987; Deamer and Nichols, 1989).

The internal [i.e., cytoplasmic pH (pH_i)] of C6/36 cells was, therefore, determined as a function of the pH_e, using the fluorescent pH indicator Quene 1 (Rogers et al., 1983). The pH_i decreases dramatically, only in infected cells and in the presence of KNC, below a pH_e of 6.2. The threshold value of pH_e corresponds to the critical pH at which the conformational change triggering the cell–cell fusion takes place (Koblet et al., 1985). This sudden influx of protons is not connected to a general leakiness of the PM. For example, the uptake of [^{14}C]sucrose is not increased. The uptake of amino acids and nucleosides into infected cells at pH 6.0 is strongly depressed.

This decrease of pH_i at a pH_e below 6.2 is observed only in infected C6/36 cells in the presence of metabolic inhibitors. Apparently, below pH_e 6.2 proton channels are opened, allowing the passive flux of protons along a pH gradient. In the absence of ATP-depleting agents, the cells overcome the flux of protons by extruding them in a process requiring energy (Kempf et al., 1987c). Protons could be exchanged against sodium and then sodium against potassium (Na^+, K^+-ATPase) (Roos and Boron, 1981; Moolenaar, 1986; Frelin et al., 1988; Madshus, 1988).

The channels might consist of viral membrane proteins (see Section II,B), or such proteins, in turn, might influence host membrane proteins to act as channels, or channels might be the result of the formation of local cylindrical structures of the membrane lipids due to fusogenic protein import. Finally, lipases might be activated, liberating fatty acids which can act as proton carriers (Gutknecht, 1987, 1988). However, it is not known in topographical terms where the channels are formed; they could be linked to the fusion bridges described above, or they could be located in cellular areas not involved in the fusion. They might lead into the donor or acceptor compartment or both.

Again, the situation is different in infected BHK cells. pH_i was measured as a function of the pH_e, using the pH indicator fluorescein diacetate (Roos and Boron, 1981). Only in infected cells below 6.2, the pH_i suddenly drops and no inhibitors of oxidative phosphorylation are needed. Thus, the permeability of the PM of infected cells to protons is increased again at the critical pH_e required to trigger SFV-induced cell–cell fusion (Kempf et al., 1988a). We conclude that, in contrast to C6/36 cells, infected BHK cells are helpless against the sudden flux of protons and exhaust their ATP in vain in processes described to occur after acid loads (Moolenaar, 1986). The cells are eventually killed. However, the process is nonlytic.

3. Fusion and Intracellular pH with Other Viruses and Cells

In C6/36 cells infected with VSV (which also induces cell–cell fusion at an acidic pH) pH_i was determined as a function of pH_e using Quene 1. Again, in the presence of KCN at pH_e values below the critical fusogenic pH, the pH_i decreases dramatically.

GPBind4 cells, murine cells constitutively expressing the uncleaved form of the hemagglutinin (HA_0) of an influenza virus, fuse at an acidic pH after trypsin treatment (Sambrook et al., 1985). Such cells were treated with trypsin and the pH_i was again measured as a function of pH_e, using fluorescein diacetate. The pH_i decreases without the application of inhibitors of oxidative phosphorylation as soon as the critical pH_e of 5.5 is reached. In nontrypsinized control cells no such deviation of the pH_i from the straight line can be observed (Kempf et al., 1987c).

4. Fusion and Other Ion Fluxes and Concentrations

As stated in Section VI,H,2, exchange processes incited by the proton influx in infected mosquito cells at the critical pH_e might lead to an increased flux of other ions. The findings are rather complicated (Kempf et al., 1988b). (1) At pH_e of 5.7 there is a rapid (i.e., 1-minute) reduction in the intracellular potassium ion concentration from about 270 nmol of potassium ion per milligram of protein (approximately 200 mM) to about 100 nmol in C6/36 cells. (2) However, an increased ^{42}K influx into infected cells below a pH_e of 6.2 can be detected. (3) The intracellular sodium ion concentration is increased. Whereas an increased influx of ^{22}Na cannot be demonstrated as a function of the critical pH_e, a greatly increased efflux of ^{22}Na from preloaded C6/36 cells can be measured below the critical pH_e of 6.2. These findings suggest increased $Na^+–K^+$ exchange rates due to increased $Na^+–K^+$ pumping. (4) Neither the fusion nor the exchange processes are sensitive to ouabain (0.5–2 mM).

The energy-dependent coupled Na^+/HCO_3^- and Cl^- antiporter (Moolenaar, 1986; Roos and Boron, 1981) might therefore play a role in mosquito cells. This is indicated by the observations that extracellular chlorine ion is necessary for fusion to occur and that the inhibitors of anion transport DIDS (100 μM) and SITS block the low pH-induced fusion of SFV-infected C6/36 cells.

Pasternak and Micklem (1973, 1974) have reported the leakage of potassium ion from Lettrée cells infected with Sendai virus shortly after adsorption (FFWO), and a decrease of intracellular potassium ion was observed after interaction of Sendai virus with HeLa cells (Fuchs et al., 1978). Fuchs and Giberman (1973) described an enhancement of ^{42}K influx in BHK cells and chicken erythrocytes during the adsorption of Sendai virus (FFWO).

These and our data reveal for the first time a link between FFWO and FFWI, between a viral fusion at pH 7.0 and at pH 6.0. There is also an increase in the sodium ion concentration in both systems (Fuchs et al., 1978), which depends on the extracellular ion composition (Kempf et al., 1988b).

5. Fusion and Membrane Potential

An alteration of the membrane potential is expected to occur as a consequence of an augmented potassium ion influx or the hydrogen ion extrusion coupled to electrogenic ion pumping (Bashford and Pasternak, 1985, 1986).

The membrane potential of SFV- or mock-infected C6/36 cells was measured with the potential sensitive dye 1,3,3,1',3',3'-hexamethyl indodicarbocyanine (Lüdi et al., 1983) as a function of the pH_e (Kempf et al., 1988b). Below 6.2 a hyperpolarization, which is more negative on the inside of the PM of infected cells, is detected. This hyperpolarization is constant in time and does not increase or decrease when the pH_e is further lowered or the extracellular potassium ion concentration is raised. It is not dependent on the final intracellular potassium ion concentration established at a given pH_e. The new low intracellular potassium ion concentration, partially compensated by a gain in sodium ion, might invoke an exchange which is not completely electroneutral, thereby shifting the membrane potential into a new range, favoring fusion.

It is interesting that the membrane potential is increased during myogenesis (Amagi et al., 1983). Furthermore, there is a close correlation between conditions inhibiting viral infection of BHK cells by SFV and those causing depolarization of the cells, whereby fusion of the viral envelope with the endosomal membrane is inhibited (Helenius et al., 1985). Changes in membrane potential were also discussed in the fusion of liposomes and polyethylene glycol-induced cell–cell fusion (Ohki and Oshima, 1985; Krahling et al., 1978).

I. Conclusions

The following principle can be stated. Enveloped viruses (e.g., para-myxoviruses), the fusogenic protein of which undergoes conformational change, with a concomitant increase in hydrophobicity at pH 7.0, infect the cells by an FFWO. At the time of viral maturation, the cells will fuse by an FFWI and show a CPE spontaneously in form of a syncytium.

Enveloped viruses (e.g., myxoviruses, rhabdoviruses, and togaviruses), the fusogenic protein of which undergoes conformational change, with a concomitant increase in hydrophobicity at pH 6.0, infect the cells by an endosomal fusion. FFWO, and later FFWI, can only be provoked by an intentional lowering of the pH of the medium. FFWO at low pH can lead to infection; however, this is probably not a common event in nature. Without intentional lowering of the pH of the medium, a viral infection of a cell culture might remain undetected because no CPE appears. This is the case in many alphavirus-infected mosquito cell lines.

This view gives a strong analogy among the three types of viral fusion reactions: FFWO, EF, and FFWI. They are supposed to be expressions of similar mechanisms at different locations of the cell. I think that this linking concept is strengthened by the data reviewed in Section VI,H.

However, for the time being, the linking of fusion types should not be exaggerated. One linking group, as proposed here, might consist of viral fusion types in which exoplasmic surfaces of bilayers are forced into primary contact. Another, more complicated, linking group could be represented by intracellular transport vesicles, where endoplasmic surfaces are primarily involved. The two leaflets of a bilayer are not at all identical in their chemical composition (van Meer and Simons, 1988; Hoekstra et al., 1989).

Obviously, the biochemistry of fusion is extremely complicated. It involves not only complex mixtures of lipids and glycolipids, but also membrane proteins with hydrophilic and hydrophobic parts exhibiting various types of posttranslational modifications. Specific proteins are needed to induce fusion. With current methodology fusogenic proteins can be identified (Kondor-Koch et al., 1983; Florkiewicz and Rose, 1984; Paterson et al., 1985; Gething et al., 1986a; Lapidot et al., 1987; Omar and Koblet, 1988). It can be shown that a single viral protein type bound to a membrane can provoke fusion with another membrane. Viral proteins are only fusogenic under certain conditions. Endoproteolytic cleavage (e.g., paramyxoviruses) or low pH (e.g., rhabdoviruses and togaviruses) or both (e.g., myxoviruses) enforce conformational changes which lead to the extrusion of hydrophobic amino acid sequences.

A general concept of virus-induced fusions at low pH is tentatively formulated as follows: (1) Spikes are bound via the fusogenic (alphavirus E_1) protein to a surface which acts as a hydrophobic support. Thereby, the conformation of the protein changes from the free state at pH 7.0 to the bound state at the same pH. The forces are mainly electrostatic. At this stage the fusion sequence within the E_1 protein (Omar and Koblet, 1988) near the amino-terminal side (Garoff et al., 1982) would be hidden in the trimer clustering.

(2) This conformation is the prerequisite for establishing the correct fusogenic conformation, which is locked into position and stabilized, possibly by sulfhydryl–disulfide exchange, as soon as the spike is exposed to a low pH. The spikes could open as a flower and expose the hydrophobic fusion sequences. This second tight binding corresponds to a hydrophobic interaction. A cooperative effect takes place because the spikes are oligomerized and even aggregate (Doms and Helenius, 1986; Bundo-Morita et al., 1988; Wharton et al., 1988). The spike clustering persists in fused membranes of Sindbis virus (von Bonsdorff and Harrison, 1978). This could be explained by the finding that spikes form bonds across the main axes of the icosahedron. In addition, protonation at pH 6.0 reduces opposing electrostatic forces.

(3) At this stage the ectodomains of the oligomerized proteins partially lose contact (influenza: Ruigrok et al., 1988), so that proton channels are opened. Exchange processes are driven by ATP. The cells are forced to build up a membrane potential, which becomes more negative inside.

(4) This, in turn, leads to an electrophoretic import of the fusogenic parts of the molecule into the opposing membrane due to positively charged groups. By analogy, a link between membrane potential and vesicular transport has been postulated (Anner, 1987).

(5) This disturbs the lipid bilayers in such a way that the two bilayers unite. Phospholipid head groups are sensors of electric charge in membranes (Seelig et al., 1987).

Thus, the viral fusion reaction is considered to be a special case of a posttranslational protein import, whereby one end of the fusogenic protein is fixed by its transmembrane piece to its original membrane, and a fusogenic loop is integrated into the membrane of the opposing compartment. The thesis is compatible with known facts (Rietveld and de Kruijff, 1986; Jain and Zakim, 1987; Sarkar et al., 1989) and theoretical considerations (Seelig et al., 1987; Blumenthal, 1988b; Brasseur et al., 1988; von Heijne, 1988a; Jacobs and White, 1989). However, no convincing proof is available, such as hydrophobic labeling (Asano and Asano, 1984; Boulay *et al.*, 1987b; Harter *et al.*, 1988; Novick and Hoekstra, 1988), for such a protein import. What has been shown by hydrophobic labeling is that a hydrophobic interaction of the amino-terminal "fusion peptide" of influenza HA2 with liposomal membranes takes place, either at the surface or by integration (Harter *et al.*, 1989).

Finally, FFWI could be an important factor in the propagation of viral disease in the living organism [e.g., VSV: Chany *et al.*, 1987; acquired immunodeficiency syndrome (AIDS): Sodroski *et al.*, 1986; Haseltine and Sodroski, 1987]. The milieu in an organ which will be infected by a virus plays a major role (Barbey-Morel *et al.*, 1987; Tashiro *et al.*, 1987a,b; Webster and Rott, 1987). In several examples fusogenicity and infectivity are directly related. Therefore, the preventive concept of the future might be to interfere with conformational changes in fusogenic viral proteins (Kida *et al.*, 1985).

ACKNOWLEDGMENTS

Work in this laboratory discussed here was supported in part by Swiss National Science Foundation grants. I am grateful to all of my collaborators: Linda Börlin, Y. Dai, M. Elgizoli, A. Flaviano, C. Kempf, Ursula Kohler, M. R. Michel, H. Y. Naim, A. Omar, F. Reigel, C. Schoch, P. Späth, and J. Stalder. I thank D. Rüsch for carefully typing the manuscript, H. Naim for the illustration, and R. Jakob for his comments.

This review would not have been possible without the generous hospitality of Professor A. Igarashi, Head of the Department of Virology at the Institute for Tropical Medicine, Sakamoto Machi, during my sabbatical in Nagasaki, Japan, in the summer of 1987.

This review is dedicated to Professor Karl Maramorosch.

REFERENCES

Acheson, N. H., and Tamm, I. (1967). *Virology* **32**, 128–143.
Ahkong, R. F., Botham, G. M., Woodward, A. W., and Lucy, J. A. (1980). *Biochem. J.* **192**, 829–836.
Ahle, S., Mann, A., Eichelsbacher, U., and Ungewickell, E. (1988). *EMBO J.* **7**, 919–929.
Aliperti, G., and Schlesinger, M. J. (1978). *Virology* **90**, 366–369.
Amagi, Y., Iijima, M., and Kasai, S. (1983). *Jpn. J. Physiol.* **33**, 547–557.
Anderson, R. G., and Orci, L. (1988). *J. Cell Biol.* **106**, 539–543.
Anderson, R. G., Brown, M. S., and Goldstein, J. L. (1977). *Cell (Cambridge, Mass.)* **10**, 351–354.
Anner, B. M. (1987). *Perspect. Biol. Med.* **30**, 537–545.
Argos, P. (1988). *Nucleic Acids Res.* **16**, 9909–9916.
Asano, A., and Asano, K. (1984). *Tumor Res.* **19**, 1–20.
Asano, K., and Asano, A. (1985). *Biochem. Int.* **10**, 115–122.
Barbey-Morel, C. L., Oeltmann, T. N., Edwards, K. M., and Wright, P. F. (1987). *J. Infect. Dis.* **155**, 667–672.
Bartles, J. R., and Hubbard, A. L. (1988). *Trends Biochem. Sci.* **13**, 181–184.

Bashford, C. L., and Pasternak, C. A. (1985). *Eur. Biophys. J.* **12**, 229–235.

Bashford, C. L., and Pasternak, C. A. (1986). *Trends Biochem. Sci.* **11**, 113–116.

Beckers, C. J. M., and Balch, W. E. (1989). *J. Cell Biol.* **108**, 1245–1256.

Beckers, C. J. M., Black, M. R., Glick, B. S., Rothman, J. E., and Balch, W. E. (1989). *Nature (London)* **339**, 397–398.

Bell, J. W., and Waite, M. R. F. (1977). *J. Virol.* **21**, 788–791.

Bell, J. W., Garry, R. F., and Waite, M. R. F. (1978). *J. Virol.* **25**, 764–769.

Berger, M., and Schmidt, M. F. G. (1984). *J. Biol. Chem.* **259**, 7245–7252.

Berger, M., and Schmidt, M. F. G. (1985). *FEBS Lett.* **187**, 289–294.

Birdwell, C. R., and Strauss, J. H. (1974). *J. Virol.* **14**, 366–374.

Bishop, W. R., and Bell, R. M. (1988). *Annu. Rev. Cell Biol.* **4**, 579–610.

Blumenthal, R. (1987). *Curr. Top. Membr. Transp.* **29**, 203–254.

Blumenthal, R. (1988a). *Cell Biophys.* **12**, 1–12.

Blumenthal, R. (1988b). *Stud. Biophys.* **127**, 147–154.

Boege, U., Wengler, G., Wengler, G., and Wittmann-Liebold, B. (1981). *Virology* **113**, 293–303.

Boere, W. A. M., Harmsen, T., Vinjé, J., Benaissa, B. J., Kraaijeveld, C. A., and Snippe, H. (1984). *J. Virol.* **52**, 575–582.

Boggs, W. M., Hahn, C. S., Strauss, E. G., Strauss, J. H., and Griffin, D. E. (1989). *Virology* **169**, 485–488.

Bonatti, S., Cancedda, R., and Blobel, G. (1979). *J. Cell Biol.* **80**, 219–224.

Bonatti, S., Migliaccio, G., and Simons, K. (1989). *J. Biol. Chem.* **264**, 12590–12595.

Boron, W. F. (1983). *J. Membr. Biol.* **72**, 1–16.

Boulay, F., Doms, R. W., Wilson, I., and Helenius, A. (1987a). *EMBO J.* **6**, 2643–2650.

Boulay, F., Doms, R. W., and Helenius, A. (1987b). *In* "Positive Strand RNA Viruses" (M. A. Brinton and R. R. Rueckert, eds.), pp. 103–112. Liss, New York.

Boulay, F., Doms, R. W., Webster, R., and Helenius, A. (1988). *J. Cell Biol.* **106**, 629–639.

Bourne, H. R. (1988). *Cell (Cambridge, Mass.)* **53**, 669–671.

Bracha, M., and Schlesinger, M. J. (1976). *Virology* **74**, 441–449.

Braell, W. A. (1987). *Proc. Natl. Acad. Sci. U.S.A.* **84**, 1137–1141.

Brasseur, R., Cornet, B., Burny, A., Vandenbranden, M., and Ruysschaert, J. M. (1988). *AIDS Res. Hum. Retroviruses* **4**, 83–89.

Bratt, M. A., and Gallaher, W. R. (1969). *Proc. Natl. Acad. Sci. U.S.A.* **64**, 536–543.

Bratt, M. A., and Gallaher, W. R. (1970). *In Vitro* **6**, 3–14.

Brodsky, F. M. (1988). *Science* **242**, 1396–1402.

Brown, D. T. (1984). *In* "Vectors in Virus Biology" (M. A. Mayo and K. A. Harrap, eds.), pp. 113–133. Academic Press, London.

Brown, D. T., and Gliedman, J. B. (1973). *J. Virol.* **12**, 1534–1539.

Brown, D. T., Waite, M. R. F., and Pfefferkorn, E. R. (1972). *J. Virol.* **10**, 524–536.

Brown, D. T., Smith, J. F., Gliedman, J. B., Riedel, B., Filtzer, D., and Renz, D. (1976). *In* "Invertebrate Tissue Cultures" (E. Kurstak and K. Maramorosch, eds.), pp. 35–48. Academic Press, New York.

Brown, W. J., Goodhouse, J., and Farquhar, M. G. (1986). *J. Cell Biol.* **103**, 1235–1247.

Bundo-Morita, K., Gibson, S., and Lenard, J. (1988). *Virology* **163**, 622–624.

Burge, B. W., and Strauss, J. H. (1970). *J. Mol. Biol.* **47**, 449–466.

Burgess, T. L., and Kelly, R. B. (1987). *Annu. Rev. Cell Biol.* **3**, 243–293.

Butters, T. D., and Hughes, R. C. (1981). *Biochim. Biophys. Acta* **640**, 655–671.

Butters, T. D., Hughes, R. C., and Vischer, P. (1981). *Biochim. Biophys. Acta* **640**, 672–686.

Cain, C. C., and Murphy, R. F. (1988). *J. Cell Biol.* **106**, 269–277.

Calisher, C. H., Shope, R. E., Brandt, W., Casals, J., Karabatsos, N., Murphy, F. A., Tesh, R. B., and Wiebe, M. E. (1980). *Intervirology* 14, 229–232.

Cameron, J. M., Clemens, M. J., Gray, M. A., Menzies, D. E., Mills, B. J., Warren, A. P., and Pasternak, C. A. (1986). *Virology* 155, 534–544.

Cancedda, R., Villa-Komaroff, L., Lodish, H. F., and Schlesinger, M. J. (1975). *Cell (Cambridge, Mass.)* 6, 215–222.

Caplan, M. J., Rosenzweig, S. A., and Jamieson, J. D. (1986). *In* "Physiology of Membrane Disorders" (T. E. Andreoli, J. F. Hoffman, D. D. Fanestil, and S. G. Schultz, eds.), pp. 273–281. Plenum, New York.

Caplan, M. J., Stow, J. L., Newman, A. P., Madri, J. A., Anderson, H. C., Farquhar, M. G., Palade, G. E., and Jamieson, J. D. (1987). "Molecular Mechanisms in the Regulation of Cell Behavior," pp. 179–185. Liss, New York.

Carrasco, L. (1977). *FEBS Lett.* 76, 11–15.

Carrasco, L. (1978). *Nature (London)* 272, 694–699.

Cassell, S., Edwards, J., and Brown, D. T. (1984). *J. Virol.* 52, 857–864.

Chanas, A. C., Ellis, D. S., Stamford, S., and Gould, E. A. (1982a). *Antiviral Res.* 2, 191–201.

Chanas, A. C., Gould, E. A., Clegg, J. C. S., and Varma, M. G. R. (1982b). *J. Gen. Virol.* 58, 37–46.

Chang, G. J. J., and Trent, D. W. (1987). *J. Gen. Virol.* 68, 2129–2142.

Chany, C., Chany-Fournier, F., and Robain, O. (1987). *Nature (London)* 326, 250.

Chirico, W. J., Waters, G., and Blobel, G. (1988). *Nature (London)* 332, 805–810.

Clague, M. J., Schoch, C., Zech, L., and Blumenthal, R. (1990). *Biochemistry* 29, 1303–1308.

Clarke, S., and Harrison, L. C. (1983). *J. Biol. Chem.* 258, 11434–11437.

Clegg, J. C. S. (1975). *Nature (London)* 254, 454–455.

Clegg, J. C. S., and Kennedy, S. I. T. (1975). *J. Mol. Biol.* 97, 401–411.

Collett, M. S., Moennig, V., and Horzinek, M. C. (1989). *J. Gen. Virol.* 70, 253–266.

Condreay, L. D., Adams, R. H., Edwards, J., and Brown, D. T. (1988). *J. Virol.* 62, 2629–2635.

Coombs, K., and Brown, D. T. (1987a). *Virus Res.* 7, 131–149.

Coombs, K., and Brown, D. T. (1987b). *J. Mol. Biol.* 195, 359–371.

Coombs, K., and Brown, D. T. (1987c). *In* "Mechanisms of Viral Toxicity in Animal Cells" (L. Carrasco, ed.), pp. 5–19. CRC Press, Boca Raton, Florida.

Copeland, C. S., Doms, R. W., Bolzau, E. M., Webster, R. G., and Helenius, A. (1986). *J. Cell Biol.* 103, 1179–1192.

Copeland, C. S., Zimmer, K. P., Wagner, K., Healey, G. A., Mellman, I., and Helenius, A. (1988). *Cell (Cambridge, Mass.)* 53, 197–209.

Cutler, D. F. (1988). *J. Cell Sci.* 91, 1–4.

Cutler, D. F., and Garoff, H. (1986). *J. Cell Biol.* 102, 889–901.

Cutler, D. F., Melançon, P., and Garoff, H. (1986). *J. Cell Biol.* 102, 902–910.

Dalgarno, L., Rice, C. M., and Strauss, J. H. (1983). *Virology* 129, 170–187.

Dalrymple, J. M., Schlesinger, S., and Russell, P. K. (1976). *Virology* 69, 93–103.

Davey, J. (1987). *Biosci. Rep.* 7, 299–306.

Davey, M. W., and Dalgarno, L. (1974). *J. Gen. Virol.* 24, 453–463.

Davey, M. W., Dennett, D. P., and Dalgarno, L. (1973). *J. Gen. Virol.* 20, 225–232.

Davidson, H. W., Rhodes, C. J., and Hutton, J. C. (1988). *Nature (London)* 333, 93–96.

Davoust, J., Gruenberg, J., and Howell, K. E. (1987). *EMBO J.* 6, 3601–3609.

Deamer, D. W. (1987). *J. Bioenerg. Biomembr.* 19, 457–479.

Deamer, D. W., and Bramhall, J. (1986). *Chem. Phys. Lipids* 40, 167–188.

Deamer, D. W., and Nichols, J. W. (1989). *J. Membr. Biol.* 107, 91–103.

De Curtis, I., and Simons, K. (1988). *Proc. Natl. Acad. Sci. U.S.A.* **85,** 8052–8056.

De Curtis, I., and Simons, K. (1989). *Cell (Cambridge, Mass.)* **58,** 719–727.

Diaz, R., Mayorga, L. S., Weidman, P. J., Rothman, J. E., and Stahl, P. D. (1989). *Nature (London)* **339,** 398–400.

Ding, M., and Schlesinger, M. J. (1989). *Virology* **171,** 280–284.

Doms, R. W., and Helenius, A. (1986). *J. Virol.* **60,** 833–839.

Doms, R. W., Helenius, A. H., and White, J. M. (1985). *J. Biol. Chem.* **260,** 2973–2981.

Doms, R. W., Keller, D. S., Helenius, A., and Balch, W. E. (1987). *J. Cell Biol.* **105,** 1957–1969.

Doms, R. W., Ruusala, A., Machamer, C., Helenius, J., Helenius, A., and Rose, J. K. (1988). *J. Cell Biol.* **107,** 89–99.

Dreyfuss, G., Philipson, L., and Mattaj, I. W. (1988). *J. Cell Biol.* **106,** 1419–1425.

Dubin, D. T., Timko, K., Gillies, S., and Stollar, V. (1979). *Virology* **98,** 131–141.

Dubovi, E. J., and Wagner, R. R. (1977). *J. Virol.* **22,** 500–509.

Duncan, J. R., and Kornfeld, S. (1988). *J. Cell Biol.* **106,** 617–628.

Dunphy, W. G., Fries, E., Urbani, L. J., and Rothman, J. E. (1981). *Proc. Natl. Acad. Sci. U.S.A.* **78,** 7453–7457.

Durbin, R. K., and Stollar, V. (1986). *Virology* **154,** 135–143.

Eaton, B. T. (1978). *In* "Viruses and Environment" (E. Kurstak and K. Maramorosch, eds.), pp. 181–193. Academic Press, New York.

Eaton, B. T., and Regnery, R. L. (1975). *J. Gen. Virol.* **29,** 35–49.

Edwards, J., and Brown, D. T. (1984). *Virus Res.* **1,** 705–711.

Edwards, J., Mann, E., and Brown, D. T. (1983). *J. Virol.* **45,** 1090–1097.

Eilers, M., and Schatz, G. (1988). *Cell (Cambridge, Mass.)* **52,** 481–483.

Elgizoli, M., Dai, Y., Kempf, C., Koblet, H., and Michel, M. R. (1989). *J. Virol.* **63,** 2921–2928.

Enzmann, P.-J. (1987a). *In* "Arboviruses in Arthropod Cells in Vitro" (C. E. Yunker, ed.), Vol. II, pp. 53–66. CRC Press, Boca Raton, Florida.

Enzmann, P.-J. (1987b). *In* "Arboviruses in Arthropod Cells in Vitro" (C. E. Yunker, ed.), Vol. II, pp. 67–76. CRC Press, Boca Raton, Florida.

Enzmann, P.-J., and Weiland, F. (1979). *Virology* **95,** 501–510.

Erwin, C., and Brown, D. T. (1983). *J. Virol.* **45,** 792–799.

Evered, D., and Whelan, J., eds. (1984). *Ciba Found. Symp.* **103.**

Fallon, A. M., and Stollar, V. (1987). *Adv. Cell Cult.* **5,** 97–137.

Fisher, J. M., and Scheller, R. H. (1988). *J. Biol. Chem.* **263,** 16515–16518.

Florkiewicz, R. Z., and Rose, J. K. (1984). *Science* **225,** 721–722.

Frelin, C., Vigne, P., Ladoux, A., and Lazdunski, M. (1988). *Eur. J. Biochem.* **174,** 3–14.

Friedman, R. M., Levin, J. G., Grimley, P. M., and Berezesky, I. K. (1972). *J. Virol.* **10,** 504–515.

Fries, E., and Helenius, A. (1979). *Eur. J. Biochem.* **97,** 213–220.

Froshauer, S., Kartenbeck, J., and Helenius, A. (1988). *J. Cell Biol.* **107,** 2075–2086.

Fuchs, P., and Giberman, E. (1973). *FEBS Lett.* **31,** 127–130.

Fuchs, P., Spiegelstein, M., Haimsohn, M., Gitelman, J., and Kohn, A. (1978). *J. Cell. Physiol.* **95,** 223–234.

Fuchs, R., Schmid, S., Male, P., Helenius, A., and Mellman, I. (1986). *J. Cell Biol.* **103,** 439a (abstr.).

Fuller, S. D. (1987). *Cell (Cambridge, Mass.)* **48,** 923–934.

Fuller, S. D., and Argos, P. (1987). *EMBO J.* **6,** 1099–1105.

Gallaher, W. R. (1987). *Cell (Cambridge, Mass.)* **50,** 327–328.

Gallaher, W. R., Levitan, D. B., Kirwin, K. S., and Blough, H. A. (1980). *In* "Cell Membranes and Viral Envelopes" (H. A. Blough and J. M. Tiffany, eds.), Vol. I, pp. 395–457. Academic Press, London.

Garoff, H. (1985). *Annu. Rev. Cell Biol.* **1**, 403–445.
Garoff, H., and Simons, K. (1974). *Proc. Natl. Acad. Sci. U.S.A.* **71**, 3988–3992.
Garoff, H., and Söderlund, H. (1978). *J. Mol. Biol.* **124**, 535–549.
Garoff, H., Frischauf, A. M., Simons, K., Lehrach, H., and Delius, H. (1980a). *Nature (London)* **288**, 236–241.
Garoff, H., Frischauf, A. M., Simons, K., Lehrach, H., and Delius, H. (1980b). *Proc. Natl. Acad. Sci. U.S.A.* **77**, 6376–6380.
Garoff, H., Kondor-Koch, C., and Riedel, H. (1982). *Curr. Top. Microbiol. Immunol.* **99**, 1–50.
Garoff, H., Kondor-Koch, C., Pettersson, R., and Burke, B. (1983). *J. Cell Biol.* **97**, 652–658.
Garry, R. F. (1988). *Biosci. Rep.* **8**, 35–48.
Garry, R. F., and Bostick, D. A. (1987). *Virus Res.* **8**, 245–259.
Garry, R. F., Bishop, J. M., Parker, S., Westbrook, K., Lewis, G., and Waite, M. R. F. (1979a). *Virology* **96**, 108–120.
Garry, R. F., Westbrook, K., and Waite, M. R. F. (1979b). *Virology* **99**, 179–182.
Gates, D., Brown, A., and Wust, C. J. (1982). *Infect. Immun.* **35**, 248–255.
Gething, M. J., White, J., and Waterfield, M. (1978). *Proc. Natl. Acad. Sci. U.S.A.* **75**, 2737–2740.
Gething, M. J., Doms, R. W., York, D., and White, J. (1986a). *J. Cell Biol.* **102**, 11–23.
Gething, M. J., McCammon, K., and Sambrook, J. (1986b). *Cell (Cambridge, Mass.)* **46**, 939–950.
Geuze, H. J., Slot, J. W., Strous, G. J., Lodish, H. F., and Schwartz, A. L. (1983). *Cell (Cambridge, Mass.)* **32**, 277–287.
Geuze, H. J., Slot, J. W., and Schwartz, A. L. (1987). *J. Cell Biol.* **104**, 1715–1723.
Gibson, S., Bundo-Morita, K., Portner, A., and Lenard, J. (1988). *Virology* **163**, 226–229.
Gillies, S., and Stollar, V. (1980). *Virology* **107**, 509–519.
Gillies, S., and Stollar, V. (1982). *Mol. Cell. Biol.* **2**, 66–75.
Glanville, N., Ranki, M., Morser, J., Kääriäinen, L., and Smith, A. E. (1976). *Proc. Natl. Acad. Sci. U.S.A.* **73**, 3059–3063.
Glick, B. S., and Rothman, J. E. (1987). *Nature (London)* **326**, 309–312.
Gliedman, J. B., Smith, J. F., and Brown, D. T. (1975). *J. Virol.* **16**, 913–926.
Gomatos, P., Kääriäinen, L., Keränen, S., Ranki, M., and Sawicki, D. L. (1980). *J. Gen. Virol.* **49**, 61–69.
Gould, E. A., and Clegg, J. C. S. (1985). In "Virology: A Practical Approach" (B. W. J. Mahy, ed.), pp. 43–78. IRL Press, Oxford, England.
Green, J., Griffiths, G., Louvard, D., Quinn, P., and Warren, G. (1981). *J. Mol. Biol.* **152**, 663–698.
Griffiths, G., and Hoppeler, H. (1986). *J. Histochem. Cytochem.* **34**, 1389–1398.
Griffiths, G., and Simons, K. (1986). *Science* **234**, 438–443.
Griffiths, G., Warren, G., Quinn, P., Mathieu-Costello, O., and Hoppeler, H. (1984). *J. Cell Biol.* **98**, 2133–2141.
Griffiths, G., Hoflack, B., Simons, K., Mellman, I., and Kornfeld, S. (1988). *Cell (Cambridge, Mass.)* **52**, 329–341.
Griffiths, G., Fuller, S. D., Back, R., Hollinshead, M., Pfeiffer, S., and Simons, K. (1989). *J. Cell Biol.* **108**, 277–297.
Grimley, P. M., Berezesky, I. K., and Friedman, R. M. (1968). *J. Virol.* **2**, 1326–1338.
Grimstad, P. R. (1983). *Adv. Virus Res.* **28**, 357–438.
Guan, J.-L., Ruusala, A., Cao, H., and Rose, J. K. (1988). *Mol. Cell. Biol.* **8**, 2869–2874.
Gutknecht, J. (1987). *Proc. Natl. Acad. Sci. U.S.A.* **84**, 6443–6446.
Gutknecht, J. (1988). *J. Membr. Biol.* **106**, 83–93.

Hahn, C. S., Lustig, S., Strauss, E. G., and Strauss, J. H. (1988). *Proc. Natl. Acad. Sci. U.S.A.* **85**, 5997–6001.

Hahn, Y. S., Grakoui, A., Rice, C. M., Strauss, E. G., and Strauss, J. H. (1989a). *J. Virol.* **63**, 1194–1202.

Hahn, Y. S., Strauss, E. G., and Strauss, J. H. (1989b). *J. Virol.* **63**, 3142–3150.

Hahn, C. S., Rice, C. M., Strauss, E. G., Lenches, E. M., and Strauss, J. H. (1989c). *J. Virol.* **63**, 3459–3465.

Hardy, W. R., and Strauss, J. H. (1988). *J. Virol.* **62**, 998–1007.

Harter, C., Bächi, T., Semenza, G., and Brunner, J. (1988). *Biochemistry* **27**, 1856–1864.

Harter, C., James, P., Bächi, T., Semenza, G., and Brunner, J. (1989). *J. Biol. Chem.* **264**, 6459–6464.

Hase, T., Summers, P. L., Eckels, K. H., and Baze, W. B. (1987). *Arch. Virol.* **96**, 135–151.

Hase, T., Summers, P. L., and Cohen, W. H. (1989). *Arch. Virol.* **108**, 101–114.

Haseltine, W. A., and Sodroski, J. G. (1987). *Ann. Inst. Pasteur, Paris: Virol.* **138**, 83–92.

Hashimoto, K., Erdei, S., Keränen, S., Saraste, J., and Kääriäinen, L. (1981). *J. Virol.* **38**, 34–40.

Hedman, K., Goldenthal, K. L., Rutherford, A. V., Pastan, I., and Willingham, M. C. (1987). *J. Histochem. Cytochem.* **35**, 233–243.

Helenius, A. (1984). *Biol. Cell.* **51**, 181–186.

Helenius, A., and Kartenbeck, J. (1980). *Eur. J. Biochem.* **106**, 613–618.

Helenius, A., and Marsh, M. (1982). *Ciba Found. Symp.* **92**, 59–76.

Helenius, A., Kartenbeck, J., Simons, K., and Fries, E. (1980). *J. Cell Biol.* **84**, 404–420.

Helenius, A., Marsh, M., and White, J. (1982). *J. Gen. Virol.* **58**, 47–61.

Helenius, A., Mellman, I., Wall, D., and Hubbard, A. (1983). *Trends Biochem. Sci.* **8**, 245–250.

Helenius, A., Kielian, M., Wellsteed, J., Mellman, I., and Rudnick, G. (1985). *J. Biol. Chem.* **260**, 5691–5697.

Herrler, G., Dürkop, I., Becht, H., and Klenk, H.-D. (1988). *J. Gen. Virol.* **69**, 839–846.

Hoekstra, D. (1988). *Indian J. Biochem. Biophys.* **25**, 76–84.

Hoekstra, D., and Kok, J. W. (1989). *Biosci. Rep.* **9**, 273–305.

Hoekstra, D., Klappe, K., Stegmann, T., and Nir, S. (1988). *In* "Molecular Mechanisms of Membrane Fusion" (S. Ohki, D. Doyle, T. D. Flanagan, S. W. Hui, and E. Mayhew, eds.), pp. 399–412. Plenum, New York.

Hoekstra, D., Eskelinen, S., and Kok, J. W. (1989). *In* "Organelles in Eukaryotic Cells: Molecular Structure and Interactions" (J. M. Tager, A. Azzi, S. Papa, and F. Guerrieri, eds.), pp. 59–83. Plenum, New York.

Horzinek, M. C., and Mussgay, M. (1969). *J. Virol.* **4**, 514–520.

Hsieh, P., and Robbins, P. W. (1984). *J. Biol. Chem.* **259**, 2375–2382.

Hsu, M.-C., Scheid, A., and Choppin, P. W. (1981). *J. Biol. Chem.* **256**, 3557–3563.

Hsu, S. H., Li, S. Y., and Cross, J. H. (1972). *J. Med. Entomol.* **9**, 86–96.

Hubbard, S. C., and Ivatt, R. J. (1981). *Annu. Rev. Biochem.* **50**, 555–583.

Hubbard, A. L., Stieger, B., and Bartles, J. R. (1989). *Annu. Rev. Physiol.* **51**, 755–770.

Hynes, R. O. (1987). *Cell (Cambridge, Mass.)* **48**, 549–554.

Igarashi, A. (1978). *J. Gen. Virol.* **40**, 531–544.

Igarashi, A. (1985). *Adv. Virus Res.* **30**, 21–42.

Igarashi, A., Koo, R., and Stollar, V. (1977). *Virology* **82**, 69–83.

Ishihama, A., and Nagata, K. (1988). *CRC Crit. Rev. Biochem.* **23**, 27–76.

Jacobs, R. E., and White, S. H. (1989). *Biochemistry* **28**, 3421–3437.

Jain, M. K., and Zakim, D. (1987). *Biochim. Biophys. Acta* **906**, 33–68.

Jalanko, A. (1985). *FEBS Lett.* **186**, 59–64.

Johnson, D. C., and Schlesinger, M. J. (1980). *Virology* **103**, 407–424.

Jones, K. J., Waite, M. R. F., and Bose, H. R. (1974). *J. Virol.* **13,** 809–817.
Kääriäinen, L., and Ranki, M. (1984). *Annu. Rev. Microbiol.* **38,** 91–109.
Kääriäinen, L., and Söderlund, H. (1978). *Curr. Top. Microbiol. Immunol.* **82,** 15–69.
Kääriäinen, L., Takkinen, K., Keränen, S., and Söderlund, H. (1987). *J. Cell Sci., Suppl.* **7,** 231–250.
Kaluza, G., and Pauli, G. (1980). *Virology* **102,** 300–309.
Kaluza, G., Rott, R., and Schwarz, R. T. (1980). *Virology* **102,** 286–299.
Kempf, C., Michel, M. R., Kohler, U., and Koblet, H. (1987a). *Arch. Virol.* **95,** 111–122.
Kempf, C., Michel, M. R., Kohler, U., and Koblet, H. (1987b). *Arch. Virol.* **95,** 283–289.
Kempf, C., Michel, M. R., Kohler, U., and Koblet, H. (1987c). *Biosci. Rep.* **7,** 761–769.
Kempf, C., Michel, M. R., Kohler, U., and Koblet, H. (1988a). *Arch. Virol.* **99,** 111–115.
Kempf, C., Michel, M. R., Kohler, U., Koblet, H., and Oetliker, H. (1988b). *Biosci. Rep.* **8,** 241–254.
Kennedy, S. I. T. (1976). *J. Mol. Biol.* **108,** 491–511.
Keränen, S., and Kääriäinen, L. (1975). *J. Virol.* **16,** 388–396.
Kida, H., Yoden, S., Kuwabara, M., and Yanagawa, R. (1985). *Vaccine, Suppl.* **3,** 219–222.
Kielian, M. C. (1987). *In* "Molecular Mechanisms in the Regulation of Cell Behavior," pp. 197–202. Liss, New York.
Kielian, M. C., and Helenius, A. (1984). *J. Virol.* **52,** 281–283.
Kielian, M. C., and Helenius, A. (1985). *J. Cell Biol.* **101,** 2284–2291.
Kielian, M. C., Marsh, M., and Helenius, A. (1986). *EMBO J.* **5,** 3103–3109.
Kimura, T., and Ohyama, A. (1988). *J. Gen. Virol.* **69,** 1247–1254.
Kinney, R. M., Johnson, R. J. B., Brown, V. L., and Trent, D. W. (1986). *Virology* **152,** 400–413.
Klausner, R. D. (1989). *Cell (Cambridge, Mass.)* **57,** 703–706.
Klenk, H.-D., and Schwarz, R. T. (1982). *Antiviral Res.* **2,** 177–190.
Klip, A., Grinstein, S., and Semenza, G. (1979). *Biochim. Biophys. Acta* **558,** 233–245.
Knipfer, M. E., and Brown, D. T. (1989). *Virology* **170,** 117–122.
Knutton, S. (1980). *J. Cell Sci.* **43,** 103–118.
Kobata, A. (1984). *In* "Biology of Carbohydrates" (V. Ginsburg and P. W. Robbins, eds.), Vol. II, Ch. 2. Wiley, New York.
Koblet, H., Kempf, C., Kohler, U., and Omar, A. (1985). *Virology* **143,** 334–336.
Koblet, H., Omar, A., and Kempf, C. (1987). *In* "Arboviruses in Arthropod Cells in Vitro" (C. E. Yunker, ed.), Vol. II, pp. 77–90. CRC Press, Boca Raton, Florida.
Koblet, H., Omar, A., Kohler, U., and Kempf, C. (1988). *In* "Invertebrate and Fish Tissue Culture" (Y. Kuroda, E. Kurstak, and K. Maramorosch, eds.), pp. 140–143. Jpn. Sci. Soc. Press, Tokyo.
Kohn, A. (1979). *Adv. Virus Res.* **24,** 223–276.
Kondor-Koch, C., Riedel, H., Söderberg, K., and Garoff, H. (1982). *Proc. Natl. Acad. Sci. U.S.A.* **79,** 4525–4529.
Kondor-Koch, C., Burke, B., and Garoff, H. (1983). *J. Cell Biol.* **97,** 644–651.
Konigsberg, I. R. (1971). *Dev. Biol.* **26,** 133–152.
Kornfeld, S. (1986). *J. Clin. Invest.* **77,** 1–6.
Kornfeld, S. (1987). *FASEB J.* **1,** 462–468.
Kornfeld, R., and Kornfeld, S. (1985). *Annu. Rev. Biochem.* **54,** 631–664.
Krahling, H., Schinkewitz, U., Barker, A., and Hulser, D. F. (1978). *Cytobiologie* **17,** 51–61.
Kräusslich, H.-G., and Wimmer, E. (1988). *Annu. Rev. Biochem.* **57,** 701–754.
Kreis, T. E., and Lodish, H. F. (1986). *Cell (Cambridge, Mass.)* **46,** 929–937.
Kuroda, Y., Kurstak, E., and Maramorosch, K. (eds.) (1988). "Invertebrate and Fish Tissue Culture." Jpn. Sci. Soc. Press, Tokyo.

Kurstak, E., and Maramorosch, K. (1976). "Invertebrate Tissue Culture." Academic Press, New York.

Kurstak, E., Maramorosch, K., and Dübendorfer, A. (1980). "Invertebrate Systems in Vitro." Elsevier, Amsterdam.

Kurtti, T. J., and Munderloh, U. G. (1984). *Adv. Cell Cult.* **3**, 259–302.

Kyte, J., and Doolittle, R. F. (1982). *J. Mol. Biol.* **157**, 105–132.

Lachmi, B., Glanville, N., Keränen, S., and Kääriäinen, L. (1975). *J. Virol.* **16**, 1615–1629.

Laine, R., Söderlund, H., and Renkonen, O. (1973). *Intervirology* **1**, 110–118.

Lambrecht, B., and Schmidt, M. F. G. (1986). *FEBS Lett.* **202**, 127–132.

Lapidot, M., Nussbaum, O., and Loyter, A. (1987). *J. Biol. Chem.* **262**, 13736–13741.

Leader, D. P., and Katan, M. (1988). *J. Gen. Virol.* **69**, 1441–1464.

Lesser, G. J., Lee, R. H., Zehfus, M. H., and Rose, G. D. (1987). *In* "Protein Engineering," pp. 175–179. Liss, New York.

Linden, C. D., and Roth, T. F. (1983). *In* "Receptor-Mediated Endocytosis" (P. Cuatrecasas and T. F. Roth, eds.), Vol. 15, Ser. B, pp. 21–44. Chapman & Hall, London.

Linderman, J. J., and Lauffenburger, D. A. (1988). *J. Theor. Biol.* **132**, 203–245.

Lloyd, J. B., and Forster, S. (1986). *Trends Biochem. Sci.* **11**, 365–368.

Lodish, H. F. (1988). *J. Biol. Chem.* **263**, 2107–2110.

Lüdi, H., Oetliker, H., Brodbeck, U., Ott, P., Schwendimann, B., and Fulpius, B. W. (1983). *J. Membr. Biol.* **74**, 75–84.

Luukkonen, A., Kääriäinen, L., and Renkonen, O. (1976). *Biochim. Biophys. Acta* **450**, 109–120.

Luukkonen, A., von Bonsdorff, C.-H., and Renkonen, O. (1977). *Virology* **78**, 331–335.

Maassen, J. A., and Terhorst, C. (1981). *Eur. J. Biochem.* **115**, 153–158.

Machamer, C. E., and Rose, J. K. (1988a). *J. Biol. Chem.* **263**, 5948–5954.

Machamer, C. E., and Rose, J. K. (1988b). *J. Biol. Chem.* **263**, 5955–5960.

Madshus, I. H. (1988). *Biochem. J.* **250**, 1–8.

Mann, E., Edwards, J., and Brown, D. T. (1983). *J. Virol.* **45**, 1083–1089.

Maramorosch, K., and Mitsuhashi, J. (1982). "Invertebrate Cell Culture Applications." Academic Press, New York.

Marsh, M. (1984). *Biochem. J.* **218**, 1–10.

Marsh, M., and Helenius, A. (1980). *J. Mol. Biol.* **142**, 439–454.

Marsh, M., and Helenius, A. (1989). *Adv. Virus Res.* **36**, 107–151.

Marsh, M., Wellsteed, J., Kern H., Harms, E., and Helenius, A. (1982). *Proc. Natl. Acad. Sci. U.S.A.* **79**, 5297–5301.

Marsh, M., Bolzau, E., and Helenius, A. (1983). *Cell (Cambridge, Mass.)* **32**, 931–940.

Marsh, M., Kern, H., Harms, E., Schmid, S., Mellman, I., and Helenius, A. (1988). *In* "Cell-Free Analysis of Membrane Traffic" (D. J. Morré, K. E. Howell, G. M. W. Cook, and W. H. Evans, eds.), pp. 21–33. Liss, New York.

Matlin, K. S. (1986). *J. Cell Biol.* **103**, 2565–2568.

Matlin, K., Reggio, H., Helenius, A., and Simons, K. (1981). *J. Cell Biol.* **91**, 601–613.

May, J. M. (1985). *J. Biol. Chem.* **260**, 462–467.

Mayne, J. T., Rice, C. M., Strauss, E. G., Hunkapiller, M. W., and Strauss, J. H. (1984). *Virology* **134**, 338–357.

McDowell, W., Romero, P. A., Datema, R., and Schwarz, R. T. (1987). *Virology* **161**, 37–44.

Melançon, P., and Garoff, H. (1986). *EMBO J.* **7**, 1543–1550.

Melançon, P., and Garoff, H. (1987). *J. Virol.* **61**, 1301–1309.

Mellman, I., Fuchs, R., and Helenius, A. (1986). *Annu. Rev. Biochem.* **55**, 663–700.

Mi, S., Durbin, R., Huang, H. V., Rice, C. M., and Stollar, V. (1989). *Virology* **170**, 385–391.

Michel, M. R., and Gomatos, P. J. (1973). *J. Virol.* **11**, 900–914.

Mitsuhashi, J. (1982). *Adv. Cell Cult.* **2**, 133–196.

Mitsuhashi, J., and Maramorosch, K. (1964). *Contrib. Boyce Thompson Inst.* **22**, 435–460.

Mitsuhashi, J., Nakasone, S., and Horie, Y. (1983). *Cell Biol. Int. Rep.* **7**, 1057–1062.

Moolenaar, W. H. (1986). *Trends Biochem. Sci.* **11**, 141–143.

Mooney, J. J., Dalrymple, J. M., Alving, C. R., and Russell, P. K. (1975). *J. Virol.* **15**, 225–231.

Morgan, M. S., Darrow, R. M., Nafz, M. A., and Varandani, P. T. (1985). *Biochem. J.* **225**, 349–356.

Muñoz, A., Castrillo, J. L., and Carrasco, L. (1985). *Virology* **146**, 203–212.

Naim, H., and Koblet, H. (1988). *Arch. Virol.* **102**, 73–89.

Naim, H., and Koblet, H. (1990). *Arch. Virol.* **110**, 221–237.

Nelson, N. (1987). *BioEssays* **7**, 251–254.

Nestorowicz, A., Laver, G., and Jackson, D. C. (1985). *J. Gen. Virol.* **66**, 1687–1695.

Ng, M. L. (1987). *J. Gen. Virol.* **68**, 577–582.

Ng, M. L., and Westaway, E. G. (1979). *J. Gen. Virol.* **43**, 91–101.

Nilsson, T., Jackson, M., and Peterson, P. A. (1989). *Cell (Cambridge, Mass.)* **58**, 707–718.

Njus, D., Kelley, P. M., and Harnadek, G. J. (1986). *Biochim. Biophys. Acta* **853**, 237–265.

Novick, S. L., and Hoekstra, D. (1988). *Proc. Natl. Acad. Sci. U.S.A.* **85**, 7433–7437.

Ohki, S., and Oshima, H. (1985). *Biochim. Biophys. Acta* **812**, 147–154.

Okada, Y. (1958). *Biken J.* **1**, 103–110.

Okada, Y. (1962a). *Exp. Cell Res.* **26**, 98–107.

Okada, Y. (1962b). *Exp. Cell Res.* **26**, 119–129.

Olson, E. N. (1988). *Prog. Lipid Res.* **27**, 177–197.

Omar, A., and Koblet, H. (1988). *Virology* **166**, 17–23.

Omar, A., and Koblet, H. (1989a). *Virology* **168**, 177–179.

Omar, A., and Koblet, H. (1989b). *In* "Invertebrate Cell System Applications" (J. Mitsuhashi, ed.), Vol. II, pp. 151–155. CRC Press, Boca Raton, Florida.

Omar, A., Flaviano, A., Kohler, U., and Koblet, H. (1986). *Arch. Virol.* **89**, 145–159.

Orci, L., Ravazzola, M., Storch, M.-J., Anderson, R. G. W., Vassalli, J.-D., and Perrelet, A. (1987). *Cell (Cambridge, Mass.)* **49**, 865–868.

Ou, J. H., Strauss, E. G., and Strauss, J. H. (1981). *J. Virol.* **109**, 281–289.

Ozawa, M., Asano, A., and Okada, Y. (1979). *Virology* **99**, 197–202.

Pakkanen, R., von Bonsdorff, C.-H., Turunen, O., Wahlström, T., and Vaheri, A. (1988). *Eur. J. Cell Biol.* **46**, 435–443.

Pasternak, C. A. (1987a). *Arch. Virol.* **93**, 169–184.

Pasternak, C. A. (1987b). *BioEssays* **6**, 14–19.

Pasternak, C. A., and Micklem, K. J. (1973). *J. Membr. Biol.* **14**, 293–303.

Pasternak, C. A., and Micklem, K. J. (1974). *Biochem. J.* **140**, 405–411.

Paterson, R. G., and Lamb, R. A. (1987). *Cell (Cambridge, Mass.)* **48**, 441–452.

Paterson, R. G., Hiebert, S. W., and Lamb, R. A. (1985). *Proc. Natl. Acad. Sci. U.S.A.* **82**, 7520–7524.

Pearse, B. M. F. (1982). *Proc. Natl. Acad. Sci. U.S.A.* **79**, 451–455.

Pearse, B. M. F. (1987). *EMBO J.* **6**, 2507–2512.

Pedersen, P. L., and Carafoli, E. (1987). *Trends Biochem. Sci.* **12**, 146–150.

Pelham, H. R. (1988). *EMBO J.* **7**, 913–918.

Pesonen, M., and Renkonen, O. (1976). *Biochim. Biophys. Acta* **455**, 510–525.

Pesonen, M., Saraste, J., Hashimoto, K., and Kääriäinen, L. (1981). *Virology* **109**, 165–173.

Petterson, R. F., Söderlund, H., and Kääriäinen, L. (1980). *Eur. J. Biochem.* **105**, 435–443.

Pfeffer, S. R., and Rothman, J. E. (1987). *Annu. Rev. Biochem.* **56**, 829–852.

Pierce, J. S., Strauss, E. G., and Strauss, J. H. (1974). *J. Virol.* **13**, 1030–1036.

Poste, G., and Nicolson, G. L. (1978). "Membrane Fusion." North-Holland, Amsterdam.

Presley, J. F., and Brown, D. T. (1989). *J. Virol.* **63**, 1975–1980.

Pressman, B. C. (1976). *Annu. Rev. Biochem.* **45**, 501–530.

Primakoff, P., Hyatt, H., and Tredick-Kline, J. (1987). *J. Cell Biol.* **104**, 141–149.

Pudney, M., Leake, C. J., and Buckley, S. (1982). *In* "Invertebrate Cell Culture Applications" (K. Maramorosch and J. Mitsuhashi, eds.), pp. 159–194. Academic Press, New York.

Quinn, P., Griffiths, G., and Warren, G. (1983). *J. Cell Biol.* **96**, 851–856.

Quinn, P., Griffiths, G., and Warren, G. (1984). *J. Cell Biol.* **98**, 2142–2147.

Rademacher, T. W., Parekh, R. B., and Dwek, R. A. (1988). *Annu. Rev. Biochem.* **57**, 785–838.

Raghow, R. S., Davey, N. W., and Dalgarno, L. (1973a). *Arch. Gesamte Virusforsch.* **43**, 165–168.

Raghow, R. S., Grace, T. D. C., Filshie, B. K., Bartley, W., and Dalgarno, L. (1973b). *J. Gen. Virol.* **21**, 109–122.

Rasilo, M.-L., and Renkonen, O. (1979). *Biochim. Biophys. Acta* **582**, 307–321.

Reigel, F., and Koblet, H. (1981). *J. Virol.* **39**, 321–324.

Rice, C. M., and Strauss, J. H. (1981). *Proc. Natl. Acad. Sci. U.S.A.* **78**, 2062–2066.

Rice, C. M., and Strauss, J. H. (1982). *J. Mol. Biol.* **154**, 325–348.

Richardson, M. A., Boulton, R. W., Raghow, R. S., and Dalgarno, L. (1980). *Arch. Virol.* **64**, 263–274.

Riedel, H., Lehrach, H., and Garoff, H. (1982). *J. Virol.* **42**, 725–729.

Rietveld, A., and de Kruijff, B. (1986). *Biosci. Rep.* **6**, 775–782.

Roberts, D. D., and Goldstein, I. J. (1984). *J. Biol. Chem.* **259**, 909–914.

Robinson, A., and Austen, B. (1987). *Biochem. J.* **246**, 249–261.

Rodriguez, J. F., Paez, E., and Esteban, M. (1987). *J. Virol.* **61**, 395–404.

Rodriguez-Boulan, E., Misek, D. E., Vega de Salas, D., Salas, P. J., and Bard, E. (1985). *Curr. Top. Membr. Transp.* **24**, 251–294.

Roehrig, J. T., Hunt, A. R., Kinney, R. M., and Mathews, J. H. (1988). *Virology* **165**, 66–73.

Rogers, J., Hesketh, T. R., Smith, G. A., and Metcalfe, J. C. (1983). *J. Biol. Chem.* **258**, 5994–5997.

Roos, A., and Boron, W. F. (1981). *Physiol. Rev.* **61**, 296–434.

Rose, J. K., and Doms, R. W. (1988). *Annu. Rev. Cell Biol.* **4**, 257–288.

Rossmann, M. G., and Johnson, J. E. (1989). *Annu. Rev. Biochem.* **58**, 533–573.

Roth, M. G. (1989) *Annu. Rev. Physiol.* **51**, 797–810.

Roth, J., Taatjes, D. J., Lecocq, J. M., Weinstein, M., and Paulson, J. C. (1985). *Cell (Cambridge, Mass.)* **43**, 287–295.

Rothman, J. E. (1987). *Cell (Cambridge, Mass.)* **50**, 521–522.

Rott, R. (1982). *Hoppe-Seyler's Z. Physiol. Chem.* **363**, 1273–1282.

Rott, R., Klenk, H.-D., and Scholtissek, C. (1984). *Zentralbl. Bakteriol., Parasitenkd., Infektionskr. Hyg., Abt. 1: Orig., Reihe A* **258**, 337–349.

Ruigrok, R. W. H., Aitken, A., Calder, L. J., Martin, S. R., Skehel, J. J., Wharton, S. A., Weis, W., and Wiley, D. C. (1988). *J. Gen. Virol.* **69,** 2785–2795.

Russell, D. L., Dalrymple, J. M., and Johnston, R. E. (1989). *J. Virol.* **63,** 1619–1629.

Sambrook, J., Rogers, L., White, J., and Gething, M.-J. (1985). *EMBO J.* **4,** 91–103.

Saraste, J., von Bonsdorff, C.-H., Hashimoto, K., Kääriäinen, L., and Keränen, S. (1980). *Virology* **100,** 229–245.

Sarkar, D. P., Morris, S. J., Eidelman, O., Zimmerberg, J., and Blumenthal, R. (1989). *J. Cell Biol.* **109,** 113–122.

Sarver, N., and Stollar, V. (1977). *Virology* **80,** 390–400.

Sawicki, D., and Sawicki, S. (1980). *J. Virol.* **34,** 108–118.

Sawicki, D. L., Barkhimer, D. B., Sawicki, S. G., Rice, C. M., and Schlesinger, S. (1990). *Virology* **174,** 43–52.

Scheefers-Borchel, U., Scheefers, H., Edwards, J., and Brown, D. T. (1981). *Virology* **110,** 292–301.

Schlesinger, R. W. (ed.) (1980). "The Togaviruses." Academic Press, New York.

Schlesinger, M. J., and Malfer, C. (1982). *J. Biol. Chem.* **257,** 9887–9890.

Schlesinger, M., and Schlesinger, S. (1973). *J. Virol.* **11,** 1013–1016.

Schlesinger, S., and Schlesinger, M. J. (eds.) (1986). "The Togaviridae and Flaviviridae." Plenum, New York.

Schmaljohn, A. L., Kokubun, K. M., and Cole, G. A. (1983). *Virology* **130,** 144–154.

Schmid, S. L., Fuchs, R., Male, P., and Mellman, I. (1988). *Cell (Cambridge, Mass.)* **52,** 73–83.

Schmid, S., Fuchs, R., Kielian, M., Helenius, A., and Mellman, I. (1989). *J. Cell Biol.* **108,** 1291–1300.

Schmidt, M. F. G. (1982). *Virology* **116,** 327–338.

Schmidt, M. F. G. (1984). *EMBO J* **3,** 2295–2300.

Schmidt, M. F. G. (1989). *Biochim. Biophys. Acta* **988,** 411–426.

Schmidt, M. F. G., and Schlesinger, M. J. (1980). *J. Biol. Chem.* **255,** 3334–3339.

Schmidt, M. F. G., Bracha, M., and Schlesinger, M. J. (1979). *Proc. Natl. Acad. Sci. U.S.A.* **76,** 1687–1691.

Schmidt, M., Schmidt, M. F. G., and Rott, R. (1988). *J. Biol. Chem.* **263,** 18635–18639.

Schultz, A. M., Henderson, L. E., and Oroszlan, S. (1988). *Annu. Rev. Cell Biol.* **4,** 611–647.

Schwarz, R. T., and Datema, R. (1980). *Trends Biochem. Sci. (Pers. Ed.)* **5,** 65–67.

Sechoy, O., Philippot, J. R., and Bienvenue, A. (1987). *J. Biol. Chem.* **262,** 11519–11523.

Seelig, J., Macdonald, P. M., and Scherer, P. G. (1987). *Biochemistry* **26,** 7535–7541.

Sefton, B. M., and Buss, J. E. (1987). *J. Cell Biol.* **104,** 1449–1453.

Shaw, A. S., Rottier, P. J. M., and Rose, J. K. (1988). *Proc. Natl. Acad. Sci. U.S.A.* **85,** 7592–7596.

Silva, A. M., Cachau, R. E., and Goldstein, D. J. (1987). *Biophys. J.* **52,** 595–602.

Silverstein, S. C. (ed.) (1978). *Life Sci. Res. Rep.* **11.**

Simizu, B., and Maeda, S. (1981). *J. Gen. Virol.* **56,** 349–361.

Simizu, B., Yamamoto, K., Hashimoto, K., and Ogata, T. (1984). *J. Virol.* **51,** 254–258.

Simmons, D. T., and Strauss, J. H. (1972). *J. Mol. Biol.* **71,** 599–613.

Simons, K., and van Meer, G. (1988). *Biochemistry* **27,** 6197–6202.

Simons, K., Keränen, S., and Kääriäinen, L. (1973). *FEBS Lett.* **29,** 87–91.

Singer, S. J., Maher, P. A., and Yaffe, M. P. (1987a). *Proc. Natl. Acad. Sci. U.S.A.* **84,** 1015–1019.

Singer, S. J., Maher, P. A., and Yaffe, M. P. (1987b). *Proc. Natl. Acad. Sci. U.S.A.* **84,** 1960–1964.

Skehel, J. J., and Waterfield, M. D. (1975). *Proc. Natl. Acad. Sci. U.S.A.* **72,** 93–97.

Skehel, J. J., Bayley, P. M., Brown, E. B., Martin, S. R., Waterfield, M. D., White, J. M., Wilson, I. A., and Wiley, D. C. (1982). *Proc. Natl. Acad. Sci. U.S.A.* **79**, 968–972.

Smith, J. F., and Brown, D. T. (1977). *J. Virol.* **22**, 662–678.

Söderlund, H., and Ulmanen, I. (1977). *J. Virol.* **24**, 907–909.

Söderlund, H., Kääriäinen, L., von Bonsdorff, C.-H., and Weckström, P. (1972). *J. Virol.* **47**, 753–760.

Söderlund, H., Kääriäinen, L., and von Bonsdorff, C.-H. (1975). *Med. Biol.* **53**, 412–417.

Söderlund, H., von Bonsdorff, C.-H., and Ulmanen, I. (1979). *J. Gen. Virol.* **45**, 15–26.

Sodroski, J. G., Goh, W. C., Rosen, C., Campbell, K., and Haseltine, W. A. (1986). *Nature (London)* **322**, 470–474.

Sowers, A. E. (ed.) (1987). "Cell Fusion." Plenum, New York.

Späth, P., and Koblet, H. (1980). *In* "Invertebrate Systems in Vitro" (E. Kurstak, K. Maramorosch, and A. Dübendorfer, eds.), pp. 375–388. Elsevier, Amsterdam.

Spear, P. G. (1987). *In* "Cell Fusion" (A. E. Sowers, ed.), pp. 3–32. Plenum, New York.

Stalder, J., Reigel, F., and Koblet, H. (1983). *Virology* **129**, 247–254.

Stanley-Samuelson, D. W., Jurenka, R. A., Cripps, C., Blomquist, G. J., and de Renobales, M. (1988). *Arch. Insect Biochem. Physiol.* **9**, 1–33.

Steele, J. H. (1981a). "Viral Zoonoses," CRC Handb., Ser. Zoonoses, Vol. I. CRC Press, Boca Raton, Florida.

Steele, J. H. (1981b). "Viral Zoonoses," CRC Handb., Ser. Zoonoses, Vol. II. CRC Press, Boca Raton, Florida.

Stegmann, T., Booy, F. P., and Wilschut, J. (1987). *J. Biol. Chem.* **262**, 17744–17749.

Stegmann, T., Doms, R. W., and Helenius, A. (1989). *Annu. Rev. Biophys. Biophys. Chem.* **18**, 187–211.

Stephens, E. B., and Compans, R. W. (1988). *Annu. Rev. Microbiol.* **42**, 489–516.

Stewart, W. W. (1981). *Nature (London)* **292**, 17–21.

Stollar, V. (1980). *In* "The Togaviruses" (R. W. Schlesinger, ed.), pp. 583–621. Academic Press, New York.

Stollar, V. (1987a). *Adv. Virus Res.* **55**, 327–365.

Stollar, V. (1987b). *In* "Arboviruses in Arthropod Cells in Vitro" (C. E. Yunker, ed.), Vol. II, pp. 91–110. CRC Press, Boca Raton, Florida.

Stollar, V., Stollar, B. D., Koo, R., Harrap, K. A., and Schlesinger, R. W. (1976). *Virology* **69**, 104–115.

Stoorvogel, W., Geuze, H. J., Griffith, J. M., and Strous, G. J. (1988). *J. Cell Biol.* **106**, 1821–1829.

Strauss, E. G., and Strauss, J. H. (1983). *Curr. Top. Microbiol. Immunol.* **105**, 2–98.

Strauss, J. H., and Strauss, E. G. (1977). *In* "The Molecular Biology of Animal Viruses" (D. P. Nayak, ed.), pp. 111–166. Dekker, New York.

Strauss, J. H., and Strauss, E. G. (1985). *In* "Immunochemistry of Viruses" (M. H. V. van Regenmortel and A. R. Neurath, eds.), pp. 407–424. Elsevier, Amsterdam.

Strauss, E. G., Rice, C. M., and Strauss, J. H. (1983). *Proc. Natl. Acad. Sci. U.S.A.* **80**, 5271–5275.

Strauss, E. G., Rice, C. M., and Strauss, J. H. (1984). *Virology* **133**, 92–110.

Strauss, J. H., Strauss, E. G., Hahn, C. S., Hahn, Y. S., Galler, R., Hardy, W. R., and Rice, C. M. (1987). *UCLA Symp., New Ser.* **54**, 209–225.

Sullivan, P. C., Ferris, A. L., and Storrie, B. (1987). *J. Cell. Physiol.* **131**, 58–63.

Superti, F., Seganti, L., Ruggeri, F. M., Tinari, A., Donelli, G., and Orsi, N. (1987). *J. Gen. Virol.* **68**, 387–399.

Takkinen, K. (1986). *Nucleic Acids Res.* **14**, 5667–5682.

Talbot, P., and Vance, D. E. (1982). *Virology* **118**, 451–455.

Tan, K. B., and Sokol, F. (1974). *J. Virol.* **13**, 1245–1253.

Tashiro, M., Ciborowski, P., Reinacher, M., Pulverer, G., Klenk, H.-D., and Rott, R. (1987a). *Virology* **157**, 421–430.

Tashiro, M., Ciborowski, P., Klenk, H.-D., Pulverer, G., and Rott, R. (1987b). *Nature (London)* **325**, 536–537.

Tatem, J., and Stollar, V. (1986). *Virus Res.* **5**, 121–130.

Tooker, P., and Kennedy, S. I. T. (1981). *J. Virol.* **37**, 589–600.

Towler, D. R., Gordon, J. I., Adams, S. P., and Glaser, L. (1988). *Annu. Rev. Biochem.* **57**, 69–99.

Tsou, C.-L. (1988). *Biochemistry* **27**, 1809–1812.

Tuomi, K., Kääriäinen, L., and Söderlund, H. (1975). *Nucleic Acids Res.* **2**, 555–565.

Tycko, B., and Maxfield, F. R. (1982). *Cell (Cambridge, Mass.)* **28**, 643–651.

Ulmanen, I. (1978). *J. Gen. Virol.* **41**, 353–365.

Ulug, E. T., Garry, R. F., Waite, M. R. F., and Bose, H. R. (1984). *Virology* **132**, 118–130.

Ulug, E. T., Garry, R. F., and Bose, H. R. (1987). *In* "Mechanisms of Viral Toxicity in Animal Cells" (L. Carrasco, ed.), pp. 91–113. CRC Press, Boca Raton, Florida.

Ulug, E. T., Garry, R. F., and Bose, H. R., Jr. (1989). *Virology* **172**, 42–50.

van Deurs, B., Petersen, O. W., Olsnes, S., and Sandvig, K. (1987). *Exp. Cell Res.* **171**, 137–152.

van Meer, G., and Simons, K. (1988). *J. Cell. Biochem.* **36**, 51–58.

van Steeg, H., Kasperaitis, M., Voorma, H. O., and Benne, R. (1984). *Eur. J. Biochem.* **138**, 473–478.

Varsanyi, T. M., Jörnvall, H., and Norrby, E. (1985). *Virology* **147**, 110–117.

Vaux, D. J. T., Helenius, A., and Mellman, I. (1988). *Nature (London)* **336**, 36–42.

Verner, K., and Schatz, G. (1988). *Science* **241**, 1307–1313.

Vogel, R. H., Provencher, S. W., von Bonsdorff, C.-H., Adrian, M., and Dubochet, J. (1986). *Nature (London)* **320**, 533–535.

von Bonsdorff, C.-H. (1973). *Commentat. Biol.* **74**, 1–53.

von Bonsdorff, C.-H., and Harrison, S. C. (1975). *J. Virol.* **16**, 141–145.

von Bonsdorff, C.-H., and Harrison, S. C. (1978). *J. Virol.* **28**, 578–583.

von Figura, K., and Hasilik, A. (1986). *Annu. Rev. Biochem.* **55**, 167–193.

von Heijne, G. (1988a). *Biochim. Biophys. Acta* **947**, 307–333.

von Heijne, G. (1988b). *In* "Institute for Scientific Information Atlas of Science" (A. M. Grimwade, ed.), Vol. I, pp. 205–209. Grimwade, Philadelphia, PA.

von Heijne, G., and Gavel, Y. (1988). *Eur. J. Biochem.* **174**, 671–678.

Wahlberg, J. M., Boere, W. A. M., and Garoff, W. (1989). *J. Virol.* **63**, 4991–4997.

Waite, M., and Pfefferkorn, E. (1970). *J. Virol.* **5**, 60–71.

Waite, M. R. F., Lubin, M., Jones, K. J., and Bose, H. R. (1974). *J. Virol.* **13**, 244–246.

Warren, G. (1987). *Nature (London)* **327**, 17–18.

Warren, G., Woodman, P., Pypaert, M., and Smythe, E. (1988). *Trends Biochem. Sci.* **13**, 462–465.

Webster, R. G., and Rott, R. (1987). *Cell (Cambridge, Mass.)* **50**, 665–666.

Weiss, B., Nitschko, H., Ghattas, I., Wright, R., and Schlesinger, S. (1989). *J. Virol.* **63**, 5310–5318.

Weiss, E. (ed.) (1971). *Curr. Top. Microbiol. Immunol.* **55**.

Welch, W. J., and Sefton, B. M. (1979). *J. Virol.* **29**, 1186–1195.

Welch, W. J., and Sefton, B. M. (1980). *J. Virol.* **33**, 230–237.

Wellink, J., and van Kammen, A. (1988). *Arch. Virol.* **98**, 1–26.

Wengler, G. (1980). *In* "The Togaviruses" (R. W. Schlesinger, ed.), pp. 459–472. Academic Press, New York.

Wengler, G. (1987). *Arch. Virol.* **94**, 1–14.

Wengler, G., and Wengler, G. (1974). *Virology* **59**, 21–35.

Wengler, G., and Wengler, G. (1976a). *Virology* **73**, 190–199.
Wengler, G., and Wengler, G. (1976b). *J. Virol.* **17**, 10–19.
Wengler, G., Wengler, G., and Gross, H. (1979). *Nature (London)* **282**, 754–756.
Wengler, G., Wengler, G., Nowak, T., and Wahn, K. (1987). *Virology* **160**, 210–219.
West, C. M. (1986). *Mol. Cell. Biochem.* **72**, 3–20.
Westaway, E. G., Brinton, M. A., Gaidamovich, S. Y., Horzinek, M. C., Igarashi, A., Kääriäinen, L., Lvov, D. K., Porterfield, J. S., Russell, P. K., and Trent, D. W. (1985). *Intervirology* **24**, 125–139.
Wharton, S. A. (1987). *Microbiol. Sci.* **4**, 119–124.
Wharton, S. A., Martin, S. R., Ruigrok, R. W. H., Skehel, J. J., and Wiley, D. C. (1988). *J. Gen. Virol.* **69**, 1847–1857.
White, L. A. (1987). *J. Clin. Microbiol.* **25**, 1221–1224.
White, J. M., and Wilson, I. A. (1987). *J. Cell Biol.* **105**, 2887–2896.
White, J., Kartenbeck, J., and Helenius, A. (1980). *J. Cell Biol.* **87**, 264–272.
White, J., Helenius, A., and Gething, M. J. (1982). *Nature (London)* **300**, 658–659.
White, J., Kielian, M., and Helenius, A. (1983). *Q. Rev. Biophys.* **16**, 151–195.
Wickner, W. T., and Lodish, H. F. (1985). *Science* **230**, 400–407.
Wilcox, G. E., and Compans, R. W. (1982). *Virology* **123**, 312–322.
Wiley, D. C., Wilson, I. A., and Skehel, J. J. (1981). *Nature (London)* **289**, 373–378.
Wilson, D. W., Wilcox, C. A., Flynn, G. C., Chen, E., Kuang, W.-J., Henzel, W.-J., Black, M. R., Ullrich, A., and Rothman, J. E. (1989). *Nature (London)* **339**, 355–359.
Wilson, I. A., Skehel, J. J., and Wiley, D. C. (1981). *Nature (London)* **289**, 366–373.
Wirth, D. F., Katz, F., Small, B., and Lodish, H. F. (1977). *Cell (Cambridge, Mass.)* **10**, 253–263.
Wittek, R., Koblet, H., Menna, A., and Wyler, R. (1977). *Arch. Virol.* **54**, 95–106.
Wolcott, J. A., Wust, C. J., and Brown, A. (1984). *J. Virol.* **49**, 379–385.
Wust, C. J., Wolcott-Nicholas, J. A., Fredin, D., Dodd, D. C., Brideau, R. J., Lively, M. E., and Brown, A. (1989). *Virus Res.* **13**, 101–112.
Yaffe, D. (1971). *Exp. Cell Res.* **66**, 33–48.
Yamamoto, K., Suzuki, K., and Simizu, B. (1981). *Virology* **109**, 452–454.
Yamashiro, D. J., and Maxfield, F. R. (1988). *Trends Pharmacol. Sci.* **9**, 190–193.
Yewdell, J. W., Gerhard, W., and Bächi, T. (1983). *J. Virol.* **48**, 239–248.
Yoshida, T., Takao, S., Kiyotani, K., and Sakaguchi, T. (1989). *Virology* **170**, 571–574.
Yoshimura, A., and Ohnishi, S.-I. (1984). *J. Virol.* **51**, 497–504.
Yunker, C. E. (1971). *Curr. Top. Microbiol. Immunol.* **55**, 113–126.
Yunker, C. E. (ed.) (1987a). "Arboviruses in Arthropod Cells in Vitro," Vol. I. CRC Press, Boca Raton, Florida.
Yunker, C. E. (ed.) (1987b). "Arboviruses in Arthropod Cells in Vitro," Vol. II. CRC Press, Boca Raton, Florida.
Zerial, M., Huylebroeck, D., and Garoff, H. (1987). *Cell (Cambridge, Mass.)* **48**, 147–155.
Ziemiecki, A., and Garoff, H. (1978). *J. Mol. Biol.* **122**, 259–269.
Ziemiecki, A., Garoff, H., and Simons, K. (1980). *J. Gen. Virol.* **50**, 111–123.

EMERGENCE, NATURAL HISTORY, AND VARIATION OF CANINE, MINK, AND FELINE PARVOVIRUSES

Colin R. Parrish

James A. Baker Institute
New York State College of Veterinary Medicine
Cornell University
Ithaca, New York 14853

I. Introduction

During 1978 a new parvovirus of dogs [canine parvovirus (CPV)] was recognized simultaneously as the cause of new diseases of dogs throughout the world. Within 2 years the virus had spread into and infected virtually every population of domestic and wild dogs which has been examined. The diseases were new syndromes for dogs, and serological testing has shown that the virus was not present in dogs before the mid-1970s. The sudden appearance of CPV has raised a number of important questions about the mutability and variation of the parvoviruses, and about its relationships to the closely related viruses of cats and minks. At a fundamental level the mechanisms which define the host range of the viruses were in need of examination. This chapter examines the emergence of CPV, the evidence concerning the previous emergence of mink enteritis virus (MEV) as the cause of a new disease in minks in the 1940s, and the mechanisms which determine the host ranges and other specific properties of the viruses of cats, minks, and dogs.

The host range of animal viruses is one of their most fundamental characteristics, as it determines the range of susceptible animals. Most viruses are restricted in host range and infect only a defined spectrum of susceptible species. The emergence of new viruses and the acquisition of new natural host ranges are unusual events. However, such new viruses have the potential to cause pandemics in their new nonimmune hosts. The emergence of viruses from other animal reservoirs is clearly involved in the evolution of new pandemic strains of influenza in humans (Webster *et al.*, 1982; Palese and Young, 1982) and could be the origin of human immunodeficiency virus. It is likely that such events have occurred in the past to give rise to the present diversity of animal viruses.

Feline Panleukopenia Virus, Mink Enteritis Virus, and Canine Parvovirus

The viruses are classified as the feline parvovirus subgroup of the genus *Parvovirus,* within the family Parvoviridae (Siegl *et al.*, 1985). Feline panleukopenia virus (FPV), MEV, and CPV are classified as "host range variants." In addition to the viruses of cats, minks, and dogs, similar viruses naturally infect many species within the families Felidae, Canidae, Procyonidae, Mustelidae, and possibly the Viverridae. Here, canine parvovirus is referred to as CPV, except when it is necessary to distinguish between the two antigenic forms of the virus, which are referred to as CPV type 2 (CPV-2) and CPV type 2a (CPV-2a) (see Section V,B).

II. Historical Background

A. Feline Panleukopenia Virus

A disease similar to that caused by FPV has been known in both large and small cats for many years. The agent was shown to be a filterable virus during the 1920s (Verge and Cristoforoni, 1928) and was the subject of considerable study during the 1930s and 1940s (Hindle and Findlay, 1932; Lawrence and Syverton, 1938; Lawrence et al., 1940, 1943; Hammon and Enders, 1939a,b; Kikuth et al., 1940; Torres, 1941; Lucas and Riser, 1945). In these studies it was determined that a number of diseases, which had been variously described as feline distemper, spontaneous agranulocytosis, feline infectious enteritis, or malignant panleukopenia, were all caused by the same agent (Hammon and Enders, 1939a). Similar diseases in common raccoons (*Procyon lotor*) and Arctic (blue) foxes (*Alopex lagopus*) were known before 1940, the viruses involved being considered the same as FPV (Waller, 1940; Phillips, 1943; Goss, 1948).

B. Mink Enteritis Virus

In 1947 a new disease was observed among minks in the area of Fort William, Ontario, Canada (Schofield, 1949; reviewed by Burger and Gorham, 1970). Initially named Fort William disease and later called mink viral enteritis, the agent was described as MEV. The relationships between mink viral enteritis and FPV were recognized early, due to the close similarities between the clinical signs and the characteristic pathology of the two diseases (Wills, 1952), and the antigenic similarity between the viruses was demonstrated by cross-protection studies during the early 1950s (Wills, 1952; Wills and Belcher, 1956). Early studies indicated that the experimental inoculation of minks with tissues from FPV-infected cats gave a similar disease to that caused by MEV (MacPherson, 1956), although in preliminary studies inoculations of kittens with MEV-containing tissues did not show any clinical disease. More recent studies have examined the *in vivo* host ranges and virulence of the viruses in more detail (see Section II,D).

The origin of MEV in Ontario has not been determined. Within 10–12 years the disease had been observed among farmed mink throughout the United States, Europe, and Scandinavia (Knox, 1960; Tuomi and Kangas, 1963; Kull, 1966). Studies of MEV were hampered until the mid-1960s by the inability to grow the virus *in vitro* and by a lack of sensitive or specific serological tests. It is therefore not known whether MEV was a long-existing virus which had been present in minks many years before 1947 or whether it had only recently acquired the host range for minks at that time. FPV and MEV were both isolated in tissue culture during the 1960s (Johnson, 1964, 1965; Lust *et al.*, 1965; King and Croghan, 1965). The delay in isolating the viruses was probably due to the fact that they replicate only in dividing cells and cause a slow and nondescript cytopathic effect in infected cell cultures. However, the identification of characteristic Feulgen test-positive intranuclear inclusion bodies and fluorescent antibody staining methods for identifying infected cells allowed the viruses and antibody responses to be analyzed and titrated (Johnson, 1965, 1967c; Lust *et al.*, 1965; Gorham *et al.*, 1966; Scott *et al.*, 1970).

During the 1960s and 1970s studies of FPV and MEV revealed them to be parvoviruses and to be closely related biochemically and serologically (Johnson, 1967a; Johnson *et al.*, 1974; Flagstad, 1975, 1977). Effective vaccines were developed which controlled the diseases in cats and minks (Wills and Belcher, 1956; Pridham and Wills, 1959, 1960; Burger *et al.*, 1963; Gorham *et al.*, 1965; King and Gutekunst, 1970; Davis *et al.*, 1970).

C. Canine Parvovirus

During 1978 and 1979 outbreaks of previously unrecognized disease syndromes were observed in dogs in a number of countries (Eugster *et al.*, 1978; Appel *et al.*, 1978, 1979a; Burtonboy *et al.*, 1979; Gagnon and Povey, 1979; Horner *et al.*, 1979; Johnson and Spradbrow, 1979). The diseases were characterized by death due to acute or chronic nonsuppurative myocarditis in pups 3–16 weeks of age or pyrexia accompanied by vomiting and/or diarrhea in older animals.

The enteritis and pathological signs observed were similar to those of FPV in cats or MEV in minks, with necrosis of the intestinal crypt epithelium, eventual loss of the villus structure in the small intestine, and necrosis and depletion of the lymphocytes in various lymphoid tissues. Within a few months of the first observation of the diseases, a parvovirus resembling FPV was isolated or observed by electron microscopy in samples from dogs with myocardial or enteric forms of the disease. The ability of the virus from a case of myocarditis to cause enteritis was also demonstrated (Robinson *et al.*, 1979a, 1980b). The close relationship of this virus to FPV was recognized early (Appel *et al.*, 1978, 1979a; Johnson and Spradbrow, 1979), and effective vaccines were developed within a short period (Eugster, 1980; Carmichael *et al.*, 1981, 1983).

The global spread of CPV has been examined by the analysis of stored sera collected from dogᵉ during the 1970s. The earliest CPV-positive sera reported were collected in Greece during 1974, and the reactivities of the three positive sera of 28 dogs examined from that year were confirmed by virus neutralization assays (Koptopoulos *et al.*, 1986). Positive sera were collected in Belgium and The Netherlands in late 1976 and during 1977 (Schwers *et al.*, 1979; Osterhaus *et al.*, 1980). The virus spread rapidly around the world during 1978, the first positive sera reported being collected in Australia during May 1978, in the United States during June 1978, in Denmark between January and June 1978, in New Zealand between July and October 1978, and in Japan during July 1978 (Walker *et al.*, 1980; Carmichael *et al.*, 1980; Have and Andersen, 1982; Jones *et al.*, 1982; Azetaka *et al.*, 1981; Mohri *et al.*, 1982).

Studies of sera collected from 1184 wild coyotes (*Canis latrans*) in three regions of the United States between 1972 and 1983 showed the first positive sera to be present during 1979, and by 1981 the majority of the animals in all areas were seropositive (Thomas *et al.*, 1984). The first positive sera from a population of wild wolves in Alaska were collected during 1980 (Zarnke and Ballard, 1987).

The rapid global spread of the virus was most likely a consequence of the high titers of virus shed in the feces of infected dogs and the resistance of these parvoviruses to inactivation in the environment, which would have allowed its transport, probably on inanimate objects, even into countries with strict quarantine procedures for dogs (Bouillant and Hanson, 1965; Carmichael *et al.*, 1981; Pollock, 1982; Meunier *et al.*, 1985a).

III. Diseases

The pathogenesis of diseases in animals of various ages is influenced primarily by the requirement of autonomous parvoviruses for actively dividing cells (Margolis and Kilham, 1965; Tennant *et al.*, 1969; Tattersall, 1972). It is also likely that this requirement determines many of the differences seen between infections of fetal, neonatal, or older animals. However, among the dividing cells in any animal, not all are permissive for virus replication. For example, the lymphoid and intestinal epithelial tissues, which contain rapidly dividing cell populations, are targets for virus replication by FPV, MEV, and CPV, but appear to be resistant to infections by many parvoviruses in other species, such as the minute virus of mice (MVM) and H1 virus in hamsters, indicating that developmentally regulated properties of the different dividing cell populations can restrict parvovirus replication at the cellular level (reviewed by Cotmore and Tattersall, 1987). Although the relationships between cell replication rates and the susceptibility of tissues to parvovirus infection and replication in dogs, minks, or cats have not been examined in detail, it is clear that there are differences between the infections of animals at different ages.

A. *Older Animals*

In animals older than 6 weeks at the time of infection, the pathogeneses of infections of cats with FPV, minks with MEV, or dogs with CPV are similar, and the diseases in all species are therefore described together. Serological studies indicate that many infections by CPV in dogs (and probably by FPV and MEV in their respective hosts) are mild or subclinical (Smith *et al.*, 1980; Meunier *et al.*, 1980; Parrish *et al.*, 1982a). Signs seen in affected animals include pyrexia, diarrhea, and vomiting, along with leukopenia and/or a relative lymphopenia. The incidence of leukopenia or lymphopenia varies among the different viruses, panleukopenia being a striking feature of FPV infections of cats, but is uncommon in CPV infections, although a relative lymphopenia is often observed in the latter case (FPV: Hammon and Enders, 1939a,b; Lawrence *et al.*, 1940; Rohovsky and Griesemer, 1967; Larsen *et al.*, 1976; Ichijo *et al.*, 1976; Carlson *et al.*, 1977; MEV: Reynolds, 1969; CPV: Robinson *et al.*, 1980b; Carmichael *et al.*, 1981; Pollock, 1982; Macartney *et al.*, 1984a). Whether this difference reflects differences in the virus or the host is not known. Death of severely affected animals is likely to be a consequence of the extensive destruction of the gut epithelium, with dehydration and endotoxic shock contributing to the disease.

Although the prominent clinical sign observed after infection is most likely to be enteritis, the virus replicates in many organs of the infected host. The course of infection is rapid, with little virus being recovered from tissues or feces after 10–14 days postinfection (FPV: Csiza *et al.*, 1971a; CPV: Macartney *et al.*, 1984b). Natural infection is most likely via the oronasal route (FPV; Csiza *et al.*, 1971a; MEV: Myers *et al.*, 1959; Reynolds, 1970; CPV: Appel *et al.*, 1979b; Carman and Povey, 1982; Pollock, 1982; Macartney *et al.*, 1984a), although animals can be infected by most parenteral routes. After oral or oronasal inoculation initial virus replication occurs in the tonsils and the regional and mesenteric lymph nodes, and shortly thereafter in the thymus and the spleen (FPV: Csiza *et al.*, 1971a; Carlson *et al.*, 1977, 1978; CPV: Macartney *et al.*, 1984b; Meunier *et al.*, 1985a; Carman and Povey, 1985b). Between 4 and 6 days after oral inoculation, infected cells are present in the intestine, and after day 5 postinoculation virus is excreted in large amounts in the feces; in the case of CPV between 10^7 and 10^9 infectious units may be excreted per gram of feces (Carmichael *et al.*, 1981; Meunier *et al.*, 1985a,b; Carman and Povey, 1985a).

In all species, infection of the lymphoid tissues (e.g., the thymus, lymph nodes, and spleen) results in lymphocytolysis, cellular depletion, and, subsequently, tissue regeneration in surviving animals. Virus replication and cell destruction in lymphoid tissues occur mostly in areas of dividing cells, including germinal centers of the lymph nodes and the thymus cortex (Fig. 1) (FPV: Hammon and Enders, 1939a,b; Lawrence *et al.*, 1940; Kikuth *et al.*, 1940; Carlson *et al.*, 1977, 1978; MEV: Reynolds, 1970; Krunajevic, 1970; CPV: Cooper *et al.*, 1979; Robinson *et al.*, 1980b; Macartney *et al.*, 1984a,b; Carman and Povey, 1985a,b).

The bone marrow can be affected severely, with a marked decrease in cellularity and decreases in cells of the myeloid, erythroid, and megakaryocytic series (FPV: Hammon and Enders, 1939a,b; Lawrence *et al.*, 1940; CPV: Robinson *et al.*, 1980b; Boosinger *et al.*, 1982; Macartney *et al.*, 1984a; Carman and Povey, 1985b). Although endotoxemia could play a role in the development of severe bone marrow lesions, both virus and infected cells are present in the bone marrow of CPV-infected dogs (Macartney *et al.*, 1984b; O'Sullivan *et al.*, 1984; Meunier *et al.*, 1985a).

In the gut the virus replicates in and destroys the rapidly dividing cells in the intestinal crypts. The subsequent loss of newly formed epithelium results in a flattened attenuated epithelium, with shortening of the intestinal villi (Figs. 2 and 3) (FPV: Rohovsky and Griesemer, 1967; Larsen *et al.*, 1976; Okaniwa *et al.*, 1976; MEV: Landsverk and Nordstoga, 1978; CPV: Cooper *et al.*, 1979; Yasoshima *et al.*, 1982; Macartney *et al.*, 1984c). These changes might lead to a loss of osmotic regulation, resulting in diarrhea, often with blood and mucus, as well as vomiting. Animals might become dehydrated and/or pyrexic, possibly because of endotoxin uptake from the gut.

The rate of intestinal epithelial cell turnover affects the severity of the lesions, probably by determining the number of cells able to replicate virus. This effect was demonstrated during infections of germ-free cats (Rohovsky and Griesemer, 1967; Carlson *et al.*, 1977), or after the treatment of cats with acid enemas which increased cell replication and

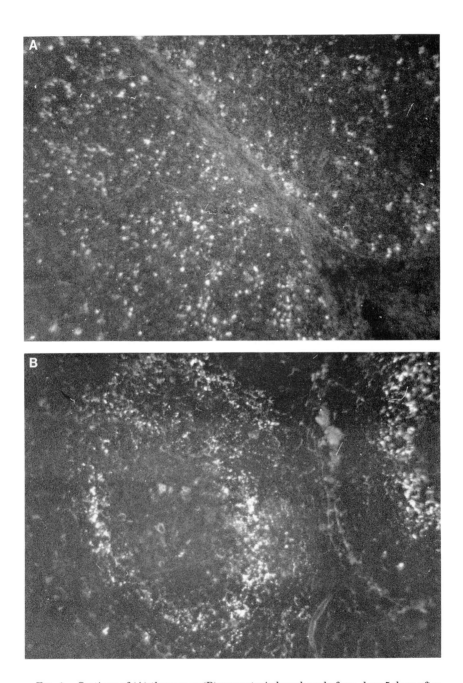

Fig. 1. Sections of (A) thymus or (B) mesenteric lymph node from dogs 5 days after parenteral inoculation with CPV and staining with a fluorescent antibody against CPV. The distribution of antigen in the thymus cortex and surrounding the lymph node germinal centers is shown. Infected cells contain intranuclear antigen, while many cells containing only cytoplasmic antigen most likely contain phagocytosed proteins.

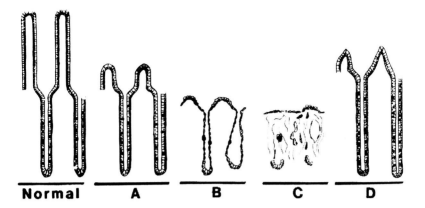

FIG. 2. Lesions in the small intestine of a dog after inoculation with CPV. (A) The observed patterns of intestinal lesions. (B) Duodenal mucosa with lesions of pattern B. Villi are stunted and occasionally fused (arrow). The crypts are dilated and lined with attenuated epithelial cells. Neutrophils (n) are present in the mucus overlying the affected mucosa. (C) Jejunal mucosa with lesions of pattern D: regenerating villi. The villi are short and covered with cuboidal epithelium, and the crypts are elongated and lined with cuboidal epithelial cells. From Macartney et al. (1984b) with permission.

FIG. 2. (cont.)

subsequent FPV replication in the colon (Shindel *et al.*, 1978). A correlation was observed between the viral titers in serum and feces and the severity of the disease observed in dogs inoculated with CPV (Meunier *et al.*, 1985b). Intestinal parasites have been suggested to play a role in the natural pathogenesis of CPV infections (Chalifoux *et al.*, 1981; Pollock, 1982).

B. Fetal or Neonatal Infections

1. Feline Ataxia

In 1967 FPV was identified as the cause of an uncharacterized feline cerebellar ataxia, which had previously been thought to be a genetically inherited disease (Johnson *et al.*, 1967; Kilham *et al.*, 1967). When kittens were infected either *in utero* or shortly after birth, viral

FIG. 3. Scanning electron micrographs of (A) the intestinal epithelium of the jejunum of a normal dog and (B) a dog killed 6 days after oral inoculation with CPV. In the latter case severely affected areas have completely lost the surface epithelium. Numerous crypt openings (c) are seen in the specimen from the control dog. From Macartney *et al.* (1984c) with permission.

replication in the cells of the external germinal epithelium of the developing cerebellum resulted in cerebellar hypoplasia (Csiza *et al.*, 1971a; Kilham *et al.*, 1971). While some kittens developed a widespread infection, leading to fetal death (Kilham *et al.*, 1967), most viable kittens subsequently suffered from ataxia (Kilham *et al.*, 1967; Csiza *et al.*, 1971a,b). Although ferrets are probably not a normal host for FPV (Parrish *et al.*, 1987; Veijalainen, 1986), ferrets inoculated with FPV prenatally or shortly after birth by the intracranial or other parenteral routes develop similar cerebellar lesions (Kilham *et al.*, 1967; Duenwald *et al.*, 1971).

2. Canine Myocarditis

Initial reports of CPV disease described outbreaks of acute myocarditis in puppies, mostly between 3 and 8 weeks old, although some animals were up to 16 weeks of age at the time of death (Kelly and Atwell, 1979; Huxtable *et al.*, 1979; Robinson *et al.*, 1979a,b, 1980a; Hayes *et al.*, 1979; Jesyk *et al.*, 1979; Carpenter *et al.*, 1980). Mortality in litters varied between 20% and 100%, and disease onset in apparently normal puppies was rapid, with pups dying of acute heart failure characterized by cardiac arrhythmia, dyspnea, and pulmonary edema (Robinson *et al.*, 1979b, 1980a; Hayes *et al.*, 1979; Jesyk *et al.*, 1979; Parrish *et al.*, 1982a). Electrocardiographically, affected pups showed a variety of subclinical abnormalities, and death appeared to result from progressive muscle necrosis, leading to ventricular fibrillation (Robinson *et al.*, 1979b; Carpenter *et al.*, 1980).

The primary pathological lesion observed was multifocal necrosis of the myocardium, often with a mononuclear cell infiltrate. Myocardial cells often contained intranuclear Feulgen test-positive amphophilic inclusion bodies. Lungs were diffusely edematous, with peribronchial and perivascular edema, most likely secondary to the heart failure (Robinson *et al.*, 1980a; Jesyk *et al.*, 1979; Carpenter *et al.*, 1980).

In a prospective study of CPV infections in a dog colony (Parrish *et al.*, 1982a) and in experimental infections of pups (Meunier *et al.*, 1984), it was shown that pups that developed myocarditis were infected either *in utero* or shortly after birth, often several weeks before the acute deaths due to heart failure, suggesting a chronic progression of the heart muscle damage. The age dependence of susceptibility of the myocarditis is probably related to the active cell division of the myocardial cells, which appears to occur only in pups under 15 days of age (Bishop, 1972).

It is of interest that there are no reports of cerebellar hypoplasia in dogs or myocarditis in cats infected with CPV or FPV, respectively. Although virus-infected cells have been observed within the heart muscle cells of neonatal kittens after experimental FPV infections (Csiza *et al.*, 1971b), disease such as that in the puppies infected with CPV has not been described. It is not known whether the difference in the pathogeneses in neonatal cats and dogs is a consequence of differences in the susceptibility of the host cells and tissues or to some other difference in virus tissue tropism.

Occasional neonatal infections give rise to a generalized infection, lesions being observed in many different tissues (Lenghaus and Studdert, 1982; Johnson and Castro, 1984). *In utero* infections in cats or Arctic foxes by FPV or blue fox parvovirus (BFPV), respectively, could result in fetal death and resorption, abortion, or neonatal death (Kilham *et al.*, 1967, 1971; Veijalainen and Smeds, 1988). However, natural infections of dogs by CPV did not appear to affect reproduction (Meunier *et al.*, 1980).

C. Immunity

The functional immunity against these viruses appears to be mediated through serum antibodies. Pups or kittens which acquire maternal immunity from their dams are protected against parvovirus infection until their serum antibody titers decline to low levels [Brun *et al.*, 1979 (FPV); Parrish *et al.*, 1982a; Pollock and Carmichael, 1982; Ishibashi *et al.*, 1983]. In addition, after parenteral administration of anti-CPV antibodies to dogs, the animals were completely protected against oral challenge, and no virus replication was seen in the intestinal epithelium 4–6 days after infection (Meunier *et al.*, 1985a).

Whether secretory antibodies play any role in immunity to or recovery from disease is unclear, although it has been shown that levels of coproantibody correlated inversely with the severity of the disease in dogs naturally infected with CPV (Rice *et al.*, 1982), and the levels and classes of antibody in the jejunum collected by cannulation after either CPV infection or vaccination suggested that the antibody was being specifically secreted (Nara *et al.*, 1983).

D. Natural Host Ranges

The natural host ranges of these viruses have not been well defined. Little of the available information has been independently confirmed, and some of the results appear contradictory or unexpected. Since natural infections by contaminating viruses are an ever-present risk in experimental studies, and the viruses are not readily characterized without monoclonal antibody reagents or restriction enzyme mapping of viral DNAs, the host ranges that have been described should be regarded as tentative in many cases until further confirmations are available.

Early studies were limited by the fact that virus could then be detected only by animal infections, so that virus or viral replication was only detected if it caused clinical disease. More recent studies have enabled the virus replication to be examined in more detail.

The natural hosts of FPV, MEV, and CPV are, by definition, the cat, mink, and dog, respectively (Siegl et al., 1985). Viruses isolated from or observed in other species have not been systematically classified and are therefore named either after a virus from a related host species or after the host animal from which they were isolated; for example, raccoon parvovirus (RPV), BFPV, and raccoon dog parvoviruses (RD) have all been described. The genetic and biological relationships among the viruses from the various hosts, their natural host ranges, and pathogenic potentials are poorly understood.

All hosts are members of five families (i.e., Felidae, Canidae, Procyonidae, Mustelidae, and Viverridae) within the order Carnivora, although within the various families only certain genera or species have been reported to be susceptible. A recently described phylogeny of the Carnivora based on thermal stability of DNA/DNA hybrids is shown in Fig. 4 (Wayne et al., 1989).

Virtually all large and small cats (i.e., several genera of the family Felidae) appear to be susceptible to FPV (Hindle and Findlay, 1932; Goss, 1948; Johnson, 1964). Within the family Procyonidae common raccoons (P. lotor) have been known to be susceptible to FPV since before 1940 (Waller, 1940; Goss, 1948; Nettles et al., 1980) and are also highly susceptible to both virus replication and disease after inoculation with MEV or FPV (Barker et al., 1983). Raccoons are not susceptible to CPV (Appel and Parrish, 1982). Ringtail coatis (family Procyonidae, Nasua nasua) are also reported to be naturally susceptible to FPV (Johnson and Halliwell, 1968). Civets (family Vivveridae) are affected by a disease similar to that caused by FPV in cats (Nain et al., 1964).

Among the family Mustelidae only MEV has been described. Domesticated ferrets (Mustela putorius) appear to be susceptible to infection by FPV when inoculated in utero or by the intraperitoneal route 1–2 days after birth, and can develop cerebellar hypoplasia (Kilham et al., 1967; Duenwald et al., 1971). No such effect was seen when ferrets were 3 days of age or older (Duenwald et al., 1971; Parrish et al., 1987), and no serological evidence for natural infections by other similar parvoviruses has been reported (Veijalainen, 1986). Serological evidence for infection of wild striped skunks (Mephitis mephitis) has been reported, but a virus has not been isolated or characterized. CPV, FPV,

CARNIVORE PHYLOGENY

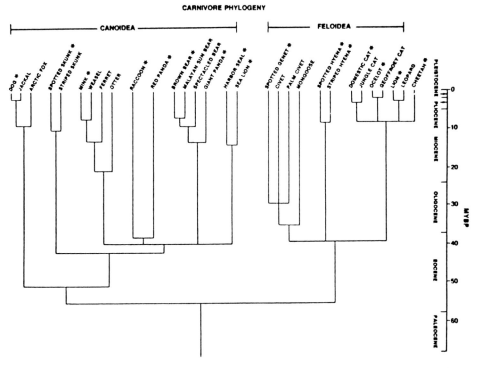

FIG. 4. Phenetic tree of members of the order Carnivora, based on the thermal stability of DNA/DNA hybrids. The time scale is based on fossil divergence times of about 40 million years before the present (MYBP) for all modern carnivore families, with the exception of the Canidae. The hosts of the FPV subgroup of parvoviruses include members of the families Canidae [represented, for example, by the dog, black-backed jackal (*Canis mesomelas*), and Arctic fox], Mustelidae (e.g., mink), Procyonidae (e.g., raccoon), Viverridae (e.g., civet), and Felidae (e.g., domestic and other cats). From Wayne *et al.* (1989) with permission.

or MEV isolates did not replicate in skunks after experimental inoculation, and the animals developed only low or undetectable antibody responses (Barker *et al.*, 1983).

Farmed American minks (*Mustela vison*) are susceptible to experimental infection by both FPV and MEV. The viruses replicate in a number of organs and are shed in the feces of many animals. Although early reports suggested that FPV isolates could cause clinical disease in mink (MacPherson, 1956), more recent studies, in which virus replication was more carefully monitored by titration, showed that only actual isolates from minks caused significant clinical disease in minks, while most infections by FPV or RPV were subclinical. In the same studies CPV isolates replicated to only low titers in minks after experimental parenteral inoculation (Barker et al., 1983; Parrish et al., 1987). It therefore appears that the principal difference between FPV and MEV in minks is manifested by a difference in virulence for minks, rather than host range.

The susceptibility of the members of the family Canidae has not been clearly defined, although it appears that, besides domestic dogs, most, if not all, members of the genus *Canis* are susceptible to CPV (coyote: Evermann et al., 1980; Thomas et al., 1984; gray wolf: Mech et al., 1986; Zarnke and Ballard, 1987). Other Canidae that are naturally susceptible to CPV include the maned wolf (*Chrysocycon brachyurus*) (Fletcher et al., 1979), the bushdog (*Speothos venaticus*) (Mann et al., 1980; Janssen et al., 1982), the crab-eating fox (*Cerdocyon thous*) (Mann et al., 1980), and the raccoon dog (*Nyctereutes procyonoides*) (Neuvonen et al., 1982; Veijalainen, 1986).

Among the foxes, which are all members of the family Canidae, the red fox (*Vulpes vulpes*) and Arctic fox (*A. lagopus*) appear to be susceptible to infection by a virus distinct from CPV, which might be FPV or a virus closely related to it (Phillips, 1943; Neuvonen et al., 1982; Barker et al., 1983; Veijalainen, 1986; Veijalainen and Smeds, 1988). In a serological survey of wild red foxes in Ontario, Canada, 79.4% had antibodies to a virus related to CPV or FPV (Barker et al., 1983). On experimental oral inoculation with CPV, MEV, or FPV isolates, red foxes did not develop clinical disease and all viruses were shed only in low titers, but the animals developed strongest serological responses against FPV (Barker et al., 1983).

IV. ANTIGENIC RELATIONSHIPS

A. Polyclonal Sera

Early studies of MEV and FPV indicated that antibodies formed to the two viruses protected animals against infection by CPV (Schofield, 1949; Wills, 1952; Wills and Belcher, 1956; Meyers et al., 1959). These studies depended on the production of clinical signs by the viruses as the indication of infection.

Cross-neutralization studies of MEV and FPV using *in vitro* culture revealed that antisera against the FPV and MEV strains examined gave similar neutralization titers against the alternative virus strain (Johnson, 1967a; Flagstad, 1975, 1977).

Several studies of the antigenic relationships between FPV or MEV and CPV were performed using polyclonal sera and revealed that the viruses all gave similar titers in neutralization or hemagglutination inhibition tests, although there was an indication of higher titers in the homologous than in the heterologous reactions (Flower *et al.*, 1980; Lenghaus and Studdert, 1980; Parrish *et al.*, 1982b; Tratschin *et al.*, 1982).

B. *Monoclonal Antibodies: Antigenic Structure and Variation among Virus Types*

The production of monoclonal antibodies (mAbs) against the various viruses has illuminated much of the complexity of the relationships among the various viruses.

Mouse mAbs against a purified 1978 CPV isolate revealed CPV-specific determinants, as well as determinants in common among CPV, FPV, RPV, and MEV (Parrish *et al.*, 1982b; Burtonboy *et al.*, 1982; Parrish and Carmichael, 1983; Surleraux *et al.*, 1987). Six of eight rat mAbs made against a 1973 FPV isolate reacted with CPV, RPV, FPV, and some MEV isolates, while two reacted more strongly with the FPV and MEV isolates than with the CPV isolates (Fig. 5) (Parrish and Carmichael, 1983).

When the mAbs were used to examine the antigenic structure of the viral capsid, the results revealed a variety of overlapping epitopes (Parrish and Carmichael, 1983). The CPV-specific mAbs appeared to recognize a single antigenic determinant, since all CPV isolates gave similar reactivities (Fig. 5). Antigenically variant neutralization escape mutants of CPV selected with two CPV-specific mAbs lost reactivity to all five CPV-specific mAbs, as well as to a group of five virus type-common mAbs (Fig. 6). Selection of mutant viruses with three of these type-common mAbs revealed two types of variation. Selection with mAb 6 or 12 gave variants which lost reactivity only to the three virus type-common mAbs prepared against CPV (mAbs 6, 12, and 18). In contrast, selecting with mAb B resulted in viruses which had lost reactivity to the same group of five virus type-common mAbs that were selected against by the CPV-specific mAbs, as well as the five CPV-specific mAbs (Fig. 6) (Parrish and Carmichael, 1983).

The two FPV-specific mAbs (i.e., G and H) recognized two different epitopes, as they reacted differently with various MEV or RPV strains (Fig. 5). In addition these mAbs selected mutant viruses with different patterns of reactivity, and the escape mutants selected by one mAb did not lose reactivity to the other (Fig. 6). mAb G cross-reacted with CPV, although to a much lower titer than to FPV (Parrish and Carmichael, 1983).

The remaining virus type-common mAbs in this study (Parrish and

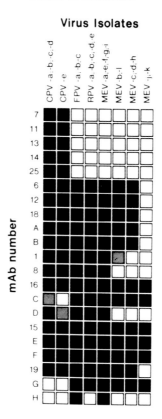

FIG. 5. Antigenic typing of FPV subgroup isolates. CPV [CPV-2 (a–d) and CPV-2a (e) strains], FPV, RPV, and MEV were titrated against mAbs, using the hemagglutination inhibition test. mAbs identified by numbers were prepared against CPV-a, and those identified by letters were prepared against FPV-b. Solid squares, at least 25% of the titer with the homologous virus; crosshatched squares, 5% to less than 25% of the titer with the homologous virus; open squares, less than 5% of the titer with the homologous virus, or no reaction. From Parrish and Carmichael (1983) with permission.

Carmichael, 1983) revealed a number of distinguishable epitopes when used to select variants. However, the patterns revealed by the analysis of naturally variant viruses (the MEV type 2 isolates, the *CPV-102/10* mutant, and the CPV-2a variant) (Parrish *et al.,* 1984, 1985; Parrish and Carmichael, 1983, 1986) suggest that these epitopes comprise only one or two distinct determinants (see Section V).

Competitive binding studies between ^{125}I-labeled and unlabeled mAbs revealed the surface of the viruses to be comprised of two non-cross-competing determinants, one recognized by mAbs 6, 12, and 18, and the other by the remaining mAbs in that panel (Fig. 7). The exten-

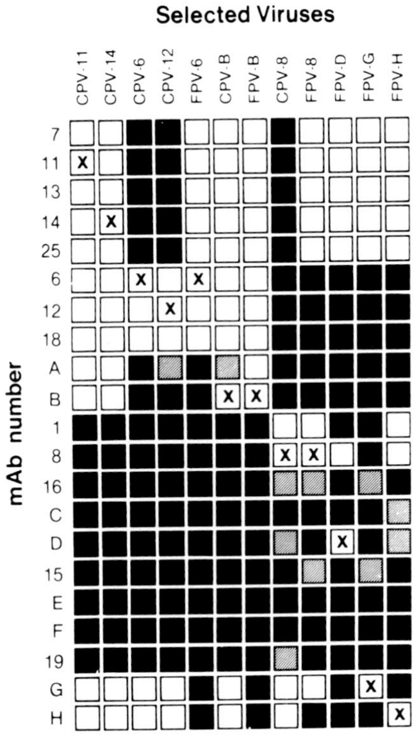

FIG. 6. Reactions of mAb-selected neutralization escape mutants of CPV-d or FPV-b in the hemagglutination inhibition assay or radioimmunoassay. X, mAb used for selection; solid squares, at least 25% of the titer with the original virus; crosshatched squares, 5% to less than 25% of the titer with the original virus; open squares, less than 5% of the titer with the original virus, or no detectable reaction. From Parrish and Carmichael (1983) with permission.

sive cross-competition between mAbs was probably a consequence of the small size of the parvovirus capsid (i.e., about 257 Å from the initial crystallographic diffraction data of CPV crystals) (Luo et al., 1988)) and the number of protein subunits (most likely about 10 VP-1 and 60 VP-2) (Tattersall et al., 1976; Paradiso et al., 1982; Paradiso, 1983), suggesting that each epitope is repeated many times. The exposed area on each repeated subunit available for antibody binding would therefore be small compared to the size of an antibody molecule.

In the study described all mAbs both inhibited hemagglutination and neutralized the virus infectivity (Parrish et al., 1982b; Parrish and Carmichael, 1983). In other studies some mAbs were nonneutralizing or were neutralizing only when an anti-rat IgG second-stage antibody was added to the neutralization reaction (Burtonboy et al., 1982;

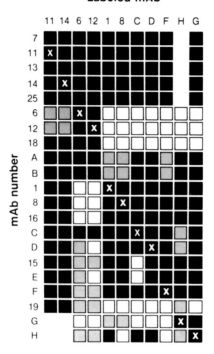

FIG. 7. Relationships between antigenic sites on CPV or FPV capsids revealed as binding of ^{125}I-labeled mAbs in the presence of dilutions of unlabeled mAbs. X, ^{125}I-labeled mAbs used in each set of tests; solid squares, efficient competition: less than 20% of the control CPM bound in the highest concentration of competing mAbs; hatched squares, partial blocking: 20–60% of the control counts per minute (CPM) bound in the highest concentration of competing mAbs; open squares, no competition: over 60% of the control CPM bound in the highest concentration of competing mAbs. From Parrish and Carmichael (1983) with permission.

Surleraux *et al.*, 1987). Thirty-nine of 47 mAbs in the latter study were either directly or indirectly neutralizing, suggesting that the determinants were mostly exposed on the surface of the intact virion. Eight of the 47 mAbs were nonneutralizing under any of the conditions tested (Surleraux *et al.*, 1987). Of the 47 mAbs 31 did not react in Western blots with proteins electrophoresed in sodium dodecyl sulfate–polyacrylamide gels and transferred to nitrocellulose sheets, and this group included the five CPV-specific mAbs in the study.

Of the 16 mAbs which reacted in Western blots, 12 reacted with both VP-2 and VP-3 in the expected molar ratio, three reacted only with VP-1, and one reacted only with VP-2. The mAb reactive only with VP-2 did not react with the blotted proteins after the digestion of full virions with trypsin, suggesting that it reacted with an amino-terminal region of the protein affected by the trypsin-cleavable site in the VP-2 of DNA-containing virions (Surleraux et al., 1987; Clinton and Hayashi, 1976; Paradiso, 1981; Paradiso et al., 1984). The mAbs with the various reactivities in Western blot reactivities included neutralizing, indirectly neutralizing, and nonneutralizing antibodies.

The antigenic relationships between a parvovirus (RD) isolated from raccoon dogs and a parvovirus isolated from an Arctic fox (BFPV) were examined by mAb analysis (Veijalainen, 1988). Of 19 mouse mAbs against the RD, two were specific in the hemagglutination inhibition test for that virus; three reacted with RD and CPV to equal titers, but to lower titers with MEV isolates; and 14 reacted to equal titers with all virus types tested. Of 20 mouse mAbs prepared against BFPV, one reacted in the hemagglutination inhibition test only with BFPV and with an MEV isolate; five reacted with BFPV, FPV, and MEV isolates, but to lower titers with a 1982 isolate of CPV; and 15 reacted to equal titers with all virus types (Veijalainen, 1988). These results suggest that RD is similar or identical to CPV, but distinct from the FPV, MEV, and BFPV isolates. Conversely, BFPV appears to be closely related antigenically to FPV and MEV but distinct from CPV.

V. Antigenic Variation and Genome Organization

A. MEV Variation

When examined with mAb panels prepared against CPV or FPV, three antigenic types of MEV were observed among the MEV isolates collected in the United States and Europe during the 1960s, 1970s, and 1980s (Parrish and Carmichael, 1983). Antigenic type 1 MEV appeared to be similar to the FPV isolates examined, while MEV types 2 and 3 differed in their reactivities to various CPV, FPV, or MEV type-common mAbs. The type 3 MEV isolates reacted with only six of 16 FPV-reactive mAbs in the panel used, while the type 2 MEVs reacted with 12 of the 16 mAbs (see Fig. 5).

The MEV types 2 and 3 isolates could be readily distinguished from the type 1 viruses by digestion of viral replicative-form DNA with various restriction endonucleases (Parrish et al., 1984). The TaqI restriction patterns are shown in Fig. 8. The epidemiological significance of the MEV antigenic variation is not known. In cross-protection studies minks immunized with inactivated preparations of any of the three MEV types were protected against parenteral challenge by any of the alternative antigenic types (Parrish et al., 1984).

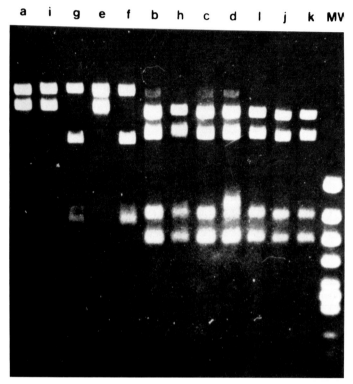

FIG. 8.   Digests of replicative-form DNA from various MEV isolates with the *Taq*I restriction enzyme. The DNA was electrophoresed in a 2% agarose gel and stained with ethidium bromide. Antigenic type 1 MEV isolates (strains a, i, g, e, and f) can be distinguished from the type 2 (strains b, h, c, d, and l) and type 3 (strains j and k) isolates. From Parrish *et al.* (1984) with permission.

B. Natural Variation of CPV Strains

Analysis of CPV isolates collected after 1978 revealed that viruses collected after 1980 did not react with several mAbs which reacted with the pre-1980 virus isolates (Parrish *et al.*, 1985). Originally described as pre-1980 and post-1980, or old and new, antigenic types, these types were later designated CPV-2 and CPV-2a, respectively (Parrish *et al.*, 1988b,c). A panel of nine mouse mAbs prepared against a 1984 isolate of CPV-2a contained three mAbs which reacted with CPV-2a isolates but not with CPV-2 isolates (Fig. 9). The CPV-2a and CPV-2 isolates also differed in the patterns of their DNA after digestion with the *Hph*I restriction endonuclease (Fig. 10) (Parrish *et al.*,

mAb	1978		1979		1980		1981		1982		1983		1984		FPV-c
	b	17	d	18	21	22	43	44	25	26	14	31	15	39	
A)															
7	512	256	512	1024	512	256	512	128	128	128	128	128	128	128	<2
14	128	64	32	64	64	64	256	128	128	256	128	256	128	256	<2
25	64	64	32	64	64	64	4	4	2	4	2	2	4	<2	2
C	128	64	64	64	128	128	<2	<2	<2	<2	<2	<2	<2	<2	64
D	256	256	256	256	256	256	2	4	4	4	2	2	2	2	256
E	256	256	256	256	128	256	8	8	8	8	8	4	8	4	128
J	128	64	128	32	128	128	<2	<2	<2	<2	<2	<2	<2	<2	64
6	64	64	128	128	128	256	256	128	128	256	64	128	256	64	64
12	64	128	64	64	64	64	64	32	64	64	32	64	64	64	64
A	32	64	64	64	128	256	256	128	128	256	256	256	256	<2	256
I	32	64	32	32	32	32	64	32	64	64	64	64	64	<2	64
1	1024	512	1024	1024	512	512	512	256	256	256	512	256	256	512	256
8	128	256	256	256	256	256	256	256	128	256	256	256	256	256	256
16	128	256	256	256	256	256	128	64	64	128	64	64	64	64	128
15	64	256	128	256	256	256	256	128	128	256	32	32	128	32	64
F	2048	1024	1024	2048	1024	2048	1024	512	1024	512	512	512	512	1024	1024
B)															
1D1	<2	<2	<2	<2	<2	<2	2048	1024	1024	1024	512	512	512	512	4
7D6	4	2	2	2	<2	<2	1024	2048	1024	1024	1024	2048	2048	1024	4
7E2	<2	<2	<2	<2	<2	<2	128	128	128	64	128	128	128	64	2
2A9	64	64	32	64	64	128	64	64	64	128	32	32	32	32	2
2E2	16	32	16	16	16	16	32	32	32	32	16	16	32	32	16
2E12	128	256	64	128	128	256	64	128	256	128	64	64	64	128	256
3G6	64	64	32	64	64	64	64	64	32	32	64	64	32	128	32
4A12	128	256	128	256	128	256	128	128	128	256	64	64	128	128	256
4E9	64	128	32	128	64	128	64	128	128	128	64	32	128	256	64

FIG. 9. Reactions of CPV isolates from 1978 to 1984 with mAbs made against either a 1978 isolate of CPV (CPV-a) or FPV-b (A) or against a 1984 CPV isolate (CPV-39). Shaded are those reactions of mAbs which distinguish CPV-2 (1978–1980) from CPV-2a (1981–1984) isolates. The reactions of FPV-c are also shown. From Parrish *et al.* (1985) with permission. Copyright 1985 by the AAAS.

1985). Typing of CPV isolates from various countries with mAbs which distinguished CPV-2 and CPV-2a showed that the CPV-2a isolates among viruses collected in Japan and the United States during 1979 and 121 of 125 CPV isolates collected in Denmark during 1980 were of the CPV-2a type (Table I). By 1981 CPV-2a was the predominant virus in all countries (Parrish *et al.*, 1988b).

Testing sera collected from wild (hence, nonvaccinated) coyotes in Utah, Texas, and Idaho (Thomas *et al.*, 1984), using an Ouchterlony agar gel immunodiffusion test, showed that the same change in virus types occurred among the CPV viruses infecting these populations between 1980 and 1981 (Parrish *et al.*, 1988b). This indicates that the replacement of CPV-2 with CPV-2a between 1979 and 1981 was most

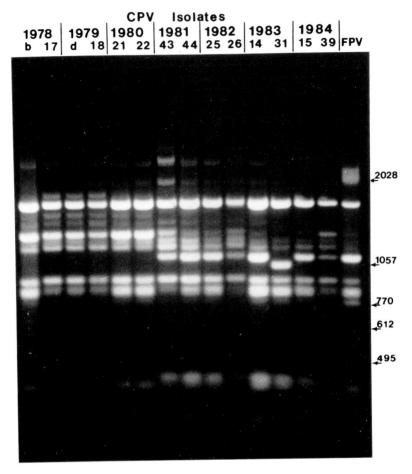

FIG. 10. Digests of replicative-form DNA from CPV isolates from 1978 to 1984 with the *Hph*I restriction enzyme, electrophoresed in a 2% agarose gel. Patterns show the difference in restriction digests between CPV-2 (1978–1980) and CPV-2a (1981–1984) isolates. Size standards (MW) are φ×174 DNA digested with *Hae*III restriction enzyme. From Parrish *et al.* (1985) with permission. Copyright 1985 by the AAAS.

likely due to some natural advantage of the latter virus, not to any human intervention, such as vaccination.

CPV isolates from Japan which were characterized by mAbs as CPV-2 or CPV-2a were examined for their abilities to hemagglutinate at 37°C or 4°C. The CPV-2 isolates were found to hemagglutinate only at 4°C, while most of the CPV-2a isolates were able to hemagglutinate at both 4°C and 37°C, although there were some virus isolates which were antigenically characterized as CPV-2a isolates, but which only

TABLE I

Antigenic Types of CPV Isolates Collected in Various Countries between 1978 and 1984 and Tested with Monoclonal Antibodies[a]

Country of origin	Year of virus isolation													
	1978		1979		1980		1981		1982		1983		1984	
	2[b]	2a[b]	2	2a	2	2a	2	2a	2	2a	2	2a	2	2a
United States	4	0	5	1	5	6	2	9	1	2	0	7	0	7
Denmark					4	121	0	110	3	84	0	3	0	2
France					0	1	0	2			0	1	0	1
Belgium	1	0									0	1	0	1
Japan			3	1	5	1	2	1	1	5	0	3	0	6
Australia	2	0	2	0			0	1	0	1			0	2

[a] From Parrish *et al.* (1988b) with permission.
[b] Antigenic type of virus determined by reactivity with monoclonal antibodies specific for either CPV-2 or CPV-2a.

hemagglutinated at 4°C (Table II). The significance of these findings in relation to any selective advantage or other properties of CPV-2a has not been defined (Senda *et al.*, 1988).

C. Mutants of Viruses

Few mutants of the FPV subgroup viruses have been described or characterized (Rivera and Sundquist, 1984; Parrish and Carmichael, 1986; Parrish *et al.*, 1988a). An isolate of MEV was reported that did not hemagglutinate pig, rhesus monkey, or horse erythrocytes at pH 6.8 at 4°C (Rivera and Sundquist, 1984). As the pH used in this study was near the upper range of the conditions reported by others for the hemagglutination of FPV (Carmichael *et al.*, 1980; Moraillon and Moraillon, 1982; Veijalainen, 1988; Parrish *et al.*, 1988c), it is not clear whether that mutant was truly nonhemagglutinating.

A nonhemagglutinating strain of CPV was derived after 12–22 passages of a 1978 isolate of the virus in the NLFK feline kidney cell line. The nonhemagglutinating virus showed no antigenic differences from wild-type CPV-2 when examined with a total of 82 mAbs prepared against CPV-2, CPV-2a, or FPV, and contained only two specific nucleotide and predicted amino acid sequence differences within the *VP-1*/*VP-2* gene, at nucleotides 1631 and 2185 in the *VP-1* gene [around 77 and 88 map units (m.u.) (Parrish *et al.*, 1988a,c). Of these two differences the sequence at nucleotide 2185 was also present in two FPV isolates, FPV-Carl and FPV-a (Carlson *et al.*, 1985; Parrish *et*

TABLE II

RELATIONSHIPS BETWEEN TEMPERATURE DEPENDENCE
OF HEMAGGLUTINATION AND THE ANTIGENIC TYPE
WITH CPV-2- OR CPV-2a-SPECIFIC mAbs[a]

Year of isolation	Number of isolates	4°C/37°C[b]	Dependence	Antigenic CPV type[c]
1978	1	≥64	+	2
1979	3	≥64	+	2
	1	2	−	2a
1980	4	≥64	+	2
	2	8	−	2
	1	2	−	2a
1981	2	8	−	2
	1	2	−	2a
1982	1	≥64	+	2
	5	2	−	2a
1983	3	1–2	−	2a
1984	7	2–8	−	2a

[a] Adapted from Senda et al. (1988) with permission.

[b] The ratio between the hemagglutination titer of the virus isolate when assayed at 4°C or 37°C.

[c] Antigenic type determined by testing with CPV-2- or CPV-2a-specific monoclonal antibodies in the hemagglutination inhibition test.

al., 1988c). It is known that FPV-a hemagglutinates under standard conditions, indicating that the functional difference in the nonhemagglutinated mutant is most likely due to the sequence at nucleotide 1631, suggesting that an amino acid affects the structure of the viral ligand which binds to the erythrocyte virus-binding receptor. Whether the nonhemagglutinating virus exhibits other differences from the wild-type virus is not known.

Another mutant of CPV-2 was derived during passage of a 1978 isolate of CPV-2 102 times in primary canine kidney cells and an additional 10 times in the NLFK feline kidney-derived cell line. During the passages in NLFK cells, a mutant (CPV-102/10) was selected which no longer replicated in dogs or in dog cells and which was variant antigenically when tested in the hemagglutination inhibition test with mAbs prepared against CPV-2 or FPV (Parrish and Carmichael, 1986). The mutation in CPV-102/10 was identified by recombination mapping, in which restriction fragments prepared from replicative-form DNA of each virus were religated to form the entire genome and then transfected into NLFK cells. By examination of seven recombinant viruses prepared between *CPV-102/10* and a wild-type CPV strain, the antigenic and host range differences of *CPV-102/10* were mapped to a 438-base pair 64–73 m.u. *Bgl*II–*Pvu*II fragment of the viral genome (Fig. 11) (Parrish and Carmichael, 1986). Within this region there were two *CPV-102/10*-specific nucleotide and amino acid sequence differences at adjacent codons (i.e., nucleotides 1400 and 1403 in the *VP-1/VP-2* gene sequence) (Parrish *et al.*, 1988c).

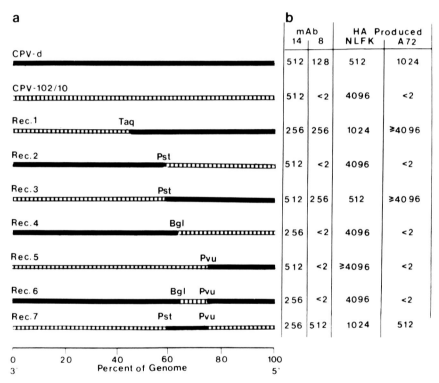

FIG. 11. (a) Analysis of recombinants (Rec.) between a wild-type CPV-2 isolate (CPV-d) and the antigenic and host range mutant *CPV-102/10*. Recombinants were formed by ligating genome fragments from these two viruses at various restriction sites (i.e., *Taq*I, *Pst*I, *Bgl*II, and *Pvu*II). Viruses produced after transfection of cell cultures were examined for recombinant genome type. (b) Antigenic and host range types of CPV-d and *CPV-102/10* and the recombinants between them. Antigenic types are represented by reactions with mAb 14 (control) or mAb 8 (which reacts with CPV-d, but not with *CPV-102/10*) in the hemagglutination inhibition assay, while host ranges are represented by NLFK (feline) or A72 (canine) cell culture supernatant hemagglutination titers 4 days after inoculation with 0.02 pfu of virus per cell. The antigenic type and host range of the mutant virus map between the *Bgl*II and *Pvu*II sites (64–73 m.u.). From Parrish and Carmichael (1986) with permission.

D. Genome Organization

The genome organization appears to be similar to that described for several other related autonomous parvoviruses, including H1, MVM, and bovine parvovirus (Rhode and Paradiso, 1983; Astell *et al.*, 1983; Chen *et al.*, 1986). The proposed genome organizations of MVM and CPV are shown in Fig. 12. The sequence contains an open reading frame DNA from a promoter at 4 m.u. (i.e., p4), which encodes the NS-1 protein which has been extensively characterized for H1 and MVM (Cotmore *et al.*, 1983; Paradiso, 1984; Cotmore and Tattersall, 1986, 1988).

An 83-kDa protein was precipitated from *in vitro* translated CPV mRNA, using rabbit antisera prepared against bacterially expressed fusion proteins from the middle portion of the *NS-1* gene of MVM. However, no translation product was precipitated by antiserum against a sequence from the amino terminus of the MVM NS-1 protein, which is contained in both the 83-kDa NS-1 and 25-kDa NS-2 proteins of MVM (Cotmore and Tattersall, 1986). The carboxy-terminal two-thirds of the FPV *NS-1* gene sequence was expressed in *Escherichia coli* as a fusion with the tryptophan *LE* gene product and the proteins used to produce antibodies in rabbits. These antibodies recognized proteins of 84, 60, 53, and 49 kDa, the most abundant product being the 60-kDa form (Carlson *et al.*, 1987). The relationships of the various proteins to each other and to the 83-kDa protein precipitated from the *in vitro* translated CPV mRNA have not been determined, but the smaller proteins most likely represent degraded forms of the 83-kDa NS-1 protein.

It is not clear whether CPV encodes a protein equivalent to the 25-kDa NS-2 protein of MVM and H1, which is translated from the MVM R2 transcript. In MVM this transcript contains 84 amino acids from the amino terminus of NS-1 in the third open reading frame, with a carboxy-terminal portion derived from 40–46 m.u. of the MVM sequence after mRNA splicing. This portion is encoded from open reading frame 2 and is followed by a second minor splice (i.e., 46–48 m.u.) to give a further short sequence encoding carboxy-terminal domains in open reading frame 2 (Morgan and Ward, 1986; Cotmore and Tattersall, 1986, 1987). An equivalent transcript or product in CPV or FPV has not been identified. The presence of a stop codon upstream from the second (minor) splice and the low homology between the translated open reading frames in this region between the sequences of MVM and CPV or FPV suggests that the latter viruses have different patterns of transcription and translation of the NS proteins from those seen for MVM and H1 (Cotmore and Tattersall, 1986).

The genomic 5' 60% of the parvovirus genome encodes the VP-1 and

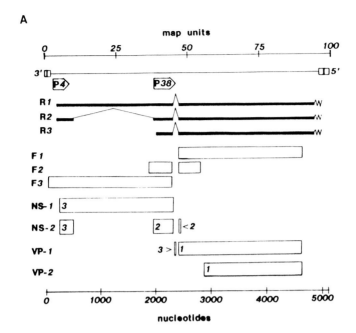

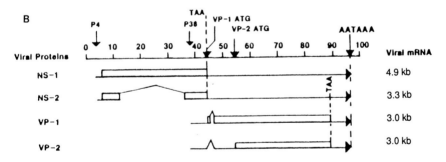

Fig. 12. Coding strategy of the minute virus of mice (MVM) genome (A) and that proposed for the the CPV genome (B). The cytoplasmic transcripts are aligned with the viral DNA strand (orientation 3′ to 5′), showing the promoters at 4 and 38–40 m.u. Major blocks of open reading frame are shown for MVM in each of the three reading frames (F1, F2, and F3) and the sequences which encode NS-1, NS-2 (which has not been identified in CPV), and the capsid proteins VP-1 and VP-2. (A) The open reading frames used to encode the various parts of each virus are shown as numbers in the transcript map. From Cotmore and Tattersall (1987) and Reed et al. (1988) with permission.

VP-2 from mRNAs transcribed from a promoter around 40 m.u. (Fig. 12) (Carlson *et al.*, 1985). The two proteins are most likely translated from differently spliced forms of the R3 transcript, as described in detail for MVM (Morgan and Ward, 1986), although it is possible that R1 and R2 transcripts can also be translated to form VP-1 or VP-2 (Cotmore and Tattersall, 1987). Messages from the CPV p40 promoter are able to direct the translation of both VP-1 and VP-2 when cloned in a bovine papillomavirus expression vector (Mazzara *et al.*, 1987).

Alternative splices predicted to generate VP-1 and VP-2 messages are most likely derived from two different splice donors. One, prior to the methionine codon thought to be the initiation codon of VP-1, would give rise to an mRNA in which the first AUG used would be that initiating VP-2 (Rhode, 1985, Carlson *et al.*, 1985; Morgan and Ward, 1986). The alternatively spliced mRNA would most likely use a predicted splice donor after the predicted ATG initiating VP-1, giving 10 codons prior to the splice, then the mRNA could be spliced to the same acceptor as for the VP-2 message. The outcome of the alternative splices would encode an additional 153 amino acids in VP-1, to give VP-1 and VP-2 of predicted sizes of 79,845 and 64,661 Da, respectively, for FPV (Carlson *et al.*, 1985) and similarly sized proteins for CPV (Rhode, 1985; Reed *et al.*, 1988). Whether these mRNA splicing reactions actually occur in CPV and FPV has not yet been determined.

The genomes of CPV, FPV, MEV, and RPV isolates contain various directly repeated units of unknown significance in the genomic 5′ (largely noncoding) region. Some of the rearrangements could have arisen during passage of the viruses in tissue culture, although some of the variable sequences are present in wild-type low-passage isolates of CPV (Fig. 13) (Rhode, 1985; Carlson *et al.*, 1985; Parrish *et al.*, 1988c). No correlation between the form of the repeat units and any biological function has been described, but all viruses examined in one study had at least one copy of each repeat, and the DNA sequences were conserved, suggesting that the sequences play some role in the virus life cycle (Cotmore and Tattersall, 1987; Parrish *et al.*, 1988c).

VI. Other Properties of FPV, MEV, and CPV

A. Hemagglutination Dependence on pH

All virus types hemagglutinate erythrocytes from a variety of species at 4°C in buffers with pH values below 7.0 (Johnson, 1971; Goto *et al.*, 1974; Goto, 1975; Mochizuki *et al.*, 1978; Senda *et al.*, 1988). FPV and MEV isolates hemagglutinated pig or rhesus monkey erythrocytes

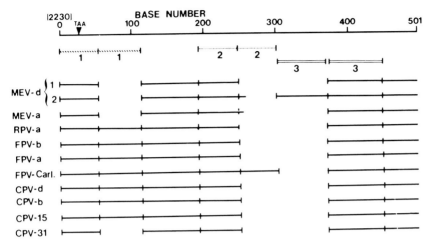

FIG. 13. The sequences between about 88 and 95 m.u. in the genomes of MEV, RPV, FPV, and CPV, isolates, showing the structure of the three repeated units of 50–60 base pairs, and the form of this region in each of the virus isolates examined. The MEV-d stock examined contained two genome forms, shown as 1 and 2. CPV-d and CPV-b are antigenic type 2 isolates, and CPV-15 and CPV-31 are type 2a isolates. From Parrish *et al.* (1988c) with permission.

only at pH values below 7.0, and the pH optimum for most strains is below pH 6.6 (Carmichael *et al.*, 1980; Moraillon and Moraillon, 1982). In contrast, CPV isolates hemagglutinated with approximately equal titers in buffers with pH values up to 8.0 (Carmichael *et al.*, 1980; Parrish *et al.*, 1988c). RD showed a hemagglutination dependence on pH similar to that of CPV isolates, while the BFPV was similar to FPV or MEV isolates, supporting the mAb antigenic typing results of these viruses (Veijalainen, 1988).

B. Experimental Host Range Properties

Isolates of CPV, FPV, and MEV all replicate in feline kidney cells and many other feline cell lines in culture (Johnson, 1965; Appel *et al.*, 1979b; Tratschin *et al.*, 1982; Parrish *et al.*, 1988c). CPV isolates replicate in canine cells in culture, but FPV and MEV isolates do not (Tratschin *et al.*, 1982; Parrish *et al.*, 1988c). Studies with RD and BFPV in tissue culture show these viruses to differ and to have host ranges similar to those for CPV and FPV, respectively (Veijalainen, 1988).

The experimental host ranges of the FPV subgroup viruses have not been well defined, and the many viruses and hosts make definitive analysis of the various host ranges difficult. In early studies feline tissues containing FPV were inoculated into dogs, ferrets, ground squirrels (*Citellus richardsonii*), mongooses, rhesus monkeys, hamsters, chick embryos, rabbits, guinea pigs, mice, and rats without producing any signs of disease (Hindle and Findlay, 1932; Hammon and Enders, 1939a; Lawrence *et al.*, 1940; Kikuth *et al.*, 1940). FPV isolates inoculated into minks in a number of studies caused little clinical disease (Myers *et al.*, 1959; Barker *et al.*, 1983; Parrish *et al.*, 1987). An FPV or RPV isolate was passaged repeatedly through minks in one study, but only one animal in each series of RPV- or FPV-inoculated minks showed any signs of clinical disease, in contrast to the MEV isolates which caused clinical disease in many of the inoculated minks (Parrish *et al.*, 1987). Inoculation of an MEV isolate into a single cat resulted in clinical disease (Higashihara *et al.*, 1981).

Inoculation of FPV isolates into dogs gave a limited replication in a number of lymphoid tissues, but the viruses could not be propagated repeatedly in this host (Parrish *et al.*, 1988c). CPV isolates inoculated into cats showed little or no replication (Goto *et al.*, 1984b; C. R. Parrish, unpublished observations).

VII. Genetic Analysis and Variation among the Viruses

A. Restriction Mapping

Comparisons of the restriction enzyme cleavage maps of CPV, FPV, and MEV replicative-form DNAs revealed a number of differences in restriction sites (McMaster *et al.*, 1981; Tratschin *et al.*, 1982). Most of the variable sites mapped to the right-hand side of the genome, within the *VP-1/VP-2* gene(s) (Fig. 14). The restriction maps of various wild-type and vaccine strains of FPV, MEV, and CPV were compared, using seven restriction enzymes (Tratschin *et al.*, 1982). A Swiss isolate of CPV (*Ka/BE*, discovered in 1979) differed from the French V1 strain of MEV in 11 of 79 restriction enzyme sites examined, allowing unambiguous identification of the CPV isolates. Low-passage MEV and FPV isolates appeared to be similar in this analysis, differing in two or fewer restriction sites (Fig. 14) (Tratschin *et al.*, 1982).

Two or three *Hin*fI sites in two of the vaccine strains of FPV were similar to CPV-specific sites in the CPV isolates (Fig. 14) (Tratschin *et al.*, 1982). Although this finding suggested a possible origin for CPV in the FPV vaccine viruses (Siegl, 1984), other studies comparing *VP-1/VP-2* gene sequences of various CPV isolates with the two FPV vaccine strains indicated that these FPV vaccine strains were not progenitors of CPV (Parrish *et al.*, 1988c; see Section IX).

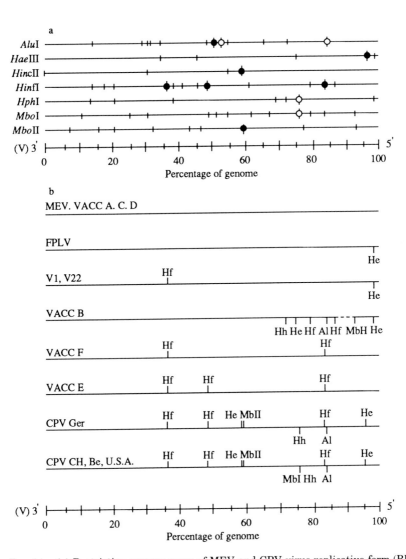

FIG. 14. (a) Restriction enzyme maps of MEV and CPV virus replicative-form (RF) DNAs, aligned with the viral genomic (−) DNA shown as the 3′–5′ orientation. Vertical lines indicate sites present in both CPV and MEV, open circles show sites present only in MEV, and closed circles indicate sites present only in CPV (Swiss isolate). (b) Comparative restriction enzyme digestion maps of the RF DNAs of various CPV, FPV, and MEV isolates, along with those of six FPV vaccine strains. The map of the MEV isolate is used for comparison. Additional restriction sites are shown above the line, while sites absent from a virus strain are shown below it. Sites common to all viruses are not shown. Live attenuated FPV vaccine strains examined were from the following manufacturers' vaccines—strain A, Iffa-Merieux; strain B, Norden; strain C, Frieosythe/Wellcome; strain E, Phillips Roxanne; and strain F, Dellen—as well as one strain (strain D) derived from MEV from Connaught. Hf, *Hin*fI; He, *Hae*III; Hh, *Hph*I; Al, *Alu*I; MbII, *Mbo*II; Hc, *Hinc*II; MbI, *Mbo*I. Comparisons of the DNA sequences of strains E and F are also shown in Fig. 17. From Tratschin *et al.* (1982) with permission.

B. DNA Sequence Analysis

An almost complete DNA sequence of one tissue culture-passaged vaccine strain of CPV isolated in 1978 has been reported (Reed *et al.*, 1988). Partial nucleotide sequences of a tissue culture-passaged vaccine strain of FPV (about 22–97 m.u.) (Carlson *et al.*, 1985) and a wild-type 1978 isolate of CPV (about 33–95 m.u.) (Rhode, 1985), as well as the *VP-1* and *VP-2* gene sequences of eight further isolates of CPV-2, CPV-2a, FPV, RPV, or MEV have also been reported (Parrish *et al.*, 1988c).

VIII. Genetic Analysis of Antigenic, Host Range, and Other Functions within the *VP-1/VP-2* Genes

Mapping CPV-Specific Properties

Recombinants between a CPV-2 isolate and an FPV isolate were prepared from purified viral replicative-form DNA, using methods described above for analysis of the *CPV-102/10* mutant (Parrish and Carmichael, 1986; Parrish *et al.*, 1988c). Recombinant genomes prepared at the *Pst*I, *Bgl*II, and *Pvu*II sites (at 59, 64, and 73 m.u. in the viral genome, respectively) were transfected into cell cultures, and recombinant viruses recovered were examined for the CPV-specific antigenic epitope, the pH dependence of hemagglutination, and the canine host range (Fig. 15 and Table III). These analyses showed that the CPV-specific pH dependence of hemagglutination and a sequence difference determining the CPV-specific antigenic epitope both mapped within the 300-base pair 59–64 m.u. region. Within that region there were only two CPV- or FPV-specific sequence differences—an asparagine–lysine change coded for by nucleotide 780 and an alanine–valine difference coded for by nucleotide 809—and therefore one or both of those differences gave rise to the differences in antigenic type and in the pH dependence of hemagglutination (Parrish *et al.*, 1988c).

The host range differences between CPV and FPV proved genetically more complex, and only those properties that allow CPV to replicate in dogs have been defined. When the 59–73 m.u. region of FPV was replaced by the same region of CPV, the resulting virus replicated in dogs to a titer as high as that for the wild-type virus, identifying this region as being responsible for the canine host range of CPV (Fig. 15 and Table III). This region contained the two sequence differences associated with the CPV-specific epitope and pH dependence of hemagglutination in the 59–64 m.u. region, as well as one further CPV-specific coding change at 71 m.u. (asparagine–aspartic acid at nu-

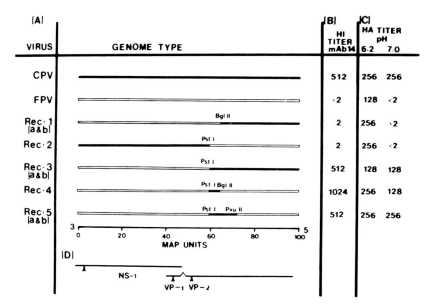

FIG. 15. Analysis of recombinants between CPV-d and FPV-b prepared from replicative-form viral DNA and transfected into NLFK cells. Viruses recovered were confirmed by restriction enzyme analysis and DNA sequencing. (A) CPV, FPV, and five types of recombinants (Rec.) between these viruses prepared using the *Pst*I, *Bgl*II, and *Pvu*II sites at 59, 64, and 73 m.u., respectively, in the virus genome. (B) Hemagglutination inhibition titers of recombinant viruses with CPV-specific mAb 14. (C) The pH dependence of hemagglutination of the parental viruses and of the recombinants. Hemagglutination assays were performed in buffers with pH values of 6.2 and 7.0. From Parrish *et al.* (1988c) with permission.

cleotide 1468 in the *VP-1/VP-2* gene) and a virus-specific noncoding difference at nucleotide 1200 (Fig. 16).

The requirements for and the relative importance of the various sequence differences within that region are not known. However, the region between 64 and 73 m.u. does not itself contain the host range function since Rec 1, which contained the 0–64 m.u. region from FPV and the 64–100 m.u. region from CPV, was not able to replicate in dogs (Fig. 15 and Table III). It appears that differences in both the 59–64 and 64–73 m.u. regions are required to endow the canine host range (Parrish *et al.*, 1988c).

A similar region within the capsid proteins genes is also important for determining the host range differences between two strains of MVM which differ in their host ranges for various differentiated cells: MVM(p), replicating in fibroblast-derived cells, and MVM(i), replicating in lymphocyte-derived cells (Engers *et al.*, 1981; Spalholz and Tattersall, 1983; Tattersall and Bratton, 1983; Kimsey *et al.*, 1986). This

TABLE III

Replication of CPV-d, FPV-b, and Recombinants
between Them in Specific Pathogen-Free Dogs

	Virus titer per 0.03 g of tissue		
Virus	Thymus	Mesenteric lymph node	Ileum
CPV-d	1.2×10^5	1.6×10^3	7.0×10^4
	3.0×10^4	1.7×10^3	2.2×10^4
FPV-b	1.4×10^3	<10	<10
	8.0×10	<10	<10
Rec 1a	10	1.0×10^2	3.0×10
	10	<10	<10
Rec 2	<10	<10	<10
	<10	<10	<10
Rec 3a	2.2×10^5	1.9×10^4	1.2×10^3
	2.8×10^4	2.4×10^4	4.0×10^2
Rec 3b	1.7×10^4	2.2×10^4	1.1×10^3
Rec 4			
Experiment 1	9.1×10^4	<10	<10
	<10	5.0×10	<10
Experiment 2	2.7×10^6	<10	10
	<10	<10	<10
Experiment 3	<10	<10	<10
	<10	<10	<10
Rec 5a	1.3×10^4	2.8×10^6	3.1×10^3
	2.7×10^6	2.0×10^3	4.1×10^2
	8.0×10^6	$>4.0 \times 10^7$	4.0×10^2
	2.2×10^6	<10	<10
Rec 5b	1.4×10^6	2.2×10^6	2.6×10^5
	6.0×10^4	4.0×10^3	1.4×10^5

[a] Virus titers in various tissues of dogs 4 days after intramuscular inoculation with FPV, CPV, or recombinants (Rec). The dogs were inoculated with preparations of Rec 4 from tissue culture (experiments 1 and 3), while the dogs in experiment 2 were inoculated with virus recovered from the thymus of a dog inoculated with Rec 4 in experiment 1. From Parrish *et al.* (1988c) with permission.

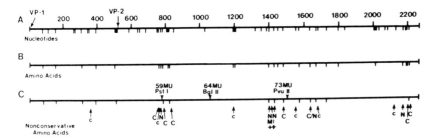

FIG. 16. Sequence variation within the *VP-1/VP-2* gene sequences, compared to a wild-type CPV type 2 sequence (CPV-d). (A) Positions of variable nucleotides. (B) Predicted amino acid sequence differences. (C) Amino acid differences which gave changes in charge or polarity. C, c, amino acid, or "silent," sequence differences which distinguished CPV from the other viruses; N, sequence differences specific for CPV-2a isolates; M, amino acid sequence difference between MEV-a and MEV-b (antigenic types 1 and 2 MEV isolates); +, mutations in *CPV-102/10* associated with host range and antigenic differences. Adapted from Parrish *et al.* (1988c) with permission.

difference in the host range mapped between 68 and 73 m.u. in these viruses and involved residues similar to those affecting the host range defined in this region for CPV or FPV (Figs. 11 and 15) (Antionetti *et al.*, 1988; Gardiner and Tattersall, 1988a,b).

Nothing is known about the specific function by which the CPV-specific differences in the *VP-1/VP-2* gene determine the canine host range. However, between MVM(i) and MVM(p) the functional difference within the *VP-1/VP-2* gene does not affect virus binding to cell receptors, but influences early events in the replication of the virus, in either mRNA transcription or DNA replication (Spalholz and Tattersall, 1983; Antionetti *et al.*, 1988; Gardiner and Tattersall, 1988a).

IX. VIRUS SEQUENCE VARIATION AND PHYLOGENETIC RELATIONSHIPS

The *VP-1/VP-2* gene sequences of a total of 10 isolates have been compared: two CPV-2, two CPV-2a, two MEV (types 1 and 2), one RPV, and three FPV isolates. An FPV and a CPV sequence were obtained from published sequences (Carlson *et al.*, 1985; Rhode, 1985). Others were obtained from M13 phage clones of viral replicative-form DNA, which were sequenced using a series of primers complementary to various positions within the *VP-1/VP-2* gene (Parrish *et al.*, 1988c). The distribution of the variant positions within the *VP-1/VP-2* gene is shown in Fig. 16. Although the differences in nucleotide sequences were evenly distributed through the genes, the coding changes were less evenly distributed. Changes that were FPV or CPV specific or that were specific for CPV-2a viruses tended to cluster into three regions within the gene: around 60 m.u., between 70 and 77 m.u., and around 85 m.u, near the carboxy terminus (Fig. 16) (Parrish *et al.*, 1988c).

The differences in nucleotide sequences within the *VP-1/VP-2* gene were used to prepare a network of the phylogenetic relationships between the viruses, assuming maximum parsimony (Fig. 17) (Fitch, 1977). Among the FPV, MEV, and RPV viruses there was a certain amount of apparently random sequence variation, but the viruses isolated from the different hosts could not be readily distinguished. However, the CPV isolates formed a separate cluster, which subdivided into the CPV-2 and the CPV-2a types that had previously been distinguished antigenically and by restriction enzyme analysis (Parrish *et al.*, 1985). It is apparent from this analysis that CPV-2 and CPV-2a most likely derived from some common ancestor virus prior to 1978 and that CPV-2a was not simply a linear descendent of CPV-2 (Parrish *et al.*, 1988c).

X. Conclusions

The emergence of MEV in minks during the 1940s and the recent emergence of CPV in dogs during the 1970s are unusual events, which raise a number of questions about the evolution of viruses and the ways in which host ranges of the viruses are determined and restricted. By examining these phenomena, we hope to derive lessons about the natural history of viruses and the interactions of these viruses with their hosts.

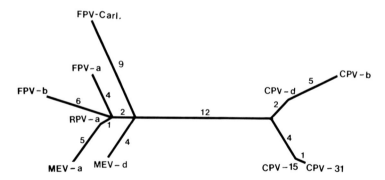

FIG. 17. An unrooted network showing the phylogenetic relationships between virus isolates based on the nucleotide sequences of the *VP-1/VP-2* gene between nucleotide 203 and the termination codon at nucleotide 2254. Isolates are feline panleukopenia virus (FPV-a, -b, and -Carl), mink enteritis virus (MEV-a and -d), raccoon parvovirus (RPV-a), canine parvovirus type 2 (CPV-d and -b), and CPV type 2a (CPV-15 and -31). FPV-Carl and FPV-a represent the Dellen Laboratories and Phillips Roxanne vaccine strains of FPV, respectively. From Parrish *et al.* (1988c) with permission.

The emergence of MEV appears to be an example not of altered host range, but was most likely due to the recognition of a long-existing virus which might have altered in virulence for minks. It is also possible that the disease was recognized at that time due to some change in the farming practices or possibly by a change in minks due to genetic selection. The differences in virulence for minks observed after inoculation of MEV or FPV suggests that there are subtle differences between FPV and MEV which have yet to be defined.

The emergence of CPV probably represents a different type of event. It is clear from the retrospective serological studies that CPV is a new virus of dogs and that it was first present in domestic dogs in Europe only during the mid-1970s. The widespread outbreaks and global spread of the virus did not occur until 1978. It is possible that this was due to some change in the virus at that time which gave it an epidemiological advantage. Genetic mapping studies indicate that only three or four sequence differences between the FPV and CPV-2 isolates within the *VP-1/VP-2* gene determine all of the specific properties of CPV that have been defined: the pH dependence of hemagglutination, the CPV-specific epitope, and the host range for canine cells and dogs. Whether these changes are all that would be required to convert FPV into a successful canine pathogen is not known, but it is clear that CPV could have been derived from FPV or some closely related virus by only a few specific changes. How this occurred and which virus was the ancestor are unknown and are subjects for future studies.

These parvoviruses, therefore, present a unique opportunity for understanding the natural evolution and variation of viruses and for determining the ways in which viruses can gain new host range and other functions. In this case the viruses apparently emerged by two different mechanisms as pathogens of the members of two families (i.e., Mustelidae and Canidae) among the Carnivora which had previously been resistant to infection or disease. Further examination of the viruses and further definition of the specific genetic properties involved in the changes will allow the details of these unusual events and their implications for virus evolution and host range stability to be understood.

REFERENCES

Antonietti, J.-P., Sahli, R., Beard, P., and Hirt, B. (1988). Characterization of the cell type-specific determinant in the genome of minute virus of mice. *J. Virol.* **62,** 552–557.

Appel, M. J. G., and Parrish, C. R. (1982). Raccoons are not susceptible to canine parvovirus. *J. Am. Vet. Med. Assoc.* **181,** 489.

Appel, M. J. G., Cooper, B. J., Greisen, H., and Carmichael, L. E. (1978). Status report: Canine viral enteritis. *J. Am. Vet. Med. Assoc.* **173,** 1516–1518.

Appel, M. J. G., Cooper, B. J., Greisen, H., Scott, F., and Carmichael, L. E. (1979a). Canine viral enteritis. I. Status report on corona- and parvo-like viral enteritides. *Cornell Vet.* **69**, 123–133.

Appel, M. J. G., Scott, F. W., and Carmichael, L. E. (1979b). Isolation and immunisation studies of a canine parvo-like virus from dogs with haemorrhagic enteritis. *Vet. Rec.* **105**, 156–159.

Astell, C., Thompson, M., Merchlinsky, M., and Ward, D. C. (1983). The complete DNA sequence of minute virus of mice, an autonomous parvovirus. *Nucleic Acids Res.* **11**, 999–1018.

Azetaka, M., Hirasawa, T., Konishi, S., and Ogata, M. (1981). Studies of canine parvovirus isolation, experimental infection and serologic survey. *Jpn. J. Vet. Sci.* **43**, 243–255.

Barker, I. K., Povey, R. C., and Voigt, D. R. (1983). Response of mink, skunk, red fox and raccoon to inoculation with mink virus enteritis, feline panleukopenia and canine parvovirus and prevalence of antibody to parvovirus in wild carnivores in Ontario. *Can. J. Comp. Med.* **47**, 188–197.

Bishop, S. P. (1972). Effects of aortic stenosis on myocardial cell growth and hyperplasia and ultrastructure in neonatal dogs. *Adv. Stud. Card. Struct. Metab.* **3**, 637–655.

Boosinger, T. R., Rebar, A. H., DeNicola, D. B., and Boon, G. D. (1982). Bone marrow alterations associated with canine parvoviral enteritis. *Vet. Pathol.* **19**, 558–561.

Bouillant, A., and Hanson, R. P. (1965). Epizootiology of mink enteritis: I. Stability of the virus in feces exposed to natural environmental factors *Can. J. Comp. Med. Vet. Sci.* **29**, 125–128.

Brun, A., Chappuis, G., Precausta, P., and Terre, J. (1979). Immunization against panleukopenia: Early development of immunity. *Comp. Immunol. Microbiol. Infect. Dis.* **1**, 335–339.

Burger, D., and Gorham, J. R. (1970). Mink virus enteritis. *In* "Infectious Diseases of Wild Mammals" (W. R. Davis, L. H. Karstad, and D. O. Trainer, eds.), pp. 76–84. Iowa State Univ. Press, Ames, Iowa.

Burger, D., Gorham, J. R., and Ott, R. L. (1963). Protection of cats against feline panleukopenia following mink virus enteritis vaccination. *Small Anim. Clin.* **3**, 611–614.

Burtonboy, G., Coignoul, F., Delferriere, N., and Pastoret, P.-P. (1979). Canine hemorrhagic enteritis: Detection of viral particles by electron microscopy. *Arch. Virol.* **61**, 1–11.

Burtonboy, G., Bazin, H., and Delferriere, N. (1982). Rat hybridoma antibodies against canine parvovirus. *Arch. Virol.* **71**, 291–302.

Carlson, J. H., and Scott, F. W. (1977). Feline panleukopenia. II. The relationship of intestinal mucosal cell proliferation to viral infection and development of lesions. *Vet. Pathol.* **14**, 173–181.

Carlson, J. H., Scott, F. W., and Duncan, J. R. (1977). Feline panleukopenia. I. Pathogenesis in germfree and specific pathogen-free cats. *Vet. Pathol.* **14**, 79–88.

Carlson, J. H., Scott, F. W., and Duncan, J. R. (1978). Feline panleukopenia. III. Development of lesions in the lymphoid tissues. *Vet. Pathol.* **15**, 383–392.

Carlson, J., Rushlow, K., Maxwell, I., Maxwell, F., Winston, S., and Hahn, W. (1985). Cloning and sequence of DNA encoding structural proteins of the autonomous parvovirus feline panleukopenia virus. *J. Virol.* **55**, 574–582.

Carlson, J. O., Lynde-Maas, M. K., and Zheng-da, S. (1987). A non-structural protein of feline panleukopenia virus: Expression in *Escherichia coli* and detection of multiple forms in infected cells. *J. Virol.* **61**, 621–624.

Carman, S., and Povey, C. (1982). Successful experimental challenge of dogs with canine parvovirus-2. *Can. J. Comp. Med.* **46**, 33–38.

Carman, P. S., and Povey, R. C. (1985a). Pathogenesis of canine parvovirus-2 in dogs: Haematology, serology and virus recovery. *Res. Vet. Sci.* **38**, 134–140.

Carman, P. S., and Povey, R. C. (1985b). Pathogenesis of canine parvovirus-2 in dogs: Histopathology and antigen identification in tissues. *Res. Vet. Sci.* **38**, 141–150.

Carmichael, L. E., Joubert, J. C., and Pollock, R. V. H. (1980). Hemagglutination by canine parvovirus: Serologic studies and diagnostic applications. *Am. J. Vet. Res.* **40**, 784–791.

Carmichael, L. E., Joubert, J. C., and Pollock, R. V. H. (1981). A modified live canine parvovirus strain with novel plaque characteristics. I. Viral attenuation and dog response. *Cornell Vet.* **71**, 408–427.

Carmichael, L. E., Joubert, J. C., and Pollock, R. V. H. (1983). A modified canine parvovirus vaccine 2. Immune response. *Cornell Vet.* **73**, 13–29.

Carpenter, J. L., Roberts, R. M., Harpster, N. K., and King, N. W. (1980). Intestinal and cardiopulmonary forms of parvovirus infection in a litter of pups. *J. Am. Vet. Med. Assoc.* **176**, 1269–1273.

Chalifoux, A., Elazhary, Y., and Frechette, J. L. (1981). Parvovirus enteritis: Possible role of endo parasites. *Med. Vet. Quebec* **11**, 66–70.

Chen, K. C., Shull, B. C., Moses, E. A., Lederman, M., Stout, E. R., and Bates, R. C. (1986). Complete nucleotide sequence and genome organization of bovine parvovirus. *J. Virol.* **60**, 1085–1097.

Clinton, G., and Hayashi, M. (1976). The parvovirus MVM: A comparison of heavy and light particle infectivity and their density conversion *in vitro*. *Virology* **74**, 57–63.

Cooper, B. J., Carmichael, L. E., Appel, M. J. G., and Greisen, H. (1979). Canine viral enteritis. II. Morphologic lesions in naturally occurring parvovirus infection. *Cornell Vet.* **69**, 134–144.

Cotmore, S. F., and Tattersall, P. (1986). Organization of the nonstructural genes of the parvovirus minute virus of mice. *J. Virol.* **58**, 724–732.

Cotmore, S. F., and Tattersall, P. (1987). The autonomously replicating parvoviruses of vertebrates. *Adv. Virus Res.* **33**, 91–174.

Cotmore, S. F., and Tattersall, P. (1988). The NS-1 polypeptide of minute virus of mice is covalently attached to the 5′ termini of duplex replicative-form DNA and progeny single strands. *J. Virol.* **62**, 851–860.

Cotmore, S. F., Sturzenbecker, L. J., and Tattersall, P. (1983). The autonomous parvovirus minute virus of mice encodes two nonstructural proteins in addition to its capsid polypeptides. *Virology* **129**, 333–343.

Csiza, C. K., Scott, F. W., de Lahunta, A., and Gillespie, J. H. (1971a). Pathogenesis of feline panleukopenia virus in susceptible newborn kittens. I. Clinical signs, hematology, serology, and virology. *Infect. Immun.* **3**, 833–837.

Csiza, C. K., de Lahunta, A., Scott, F. W., and Gillespie, J. H. (1971b). Pathogenesis of feline panleukopenia virus in susceptible newborn kittens. II. Pathology and immunofluorescence. *Infect. Immun.* **3**, 838–846.

Davis, E. V., Gregory, G. G., and Beckenhauer, W. H. (1970). Infectious feline panleukopenia (developmental report of a tissue culture origin formalin-inactivated vaccine). *VM/SAC, Vet. Med. Small Anim. Clin.* **65**, 237–242.

Duenwald, J. C., Holland, J. M., Gorham, J. R., and Ott, R. L. (1971). Feline panleukopenia: Experimental cerebellar hypoplasia produced in neonatal ferrets with live virus vaccine. *Res. Vet. Sci.* **12**, 394–396.

Engers, H. D., Louis, J. A., Zubler, R. H., and Hirt, B. (1981). Inhibition of T cell-mediated functions by MVM(i), a parvovirus closely related to minute virus of mice. *J. Immunol.* **127**, 2280–2285.

Eugster, A. K. (1980). Studies on canine parvovirus infection: Development of an inactivated vaccine. *Am. J. Vet. Res.* **41**, 2020–2024.

Eugster, A. K., Bendele, R. A., and Jones, L. P. (1978). Parvovirus infection in dogs. *J. Am. Vet. Med. Assoc.* **173**, 1340–1341.

Evermann, J. F., Foreyt, W., Maag-Miller, L., Leathers, C. W., McKeirnan, A. J., and LeaMaster, B. (1980). Acute hemorrhagic enteritis associated with canine coronavirus and parvovirus infections in a captive coyote population. *J. Am. Vet. Med. Assoc.* **177**, 784–786.

Fitch, W. M. (1977). On the problem of discovering the most parsimonious tree. *Am. Nat.* **111**, 223–257.

Flagstad, A. (1977). Feline panleukopenia virus and mink enteritis virus. A serological study. *Acta. Vet. Scand.* **18**, 1–9.

Fletcher, K. C., Eugster, A. K., Schmidt, R. E., and Hubbard, G. B. (1979). Parvovirus infection in maned wolves. *J. Am. Vet. Med. Assoc.* **175**, 897–900.

Flower, R. L. P., Wilcox, G. E., and Robinson, W. F. (1980). Antigenic differences between canine parvovirus and feline panleukopenia virus. *Vet. Rec.* **107**, 254–256.

Gagnon, A. N., and Povey, R. C. (1979). A possible parvovirus associated with an epidemic gastroenteritis in Canada. *Vet. Rec.* **104**, 263–264.

Gardiner, E. M., and Tattersall, P. (1988a). Evidence that developmentally regulated control of gene expression by a parvoviral allotropic determinant is particle mediated. *J. Virol.* **62**, 1713–1722.

Gardiner, E. M., and Tattersall, P. (1988b). Mapping of the fibrotropic and lymphotropic host range determinants of the parvovirus minute virus of mice. *J. Virol.* **62**, 2605–2613.

Gorham, J. R., Hartsough, G. R., Burger, D., Lust, S., and Sato, N. (1965). The preliminary use of attenuated feline panleukopenia virus to protect cats against panleukopenia and mink against virus enteritis. *Cornell Vet.* **55**, 559–566.

Gorham, J. R., Hartsough, G. R., Sato, N., and Lust, S. (1966). Studies of cell culture adapted feline panleukopenia virus—Virus neutralization and antigenic extinction. *VM/SAC, Vet. Med. Small Anim. Clin.* **61**, 35–40.

Goss, L. J. (1948). Species susceptibility to the viruses of Carre and feline enteritis. *Am. J. Vet. Res.* **9**, 65–68.

Goto, H. (1975). Feline panleukopenia in Japan. II. Hemagglutinability of the isolated virus. *Jpn. J. Vet. Sci.* **37**, 431–438.

Goto, H., Yachida, S., Shiahata, T., and Shimizu, K. (1974). Feline panleukopenia in Japan. I. Isolation and characterization of the virus. *Jpn. J. Vet. Sci.* **36**, 203–211.

Goto, H., Hirano, T., Uchida, E., Watanabe, K., Shinagawa, M., Ichijo, S., and Shimizu, K. (1984a). Comparative studies of physicochemical and biological properties between canine parvovirus and feline panleukopenia virus. *Jpn. J. Vet. Sci.* **46**, 519–526.

Goto, H., Uchida, S., Ichijo, S., Shimizu, K., Morohoshi, Y., and Nakano, K. (1984b). Experimental infection of canine parvovirus in specific pathogen-free cats. *Jpn. J. Vet. Sci.* **46**, 729–731.

Hammon, W. D., and Enders, J. F. (1939a). A virus disease of cats, principally characterized by aleucocytosis, enteric lesions and the presence of intranuclear inclusion bodies. *J. Exp. Med.* **69**, 327–353.

Hammon, W. D., and Enders, J. F. (1939b). Further studies on the blood and the hematapoietic tissues in malignant panleucopenia of cats. *J. Exp. Med.* **70**, 557–564.

Have, P., and Andersen, A. B. (1982). Parvovirus-diarre hos hunde I Danmark. Forkomst, samt laboratorie-diagnostic belyst ved en underjogelsae af virusudskillelse med feces. *Proc. Nord. Vet. Kongr., 14th* pp. 148–151 (in Norwegian).

Hayes, M. A., Russell, R. G., Mueller, R. W., and Lewis, R. J. (1979). Myocarditis in young dogs associated with a parvovirus-like agent. *Can. Vet. J.* **20**, 126.

Higashihara, T., Izawa, H., Onuma, M., Kodama, H., and Mikami, T. (1981). Mink enteritis in Japan. I. Isolation and characterization of the causative virus and its pathogenicity in cat. *Jpn. J. Vet. Sci.* **43**, 841–851.

Hindle, E., and Findlay, G. M. (1932). Studies on feline distemper. *J. Comp. Pathol. Ther.* **45**, 11–26.

Horner, G. W., Hunter, R., and Chisholm, E. G. (1979). Isolation of parvovirus from dogs with enteritis. *N.Z. Vet. J.* **27**, 280.

Huxtable, C. R. R., McHowell, J., Robinson, W. F., Wilcox, G. E., and Pass, D. A. (1979). Sudden death associated with a suspected viral myocarditis. *Aust. Vet. J.* **55**, 37–38.

Ichijo, S., Osame, S., Konishi, T., and Ogata, H. (1976). Clinical and hematological findings and myelograms on feline panleukopenia. *Jpn. J. Vet. Sci.* **38**, 197–205.

Ishibashi, K., Maede, Y., Oshugi, T., Onuma, M., and Mikama, T. (1983). Serotherapy for dogs infected with canine parvovirus. *Jpn. J. Vet. Sci.* **45**, 59–66.

Janssen, D. L., Bartz, C. R., Bush, M., Marchwicki, R. H., Grates, S. J., and Montali, R. J. (1982). Parvovirus enteritis in vaccinated juvenile bush dogs. *J. Am. Vet. Med. Assoc.* **181**, 1225–1227.

Jesyk, P. F., Haskins, M. E., and Jones, C. L. (1979). Myocarditis of probable viral origin in pups of weaning age. *J. Am. Vet. Med. Assoc.* **174**, 1204–1207.

Johnson, R. H. (1964). Isolation of a virus from a condition simulating feline panleukopaenia in a leopard. *Vet. Rec.* **76**, 1008–1012.

Johnson, R. H. (1965). Feline panleukopaenia. I. Identification of a virus associated with the syndrome. *Res. Vet. Sci.* **6**, 466–471.

Johnson, R. H. (1967a). Feline panleukopaenia virus—*In vitro* comparison of strains with a mink enteritis virus. *J. Small Anim. Pract.* **8**, 319–324.

Johnson, R. H. (1967b). Feline panleukopaenia virus. IV. Methods for obtaining reproducible *in vitro* results. *Res. Vet. Sci.* **8**, 256–264.

Johnson, R. H. (1971). Serologic procedures for the study of feline panleukopaenia. *J. Am. Vet. Med. Assoc.* **158**, 876–884.

Johnson, B. J., and Castro, A. E. (1984). Isolation of canine parvovirus from a dog brain with severe necrotizing vasculitis and encephalomalacia. *J. Am. Vet. Med. Assoc.* **184**, 1398–1399.

Johnson, R. H., and Halliwell, R. E. W. (1968). Natural susceptibility to feline panleukopaenia of the coatimundi. *Vet. Rec.* **82**, 582.

Johnson, R. H., and Spradbrow, P. B. (1979). Isolation from dogs with severe enteritis of a parvovirus related to feline panleukopaenia virus. *Aust. Vet. J.* **55**, 151.

Johnson, R. H., Margolis, G., and Kilham, L. (1967). Identity of feline ataxia virus with panleukopaenia virus. *Nature (London)* **214**, 175–177.

Johnson, R. H., Siegl, G., and Gautschi, M. (1974). Characteristics of feline panleukopaenia virus strains enabling definitive classification as parvoviruses. *Arch. Ges. Virusforsch.* **46**, 315–324.

Jones, B. R., Robinson, A. J., Fray, L. M., and Lee, E. A. (1982). A longitudinal serological survey of parvovirus infection in dogs. *N.Z. Vet. J.* **30**, 19–20.

Kelly, W. R., and Atwell, R. B. (1979). Diffuse subacute myocarditis of possible viral aetiology: Cause of sudden deaths in pups. *Aust. Vet. J.* **55**, 36–37.

Kikuth, V. W., Gonnert, R., and Schweickert, M. (1940). Infectiose aleukozytose der katzen. *Zentralbl. Bakteriol., Parasitenkd. Infektionskr.* **146**, 1–17 (in German).

Kilham, L., Margolis, G., and Colby, E. D. (1967). Congenital infections of cats and ferrets by feline panleukopenia virus manifested by cerebellar hypoplasia. *Lab. Invest.* **17**, 465–480.

Kilham, L., Margolis, G., and Colby, E. D. (1971). Cerebellar ataxia and its congenital transmission in cats by feline panleukopenia virus. *J. Am. Vet. Med. Assoc.* **158**, 888–901.

Kimsey, P. B., Engers, H. D., Hirt, B., and Jongeneel, C. V. (1986). Pathogenicity of fibroblast- and lymphocyte-specific variants of minute virus of mice. *J. Virol.* **59,** 8–13.

King, D. A., and Croghan, D. L. (1965). Immunofluorescence of feline panleukopenia virus in cell culture: Determination of immunological status of felines by serum neutralization. *Can. J. Comp. Med.* **29,** 85–89.

King, D. A., and Gutekunst, D. E. (1970). A new mink enteritis vaccine for immunization against feline panleukopenia. *VM/SAC, Vet. Med. Small Anim. Clin.* **70,** 377–383.

Knox, B. (1960). Outbreaks of virus enteritis in mink in Denmark. *Nord. Veterinaermed.* **12,** 166–168.

Koptopoulos, G., Papadopoulos, O., Papanastasopoulou, M., and Cornwell, H. J. C. (1986). Presence of antibody cross-reacting with canine parvovirus in the sera of dogs from Greece. *Vet. Rec.* **118,** 332–333.

Krunajevic, T. (1970). Experimental virus enteritis in mink. A pathologic–anatomical and electron microscopical study. *Acta. Vet. Scand.* **11** (Suppl. 30), 1–88.

Kull, K.-E. (1966). Virusenterit: Sverige. *Vara Palsdjur* **8,** 331 (in Swedish).

Landsverk, T., and Nordstoga, K. (1978). Virus enteritis of mink, a scanning electron microscopic study. *Acta Vet. Scand.* **19,** 569–573.

Larsen, S., Flagstad, A., and Aalbak, B. (1976). Experimental feline panleukopenia in the conventional cat. *Vet. Pathol.* **13,** 216–240.

Lawrence, J. S., and Syverton, J. T. (1938). Spontaneous agranulocytosis in the cat. *Proc. Soc. Exp. Biol. Med.* **38,** 914–918.

Lawrence, J. S., Syverton, J. T., Shaw, J. S., and Smith, F. P. (1940). Infectious feline agranulocytosis. *Am. J. Pathol.* **16,** 333–354.

Lawrence, J. S., Syverton, J. T., Ackart, R. J., Adams, W. S., Ervin, D. M., Haskins, A. L., Saunders, R. H., Stringfellow, M. B., and Wetrich, R. M. (1943). The virus of infectious feline agranulocytosis. II. Immunological relationships to other viruses. *J. Exp. Med.* **77,** 57–64.

Lenghaus, C., and Studdert, M. J. (1980). Relationships of canine panleukopaenia (enteritis) and myocarditis parvoviruses to feline panleukopaenia virus. *Aust. Vet. J.* **56,** 152–153.

Lenghaus, C., and Studdert, M. J. (1982). Generalized parvovirus disease in neonatal pups. *J. Am. Vet. Med. Assoc.* **181,** 41–45.

Lucas, A. M., and Riser, W. H. (1945). Intranuclear inclusion in panleukopenia in cats. *Am. J. Pathol.* **21,** 435–465.

Luo, M., Tsao, J., Rossmann, M. G., Basak, S., and Compans, R. W. (1988). Preliminary X-ray crystallographic analysis of canine parvovirus crystals. *J. Mol. Biol.* **200,** 209–211.

Lust, S. J., Gorham, J. R., and Sato, N. (1965). Occurrence of intranuclear inclusions in cell cultures infected with infectious feline enteritis virus. *Am. J. Vet. Res.* **26,** 1163–1166.

Macartney, L., McCandlish, I. A. P., Thompson, H., and Cornwell, H. J. C. (1984a). Canine parvovirus enteritis. 1: Clinical, haematological and pathological features of experimental infection. *Vet. Rec.* **115,** 201–210.

Macartney, L., McCandlish, I.A.P., Thompson, H., and Cornwell, H. J. C. (1984b). Canine parvovirus enteritis. 2: Pathogensis. *Vet. Rec.* **115,** 453–460.

Macartney, L., McCandlish, I. A. P., Thompson, H., and Cornwell, H. J. C. (1984c). Canine parvovirus enteritis. 3: Scanning electron microscopical features of experimental infection. *Vet. Rec.* **115,** 533–537.

MacPherson, L. W. (1956). Feline enteritis virus—Its transmission to mink under natural and experimental conditions. *Can. J. Comp. Med.* **20**, 197–202.

Mann, P. C., Bush, M., Appel, M. J. G., Beehler, B. A., and Montali, R. J. (1980). Canine parvovirus infection in South American canids. *J. Am. Vet. Med. Assoc.* **177**, 779–783.

Margolis, G., and Kilham, L. (1965). Rat virus, an agent with an affinity for dividing cells. *Monogr. Inst. Neurol. Dis. Blindness* **2**, 361–367.

Mazzara, G. P., Destree, A. T., Williams, H. W., Sue, J. M., Belanger, L. M., Panicali, D., and Parrish, C. R. (1987). Successful vaccination of dogs with empty capsids derived from canine parvovirus–bovine papillomavirus chimeric plasmids. *In* "Vaccines '87" (R. M. Chanock, R. A. Lerner, F. Brown, H. Ginsberg, eds.), pp. 419–424. Cold Spring Harbor Lab., Cold Spring Harbor, New York.

McMaster, G. K., Tratschin, J.-D., and Siegl, G. (1981). Comparison of canine parvovirus with mink enteritis virus by restriction site mapping. *J. Virol.* **38**, 368–371.

Mech, L. D., Goyal, S. M., Bota, C. N., and Seal, U. S. (1986). Canine parvovirus infection in wolves (*Canis lupus*) from Minnesota. *J. Wildl. Dis.* **22**, 104–106.

Meunier, P. C., Glickman, L. T., Appel, M. J. G., and Shin, S. J. (1980). Canine parvovirus in a commercial kennel: Epidemiologic and pathologic findings. *Cornell Vet.* **71**, 96–110.

Meunier, P. C., Cooper, B. J., Appel, M. J. G., and Slauson, D. O. (1984). Experimental viral myocarditis—Parvovirus infection of neonatal pups. *Vet. Pathol.* **21**, 509–515.

Meunier, P. C., Cooper, B. J., Appel, M. J. G., Lanieu, M. E., and Slauson, D. O. (1985a). Pathogenesis of canine parvovirus enteritis: Sequential virus distribution and passive immunization studies. *Vet. Pathol.* **22**, 617–624.

Meunier, P. C., Cooper, B. J., Appel, M. J. G., and Slauson, D. O. (1985b). Pathogenesis of canine parvovirus enteritis: The importance of viremia. *Vet. Pathol.* **22**, 60–71.

Mochizuki, M., Konishi, S., and Ogata, M. (1978). Studies on feline panleukopenia. II. Antigenicities of the virus. *Jpn. J. Vet. Sci.* **40**, 375–383.

Mohri, S., Handa, S., Wada, T., and Tokiyoshi, S. (1982). Sero-epidemiologic survey on canine parvovirus infection. *Jpn. J. Vet. Sci.* **44**, 543–545.

Moraillon, A., and Moraillon, R. (1982). Distinction des parvovirus félins et canins par hemagglutination. *Rec. Med. Vet.* **158**, 799–804 (in French).

Morgan, W. R., and Ward, D. C. (1986). Three splicing patterns are used to excise the small intron common to all minute virus of mice RNAs. *J. Virol.* **60**, 1170–1174.

Myers, W. L., Alberts, J. O., and Brandley, C. A. (1959). Certain characteristics of the virus of infectious enteritis of mink and observations on pathogenesis of the disease—Preliminary report. *Can. J. Comp. Med.* **23**, 282–287.

Nain, K. P. D., Padmainbhaiyer, R., and Venugapolar, A. (1964). An outbreak of gastroenteritis in civet cats. *Indian Vet. J.* **41**, 763–765.

Nara, P. L., Winters, K., Rice, J. B., Olsen, R. G., and Krakowka, S. (1983). Systemic and local intestinal antibody response in dogs given both infective and inactivated canine parvovirus. *Am. J. Vet. Res.* **44**, 1989–1995.

Nettles, V. F., Pearson, J. E., Gustafson, G. A., and Blue, J. L. (1980). Parvovirus infection in translocated raccoons. *J. Am. Vet. Med. Assoc.* **177**, 787–789.

Neuvonen, E., Veijalainen, P., and Kangas, J. (1982). Canine parvovirus infection in housed raccoon dogs and foxes in Finland. *Vet. Rec.* **110**, 448–449.

Okaniwa, A., Yasoshima, H., Kojima, A., and Doi, K. (1976). Fine structure of epithelial cells of Lieberkuhn's crypts in feline panleukopenia. *Natl. Inst. Health Q. (Tokyo)* **16**, 167–175.

Osterhaus, A. D. M. E., Drost, G. A., Wirahadiredja, R. M. S., and van den Ingh, T. S. G. A. M. (1980). Canine viral enteritis: Prevalence of parvo-, corona- and rotavirus infections in dogs in The Netherlands. *Vet. Q.* **2,** 181–190.

O'Sullivan, G., Durham, P. J. K., Smith, J. R., and Campbell, R. S. F. (1984). Experimentally induced severe canine parvoviral enteritis. *Aust. Vet. J.* **61,** 1–4.

Palese, P., and Young, J. F. (1982). Variation of influenza A, B, and C viruses. *Science* **215,** 1468–1474.

Paradiso, P. R. (1981). Infectious process of the parvovirus H-I: Correlation of protein content, particle density and viral infectivity. *J. Virol.* **39,** 800–807.

Paradiso, P. R. (1983). Analysis of the protein–protein interactions in the parvovirus H-1 capsid. *J. Virol.* **46,** 94–102.

Paradiso, P. R. (1984). Identification of multiple forms of the noncapsid parvovirus protein NCVP1 in H-1 parvovirus infected cells. *J. Virol.* **52,** 82–87.

Paradiso, P. R., Rhode, S. L., and Singer, I. I. (1982). Canine parvovirus: A biochemical and ultrastructural characterization. *J. Gen. Virol.* **62,** 113–126.

Paradiso, P. R., Williams, K. R., and Costantino, R. L. (1984). Mapping of the amino terminus of the H-1 major capsid protein. *J. Virol.* **52,** 77–81.

Parrish, C. R., and Carmichael, L. E. (1983). Antigenic structure and variation of canine parvovirus type-2, feline panleukopenia virus and mink enteritis virus. *Virology* **129,** 401–414.

Parrish, C. R., and Carmichael, L. E. (1986). Characterization and recombination mapping of an antigenic and hostrange mutation of canine parvovirus. *Virology* **148,** 121–132.

Parrish, C. R., Oliver, R. E., and McNiven, R. (1982a). Canine parvovirus infections in a colony of dogs. *Vet. Microbiol.* **7,** 317–324.

Parrish, C. R., Carmichael, L. E., and Antczak, D. F. (1982b). Antigenic relationships between canine parvovirus type 2, feline panleukopenia virus and mink enteritis virus using conventional antisera and monoclonal antibodies. *Arch. Virol.* **72,** 267–278.

Parrish, C. R., Gorham, J. R., Schwartz, T. M., and Carmichael, L. E. (1984). Characterization of antigenic variation among mink enteritis virus isolates. *Am. J. Vet. Res.* **45,** 2591–2599.

Parrish, C. R., O'Connell, P. H., Evermann, J. F., and Carmichael, L. E. (1985). Natural variation of canine parvovirus. *Science* **230,** 1046–1048.

Parrish, C. R., Leathers, C. W., Pearson, R., and Gorham, J. R. (1987). Comparisons of feline panleukopenia virus, canine parvovirus, raccoon parvovirus, and mink enteritis virus and their pathogenicity for mink and ferrets. *Am. J. Vet. Res.* **48,** 1429–1435.

Parrish, C. R., Burtonboy, G., and Carmichael, L. E. (1988a). Characterization of a nonhemagglutinating mutant of canine parvovirus. *Virology* **163,** 230–232.

Parrish, C. R., Have, P., Foreyt, W. J., Evermann, J. F., Senda, M., and Carmichael, L. E. (1988b). The global spread and replacement of canine parvovirus strains. *J. Gen. Virol.* **69,** 1111–1116.

Parrish, C. R., Aquadro, C. F., and Carmichael, L. E. (1988c). Canine host range and a specific epitope map along with variant sequences in the capsid protein gene of canine parvovirus and related feline, mink and raccoon parvoviruses. *Virology* **166,** 293–307.

Phillips, C. E. (1943). Haemorrhagic enteritis in the Arctic blue fox caused by the virus of feline enteritis. *Can. J. Comp. Med.* **7,** 33–35.

Pollock, R. V. H. (1982). Experimental canine parvovirus infection in dogs. *Cornell Vet.* **72,** 103–119.

Pollock, R. V. H., and Carmichael, L. E. (1982). Maternally derived immunity to canine parvovirus infection: Transfer, decline and interference with vaccination. *J. Am. Vet. Med. Assoc.* **180**, 37–42.

Pridham, T. J., and Wills, C. G. (1959). The prevention of virus enteritis in newly weaned mink kits by the use of a homologous tissue vaccine. *J. Am. Vet. Med. Assoc.* **135**, 279–282.

Pridham, T. J., and Wills, C. G. (1960). Variations in the effectiveness of commercial infectious feline enteritis vaccines in preventing virus enteritis of mink. *Can. Vet. J.* **1**, 51–56.

Reed, A. P., Jones, E. V., and Miller, T. J. (1988). Nucleotide sequence and genome organization of canine parvovirus. *J. Virol.* **62**, 266–276.

Reynolds, H. A. (1969). Some clinical and hematological features of virus enteritis of mink. *Can. J. Comp. Med.* **33**, 155–159.

Reynolds, H. A. (1970). Pathological changes in virus enteritis of mink. *Can. J. Comp. Med.* **34**, 155–163.

Rhode, S. L. (1985). Nucleotide sequence of the coat protein gene of canine parvovirus. *J. Virol.* **54**, 630–633.

Rhode, S. L., and Paradiso, P. R. (1983). Parvovirus genome: Nucleotide sequence of H-1 and mapping of its genes by hybrid-arrested translation. *J. Virol.* **45**, 173–184.

Rice, J. B., Winters, K. A., Krakowka, S., and Olsen, R. G. (1982). Comparison of systemic and local immunity in dogs with canine parvovirus gastroenteritis. *Infect. Immun.* **38**, 1003–1009.

Rivera, E., and Sundquist, B. (1984). A non-haemagglutinating isolate of mink enteritis virus. *Vet. Microbiol.* **9**, 345–353.

Robinson, W. F., Wilcox, G. E., Flower, R. L. P., and Smith, J. R. (1979a). Evidence for a parvovirus as the aetiological agent in myocarditis of puppies. *Aust. Vet. J.* **55**, 294–295.

Robinson, W. F., Huxtable, C. R. R., Pass, D. A., and Howell, J. M. (1979b). Clinical and electrocardiographic findings in suspected viral myocarditis in pups. *Aust. Vet. J.* **55**, 351–355.

Robinson, W. F., Huxtable, C. R., and Pass, D. A. (1980a). Canine parvovirus myocarditis: A morphological description of the natural disease. *Vet. Pathol.* **17**, 282–293.

Robinson, W. F., Wilcox, G. E., and Flower, R. L. P. (1980b). Canine parvoviral disease: Experimental reproduction of the enteric form with a parvovirus isolated from a case of myocarditis. *Vet. Pathol.* **17**, 589–599.

Rohovsky, M. W., and Griesemer, R. A. (1967). Experimental feline infectious enteritis in the germfree cat. *Pathol. Vet.* **4**, 391–410.

Schofield, F. W. (1949). Virus enteritis in mink. *North Am. Vet.* **30**, 651–654.

Schwers, A., Pastoret, P.-P., Burtonboy, G., and Thiry, E. (1979). Fréquence en Belgique de l'infection à parvovirus chez le chien, avant et après l'observation des premiers cas cliniques. *Ann. Vet. Med.* **123**, 561–566 (in French).

Scott, F. W., Csiza, C. K., and Gillespie, J. H. (1970). Feline viruses. V. Serum neutralization test for feline panleukopenia. *Cornell Vet.* **60**, 183–191.

Senda, M., Hirayama, N., Itoh, O., and Yamamoto, H. (1988). Canine parvovirus: Strain difference in haemagglutination activity and antigenicity. *J. Gen. Virol.* **69**, 349–354.

Shindel, N. M., van Kruiningen, H. J., and Scott, F. W. (1978). The colitis of feline panleukopenia. *J. Am. Anim. Hosp. Assoc.* **14**, 738–747.

Siegl, G. (1984). Panine parvovirus: Origin and significance of a "new" viral pathogen. *In* "The Parvoviruses" (K. I. Berns, ed.), pp. 363–388. Plenum, New York.

Siegl, G., Bates, R. C., Berns, K. I., Carter, B. J., Kelly, D. C., Kurstak, E., and Tattersall, P. (1985). Characteristics and taxonomy of Parvoviridae. *Intervirology* **23**, 61–73.

Smith, J. R., Farmer, T. S., and Johnson, R. H. (1980). Serological observations on the epidemiology of parvovirus enteritis of dogs. *Aust. Vet. J.* **56**, 149–150.

Spalholz, B. A., and Tattersall, P. (1983). Interaction of minute virus of mice with differentiated cells: Strain-dependent target cell specificity is mediated by intracellular factors. *J. Virol.* **46**, 937–943.

Surleraux, M., Bodeus, M., and Burtonboy, G. (1987). Study of canine parvovirus polypeptides by immunoblot analysis. *Arch. Virol.* **95**, 271–281.

Tattersall, P. (1972). Replication of the parvovirus minute virus of mice. I. dependence of virus multiplication and plaque formation on cell growth. *J. Virol.* **10**, 586–590.

Tattersall, P., and Bratton, J. (1983). Reciprocal productive and restrictive virus cell interactions of immunosuppressive and prototype strains of minute virus of mice. *J. Virol.* **46**, 944–955.

Tattersall, P., Cawte, P. J., Shatkin, A. J., and Ward, D. C. (1976). Three structural polypeptides coded for by minute virus of mice, a parvovirus. *J. Virol.* **20**, 273–289.

Tennant, R. W., Layman, K. R., and Hand, R. E. (1969). Effect of cell physiological state on infection by rat virus. *Virology* **11**, 872–878.

Thomas, N. J., Foreyt, W. J., Evermann, J. F., Windberg, L. A., and Knowlton, F. F. (1984). Seroprevalence of canine parvovirus in wild coyotes from Texas, Utah and Idaho (1972 to 1983). *J. Am. Vet. Med. Assoc.* **185**, 1283–1287.

Torres, E. (1941). Infectious feline gastroenteritis in wild cats. *North Am. Vet.* **22**, 297.

Tratschin, J.-D., McMaster, G. K., Kronauer, G., and Siegl, G. (1982). Canine parvovirus: Relationship to wild-type and vaccine strains of feline panleukopenia virus and mink enteritis virus. *J. Gen. Virol.* **61**, 33–41.

Tuomi, J., and Kangas, J. (1963). A fluorescent antibody technique for studies of mink virus enteritis. *Arch. Gesamte Virusforsch.* **13**, 432–434.

Veijalainen, P. (1986). A serological survey of enteric parvovirus infections in Finnish fur-bearing animals. *Acta Vet. Scand.* **27**, 159–171.

Veijalainen, P. (1988). Characterization of biological and antigenic properties of raccoon dog and blue fox parvoviruses: A monoclonal antibody study. *Vet. Microbiol.* **16**, 219–230.

Veijalainen, P. M.-L., and Smeds, E. (1988). Pathogenesis of blue fox parvovirus on blue fox kits and pregnant vixens. *Am. J. Vet. Res.* **49**, 1941–1944.

Verge, J., and Cristoforoni, N. (1928). La gastro enterite infectieuse des chats; est-elle due à un virus filterable? *C. R. Soc. Biol.* **99**, 312.

Walker, S. T., Feilen, C. P., Sabine, M., Love, D. N., and Jones, R. F. (1980). A serological survey of canine parvovirus infection in New South Wales, Australia. *Vet. Rec.* **106**, 324–325.

Waller, E. F. (1940). Infectious gastroenteritis in raccoons (Procyon lotor). *J. Am. Vet. Med. Assoc.* **96**, 266–268.

Wayne, R. K., Benveniste, R. E., Janczewski, D. N., and O'Brien, S. J. (1989). Molecular and biochemical evolution of the Carnivora. *In* "Carnivore Behavior, Ecology and Evolution" (J. L. Gittleman, ed.), pp. 465–494. Cornell Univ. Press, Ithaca, New York.

Webster, R. G., Laver, W. G., Air, G. M., and Schild, G. C. (1982). Molecular mechanisms of variation in infleunza viruses. *Nature (London)* **296**, 115–121.

Wills, C. G. (1952). Notes on infectious enteritis of mink and its relationship to feline enteritis. *Can. J. Comp. Med.* **16**, 419–420.

Wills, G., and Belcher, J. (1956). The prevention of virus enteritis of mink with commercial feline panleukopenia vaccine. *J. Am. Vet. Med. Assoc.* **128**, 559–560.

Yasoshima, A., Doi, K., Kojima, A., and Okaniwa, A. (1982). Electron microscopic findings on epithelial cells of Lieberkuhn's crypts in canine parvovirus infection. *Jpn. J. Vet. Sci.* **44**, 81–88.

Zarnke, R. L., and Ballard, W. B. (1987). Serological survey for selected microbial pathogens of wolves in Alaska, 1975–1982. *J. Wildl. Dis.* **23**, 77–85.

INDEX